Remote Sensing and Modeling

Coastal Research Library

VOLUME 9

Series Editor:

Charles W. Finkl
Department of Geosciences
Florida Atlantic University
Boca Raton, FL 33431
USA

The aim of this book series is to disseminate information to the coastal research community. The Series covers all aspects of coastal research including but not limited to relevant aspects of geological sciences, biology (incl. ecology and coastal marine ecosystems), geomorphology (physical geography), climate, littoral oceanography, coastal hydraulics, environmental (resource) management, engineering, and remote sensing. Policy, coastal law, and relevant issues such as conflict resolution and risk management would also be covered by the Series. The scope of the Series is broad and with a unique crossdisciplinary nature. The Series would tend to focus on topics that are of current interest and which carry some import as opposed to traditional titles that are esoteric and non-controversial. Monographs as well as contributed volumes are welcomed.

For further volumes:
http://www.springer.com/series/8795

Charles W. Finkl • Christopher Makowski
Editors

Remote Sensing and Modeling

Advances in Coastal and Marine Resources

Springer

Editors
Charles W. Finkl
Florida Atlantic University
Boca Raton, FL, USA

Coastal Education and Research Foundation (CERF)
Coconut Creek, FL, USA

Christopher Makowski
Florida Atlantic University
Boca Raton, FL, USA

Coastal Education and Research Foundation (CERF)
Coconut Creek, FL, USA

ISSN 2211-0577 ISSN 2211-0585 (electronic)
ISBN 978-3-319-06325-6 ISBN 978-3-319-06326-3 (eBook)
DOI 10.1007/978-3-319-06326-3
Springer Cham Heidelberg New York Dordrecht London

Library of Congress Control Number: 2014945760

Printed on acid-free paper

Springer is part of Springer Science+Business Media (www.springer.com)

Preface

Remote sensing is a very large field of study that involves many different types of sensors, platforms (videographic stations, satellites, aircraft, UAVs or drones [flying robots], etc.), and purposes (research, surveillance, warfare). Recent advances in the field focus on merging technologies in GIS (Geographic Information Systems), robotics, and numerical modeling, as these endeavors tend to reinforce each other. The coastal zone is a complex system, parts of which are difficultly accessible under the best of conditions. During inclement weather or when high-energy conditions prevail, study of coastal systems and environments often requires the acquisition of remotely sensed data that cannot be directly collected by humans. Nevertheless, even under optimal weather conditions the use of remote sensing techniques is advantageous for many reasons, not the least of which is the fact that enormous datasets can be collected over large spatial scales in relatively short time spans. Data acquisition covers a wide array of surface, intertidal, and submarine environments.

This book is not a comprehensive review of recent advances in remote sensing, as the field of study is too large to handle and such a broad view is beyond the scope of our subject area. Instead, this volume in the Coastal Research Library contains selected vignettes that make up 20 chapters. These examples of advances are considered in four parts, each with several chapters. Part I (Remote Sensing and Mapping of Coastal Biophysical Environments) contains seven chapters. The first chapter (Remote Sensing of Coastal Ecosystems and Environments), by Vic Klemas, sets the stage for the volume by providing an overview of remote sensing of coastal biophysical environments. Considered here are advances in sensor design and data analysis techniques as related to hyperspectral imagers, LiDAR, and radar systems. Chapter 2 (Advanced Techniques for Mapping Biophysical Environments on Carbonate Banks Using Laser Airborne Depth Sounding (LADS) and IKONOS Satellite Imagery) by Charles W. Finkl, Christopher Makowski, and Heather Vollmer, investigates recent advances in the mapping of seafloor environments on carbonate shelves using the example of southern Florida. Interpretation of seafloor data derived from LADS and IKONOS imagery was used to develop new cognitive mapping techniques and classification systems that are useful for the study of large areas. Chapter 3 (Terrestrial Lase Scanner Surveying in Coastal Settings) reports on advances associated with the proliferation of commercially available tripod-mounted terrestrial laser scanner (TLS) systems that use the phase difference or the time-of-flight of emitted pulses of light to rapidly acquire high-density topographic and surface reflectance data. Chapter 4 (Advances in Applied Remote Sensing to Coastal Environments Using Free Satellite Imagery), by Cristina Lira and Rui Taborda, reports on advances associated with Landsat 8, which supports improved radiometric and spectral resolutions (compared to previous Landsat platforms). Discussed here are potentialities of these new sensors for temporal coverage, frequency of coverage, radiometric resolution, and spectral resolution. Chapter 5 (Remote Sensing and Modeling of Coral Reef Resilience), by Anders Knudby, Simon J. Pittman, Joseph Maina, and Gwilym Rowlands, reviews the state of the art of coral reef resilience mapping, based on remote sensing, spatial distribution modeling, and process modeling. Case studies illustrate coarse-scale mapping of reef exposure to climate-driven disturbances, intermediate-scale mapping of water quality and its influence on coral bleaching susceptibility, and fine-scale mapping of local factors that influence the ability of reefs to resist and rebound from climate-driven disturbance. Chapter 6 (An Assessment of Physiographic Habitats, Geomorphology and Evolution of Chilika Lagoon (Odisha, India) Using Geospatial Technology), by Ashis Kr. Paul, Sk Majharul Islam, and Subrata Jana, studies the geomorphologic changes, ecologic responses, and evolution of the Chilika Lagoon using geospatial technology with temporal image data. Lastly, Chap. 7 (Foreshore Applications of X-band Radar), by G. M. Jahid Hasan and Satoshi Takewaka, employed an X-band nautical radar system to examine alongshore propagation of low frequency run-up motion, as well as estimate the morphodynamic parameters from two typhoon events in the Pacific Ocean.

Part II (Advances in the Study and Interpretation of Coastal Oceans, Estuaries, Sea-Level Variation, and Water Quality) brings together in five chapters some disparate advances under a larger umbrella with examples from the coastal ocean, estuaries, and gulfs. Chapter 8 (Digital Ocean Technological Advances), by Xin Zhang, Xiaoyi Jiang, Suixiang Shi, and Tianhe Chi, considers the Digital Ocean (DO) as a new research domain of Digital Earth. Here, DO technological advances are introduced for (1) data sources, (2) three-dimensional ocean data integration platform, (3) dynamic tide data visualization, (4) integration and sharing of remote sensing products, (5) computational ocean model data integration service, and (6) spatio-temporal model of marine disasters. In Chap. 9 (A New Statistical-Empirical Hybrid Based Model to Estimate Seasonal Sea-Level Variation in the Gulf of Paria from River Discharge) by Carol Subrath-Ali, new insight is provided for the quantitative role of the Orinoco River in South America. This chapter reports on a vertically integrated 2D numerical modeling suite that is applied to the execution of a series of experiments to ascertain variation of coastal water levels from river discharge. The modeling advance here shows how a third order model function, which is dependent only on river discharge, can estimate the average monthly river-driven water level in the Gulf of Paria. In a similar vein, Chap. 10 (Advances in Modeling of Water Quality in Estuaries) by K.I. Ascione,

F. Campuzano, G. Franz, R. Fernandes, C. Viegas, J. Sobrinho, H. De Pablo, A. Amaral, L. Pinto, M. Mateus, and R. Neves, posits that water quality models complement studies about the status of estuarine waters. This chapter serves as an exemplar showing how advanced modeling applications can be used to perform water quality studies in Portuguese estuaries. Boundary conditions for hydrodynamics and biogeochemistry, provided by the Portuguese Coast Operational Model, are downscaled by using nested domains with increasing resolution from the regional to the local scale. Chapter 11 (Advances in Video Monitoring of the Beach and Nearshore: The Long-Term Perspective), by Ana Nobre Silva and Rui Taborda, summarizes recent developments on the use of video systems in the understanding of yearly to decadal beach morphological changes and describes the application of such a video system deployed at Nazaré, Portugal. While Chap. 12 (Advances in Application of Remote Sensing Techniques to Enhance the Capability of Hydrodynamic Modeling in Estuary), by A.K.M Azad Hossain, Yafei Jia, Xiaobo Chao, and Mustafa Altinakar, provides evidence that the application of remote sensing techniques for estuarine water quality studies can be advanced by integrating them with numerical models.

Part III (Advances in Coastal Modeling Using Field Data, Remote Sensing, GIS and Numerical Simulations) contains five chapters that consider integrated approaches to coastal modeling. Chapter 13 (Developments in Salt Marsh Topography Analysis Using Airborne Infrared Photography), by Francisco Andrade, Jackson Blanton, M. Adelaide Ferreira, and Julie Amft, shows how only recently have remote-sensing techniques become widely available to obtain high-resolution topographic data in salt marshes. These authors describe how a detailed digital elevation model (DEM) of the Duplin River (Georgia, southeastern USA) with a 1 m^2 resolution was constructed through the classification and analysis of a time-series of 7 IR (infrared) aerial photography mosaics taken at 1 h intervals from low- to high-water during a rising tide. In Chap. 14 (Examining Material Transport in Dynamic Coastal Environments: An Integrated Approach Using Field Data, Remote Sensing and Numerical Modeling), Richard L. Miller, Ramón López, Ryan P. Mulligan, Robert E. Reed, Cheng-Chien Liu, Christopher J. Buonassissi, and Matthew M. Brown describe an integrated approach based on field measurements, remote sensing and numerical modeling that examines the transport of dissolved (colored dissolved organic matter (CDOM), dissolved organic carbon (DOC)) and particulate material (total suspended matter (TSM)) within a complex coastal system, the Albemarle-Pamlico Estuarine System (APES), North Carolina, USA. The advanced Delft3D numerical model is used to simulate freshwater and DOC transport following major rain events. Chapter 15 (Simulated Management Systems Developed by the Northern Gulf Coastal Hazards Collaboratory (NG-CHC): An Overview of Cyberinfrastructure to Support the Coastal Modeling Community in the Gulf of Mexico), by a team composed of Robert R. Twilley, Steve Brandt, Darlene Breaux, John Cartwright, Jim Chen, Greg Easson, Patrick Fitzpatrick, Kenneth Fridley, Sara Graves, Sandra Harper, Carola Kaiser, Alexander Maestre, Manil Maskey, William H. McAnally, John McCorquodale, Ehab Meselhe, Tina Miller-Way, Kyeong Park, Joao Pereira, Thomas Richardson, Jian

Tao, Amelia Ward, Jerry Wiggert, and Derek Williamson, explains how a collaboratory was established to catalyze collaborative research via enhanced CI (cyberinfrastructure) to reduce regional vulnerability to natural and human disasters by facilitating high performance modeling to test hypotheses focused on engineering design, coastal system response, and risk management of coastal hazards. This advanced technology is used to promote collaborative environmental modeling in coastal systems. Chapter 16 (Advancement of Technology for Detecting Shoreline Changes in East Coast of India and Comparison with Prototype Behavior), by Ramasamy Manivanan, discusses how the information predicted by cross-shore and longshore impact mathematical model match the information shown by satellite imagery. Thus, satellite information can be useful for the overall calibration of the mathematical models. Chapter 17 (Advances in Remote Sensing of Coastal Wetlands: LiDAR, SAR, and Object-Oriented Case Studies from North Carolina), by Thomas R. Allen, reviews the different advancements from the use of Light Detection and Ranging (LiDAR), space-borne Synthetic Aperture Radar (SAR), and multi-sensor and object-oriented image analysis techniques, which aid the inventorying, monitoring, and management of coastal wetlands.

Part IV (Advances in the Management of Coastal Resources Using Remote Sensing Data and GIS) contains three chapters that extol the virtues of numerical simulations and satellite remote sensing tools for research and management. Chapter 18 (Numerical Simulations and Satellite Remote Sensing as Tools for Research and Management of Marine Fishery Resources), by Grinson George, discusses modeled and satellite remote sensing data that support research, technology-development, and management of marine fishery resources. Of interest here is the fact that numerical simulations and remote sensing data of the marine environment provides sufficient cues in the form of surrogate databases that support monitoring, surveillance, and management of marine fishery resources in the context of an ecosystem approach. Chapter 19 (Identifying Suitable Sites of Shrimp Culture in Southwest Bangladesh Using GIS and Remote Sensing Data), by Shak Md. Bazlur Rahaman, Khandaker Anisul Huq, and Md. Mujibor Rahman, shows how satellite imagery and GIS data (e.g. water and soil quality, shrimp culture area, method and production, source and seasonal availability of water, drainage system, water logging, disease outbreak, sanitation facility, road communication, electricity supply, land use pattern, land elevation, hazard frequency, fisheries statistics, and population census data) were collected in studies of site suitability. The advanced methodology in this study shows how to formulate shrimp culture policy for sustainable development. And in Chap. 20 (A Multi-Criteria Approach for Erosion Risk Assessment Using a New Concept of Spatial Unit Analysis, Wave Model, and High Resolution DEMs), by Helena Granja, José Pinho, and João Mendes, field data and model outputs were integrated, processed, and analyzed within a GIS interface in order to assess the vulnerability to erosion and to produce associated risk maps using a multi-criteria approach.

When discussing advancements in the remote sensing and modeling of biophysical coastal systems, researchers have made many great strides in recent years. This volume of the Coastal Research Library (CRL), while presenting a wide range of topics related to the innovative technologies associated with the development and improvement of coastal remote sensing and modeling, also offers a look into the many parts of the world where these advancements are being implemented. For example, a new seasonal sea-level variation model is introduced along the Orinoco River in South America, the advancement of water quality models is presented for estuaries in Portugal, the use of remote sensing and GIS data is shown to further the advancement of shrimp fisheries in Bangladesh, and by incorporating Laser Airborne Depth Sounding (LADS) and IKONOS satellite imagery, a new level of mapping biophysical environments on carbonate banks is achieved in south Florida. The following chapters establish the ongoing advancements in the fields of remote sensing and modeling, as well as provide a comprehensive look into diverse coastal environments where these studies are being conducted.

Coconut Creek, FL, USA — Charles W. Finkl
Boca Raton, FL, USA — Christopher Makowski

Contents

Contributors

Thomas R. Allen Department of Geography, Planning, & Environment, East Carolina University, Greenville, NC, USA

Mustafa Altinakar National Center for Computational Hydroscience and Engineering (NCCHE), The University of Mississippi, Oxford, MS, USA

Andreia Amaral Marine Environment Technology Center (MARETEC), Instituto Superior Técnico, Universidade de Lisboa, Lisbon, Portugal

Julie Amft Skidaway Institute of Oceanography, Savannah, GA, USA

Francisco Andrade Centro de Oceanografia – Laboratório Marítimo da Guia, Faculdade de Ciências da Universidade de Lisboa, Cascais, Portugal

Isabella Ascione Kenov Marine Environment Technology Center (MARETEC), Instituto Superior Técnico, Universidade de Lisboa, Lisbon, Portugal

Jackson Blanton Skidaway Institute of Oceanography, Savannah, GA, USA

Steve Brandt Center for Computation and Technology, Louisiana State University, Baton Rouge, LA, USA

Darlene Breaux Department of Civil Engineering, University of Louisiana at Lafayette, Lafayette, LA, USA

Matthew M. Brown Department of Geological Sciences, East Carolina University, Greenville, NC, USA

Christopher J. Buonassissi Institute for Coastal Science and Policy, Coastal Resource Management PhD Program, East Carolina University, Greenville, NC, USA

Francisco Campuzano Marine Environment Technology Center (MARETEC), Instituto Superior Técnico, Universidade de Lisboa, Lisbon, Portugal

John Cartwright Geosystems Research Institute, Mississippi State University, Mississippi State, MS, USA

Xiaobo Chao National Center for Computational Hydroscience and Engineering (NCCHE), The University of Mississippi, Oxford, MS, USA

Jim Chen Center for Computation and Technology, Louisiana State University, Baton Rouge, LA, USA

Department of Civil and Environmental Engineering, Louisiana State University, Baton Rouge, LA, USA

Tianhe Chi State Key Laboratory of Remote Sensing Science, Institute of Remote Sensing and Digital Earth, Chinese Academy of Sciences, Beijing, China

Hilda de Pablo Marine Environment Technology Center (MARETEC), Instituto Superior Técnico, Universidade de Lisboa, Lisbon, Portugal

Greg Easson Department of Geology and Geological Engineering, University of Mississippi, Oxford, MS, USA

Rodrigo Fernandes Marine Environment Technology Center (MARETEC), Instituto Superior Técnico, Universidade de Lisboa, Lisbon, Portugal

M. Adelaide Ferreira Laboratório Marítimo da Guia, Instituto do Mar, Cascais, Portugal

Charles W. Finkl Coastal Education and Research Foundation (CERF), Coconut Creek, FL, USA

Florida Atlantic University, Boca Raton, FL, USA

Patrick Fitzpatrick Northern Gulf Institute, Mississippi State University Science and Technology Center, Mississippi State, MS, USA

Guilherme Franz Marine Environment Technology Center (MARETEC), Instituto Superior Técnico, Universidade de Lisboa, Lisbon, Portugal

Kenneth Fridley Department of Civil, Construction and Environmental Engineering, University of Alabama, Tuscaloosa, AL, USA

Grinson George Fishery Resources Assessment Division, Central Marine Fisheries Research Institute, Kochi, India

Helena Granja Interdisciplinary Centre of Marine and Environmental Research (CIIMAR/CIMAR), University of Porto, Porto, Portugal

Earth Sciences Department, University of Minho, Braga, Portugal

Sara Graves Information Technology and Systems Center, University of Alabama in Huntsville, Huntsville, AL, USA

Sandra Harper Information Technology and Systems Center, University of Alabama in Huntsville, Huntsville, AL, USA

G.M. Jahid Hasan Department of Civil and Environmental Engineering, Shahjalal University of Science and Technology, Sylhet, Bangladesh

Department of Civil Engineering, Military Institute of Science and Technology, Dhaka, Bangladesh

AKM Azad Hossain National Center for Computational Hydroscience and Engineering (NCCHE), The University of Mississippi, Oxford, MS, USA

Khandaker Anisul Huq Fisheries and Marine Resource Technology Discipline, Khulna University, Khulna, Bangladesh

Sk. Majharul Islam Department of Geography and Environment Management, Vidyasagar University, Midnapore, West Bengal, India

Subrata Jana Department of Geography and Environment Management, Vidyasagar University, Midnapore, West Bengal, India

Yafei Jia National Center for Computational Hydroscience and Engineering (NCCHE), The University of Mississippi, Oxford, MS, USA

Xiaoyi Jiang National Marine Data & Information Service, State Oceanic Administration, Tianjin, China

Carola Kaiser Department of Oceanography and Coastal Sciences, School of the Coast and Environment, Louisiana State University, Baton Rouge, LA, USA

Center for Computation and Technology, Louisiana State University, Baton Rouge, LA, USA

Victor V. Klemas School of Marine Science and Policy, University of Delaware, Newark, DE, USA

Anders Knudby Department of Geography, Simon Fraser University, Burnaby, BC, Canada

Cristina Lira Instituto Dom Luiz, Centro de Geologia, Geology Department, Faculdade de Ciências da Universidade de Lisboa, Lisbon, Portugal

Cheng-Chien Liu Department of Earth Sciences, National Cheng Kung University, Tainan, Taiwan R.O.C

Ramón López Institute for Coastal Science and Policy, East Carolina University, Greenville, NC, USA

Alexander Maestre Department of Civil, Construction and Environmental Engineering, University of Alabama, Tuscaloosa, AL, USA

Joseph Maina ARC Centre of Excellence for Environmental Decisions, The University of Queensland, QLD, Australia

Wildlife Conservation Society, Bronx, NY, United States of America

Christopher Makowski Coastal Education and Research Foundation (CERF), Coconut Creek, FL, USA

Florida Atlantic University, Boca Raton, FL, USA

Ramasamy Manivanan Mathematical Modeling for Coastal Engineering (MMCE), Central Water and Power Research Station, Khadakwasla, Pune, India

Manil Maskey Information Technology and Systems Center, University of Alabama in Huntsville, Huntsville, AL, USA

Marcos Mateus Marine Environment Technology Center (MARETEC), Instituto Superior Técnico, Universidade de Lisboa, Lisbon, Portugal

William H. McAnally Geosystems Research Institute, Mississippi State University, Mississippi State, MS, USA

John McCorquodale Department of Civil and Environmental Engineering, and Pontchartrain Institute for Environmental Sciences, University of New Orleans, New Orleans, LA, USA

João Mendes C-TAC/Department of Civil Engineering, University of Minho, Braga, Portugal

Ehab Meselhe Department of Civil Engineering, University of Louisiana at Lafayette, Lafayette, LA, USA

The Water Institute of the Gulf, Baton Rouge, LA, USA

Richard L. Miller Department of Geological Sciences, East Carolina University, Greenville, NC, USA

Institute for Coastal Science and Policy, East Carolina University, Greenville, NC, USA

Tina Miller-Way Department of Marine Sciences, University of South Alabama, Dauphin Island, AL, USA

Ryan P. Mulligan Department of Civil Engineering, Queen's University, Kingston, ON, Canada

Ramiro Neves Marine Environment Technology Center (MARETEC), Instituto Superior Técnico, Universidade de Lisboa, Lisbon, Portugal

Michael A. O'Neal Department of Geological Sciences, University of Delaware, Newark, USA

Kyeong Park Department of Marine Sciences, University of South Alabama, Dauphin Island, AL, USA

Ashis Kr. Paul Department of Geography and Environment Management, Vidyasagar University, Midnapore, West Bengal, India

Joao Pereira Department of Civil and Environmental Engineering, and Pontchartrain Institute for Environmental Sciences, University of New Orleans, New Orleans, LA, USA

José Pinho C-TAC/Department of Civil Engineering, University of Minho, Braga, Portugal

Ligia Pinto Marine Environment Technology Center (MARETEC), Instituto Superior Técnico, Universidade de Lisboa, Lisbon, Portugal

Simon J. Pittman Biogeography Branch, National Oceanic and Atmospheric Administration, Silver Spring, MD, USA

Centre for Marine and Coastal Policy Research, Marine Institute, Plymouth University, Plymouth, UK

Shak Md. Bazlur Rahaman Fisheries and Marine Resource Technology Discipline, Khulna University, Khulna, Bangladesh

Md. Mujibor Rahman Environmental Science Discipline, Khulna University, Khulna, Bangladesh

Robert E. Reed Center for Applied Aquatic Ecology, North Carolina State University, Raleigh, NC, USA

Thomas Richardson Jackson State University, Jackson, MS, USA

Gwilym Rowlands Oceanographic Center, Nova Southeastern University, Fort Lauderdale, FL, USA

Khaled bin Sultan Living Oceans Foundation, Hyattsville, MD, USA

Suixiang Shi National Marine Data & Information Service, State Oceanic Administration, Tianjin, China

Ana Nobre Silva Instituto Dom Luiz, Geology Department, Science Faculty, University of Lisbon, Campo Grande, Lisbon, Portugal

João Sobrinho Marine Environment Technology Center (MARETEC), Instituto Superior Técnico, Universidade de Lisboa, Lisbon, Portugal

Carol Subrath-Ali Department of Civil and Environmental Engineering, Faculty of Engineering, The University of the West Indies, St. Augustine, Trindad and Tobago

Rui Taborda Instituto Dom Luiz, Geology Department, Faculdade de Ciências da Universidade de Lisboa, Lisbon, Portugal

Satoshi Takewaka Department of Engineering Mechanics and Energy, Graduate School of Systems and Information Engineering, University of Tsukuba, Tsukuba, Ibaraki, Japan

Jian Tao Center for Computation and Technology, Louisiana State University, Baton Rouge, LA, USA

Robert R. Twilley Department of Oceanography and Coastal Sciences, School of the Coast and Environment, Louisiana State University, Baton Rouge, LA, USA

Claudia Viegas Marine Environment Technology Center (MARETEC), Instituto Superior Técnico, Universidade de Lisboa, Lisbon, Portugal

Heather Vollmer Coastal Education and Research Foundation (CERF), Asheville, NC, USA

Amelia Ward Center for Freshwater Studies, University of Alabama, Tuscaloosa, AL, USA

Jerry Wiggert Department of Marine Sciences, University of Southern Mississippi, Hattiesburg, MS, USA

Derek Williamson Department of Civil, Construction and Environmental Engineering, University of Alabama, Tuscaloosa, AL, USA

Xin Zhang State Key Laboratory of Remote Sensing Science, Institute of Remote Sensing and Digital Earth, Chinese Academy of Sciences, Beijing, China

Hebei University of Engineering, Handan, Hebei Province, China

Part I
Remote Sensing, Mapping and Survey of Coastal Biophysical Environments

Chapter 1
Remote Sensing of Coastal Ecosystems and Environments

Victor V. Klemas

Abstract Advances in sensor design and data analysis techniques are making remote sensing systems suitable for monitoring coastal ecosystems and their changes. Hyperspectral imagers, LiDAR and radar systems are available for mapping coastal marshes, submerged aquatic vegetation, coral reefs, beach profiles, algal blooms, and concentrations of suspended particles and dissolved substances in coastal waters. Since coastal ecosystems have high spatial complexity and temporal variability, they benefit from new satellites, carrying sensors with fine spatial (0.4–4 m) or spectral (200 narrow bands) resolution. Imaging radars are sensitive to soil moisture and inundation and can detect hydrologic features beneath the vegetation canopy. Multi-sensor and multi-seasonal data fusion techniques are significantly improving coastal land cover mapping accuracy and efficiency. Using time-series of images enables scientists to study coastal ecosystems and to determine long- term trends and short- term changes.

1.1 Introduction

Coastal ecosystems, including marshes, mangroves, seagrasses and coral reefs, are highly productive and act as critical habitats for a wide variety of plants, fish, shellfish, and other wildlife. For instance, coastal wetlands provide flood protection, protection from storm and wave damage, water quality improvement through filtering of agricultural and industrial waste, and recharge of aquifers (Morris et al. 2002; Odum 1993). Since more than half of the U.S. population lives in the coastal zone, coastal ecosystems have been exposed to a wide range of stress-inducing alterations, including dredge and fill operations, hydrologic modifications,

V.V. Klemas (✉)
School of Marine Science and Policy, University of Delaware, Newark, DE 19716, USA
e-mail: klemas@udel.edu

C.W. Finkl and C. Makowski (eds.), *Remote Sensing and Modeling: Advances in Coastal and Marine Resources*, Coastal Research Library 9,
DOI 10.1007/978-3-319-06326-3_1, © Springer International Publishing Switzerland 2014

pollutant run-off, eutrophication, impoundments and fragmentation by roads and ditches (Waycott et al. 2009). Furthermore, with events such as the hurricanes of 2004, 2005 and 2012 annual losses to coastal communities can total billions of dollars. Environmental impacts from coastal storms include beach erosion, wetland destruction, excessive nutrient loading, algal blooms, hypoxia and anoxia, fish kills, releases of pollutants, spread of pathogens, and bleaching of coral reefs.

Over the long term, coastal communities are also facing a rising sea level. The substantial sea level rise and more frequent storms predicted for the next 50–100 years will affect coastal towns and roads, coastal economic development, beach erosion control strategies, salinity of estuaries and aquifers, coastal drainage and sewage systems, and coastal wetlands and coral reefs (Gesch 2009; IPCC 2007; NOAA 1999). Coastal areas such as barrier islands, beaches, and wetlands are especially sensitive to sea-level changes. A major hurricane can devastate a wetland (Klemas 2009). Rising seas will intensify coastal flooding and increase the erosion of beaches, bluffs and wetlands, as well as threaten jetties, piers, seawalls, harbors, and waterfront property. Along barrier islands, the erosion of beachfront property by flooding water will be severe, leading to greater probability of overwash during storm surges (NOAA 1999).

Since coastal ecosystems have high spatial complexity and temporal variability, they require high spatial, spectral and temporal resolutions. Recent advances in sensor design and data analysis techniques are making remote sensing systems practical and cost-effective for monitoring natural and man-made changes impacting coastal ecosystems. High resolution multispectral and hyperspectral imagers, LiDAR and radar systems are available for monitoring changes in coastal marshes, submerged aquatic vegetation, coral reefs, beach profiles, algal blooms, and concentrations of suspended particles and dissolved substances in coastal waters. Some of the ecosystem health indicators that can be mapped with new high-resolution remote sensors include natural vegetation cover, wetland loss and fragmentation, wetland biomass change, percent of impervious watershed area, buffer degradation, changes in hydrology, water turbidity, chlorophyll concentration, eutrophication level, salinity, *etc.* (Lathrop et al. 2000; Martin 2004; Wang 2010).

With the rapid development of new remote sensors, data bases and image analysis techniques, potential users need guidance in choosing remote sensors and data analysis methods that are most appropriate for each specific coastal application (Yang 2009). The objective of this paper is to review those remote sensing techniques that are cost-effective and practical for shoreline delineation, wetland mapping and other coastal applications.

1.2 Wetland Mapping

For more than three decades remote sensing techniques have been used by researchers and government agencies to map and monitor wetlands (Dahl 2006; Tiner 1996). Traditionally, in addition to airborne sensors, the Landsat Thematic Mapper (TM) and the French SPOT satellite have been reliable data sources for

Fig. 1.1 The Moderate Resolution Imaging Spectroradiometer (MODIS) on NASA's Terra satellite captured this image on September 26, 2008, 13 days after Hurricane Ike came ashore. The *brown* areas in the image are the result of a massive storm surge that Ike pushed far inland over Texas and Louisiana causing a major marsh dieback (Color figure online) (Credits: NASA/GSFC)

wetland and land cover mapping (Klemas 2011). Their 30 m and 10–20 m respective spatial resolutions and spectral bands have proven cost-effective for mapping land cover and changes in large coastal watersheds (Harvey and Hill 2001; Houhoulis and Michener 2000; Jensen 2007; Lunetta and Balogh 1999). Landsat TM and ETM+ imagery have also been used to study water turbidity and depth in marshes as well as the seasonal dynamics of inundation, turbidity, and vegetation cover (Bustamante et al. 2009; Ward et al. 2012).

More recently other medium spatial resolution satellite sensors, such as MODIS on NASA's Terra and Aqua satellites, have been used to map wetlands and study their interaction with storm surges. This is illustrated in Fig. 1.1, which shows an image of the Texas coast captured by MODIS on NASA's Terra satellite 13 days after Hurricane Ike made landfall on September 13, 2008. The storm's surge covered hundreds of kilometers of the Gulf Coast because Ike was a large storm, with tropical-storm-strength winds stretching more than 400 km from the center of the storm. Most of the shoreline in this region is coastal wetland. One can clearly distinguish the red-brown areas in the image which are the result of the massive storm surge that Ike had pushed far inland over Texas and Louisiana, causing a major marsh dieback. The salty water burned the plants, leaving them wilted and brown. The brown line corresponds with the location and extent of the wetlands. North of the brown line, the vegetation gradually transitions to pale green farmland

and dark green natural vegetation untouched by the storm's surge. The powerful tug of water returning to the Gulf also stripped marsh vegetation and soil off the land. Therefore, some of the brown seen in the wetlands may be deposited sediment. Plumes of brown water are visible as sediment-laden water drains from rivers and the coast in general. The muddy water slowly diffuses, turning pale green, green, and finally blue as it blends with clearer Gulf water (NASA/GSFC 2010; Ramsey and Rangoonwala 2005).

Many coastal ecosystems are patchy and exhibit considerable variations in their extent, spatial complexity, and temporal variability (Dahl 2006). Protecting them requires the ability to monitor their biophysical features and controlling processes at high spatial and temporal resolutions, such as that provided by aircraft and high spatial resolution satellite sensors (Adam et al. 2010; Klemas 2011). More recently, the availability of high spatial and spectral resolution satellite data has significantly improved the capacity for mapping salt marshes and other coastal ecosystems (Jensen et al. 2007; Laba et al. 2008; Ozesmi and Bauer 2002; Wang et al. 2010). High resolution imagery (0.4–4 m) can now be obtained from satellites, such as IKONOS and QuickBird. Major plant species within a complex, heterogeneous tidal marsh have been classified using multitemporal high-resolution QuickBird images, field reflectance spectra and LiDAR height information. *Phragmites*, *Typha* spp. and *S. patens* were spectrally distinguishable at particular times of the year, likely due to differences in biomass and pigments and the rate at which change occurred throughout the growing season. For instance, classification accuracies for *Phragmites* were high due to the uniquely high near-infrared reflectance and height of this plant in the early fall (Ghioca-Robrecht et al. 2008; Gilmore et al. 2010).

High resolution imagery is more sensitive to within- class spectral variance, making separation of spectrally mixed land cover types more difficult than when using medium resolution imagery. Therefore, pixel-based techniques are sometimes replaced by object-based methods, which incorporate spatial neighborhood properties, by segmenting/partitioning the image into a series of closed objects which coincide with the actual spatial pattern, and then proceed to classify the image. "Region growing" is among the most commonly used segmentation methods. This procedure starts with the generation of seed points over the whole scene, followed by grouping neighboring pixels into an object under a specific homogeneity criterion. Thus the object keeps growing until its spectral closeness metric exceeds a predefined break-off value (Kelly and Tuxen 2009; Shan and Hussain 2010; Wang et al. 2004).

Small wetland sites are often mapped and studied using airborne sensors (Jensen 2007; Klemas 2011). Airborne georeferenced digital cameras, providing color and color infrared digital imagery are particularly suitable for accurate wetland mapping and interpreting satellite data. Most digital cameras are capable of recording reflected visible to near-infrared light. A filter is placed over the lens that transmits only selected portions of the wavelength spectrum. For a single camera operation, a filter is chosen that generates natural color (blue-green-red wavelengths) or color-infrared (green-red-near IR wavelengths) imagery. For multiple camera operation, filters that transmit narrower bands are chosen (Ellis and Dodd 2000).

Digital camera imagery can be integrated with GPS position information and used as layers in a GIS for a wide range of modeling applications (Lyon and McCarthy 1995). Small aircraft flown at low altitudes (*e.g*. 200–500 m) can also be used to guide field data collection (McCoy 2005). However, cost becomes excessive if the study site is larger than a few hundred square kilometers, and in that case, medium resolution multispectral sensors, such as Landsat TM (30 m) and SPOT (20 m), become more cost-effective (Klemas 2011).

1.3 Hyperspectral Remote Sensing of Wetlands

Airborne hyperspectral imagers, such as the Advanced Visible Infrared Imaging Spectrometer (AVIRIS) and the Compact Airborne Spectrographic Imager (CASI) have been used for mapping coastal wetlands and shallow water substrate (Fearns et al. 2011; Lesser and Mobley 2007; Li et al. 2005; Rosso et al. 2005; Ozesmi and Bauer 2002; Schmidt and Skidmore 2003; Thomson et al. 1998). Hyperspectral imagers may contain hundreds of narrow spectral bands located in the visible, near-infrared, mid-infrared, and sometimes thermal portions of the EM spectrum (Jensen et al. 2007).

The advantages and problems associated with hyperspectral mapping have been clearly demonstrated by Hirano et al. (2003) who used AVIRIS hyperspectral data to map vegetation for a portion of Everglades National Park in Florida. The AVIRIS provides 224 spectral bands from 0.4 to 2.45 μm, each with 0.01 μm bandwidth, 20 m spatial resolution, and a swath width of 10.5 km. Hirano et al. compared the geographic locations of spectrally pure pixels in the AVIRIS image with dominant vegetation polygons of the Everglades Vegetation Database and identified spectrally pure pixels as ten different vegetation classes, plus water and mud. An adequate number of pure pixels was identified to permit the selection of training samples used in the automated classification procedure. The spectral signatures from the training samples were then matched to the spectral signatures of each individual pixel. Image classification was undertaken using the ENVI spectral angle mapper (SAM) classifier in conjunction with the spectral library created for the Everglades study area. The SAM classifier examines the digital numbers (DNs) of all bands from each pixel in the AVIRIS data set to determine similarity between the angular direction of the spectral signature (*i.e.* color) of the image pixel and that of a specific class in the spectral library. A coincident or small spectral angle between the vector for the unknown pixel and that for a vegetation class training sample indicates that the image pixel likely belongs to that vegetation class. In the case of spectrally mixed pixels, the relative probability of membership (based on the spectral angle) to all vegetation classes is calculated. Mixed pixels are then assigned to the class of the greatest probability of membership (Hirano et al. 2003).

The hyperspectral data proved effective in discriminating spectral differences among major Everglades plants such as red, black and white mangrove communities and enabled the detection of exotic invasive species (Hirano et al. 2003). The overall classification accuracy for all vegetation pixels was 65.7 %, with different mangrove tree species ranging from 73.5 to 95.7 % correct. Limited spatial resolution was a problem, resulting in too many mixed pixels. Another problem was the complexity of image-processing procedures that are required before the hyperspectral data can be used for automated classification of wetland vegetation. The tremendous volume of hyperspectral image data necessitated the use of specific software packages, large data storage, and extended processing time (Hirano et al. 2003). A detailed accuracy assessment of airborne hyperspectral data for mapping plant species in freshwater coastal wetlands has been performed by Lopez et al. (2004).

A number of advanced new techniques have been developed for mapping wetlands and even identifying wetland types and plant species (Schmidt et al. 2004; Jensen et al. 2007; Klemas 2011; Yang et al. 2009). For instance, using LiDAR, hyperspectral and radar imagery, and narrow-band vegetation indices, researchers have been able not only discriminate some wetland species, but also make progress on estimating biochemical and biophysical parameters of wetland vegetation, such as water content, biomass and leaf area index (Adam et al. 2010; Artigas and Yang 2006; Filippi and Jensen 2006; Gilmore et al. 2010; Ozesmi and Bauer 2002; Simard et al. 2010; Wang 2010). The integration of hyperspectral imagery and LiDAR-derived elevation has also significantly improved the accuracy of mapping salt marsh vegetation. The hyperspectral images help distinguish high marsh from other salt marsh communities due to its high reflectance in the near-infrared region of the spectrum, and the LiDAR data help separate invasive *Phragmites* from low marsh plants (Yang and Artigas 2010).

Hyperspectral imaging systems are now available not only for airborne applications, but also in space, such as the satellite-borne Hyperion system, which can detect fine differences in spectral reflectance, assisting in species discrimination on a global scale (Christian and Krishnayya 2009; Pengra et al. 2007). The Hyperion sensor provides imagery with 220 spectral bands at a spatial resolution of 30 m. Although there have been few studies using satellite-based hyperspectral remote sensing to detect and map coastal vegetation species, results so far have shown that discrimination between multiple species is possible (Blasco et al. 2005; Heumann 2011).

1.4 Wetland Applications of Synthetic Aperture Radar (SAR)

Imaging radars provide information that is fundamentally different from sensors that operate in the visible and infrared portions of the electromagnetic spectrum. This is primarily due to the much longer wavelengths used by SAR sensors and the fact that they send out and receive their own energy (*i.e.*, active sensors). One of the most common types of imaging radar is Synthetic Aperture Radar (SAR). SAR technology provides the increased spatial resolution that is necessary in regional wetland mapping and SAR data have been used extensively for this purpose (Lang and McCarty 2008; Novo et al. 2002).

When mapping and monitoring wetland ecosystems, imaging radars have some advantages over sensors that operate in the visible and infrared portions of the electromagnetic spectrum. Microwave energy is sensitive to variations in soil moisture and inundation, and is only partially attenuated by vegetation canopies, especially in areas of lower biomass (Baghdadi et al. 2001; Kasischke et al. 1997a, b; Lang and Kasischke 2008; Rosenqvist et al. 2007; Townsend 2000, 2002; Townsend and Walsh 1998) or when using data collected at longer wavelengths (Hess et al. 1990; Martinez and Le Toan 2007).

The sensitivity of microwave energy to water and its ability to penetrate vegetative canopies, make SAR ideal for the detection of hydrologic features below the vegetation (Kasischke et al. 1997a; Kasischke and Bourgeau-Chavez 1997; Phinn et al. 1999; Rao et al. 1999; Wilson and Rashid 2005). The presence of standing water interacts with the radar signal differently depending on the dominant vegetation type/structure (Hess et al. 1995) as well as the biomass and condition of vegetation (Costa and Telmer 2007; Töyrä et al. 2002). When exposed to open water without vegetation, specular reflection occurs and a dark signal (weak or no return) is observed (Dwivedi et al. 1999). The radar signal is often reduced in wetlands dominated by lower biomass herbaceous vegetation when a layer of water is present due largely to specular reflectance (Kasischke et al. 1997a). Conversely, the radar signal is often increased in forested wetlands when standing water is present due to the double-bounce effect (Harris and Digby-Arbus 1986; Dwivedi et al. 1999). This occurs in flooded forests when the radar pulse is reflected strongly by the water surface away from the sensor (specular reflectance) but is then redirected back towards the sensor by a second reflection from a nearby tree trunk. The use of small incidence angles (closer to nadir) enhances the ability to map hydrology beneath the forest canopy due to increased penetration of the canopy (Bourgeau-Chavez et al. 2001; Hess et al. 1990; Lang and McCarty 2008; Töyrä et al. 2001).

1.5 Wetland Change Detection

Many coastal wetlands, such as the tidal salt marshes along the Louisiana coast, are generally within fractions of a meter of sea level and will be lost, especially if the impact of sea level rise is amplified by coastal storms. Man-made modifications of wetland hydrology and extensive urban development will further limit the ability of wetlands to survive sea level rise. To identify long-term trends and short term variations, such as the impact of rising sea levels and storm surges on wetlands, one needs to analyze time-series of remotely sensed imagery. High temporal resolution, precise spectral bandwidths, and accurate georeferencing procedures are factors that contribute to the frequent use of satellite image data for change detection analysis (Baker et al. 2007; Coppin et al. 2004; Shalabi and Tateishi 2007). A good example is the study of the onset and progression of marsh dieback performed by Ramsey and Rangoonwala (2010).

The acquisition and analysis of time-series of multi-spectral imagery is a challenging task. The imagery must be acquired under similar environmental conditions (*e.g.* same time of year, sun angle, *etc.*) and in the same or similar spectral bands. There will be changes in both, time and spectral content. One way to approach this problem is to reduce the spectral information to a single index, reducing the multispectral imagery into one single field of the index for each time step. In this way the problem is simplified to the analysis of a time-series of a single variable, one for each pixel of the images.

The most common index used is the Normalized Difference Vegetation Index (NDVI), which is expressed as the difference between the red and near infrared (NIR) reflectances divided by their sum (Jensen 2007). These two spectral bands represent the most detectable spectral characteristic of green plants. This is because the red (and blue) radiation is absorbed by the chlorophyll in the surface layers of the plant (*Palisade parenchyma*) and the NIR is reflected from the inner leaf cell structure (*Spongy mesophyll*) as it penetrates several leaf layers in a canopy. Thus the NDVI can be related to plant biomass or stress, since the NIR reflectance depends on the abundance of plant tissue, whereas the red reflectance indicates the surface condition of the plant. It has been shown by researchers that time-series remote sensing data can be used effectively to identify long term trends and subtle changes of NDVI by means of Principal Component Analysis (Jensen 2007; Young and Wang 2001; Yuan et al. 1998).

The pre-processing of multi-date sensor imagery when absolute comparisons between different dates are to be carried out, is much more demanding than the single-date case. It requires a sequence of operations, including calibration to radiance or at-satellite reflectance, atmospheric correction, image registration, geometric correction, mosaicking, sub-setting, and masking out clouds and irrelevant features (Coppin et al. 2004; Lunetta and Elvidge 1998).

Detecting the actual changes between two registered and radiometrically corrected images from different dates can be accomplished by employing one of several techniques, including post-classification comparison (PCC), spectral image differencing (SID), and change vector analysis (CVA). In PCC change detection, two images from different dates are independently classified. The two classified maps are then compared on a pixel-by-pixel basis. One disadvantage is that every error in the individual date classification maps will also be present in the final change detection map (Jensen 1996; Lunetta and Elvidge 1998).

Spectral image differencing (SID) is the most widely applied change detection algorithm. SID techniques rely on the principle that land cover changes result in changes in the spectral signature of the affected land surface. SID techniques involve the transformation of two original images to a new single-band or multiband image in which the areas of spectral change are highlighted. This is accomplished by subtracting one date of raw or transformed (*e.g.* vegetation indices, albedo, *etc.*) imagery from a second date, which has been precisely registered to the image of the first date. Pixel difference values exceeding a selected threshold are considered as changed. This approach eliminates the need to identify land cover changes in areas where no significant spectral change has occurred between the two dates of imagery (Coppin et al. 2004; Jensen 1996). A comparison of the SID and the PCC change detection algorithms is provided by Macleod and Congalton (1998). The SID and the PCC based change detection methods are often combined in a hybrid approach. For instance, SID can be used to identify areas of significant spectral change, and then PCC can be applied within areas where spectral change was detected in order to obtain class-to-class change information.

The changeable nature of wetland ecosystems sometimes requires a more dynamic change detection procedure. These ecosystems can exhibit a variety of vegetative or hydrologic changes that might not be detected when using only one or two spectral bands. Change vector analysis (CVA) is a change detection technique that can measure change in more than two spectral bands, giving it an advantage when mapping rapidly changing and highly diverse wetlands (Baker et al. 2007; Coppin et al. 2004; Mitsch and Gosselink 2000). CVA determines the direction and magnitude of changes in multi-dimensional spectral space (Houhoulis and Michener 2000). CVA concurrently analyzes change in all data layers, instead of a few selected spectral bands (Coppin et al. 2004). The CVA method identifies a change magnitude threshold that is used to separate actual land cover changes from subtle changes due to the variability within land cover classes, as well as radiometric changes associated with instrument and atmospheric variations (Hame et al. 1998; Johnson and Kasischke 1998). Defining spectral threshold values to separate true landscape changes from inherent spectral variation is particularly beneficial for studies of broadly diverse ecosystems, such as wetlands (Houhoulis and Michener 2000). Human interpretation and sometimes an empirical threshold method need to be applied for interpretation of CVA results to obtain accurate information on wetland changes.

1.6 Submerged Aquatic Vegetation (SAV)

Seagrass beds provide essential habitat for many aquatic species, stabilize and enrich sediments, dissipate turbulence, reduce current flow, cycle nutrients, and improve water quality (Hughes et al. 2009). However, in many parts of the world, the health and quantity of seagrass beds has been declining (Orth et al. 2006; Waycott et al. 2009). The decline of coral reefs and SAV is closely linked to human activity since the coastlines and estuaries that host them are often heavily populated. Specifically, the declines have been attributed to reduction in water clarity, alteration of sediment migration via dredging, destruction from coastal engineering, boating and commercial fishing. High concentrations of nutrients exported from agriculture or urban sprawl in coastal watersheds are causing algal blooms in many estuaries and coastal waters (Klemas 2012). Algal blooms are harmful in that they cause eutrophic conditions, depleting oxygen levels needed by organic life and limiting aquatic plant growth by reducing water transparency.

Submerged aquatic plants and their properties are not as easily detectable as terrestrial vegetation. The spectral response of aquatic vegetation resembles that of terrestrial vegetation, yet the submerged or flooded conditions introduce factors that alter its overall spectral characteristics (Fyfe 2003; Han and Rundquist 2003; Pinnel et al. 2004; Williams et al. 2003). Thus the main challenge for remote sensing of submerged aquatic plants is to isolate the weakened plant signal from the interference of the water column, the bottom and the atmosphere. In addition to bottom reflectance, optically active materials, such as phytoplankton, suspended sediments and dissolved organics, affect the scattering and absorption of the radiation. A careful correction of atmospheric effects is important prior to the analysis of submerged vegetation imagery derived from satellite or high altitude airborne data (Silva et al. 2008).

More recently water column optical models have been used to correct water and bottom effects by including bathymetric information as one of the variables (Dierssen et al. 2003; Heege et al. 2003). Paringit et al. (2003) developed a seagrass canopy model to predict the spectral response of submerged macrophytes in shallow waters. The model considers not only the effects of the water column through radiative transfer modeling, but also viewing and illumination conditions, leaf and bottom reflectance, leaf area index and the vertical distribution of biomass. By inverting the model, the authors were able to estimate plant coverage and abundance with IKONOS satellite imagery, and compare the remotely sensed results with field measurements (Silva et al. 2008). In several other studies digital elevation models and bathymetric data have also been successfully incorporated in the SAV classification approach in order to relate the change in the SAV to water depth (Valta-Hulkkonen et al. 2003, 2004; Wolter et al. 2005).

Since SAV communities have high spatial complexity and temporal variability, standard methods for determining seagrass status and trends have been based on high resolution aerial color photography taken from low to medium altitude flights (Ferguson et al. 1993; NOAA-CSC 2001; Pulich et al. 1997). The color photos are traditionally analyzed by photo-interpreting the 9×9 in. positive photo-transparencies to map the SAV distribution. This is often followed by digitization of the seagrass polygons from map overlays and compilation of digital data into a spatial GIS database. Using airborne color and color infrared video imagery researchers have been able to distinguish between water hyacinth and hydrilla with an accuracy of 87.7 % (Everitt et al. 1999). Good mapping results have also been obtained with recently available airborne digital cameras (Kolasa and Craw 2009).

Large SAV beds and other benthic habitats have been mapped using Landsat TM with limited accuracies ranging from 60 to 74 %. Eight bottom types could be spectrally separated using supervised classification: sand, dispersed communities over sand, dense seagrass, dispersed seagrass over sand, reef communities, mixed vegetation over muddy bottom, and deep water (Gullstrom et al. 2006; Nobi and Thangaradjou 2012; Schweitzer et al. 2005; Wabnitz et al. 2008). SAV biomass has been mapped with Landsat TM using regression analysis between the principal components and biomass, after eigenvector rotation of four TM bands (Armstrong 1993; Zhang 2010). Changes in eelgrass and other seagrass beds have also been mapped with TM data with accuracies of about 66 %, including a study which showed that image differencing was more effective than post-classification or principal component change detection analysis (Macleod and Congalton 1998; Gullstrom et al. 2006).

The mapping of submerged aquatic vegetation (SAV), coral reefs and general bottom characteristics from satellites has become more accurate since high resolution (0.4–4 m) multispectral imagery became available (Mumby and Edwards 2002; Purkis et al. 2002; Purkis 2005). Coral reef ecosystems usually exist in clear water and can be classified to show different forms of coral reef, dead coral, coral rubble, algal cover, sand, lagoons, different densities of seagrasses, *etc*. SAV often grows in somewhat turbid waters and thus is more difficult to map. Aerial hyperspectral scanners and high resolution multispectral satellite imagers, such as IKONOS and QuickBird, have been used to map SAV with accuracies of about 75 % for classes including high-density seagrass, low-density seagrass, and unvegetated bottom (Dierssen et al. 2003; Mishra et al. 2006; Wolter et al. 2005).

Hyperspectral imagers have improved SAV and coral reef mapping results by being able to identify more estuarine and intertidal habitat classes (Garono et al. 2004; Mishra et al. 2006; Phinn et al. 2008; Pu et al. 2012; Purkis et al. 2008; Nayegandhi et al. 2009). Figure 1.2 shows a 2006 hyperspectral image of seagrass communities in the St. Joseph Bay Aquatic Preserve in Florida. The maps produced from such images showed coverage and extent of seagrass communities in the bay, and provided an indicator of the bay's health. The maps were used to identify "good" areas to be targeted for protection, and "poor" areas to be targeted for restoration (CAMA 2009).

SAV has been mapped with high accuracies using airborne hyperspectral imagers and regression models, binary decision trees incorporating spectral mixture analysis, spectral angle mapping, and band indexes (Hestir et al. 2008; Peneva et al. 2008). Airborne LiDARS have also been used with multispectral or hyperspectral imagers to map coral reefs and SAV (Brock et al. 2004, 2006; Brock and Purkis 2009).

Acoustic techniques have been used for rapid detection of submerged aquatic vegetation in turbid waters. The acoustic impedance (density difference between the plant and surrounding water) which produces the reflections, is thought to result primarily from the gas within the plant, since the more buoyant species (with more gas) reflect acoustic signals more strongly. Hydroacoustic techniques include horizontally-aimed side scanning sonar systems and vertically-aimed echo sounders. Side-scan sonar systems provide complete bottom coverage and generate an image. They have been effective for delineating seagrass beds (Moreno et al. 1998; Sabol et al. 2002). The horizontal orientation of the acoustic beam results in a stronger reflected signal from the vertically oriented grass blades.

Echo sounders are pointed vertically downward and traverse a path generating an analog strip chart, with the horizontal axis equal to distance, vertical axis equal to depth, and echo intensity shown as gray scale. Numerous researchers using echo sounders have reported success in detecting and qualitatively characterizing seagrass beds (Miner 1993; Hundley 1994). For instance, Sabol et al. (2002) used high resolution digital echo sounders linked with GPS equipment. The acoustic reflectivity of SAV allowed for detection and measurement of canopy geometry, using digital signal processing algorithms. Comparison with field data showed good detection and measurement performance over a wide range of conditions.

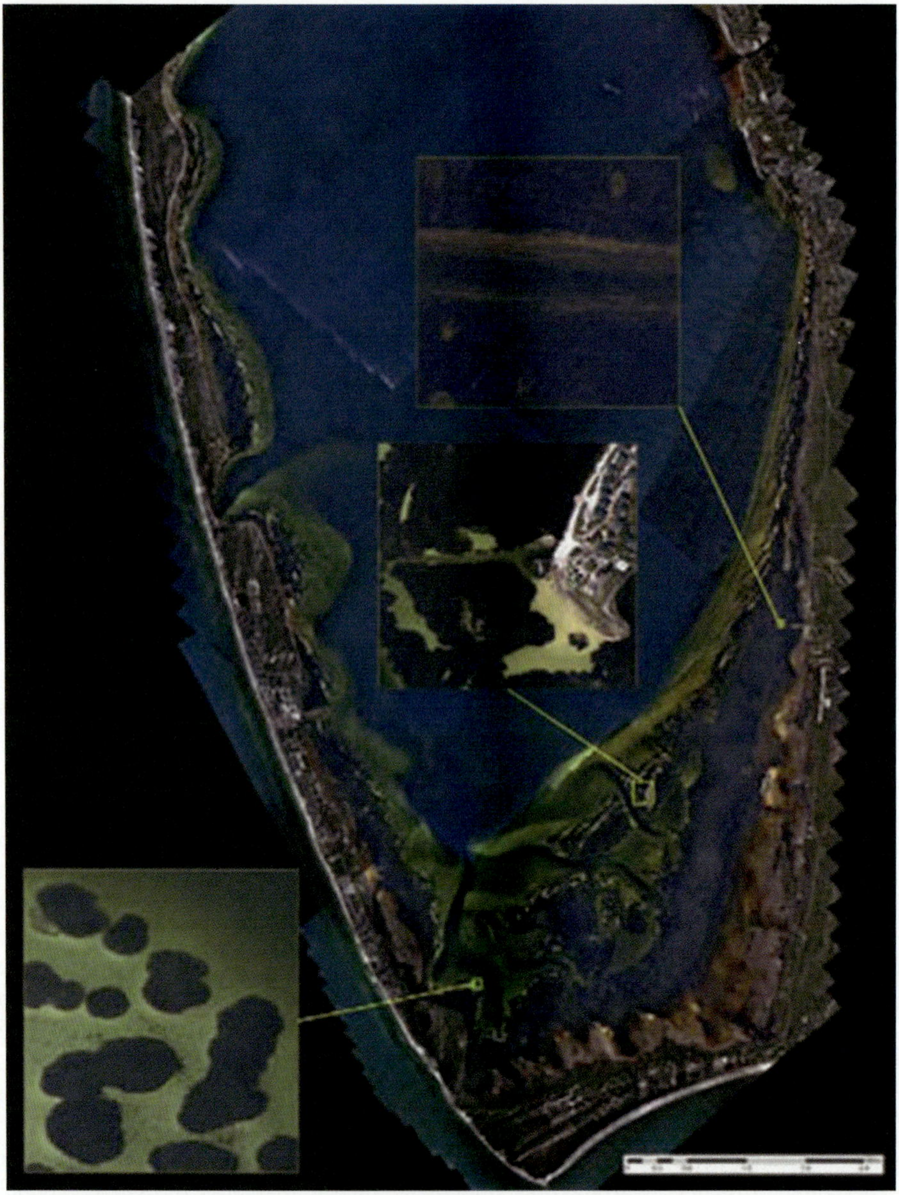

Fig. 1.2 Hyperspectral imagery of seagrass communities in St. Joseph Bay, FL (CAMA 2009)

1.7 Beach Profiling and Shoreline Change Detection

Information on beach profiles and coastal bathymetry is needed for studies of near-shore geomorphology, hydrology and sedimentary processes (Finkl et al. 2005a; Lidz et al. 1997). In order to plan sustainable coastal development

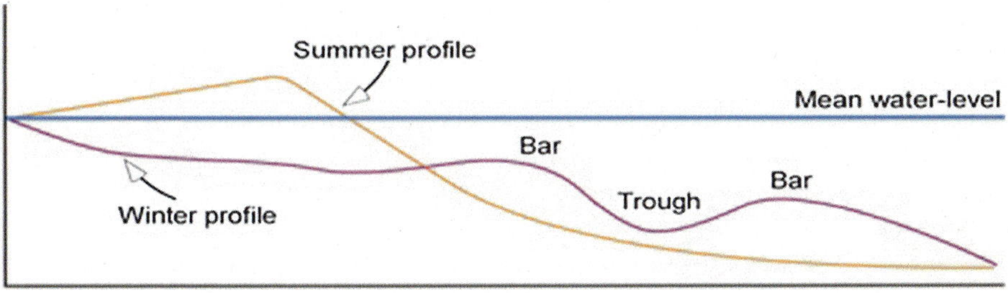

Fig. 1.3 Changes in beach profiles between summer and winter due to changes in wave climate. During winter storms the beach is eroded and seaward cross-shore sediment transport results in the formation of off-shore bars (Purkis and Klemas 2011)

and implement effective beach erosion control, flood zone delineation and ecosystem protection, coastal managers and researchers need information on long-term and short-term changes taking place along the coast, including changes in beach profiles due to erosion by storms and littoral drift, wetlands changes due to inundation, *etc*. (Gesch 2009; West et al. 2001).

Beach profiles and shoreline positions can change rapidly with seasons and after storms, in addition to exhibiting slower changes due to littoral drift and sea level rise (Stockdon et al. 2009). During winter storms waves remove sand from the beach and deposit it offshore, typically in bar formation (Fig. 1.3). During summer, milder wave formations move the bars onshore and rebuild the wider berm for the "summer beach". Long-term changes of shorelines due to littoral drift or sea level rise can be aggravated by man-made structures such as jetties, seawalls and groins (Finkl 1996; Irish and White 1998; Khalil and Finkl 2007; Klemas 2009; Wang 2010).

Before the advent of the Global Positioning System (GPS) and Light Detection and Ranging (LiDAR) systems, shoreline position analysis and beach profiling were based on historical aerial photographs and topographical sheets (Morton and Miller 2005). To map long-term changes of shorelines due to beach erosion, time-series of historical aerial photographs are available dating back to the 1930s and topographic maps exist to extend the record of shoreline change to the mid-late 1800s. Such data are held by local, state and federal agencies, including the U.S. Geological Survey and the USDA Soil Conservation Service. These agencies also have various maps, including planimetric, topographic, quadrangle, thematic, ortho-photo, satellite and digital maps (Jensen 2007; Rasher and Weaver 1990). Time series of high resolution satellite images have also been used to map shoreline changes, but with accuracies of the order of several meters.

To perform a shoreline position analysis, the shoreline can be divided into segments which are uniformly eroding or accreting. Then the change in the distance of the waterline can be measured in reference to some stable feature like a coastal highway. The instantaneous water line in the image is not a temporally representative shoreline. The high water line, also referred to as the wet/dry line, is a commonly used indicator because it is visible in most images. Other indicators include the vegetation line, bluff line, or man-made shore vestments (Boak and Turner 2005; Thieler and Danforth 1994).

Topographical and depth data can now be effectively acquired at various spatial scales by airborne laser surveys using LiDAR techniques (Guenther et al. 1996; Ackermann 1999; Krabill et al. 2000; Lillycrop et al. 2002). A laser transmitter/receiver mounted on an aircraft transmits a laser pulse that travels to the land surface or the air-water interface, where a portion of this energy reflects back to the receiver. The land topography is obtained from the LiDAR pulse travel-time. On water, some of the energy propagates through the water column and reflects off the sea bottom. The water depth is calculated from the time lapse between the surface return and the bottom return (Hapke 2010; Purkis and Klemas 2011).

Global Positioning Systems (GPS), combined with airborne LiDAR techniques, make it now possible to obtain accurate topographical maps, including shoreline positions (Jensen 2007; Morton and Miller 2005; Schmid et al. 2011). A particularly effective approach for studying sand dynamics along coastlines includes the combined use of airborne hyperspectral data and airborne LiDAR data. Airborne LiDAR surveying has been significantly enhanced by kinematic differential GPS methods which enable the positioning of small aircraft to within several centimeters. Inertial navigation systems provide three-dimensional aircraft orientation to within 0.005° making aero-triangulation with ground data points unnecessary. LiDAR transmitters can provide elevation measurements at over 2,000 points per second from altitudes of 500–1,000 m with vertical accuracies of 10–15 cm (Brock and Sallenger 2000; Cracknell and Hayes 2007; Finkl et al. 2005a; Hapke 2010).

A typical beach profiling procedure using LiDAR may include cross-shore profiles every 10 m. Beach slope and location, elevation of the berm, dune base and dune crest can also be determined from these beach profiles. One can use a known vertical datum to remove the subjective nature of identifying the shoreline. The water line is then readily identified, because laser returns from the sea are noisy (Stockdon et al. 2002).

A LiDAR aircraft mapping configuration usually includes a light aircraft equipped with a LiDAR instrument and GPS, which is operated in tandem with a GPS base station. In coastal applications, the aircraft flies along the coast at heights of about 200–500 m, surveying a ground swath directly below the aircraft. The aircraft position throughout the flight is recorded by an onboard GPS receiver. The aircraft GPS signals are later combined with signals concurrently collected by a nearby GPS base station. Differential kinematic GPS post-processing determines the aircraft flight trajectory to within about 5 cm (Cracknell and Hayes 2007; Wang 2010). Although airborne laser mapping may be carried out at night, flight safety dictates that coastal LiDAR operations are normally confined to daylight hours and timed to coincide with low tide to maximize coverage of the beach face.

LiDAR surveys can now produce a 10 cm vertical accuracy at spatial densities greater than one elevation measurement per square meter. Such performance satisfies the needs of various coastal applications, including flood zone delineation, monitoring beach nourishment projects and mapping changes along barrier island beaches and other sandy coasts (Brock and Purkis 2009; Deronde et al. 2006; Gares et al. 2006; Raber et al. 2007; Webster et al. 2004; Wozencraft and Millar 2005). The ability of LiDAR to rapidly survey long, narrow strips of terrain is important, because beaches are elongate, highly dynamic sedimentary environments that undergo seasonal and long-term erosion or accretion and are also impacted by severe storms (Kempeneers et al. 2009; Krabill et al. 2000; Stockdon et al. 2002; Zhou 2010).

In order to develop digital flood insurance maps and data for habitat studies and vegetation identification, in 2005 the State of Delaware contracted with USGS and NASA to produce high detail elevation data using NASA's Experimental Advanced Airborne Research LiDAR (EAARL), which was specifically designed to measure submerged topography and adjacent coastal land elevations. Emergency managers have been able to use this data to develop statewide inundation maps and to incorporate this data into flood and storm surge models to create an early-flood-warning system (Carter and Scarborough 2010).

1.8 Bathymetry

The lack of accurate near-shore bathymetric data has been identified as a key limitation in the application of geospatial data to coastal environments (Malthus and Mumby 2003). Remote sensing techniques that have been successfully used to map coastal water depth include LiDAR and acoustic depth sounding. In LiDAR bathymetry, a laser transmitter/receiver mounted on an aircraft transmits a pulse that reflects off the air-water interface and the sea bottom. Since the velocity of the light pulse is known, the water depth can be calculated from the time lapse between the surface return and the bottom return. Because laser energy is lost due to refraction, scattering and absorption at the water surface, the sea bottom and inside the water column, these effects limit the strength of the bottom return and limit the maximum detectable depth.

Examples of LiDAR applications include regional mapping of changes along sandy coasts due to storms or long-term sedimentary processes and in the analysis of shallow benthic environments (Bonisteel et al. 2009; Guenther et al. 1996; Irish and Lillycrop 1997; Gutierrez et al. 1998; Kempeneers et al. 2009; Sallenger et al. 1999). Bertels et al. (2012) used integrated optical and acoustic remote sensing data over the backshore-foreshore-nearshore continuum to study sediment dynamics along the Belgian coastline. To accomplish this, the authors combined airborne hyperspectral imaging spectroscopy, airbore laser scanning and seaborne sonar. The LiDAR and hyperspectral data were combined with side-scan sonar and single- and multibeam depth and backscatter data to construct integrated sedimentological and morphological maps (Bertels et al. 2012).

Another example is the use of airborne laser bathymetry (ALB) to rapidly acquire large, high-quality data sets along the Continental Shelf of southeast Florida by Finkl et al. (2005b). The ALB provided a contiguous set of data for 160 km of coast from onshore to 6 km offshore. Image enhancement of the ALB digital data facilitated recognition of numerous seafloor features and bathymetric patterns. Bathymetric analysis of the 600 sq km survey area allowed for the first time an assessment of links between the influence of seabed morphology on wave transformation patterns and beach morphodynamics in southeast Florida.

To maximize water penetration, bathymetric LiDARs employ a blue-green laser with a typical wavelength of 530 nm to range the distance to the seabed. With the near-exponential attenuation of electromagnetic energy by water with increasing wavelength, a pure blue laser with a wavelength shorter than 500 nm would offer greater penetration. However, this wavelength is not used because, first, blue light interacts much more strongly with the atmosphere than longer wavelengths; and second, creating a high-intensity blue laser is energetically less efficient than blue-green and consumes a disproportionately large amount of instrument power.

Conversely, terrestrial topographical LiDARs typically utilize near-infrared (NIR) lasers with wavelengths of 1,064 nm. As is the case for the blue-green laser used for hydrography, this NIR wavelength is focused and easily absorbed by the eye. Therefore, the maximum power of the LiDAR system is limited by the need to make them eye-safe. While bathymetric lasers are limited in their accuracy by water column absorption, terrestrial infrared lasers suffer from null or poor returns from certain materials and surfaces such as water, asphalt, tar, clouds and fog, all of which absorb NIR wavelengths.

Because they do not penetrate water predictably, NIR topographical lasers cannot be used to assess bathymetry. Dual-wavelength LiDAR provides both bathymetric and topographical LiDAR mapping capability by carrying both an NIR and a blue-green laser. The NIR laser is not redundant over water, because it reflects off the air-water interface and can be used to refine the altitude above the sea surface as well as to distinguish dry land from water using the signal polarization (Guenther 2007). In addition, specific LiDAR systems, like the Scanning Hydrographic Operational Airborne LiDAR System (SHOALS), record the red wavelength Raman signal (647 nm). The Raman signal comes from interactions between the blue-green laser and water molecules, causing part of the energy to be backscattered while changing wavelength (Guenther et al. 1994; Irish and Lillycrop 1999). A detailed description of the SHOALS system is provided by Lillycrop et al. (1997).

By employing a very high scan-rate, state-of-the-art systems such as the Experimental Advanced Airborne Research LiDAR (EAARL) have been used to measure both, topography and bathymetry, from the return time of a single blue-green laser (Bonisteel et al. 2009; McKean et al. 2009; Nayegandhi et al. 2009). Operating in the blue-green portion of the electromagnetic spectrum, the EAARL is specifically designed to measure submerged topography and adjacent coastal land elevations seamlessly in a single scan of transmitted laser pulses. Figure 1.4 shows such a bathymetric-topographical DEM of a section of the Assateague Island National Seashore, captured by the EAARL. Assateague Island National Seashore consists of a 37-mile-long barrier island along the Atlantic coasts of Maryland and Virginia. This experimental advance signaled the future move towards commercial implementation of dual-application but single-wavelength instruments (Krabill et al. 2000; Wozencraft and Lillycrop 2003).

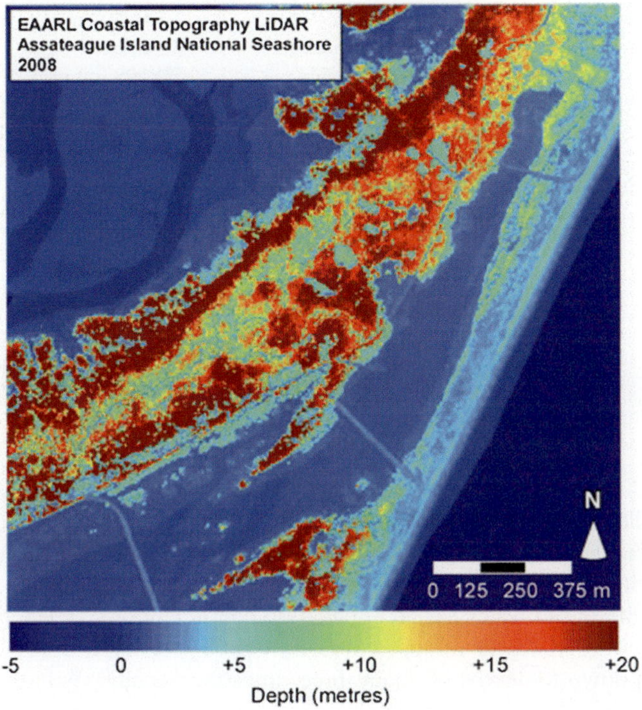

Fig. 1.4 Coastal topography for a section of the Assateague Island National Seashore acquired using the Experimental Advanced Airborne Research LiDAR (EAARL) (Bonisteel et al. 2009) (Credit: USGS)

While the EAARL and dual-wavelength LiDARs offer nearly seamless profiles between bathymetry and terrestrial terrain, neither bathymetric system can acquire dependable bathymetric data in very shallow depths or over white water in the surf zone. When white-caps are present, the laser does not penetrate the water column. Furthermore, if the depth is less than 2 m, even in clear water it becomes difficult to separate the laser pulse returning from the water surface from the one reflected by the bottom bed (Parson et al. 1997; Philpot 2007; Bonisteel et al. 2009). For coastal mapping, both problems are obviated by combining successive flights at low tide with a topographical LiDAR, and at high tide with a bathymetric LiDAR (Pastol et al. 2007; Sinclair 2008; Stoker et al. 2009). Such a strategy is not possible for coastal areas that do not have large tidal variations, or for non-tidal inland water bodies.

Optical water clarity is the most limiting factor for LiDAR depth detection, so it is important to conduct the LiDAR overflights during tidal and current conditions that minimize the water turbidity due to sediment re-suspension and river inflow (Sinclair 2008). LiDAR overflights should not be conducted during high wind conditions, since the rough water surface will scatter the LiDAR pulse and make it difficult to detect (Brock and Sallenger 2000; Irish and Lillycrop 1999).

Table 1.1 LiDAR flight parameters

Flying height	200–500 m (400 m typical)
Vertical accuracy	±15 cm
Horizontal accuracy	DGPS = 3 m; KGPS = 1 m
Max mapping depth	60 m (clear water)
Typical *kd* product	4
Coastal *k*	0.2–0.8 (d = 5–20 m)
Estuarine *k*	1.0–4.0 (d = 1–4 m)
Sounding density	3–15 m
Sun angle	18°–25° (to minimize glare)
Scan geometry	Circular (220 m swath typical)
Sea state	Low (0–1 Beaufort scale)
Water penetration	Green LiDAR (532 nm) used
Aircraft height	Infrared LiDAR (1,064 nm) used

DGPS differential GPS mode, *KGPS* kinematic GPS mode

Typical flight parameters for airborne LiDARs used in bathymetry are shown in Table 1.1. The LiDAR system must have a *kd* factor large enough to accommodate the water depth and water turbidity at the study site (k = attenuation coefficient; d = max. water penetration depth). For instance, if a given LiDAR system has a $kd = 4$ and the turbid water has an attenuation coefficient of $k = 1$, the system will be effective only to depths of approximately 4 m. Typically, a LiDAR sensor may collect data down to depths of about three times the Secchi (visible) depth (Estep et al. 1994; Sinclair 2008). Beyond that depth, one may have to use acoustic echo-sounding profilers or side-scan imaging sonars (Brock and Sallenger 2000).

Echo-sounding profilers, which measure water depth and changes in bottom topography, send out pulses of acoustic energy from beneath the boat or other platform. The acoustic "ping" is reflected off the seabottom and submerged objects and recorded by the transceiver. The depth to target calculation is based on how long it took the reflected pulse to return to the surface and the speed of sound in water under prevailing environmental conditions. The earliest sounders used single beams, but the newer systems use multiple beams, with a large array of beams measuring bottom depths across a wide swath (Bergeron et al. 2007; Cracknell and Hayes 2007).

Side-scan imaging sonars emit acoustic pulses in the form of a very wide fan-shaped beam to both sides and at right angles to the track, to produce an image of the seabottom from the backscattered acoustic energy. Sonar echo-sounders and side-scan sonars are frequently housed in a torpedo-shaped "fish", which is towed by cable behind the survey ship at a predetermined height off the bottom (Avery and Berlin 1992; Pittenger 1989; Thompson and Schroeder 2010). More recently, various acoustic sensors have been housed in Remotely Operated Vehicles (ROVs) or Autonomous Underwater Vehicles (AUV's) and used to monitor features on the bottom and in the water column (Chadwick 2010).

1.9 Summary and Conclusions

Coastal ecosystems are highly productive and act as critical habitats for a wide variety of plants, fish, shellfish, and other wildlife. Coastal wetlands and estuaries have been exposed to a wide range of natural and man-induced alterations, including dredge and fill operations, hydrologic modifications, pollutant run-off, eutrophication, impoundments and fragmentation by roads and ditches. Environmental impacts from sea level rise and more frequent coastal storms are causing increased beach erosion, wetland destruction, algal blooms, hypoxia and anoxia, fish kills, releases of pollutants, and spread of pathogens.

Since coastal ecosystems have high spatial complexity and temporal variability, they require high spatial, spectral and temporal resolutions. Advances in sensor design and data analysis techniques are making remote sensing systems attractive for monitoring and mapping changes in the coastal zone. High resolution multispectral and hyperspectral imagers, LiDAR and radar systems are available for monitoring changes in coastal marshes, submerged aquatic vegetation, coral reefs, and beach profiles. Imaging radars are sensitive to soil moisture and inundation and can detect hydrologic features beneath the vegetation canopy.

Submerged aquatic plants and their properties are not as easily detectable as terrestrial vegetation. The main challenge for remote sensing of submerged aquatic plants is to isolate the weakened plant signal from the interference of the water column, the bottom and the atmosphere. Airborne hyperspectral imagers have improved SAV and coral reef mapping results by being able to identify more estuarine and intertidal habitat classes. The mapping of submerged aquatic vegetation (SAV), coral reefs and general bottom characteristics from satellites has become more accurate since high resolution multispectral imagery became available.

To identify long-term trends and short term variations, such as the impact of rising sea levels and hurricanes on wetlands, one needs to analyze time-series of remotely sensed imagery. High temporal resolution, precise spectral bandwidths, and accurate georeferencing procedures are factors that contribute to the frequent use of satellite image data for change detection analysis. Detecting the actual changes between two registered and radiometrically corrected images from different dates can be accomplished by employing one of several techniques, including post-classification comparison (PCC), spectral image differencing (SID), and change vector analysis (CVA). A comparison of the SID and the PCC change detection algorithms is provided by Macleod and Congalton (1998).

Global Positioning Systems (GPS), combined with airborne LiDAR techniques, make it now possible to obtain accurate topographical maps, including shoreline positions. LiDAR surveys can produce a 10 cm vertical accuracy at spatial densities greater than one elevation measurement per square meter. Such performance satisfies the needs of most coastal applications. Airborne LiDAR and ship-borne acoustic depth profilers are being used to map coastal water depth and bottom topography. LiDAR profilers, employing blue-green lasers, can penetrate the water column down to depths of about three times the Secchi depth.

Remote sensors can monitor and assess long-term trends and short-term changes of vegetation and hydrology faster, more completely and at lower cost per unit area than field or ship surveys alone. Multi-sensor and multi-seasonal data fusion techniques are significantly improving coastal land cover mapping accuracy and efficiency. Combinations of different satellite and airborne sensors can provide data that enhances the research and management of coastal ecosystems.

Future research priorities should include investigations of best approaches for processing hyperspectral data. Hyperspectral sensors need to be tested for SAV and bottom type discrimination using data obtained from satellites. The question of how differing levels of tidal inundation affect the reflectance characteristics of emergent marsh vegetation still needs to be better documented. Finally there is a need to investigate improvements to be gained from synergistic use of multi-wavelength remote sensing approaches, change detection techniques and multi-temporal comparisons and knowledge-based methods for improving classification accuracy (Malthus and Mumby 2003).

References

Ackermann F (1999) Airborne laser scanning – present status and future expectations. ISPRS J Photogramm Remote Sen 54:64–67

Adam E, Mutanga O, Rugege D (2010) Multispectral and hyperspectral remote sensing for identification and mapping of wetland vegetation: a review. Wet Ecol Manag 18:281–296

Armstrong RA (1993) Remote sensing of submerged vegetation canopies for biomass estimation. Int J Remote Sens 14:621–627

Artigas FJ, Yang J (2006) Spectral discrimination of marsh vegetation types in the New Jersey meadowlands, USA. Wetlands 26:271–277

Avery TE, Berlin GL (1992) Fundamentals of remote sensing and airphoto interpretation. Macmillan Publishing Company, New York

Baghdadi N, Bernier M, Gauthier R, Neeson I (2001) Evaluation of C-band SAR data for wetlands mapping. Int J Remote Sens 22:71–88

Baker C, Lawrence RL, Montagne C, Patten D (2007) Change detection of wetland ecosystems using Landsat imagery and change vector analysis. Wetlands 27:610–619

Bergeron E, Worley CR, O'Brien T (2007) Progress in the development of shallow water mapping systems. Sea Technol 48:10–15

Bertels L, Houthuys R, Deronde B, Janssens R, Verfaillie E, Van Lancker V (2012) Integration of optical and acoustic remote sensing data over the backshore-foreshore-nearshore continuum: a case study in Ostend (Belgium). J Coast Res 28:1426–1436 (in press)

Blasco F, Aizpuru M, Din Ndongo D (2005) Mangroves remote sensing. In: Schwartz ML (ed) Encyclopedia of coastal science. Springer, Dordrecht, pp 614–617

Boak EH, Turner IL (2005) Shoreline definition and detection: a review. J Coast Res 21:688–703

Bonisteel JM, Nayegandhi A, Wright CW, Brock JC, Nagle DB (2009) Experimental Advanced Airborne Research LiDAR (EAARL) data processing manual. US Geological Survey Open-File Report, 2009-1078

Bourgeau-Chavez LL, Kasischke ES, Brunzell SM, Mudd JP, Smith KB, Frick AL (2001) Analysis of space-borne SAR data for wetland mapping in Virginia riparian ecosystems. Int J Remote Sens 22:3665–3687

Brock J, Sallenger A (2000) Airborne topographic mapping for coastal science and resource management. USGS Open-File Report 01–46

Brock JC, Purkis SJ (2009) The emerging role of LiDAR remote sensing in coastal research and resource management. J Coast Res 53:1–5, Special issue 53, Coastal Applications of Airborne LiDAR
Brock JC, Wright CW, Clayton TD, Nayegandhi A (2004) LIDAR optical rugosity of coral reefs in Biscayne National Park, Florida. Coral Reefs 23:48–59
Brock J, Wright CW, Hernandez R, Thompson P (2006) Airborne LiDAR sensing of massive stony coral colonies on patch reefs in the northern Florida reef tract. Remote Sens Environ 104:31–42
Bustamante J, Pacios F, Diaz-Delgado R, Aragones D (2009) Predictive models of turbidity and water depth in the Donana marshes using Landsat TM and ETM+ images. J Environ Manage 90:2219–2225
CAMA (2009) Mapping and monitoring seagrass communities. Office of Coastal and Aquatic Management Areas, Florida Department of Environmental Protection. Available online at: http://www.dep.state.fl.us/coastal/habitats/seagrass/management/mapping.htm. Accessed 21 June 2012
Carter D, Scarborough R (2010) Sharing Delaware's lidar lessons. NOAA Coastal Services Center report. Local Strateg Addressing Clim Chang 2:16–17
Chadwick W (2010) Remotely Operated Vehicles (ROVs) and Autonomous Underwater Vehicles (AUVs). NOAA Ocean Explorer: submarine ring of fire 2002: Background. http://oceanexplorer.noaa.gov/explorations/02fire/background/rovsauvs/rov_auv.html
Christian B, Krishnayya NSR (2009) Classification of tropical trees growing in a sanctuary using Hyperion (EO-1) and SAM algorithm. Curr Sci 96:1601–1607
Coppin P, Jonckhere I, Nackaerts K, Mays B, Lambin E (2004) Digital change detection methods in ecosystem monitoring: a review. Int J Remote Sens 25:1565–1596
Costa MPF, Telmer KH (2007) Mapping and monitoring lakes in the Brazilian Pantanal wetland using synthetic aperture radar imagery. Aquat Conserv Mar Freshw Ecosyst 17:277–288
Cracknell AP, Hayes L (2007) Introduction to remote sensing. CRC Press, New York
Dahl TE (2006) Status and trends of wetlands in the conterminous United States 1998 to 2004. U.S. Department of the Interior, Fish and Wildlife Service Publication, Washington, DC, p 112
Deronde B, Houthuys R, Debruyn W, Fransaer D, Lancker VV, Hernriet J-P (2006) Use of airborne hyperspectral data and laser scan data to study beach morphodynamics along the Belgian coast. J Coast Res 22:1108–1117
Dierssen HM, Zimmermann RC, Leathers RA, Downes V, Davis CO (2003) Ocean color remote sensing of seagrass and bathymetry in the Bahamas banks by high resolution airborne imagery. Limnol Oceanogr 48:444–455
Dwivedi R, Rao B, Bhattacharya S (1999) Mapping wetlands of the Sundarban Delta and it's environs using ERS-1 SAR data. Int J Remote Sens 20:2235–2247
Ellis JM, Dodd HS (2000) Applications and lessons learned with airborne multispectral imaging. In: Fourteenth International Conference on Applied Geologic Remote Sensing, Las Vegas, NV
Estep LL, Lillycrop WJ, Parson LE (1994) Estimation of maximum depth of penetration of a bathymetric lidar system using a Secchi depth data base. Mar Technol Soc J 28:31–36
Everitt JH, Yang C, Escobar DE, Webster CF, Lonard RI, Davis MR (1999) Using remote sensing and spatial information technologies to detect and map two aquatic macrophytes. J Aquat Plant Manag 37:71–80
Fearns PRC, Klonowski W, Babcock RC, England P, Phillips J (2011) Shallow water substrate mapping using hyperspectral remote sensing. Cont Shelf Res 31:1249–1259
Ferguson RL, Wood LL, Graham DB (1993) Monitoring spatial change in seagrass habitat with aerial photography. Photogramm Eng Remote Sens 59:1033–1038
Filippi AM, Jensen JR (2006) Fuzzy learning vector quantization for hyperspectral coastal vegetation classification. Remote Sens Environ 100:512–530
Finkl CW (1996) What might happen to America's shorelines if artificial beach replenishment is curtailed: a prognosis for southeastern Florida and other sandy regions along regressive coasts. J Coast Res 12:ii–ix
Finkl CW, Benedet L, Andrews JL (2005a) Interpretation of seabed geomorphology based on spatial analysis of high-density airborne laser bathymetry (ALB). J Coast Res 21:501–514

Finkl CW, Benedet L, Andrews JL (2005b) Submarine geomorphology of the continental shelf off southeast Florida based on interpretation of airborne laser bathymetry. J Coast Res 21:1178–1190

Fyfe SK (2003) Are seagrasses spectrally distinct? Limnol Oceanogr 48:464–479

Gares PA, Wang Y, White SA (2006) Using Lidar to monitor a beach nourishment project at Wrightsville Beach, North Carolina, USA. J Coast Res 22:1206–1219

Garono RJ, Simenstad CA, Robinson R, Ripley H (2004) Using high spatial resolution hyperspectral imagery to map intertidal habitat structure in Hood Canal Washington, USA. Can J Remote Sens 30:54–63

Gesch DB (2009) Analysis of Lidar elevation data for improved identification and delineation of lands vulnerable to sea-level rise. J Coast Res 49–58, Special issue 53, Coastal Applications of Airborne Lidar

Ghioca-Robrecht DM, Johnston CA, Tulbure MG (2008) Assessing the use of multiseason QuickBird imagery for mapping invasive species in Great Lakes coastal marsh. Wetlands 28:1028–1039

Gilmore MS, Civco DL, Wilson EH, Barrett N, Prisloe S, Hurd JD, Chadwick C (2010) Remote sensing and in situ measurements for delineation and assessment of coastal marshes and their constituent species. In: Wang J (ed) Remote sensing of coastal environment. Springer, Boca Raton

Guenther G (2007) Airborne LiDAR bathymetry digital elevation. Model technologies and applications. In: Maune D (ed) The DEM users manual. American Society for Photogrammetry and Remote Sensing, pp 253–320

Guenther G, Larocque P, Lillycrop W (1994) Multiple surface channels in SHOALS airborne LiDAR. SPIE Ocean Opt XII 2258:422–430

Guenther GC, Tomas RWL, Larocque PE (1996) Design considerations for achieving high accuracy with the SHOALS bathymetric LiDAR system. SPIE: Laser Remote Sens Nat Waters From Theory Pract 15:54–71

Gullstrom M, Lunden B, Bodin M, Kangwe J, Ohman MC, Mtolera SP, Bjork M (2006) Assessment of changes in the seagrass-dominated submerged vegetation of tropical Chwaka Bay (Zanzibar) using satellite remote sensing. Estuar Coast Shelf Sci 67:399–408

Gutierrez R, Gibeaut JC, Crawford MM, Mahoney MP, Smith S, Gutelius W, Carswell D, Macpherson E (1998) Airborne laser swath mapping of Galveston Island and Bolivar Peninsula, Texas. In: Proceedings of the fifth international conference on remote sensing for marine and coastal environments, San Diego, 1, 236–243

Hame TI, Heiler I, Miguel-Ayanz JS (1998) An unsupervised change detection and recognition system for forestry. Int J Remote Sens 19:1079–1099

Han L, Rundquist D (2003) The spectral responses of *Ceratophyllum demersum* at varying depths in an experimental tank. Int J Remote Sens 24:859–864

Hapke CJ (2010) Integration of LiDAR and historical maps to measure coastal change on a variety of time and spatial scales. In: Wang Y (ed) Remote sensing of coastal environments. Springer/CRC Press, Boca Raton

Harris J, Digby-Argus S (1986) The detection of wetlands on radar imagery. In: Proceedings of the tenth Canadian symposium on remote sensing Edmonton, AB

Harvey KR, Hill JE (2001) Vegetation mapping of a tropical freshwater swamp in the Northern Territory, Australia: a comparison of aerial photography, Landsat TM and SPOT satellite imagery. Remote Sens Environ 22:2911–2925

Heege T, Bogner A, Pinnel N (2003) Mapping of submerged aquatic vegetation with a physically based process chain. SPIE Proc Remote Sens 5233:8

Hess L, Melack J, Simonett D (1990) Radar detection of flooding beneath the forest canopy: a review. Int J Remote Sens 11:1313–1325

Hess L, Melack J, Filoso S, Wang Y (1995) Delineation of inundated area and vegetation along the Amazon floodplain with the SIR-C synthetic aperture radar. IEEE Trans Geosci Remote Sens 33:896–904

Hestir EL, Khanna S, Andrew ME, Santos MJ, Viers JH, Greenberg JA, Rajapakse SS, Ustin S (2008) Identification of invasive vegetation using hyperspectral remote sensing in the California Delta ecosystem. Remote Sens Environ 112:4034–4047
Heumann BW (2011) Satellite remote sensing of mangrove forests: recent advances and future opportunities. Prog Phys Geogr 35:87–108
Hirano A, Madden M, Welch R (2003) Hyperspectral image data for mapping wetland vegetation. Wetlands 23:436–448
Houhoulis P, Michener W (2000) Detecting wetland change: a rule-based approach using NWI and SPOT-XS data. Photogramm Eng Remote Sens 66:205–211
Hundley A (1994) Report on the use of an acoustic method for mapping seagrass density and location. Report no. 940401. Offshore Scientific Services, Sydney
Hughes AR, Williams SL, Duarte CM, Heck KL Jr, Waycott M (2009) Associations of concern: declining seagrasses and threatened dependent species. Front Ecol Environ 7:242–246
Intergovernmental Panel on Climate Change (2007) Climate change 2007: the physical science basis. WMO/UNEP, Paris, www.ipcc.ch
Irish JL, Lillycrop WJ (1997) Monitoring new pass, Florida, with high-density LiDAR bathymetry. J Coast Res 13:1130–1140
Irish JL, Lillycrop WJ (1999) Scanning laser mapping of the coastal zone: the SHOALS system. ISPRS J Photogram Remote Sens 54:123–129
Irish JL, White TE (1998) Coastal engineering applications of high resolution bathymetry. Coast Eng 35:47–71
Jensen JR (1996) Introductory digital image processing: a remote sensing perspective, 2nd edn. Pearson Prentice-Hall, Upper Saddle River
Jensen JR (2007) Remote sensing of the environment: an earth resource perspective. Pearson Prentice-Hall, Upper Saddle River
Jensen RR, Mausel P, Dias N, Gonser R, Yang C, Everitt J, Fletcher R (2007) Spectral analysis of coastal vegetation and land cover using AISA+ hyperspectral data. Geocarto Int 22:17–28
Johnson RD, Kasischke ES (1998) Change vector analysis: a technique for the multispectral monitoring of land cover and condition. Int J Remote Sens 19:411–426
Kasischke E, Bourgeau-Chavez L (1997) Monitoring South Florida wetlands using ERS-1 SAR imagery. Photogramm Eng Remote Sens 63:281–291
Kasischke E, Bourgeau-Chavez L, Smith K, Romanowicz E, Richardson C (1997a) Monitoring hydropatterns in South Florida ecosystems using ERS SAR data. In: 3rd ERS symposium on space at the service of our environment, Florence, pp 71–76
Kasischke E, Melack J, Dobson M (1997b) The use of imaging radars for ecological applications-a review. Remote Sens Environ 59:141–156
Kelly M, Tuxen K (2009) Remote sensing support for tidal wetland vegetation research and management. In: Yang X (ed) Remote sensing and geospatial technologies for coastal ecosystem assessment and management. Springer, Berlin
Kempeneers P, Deronde B, Provoost S, Houthuys R (2009) Synergy of airborne digital camera and LiDAR data to map coastal dune vegetation. J Coast Res 53:73–82
Khalil SM, Finkl CW (2007) Submarine geomorphology and coastal process zones: morphodynamics of the inner continental shelf off southeast Florida. J Coast Res 50:480–485 (special issue)
Klemas V (2009) The role of remote sensing in predicting and determining coastal storm impacts. J Coast Res 25:1264–1275
Klemas V (2011) Remote sensing of wetlands: case studies comparing practical techniques. J Coast Res 27:418–427
Klemas V (2012) Remote sensing of algal blooms: an overview with case studies. J Coast Res 28:34–43
Kolasa KV, Craw V (2009) Improving seagrass maps of Florida's Springs Coast through digital imagery. Proceedings of ASPRS 2009 annual conference, Baltimore, MD, 9–13 Mar 2009
Krabill WB, Wright CW, Swift RN, Frederick EB, Manizade SS, Yungel JK, Martin CF, Sonntag JG, Duffy M, Hulslander W, Brock JC (2000) Airborne laser mapping of Assateague National Seashore Beach. Photogramm Eng Remote Sens 66:65–71

Laba M, Downs R, Smith S, Welsh S, Neider C, White S, Richmond M, Philpot W, Baveye P (2008) Mapping invasive wetland plants in the Hudson River National Estuarine Research Reserve using Quickbird satellite imagery. Remote Sens Environ 112:286–300

Lang MW, Kasischke ES (2008) Using C-band synthetic aperture radar data to monitor forested wetland hydrology in Maryland's Coastal Plain, USA. IEEE Trans Geosci Remote Sens 46:535–546

Lang MW, Mccarty GW (2008) Remote sensing data for regional wetland mapping in the United States: trends and future prospects. In: Russo RE (ed) Wetlands: ecology, conservation and restoration. Nova Science Publishers, Inc., Hauppauge

Lathrop RG, Cole MB, Showalter RD (2000) Quantifying the habitat structure and spatial pattern of New Jersey (U.S.A.) salt marshes under different management regimes. Wetl Ecol Manag 8:163–172

Lesser MP, Mobley CD (2007) Bathymetry, water optical properties, and benthic classification of coral reefs using hyperspectral remote sensing imagery. Coral Reefs 26:819–829

Li L, Ustin SL, Lay M (2005) Application of multiple endmember spectral mixture analysis (MESMA) to AVIRIS imagery for coastal salt marsh mapping: a case study in China Camp, CA, USA. Int J Remote Sens 26:5193–5207

Lidz BH, Shinn EA, Hine AC, Locker SD (1997) Contrasts within an outlier-reef system: evidence for differential quaternary evolution, south Florida windward margin, U.S.A. J Coast Res 13 (3):711–731

Lillycrop WJ, Irish JL, Parson LE (1997) SHOALS system. Sea Technol 38:17–25

Lillycrop WJ, Pope RW, Wozencraft JM (2002) Airborne lidar hydrography: a vision for tomorrow. Sea Technol 43:27–34

Lopez RD, Edmonds CM, Slonecker TS, Jones KB, Heggem DT, Lyon JG, Jaworski E, Garofalo D, Williams D (2004) Accuracy assessment of airborne hyperspectral data for mapping opportunistic plant species in freshwater coastal wetlands. In: Lunetta RS, Lyon JG (eds) Remote sensing and GIS accuracy assessment. CRC Press, New York, pp 253–267

Lunetta RS, Balogh ME (1999) Application of multi-temporal Landsat 5 TM imagery for wetland identification. Photogramm Eng Remote Sens 65:1303–1310

Lunetta RS, Elvidge CD (1998) Remote sensing change detection: environmental monitoring methods and applications. Ann Arbor Press, Ann Arbor

Lyon JG, Mccarthy J (1995) Wetland and environmental applications of GIS. Lewis Publishers, New York

Macleod RD, Congalton RG (1998) A quantitative comparison of change detection algorithms for monitoring eelgrass from remotely sensed data. Photogramm Eng Remote Sens 64:207–216

Malthus TJ, Mumby PJ (2003) Remote sensing of the coastal zone: an overview and priorities for future research. Int J Remote Sens 24:2805–2815

Martin S (2004) An introduction to remote sensing. Cambridge University Press, Cambridge, UK

Martinez JM, Le Toan T (2007) Mapping of flood dynamics and spatial distribution of vegetation in the Amazon floodplain using multitemporal SAR data. Remote Sens Environ 108:209–223

Mccoy R (2005) Field methods in remote sensing. Guilford Press, New York

Mckean J, Nagel D, Tonina D, Bailey P, Wright CW, Bohn C, Nayegandhi A (2009) Remote sensing of channels and riparian zones with a narrow-beam aquatic-terrestrial LiDAR. Remote Sens 1:1065–1096

Miner SP (1993) Application of acoustic hydrosurvey technology to the mapping of eelgrass (Zostera marina) distribution in Humboldt Bay, California. Coastal Zone '93. Proceedings of the 8th symposium on coastal and ocean management, 19–23 July 1993, New Orleans, LA

Mishra D, Narumalani S, Rundquist D, Lawson M (2006) Benthic habitat mapping in tropical marine environments using QuickBird multispectral data. Photogramm Eng Remote Sens 72:1037–1048

Mitsch WJ, Gosselink JG (2000) The value of wetlands: importance of scale and landscape setting. Ecol Econ 35:25–33

Moreno A, Siljestrom P, Rey J (1998) Benthicphanerogam species recognition in side scan sonar images: importance of the sensor direction. In: Alippi A, Cannelli GB (eds) Proceedings 4th European conference on underwater acoustics. Italian National research Council, Rome, pp 173–178

Morris JT, Sundareshwar PV, Nietch CT, Kjerfve B, Cahoon DR (2002) Responses of coastal wetlands to rising sea level. Ecology 83:2869–2877

Morton RA, Miller TL (2005) National assessment of shoreline change: Part 2. Historical shoreline change and associated coastal land loss along the US southeast Atlantic coast. US Geological Survey Open-File Report 2005-1401

Mumby PJ, Edwards AJ (2002) Mapping marine environments with IKONOS imagery: enhanced spatial resolution can deliver greater thematic accuracy. Remote Sens Environ 82:248–257

NASA/GSFC (2010) Hurricane Ike: storm surge flooding image of the Gulf Coast. NASA image courtesy Jeff Schmaltz, MODIS Rapid Response team at NASA GSFC

Nayegandhi A, Brock JC, Wright CW (2009) Small-footprint, waveform-resolving LiDAR estimation of submerged and sub-canopy topography in coastal environments. Int J Remote Sens 30:861–878

NOAA (1999) Trends in U.S. coastal regions, 1970–1998. Addendum to the Proceedings: trends, and future challenges for U.S. National Ocean and Coastal Policy. NOAA, August 1999

NOAA-CSC (2001) Guidance for Benthic Habitat mapping: an aerial photographic approach (by M. Finkbeiner, W. Stevenson, and R. Seaman, NOAA-CSC, Charleston, SC). NOAA/CSC/20117-PUB, 73 pp

Nobi EP, Thangaradjou T (2012) Evaluation of the spatial changes in seagrass cover in the lagoons of Lakshadweep islands, India, using IRS LISS III satellite images. Geocarto Int 27:647–660

Novo EMLM, Costa MPF, Mantovani JE, Lima IBT (2002) Relationship between macrophyte stand variables and radar backscatter at L and C band, Tucurui reservoir, Brazil. Int J Remote Sens 23:1241–1260

Odum EP (1993) Ecology and our endangered life-support systems, 2nd edn. Sinauer Associates, Inc., Sunderland

Orth RJ, Carruthers TJB, Dennison WC, Duarte CM, Fourqurean JW, Heck KL Jr, Hughes AR, Kendrick GA, Kenworthy WJ, Olyarnik S, Short FT, Waycott M, Williams SL (2006) A global crisis for seagrass ecosystems. Bioscience 56:987–996

Ozesmi SL, Bauer ME (2002) Satellite remote sensing of wetlands. Wetl Ecol Manag 10:381–402

Paringit EC, Nadaoka K, Fortes MD, Harii S, Tamura H, Mistui J (2003) Multiangular and hyperspectral reflectance modeling of seagrass beds for remote sensing studies. In: Proceedings of the international geoscience and remote sensing symposium '03, vol 3, pp 21–25, New York: IEEE

Parson LE, Lillycrop WJ, Klein CJ, Ives RC, Orlando SP (1997) Use of LiDAR technology for collecting shallow bathymetry of Florida Bay. J Coast Res 13:1173–1180

Pastol Y, Le Roux C, Louvart L (2007) LITTO3D: a seamless digital terrain model. Int Hydrogr Rev 8:38–44

Peneva EI, Griffith JA, Carter GA (2008) Seagrass mapping in the Northern Gulf of Mexico using airborne hyperspectral imagery: a comparison of classification methods. J Coast Res 24:850–856

Pengra BW, Johnston CA, Loveland TR (2007) Mapping an invasive plant, Phragmitesaustralis, in coastal wetlands using the EO-! Hyperion hyperspectral sensor. Remote Sens Environ 108:74–81

Philpot W (2007) Estimating atmospheric transmission and surface reflectance from a glint-contaminated spectral image. IEEE Trans Geosci Remote Sens 45:448–457

Phinn S, Hess LL, Finlayson CM (1999) An assessment of the usefulness of remote sensing for wetland inventory and monitoring in Australia. In: Finlayson CM, Spiers AG (eds) Techniques for enhanced wetland inventory and monitoring. Supervising Scientist Report 147, Supervising Scientist, Canberra

Phinn S, Roelfsema C, Decker A, Brando V, Anstee J (2008) Mapping seagrass species, cover and biomass in shallow waters: an assessment of satellite multi-spectral and airborne hyper-spectral imaging systems in Moreton Bay (Australia). Remote Sens Environ 112:3413–3425

Pinnel N, Heege T, Zimmermann S (2004) Spectral discrimination of submerged macrophytes in lakes using hyperspectral remote sensing data. SPIE Proc Ocean Opt XVII 1:1–16
Pittenger RF (1989) Exploring and mapping the seafloor. Nat Geosci 177:61A
Pu R, Bell S, Baggett L, Meyer C, Zhao Y (2012) Discrimination of seagrass species and cover classes with in situ hyperspectral data. J Coast Res 28:1330–1334 (in press)
Pulich W Jr, Blair C, White WA (1997) Current status and historical trends of seagrass in the Corpus Christi Bay National Estuary Program study area. Publication CCBNEP-20. Texas Natural Resource Conservation Commission, Austin, p 131
Purkis SJ (2005) A 'reef-up' approach to classifying coral habitats from IKONOS imagery. IEEE Trans Geosci Remote Sens 43:1375–1390
Purkis SJ, Graham NAJ, Riegl BM (2008) Predictability of reef fish diversity and abundance using remote sensing data in Diego Garcia (Chagos Archipelago). Coral Reefs 27:167–178
Purkis S, Klemas V (2011) Remote sensing and global environmental change. Wiley-Blackwell, Oxford
Purkis SJ, Kenter JAM, Oikonomou EK, Robinson IS (2002) High-resolution ground verification, cluster analysis and optical model of reef substrate coverage on Landsat TM imagery (Red Sea, Egypt). Int J Remote Sens 23:1677–1698
Raber GT, Jensen JR, Hodgson ME, Tullis JA, Davis BA, Berglund J (2007) Impact of Lidar nominal post-spacing on DEM accuracy and flood zone delineation. Photogramm Eng Remote Sens 73:793–804
Ramsey E, Rangoonwala A (2005) Leaf optical property changes associated with the occurrence of Spartina alterniflora dieback in coastal Louisiana related to remote sensing mapping. Photogramm Eng Remote Sens 71:299–311
Ramsey E, Rangoonwala A (2010) Mapping the onset and progression of marsh dieback. In: Wang J (ed) Remote sensing of coastal environment. CRC Press, Boca Raton
Rao BRM, Dwivedi RS, Kushwaha SPS, Bhattacharya SN, Anand JB, Dasgupta S (1999) Monitoring the spatial extent of coastal wetlands using ERS-1 SAR data. Int J Remote Sens 20:2509–2517
Rasher ME, Weaver W (1990) Basic photo interpretation: a comprehensive approach to interpretation of vertical aerial photography for natural resource applications. US Department of Agriculture, Washington, DC
Rosenqvist A, Finlayson CM, Lowry J, Taylor D (2007) The potential of long-wavelength satellite-borne radar to support implementation of the Ramsar Wetland Convention. Aquat Conserv Mar Freshw Ecosyst 17:229–244
Rosso PH, Ustin SL, Hastings A (2005) Mapping marshland vegetation of San Francisco Bay, California, using hyperspectral data. Int J Remote Sens 26:5169–5191
Sabol BM, Melton REJR, Chamberlain R, Doering P, Haunert K (2002) Evaluation of a digital echo sounder system for detection of submersed aquatic vegetation. Estuaries 25:133–141
Sallenger AH, Krabill WB, Brock JC, Swift RN, Jansen M, Manizade S, Richmond B, Hampto M, Eslinger D (1999) Airborne laser study quantifies El Niño-induced coastal change. Am Geophys Union EOS Trans 80:89–93
Schmid KA, Hadley BC, Wijekoon N (2011) Vertical accuracy and use of topographic LIDAR data in coastal marshes. J Coast Res 27:116–132
Schmidt KS, Skidmore KA (2003) Spectral discrimination of vegetation types in a coastal wetland. Remote Sens Environ 85:92–108
Schmidt KS, Skidmore AK, Kloosterman EH, Van Oosten H, Kumar L, Janssen JAM (2004) Mapping coastal vegetation using an expert system and hyperspectral imagery. Photogramm Eng Remote Sens 70:703–716
Schweitzer D, Armstrong RA, Posada J (2005) Remote sensing characterization of benthic habitats and submerged vegetation biomass in Los Roques Archipelago National Park, Venezuela. Int J Remote Sens 26:2657–2667
Shalaby A, Tateishi R (2007) Remote sensing and GIS for mapping and monitoring land cover and land-use changes in the Northwestern coastal zone of Egypt. Appl Geogr 27:28–41

Shan J, Hussain E (2010) Object-based data integration and classification for high-resolution coastal mapping. In: Wang J (ed) Remote sensing of coastal environment. CRC Press, Boca Raton
Silva TSF, Costa MPF, Melack JM, Novo EMLM (2008) Remote sensing of aquatic vegetation: theory and applications. Environ Monit Assess 140:131–145
Simard M, Fatoyinbo LE, Pinto N (2010) Mangrove canopy 3D structure and ecosystem productivity using active remote sensing. In: Wang J (ed) Remote sensing of coastal environment. CRC Press, Boca Raton
Sinclair M (2008) Airborne LiDAR hydrographic survey for homeland security. Sea Technol 49:15–20
Stockdon HF, Sallenger AH, List JH, Holman RA (2002) Estimation of shoreline position and change using airborne Lidar topographic data. J Coast Res 18:502–513
Stockdon HF, Doran KS, Sallenger AH (2009) Extraction of LiDAR-based dune-crest elevations for use in examining the vulnerability of beaches to inundation during hurricanes. J Coast Res 35:59–65
Stoker JM, Tyler DJ, Turnipseed DP, Van Wilson K, Oimoen MJ (2009) In: Brock and Purkis (eds) Integrating disparate Lidar datasets for a regional storm tide inundation analysis of Hurricane Katrina. J Coast Res 66–72, Special issue 53, Coastal Applications of Airborne Lidar
Thieler ER, Danforth WW (1994) Historical shoreline mapping: improving techniques and reducing positioning errors. J Coast Res 10:539–548
Thomson AG, Fuller RM, Sparks TH, Yates MG, Eastwood JA (1998) Ground and airborne radiometry over intertidal surfaces: waveband selection for cover classification. Int J Remote Sens 19:1189–1205
Thompson RL, Schroeder AJ Jr (2010) High-definition 3-D tools for underwater surveying and inspection. Sea Technol 51:43
Tiner RW (1996) Wetlands. In: Manual of photographic interpretation, 2nd ed. American Society for Photogrammetry and Remote Sensing, Falls Church, 2440 p
Townsend PA (2000) A quantitative fuzzy approach to assess mapped vegetation classifications for ecological applications. Remote Sens Environ 72:253–267
Townsend PA (2002) Relationships between forest structure and the detection of flood inundation in forested wetlands using C-band SAR. Int J Remote Sens 23:443–460
Townsend P, Walsh S (1998) Modeling floodplain inundation using an integrated GIS with radar and optical remote sensing. Geomorphology 21:295–312
Töyrä J, Pietroniro A, Martz LW (2001) Multisensor hydrologic assessment of a freshwater wetland. Remote Sens Environ 75:162–173
Töyrä J, Pietroniro A, Martz LW, Prowse TD (2002) A multi-sensor approach to wetland flood monitoring. Hydrol Process 16:1569–1581
Valta-Hulkkonen K, Pellika P, Tanskanen H, Ustinov A, Sandman O (2003) Digital false color aerial photographs for discrimination of aquatic macrophyte species. Aquat Bot 75:71–88
Valta-Hulkkonen K, Kanninen A, Pellikka P (2004) Remote sensing and GIS for detecting changes in the aquatic vegetation of a rehabilitated lake. Int J Remote Sens 25:5745–5758
Wabnitz CC, Andrefouet S, Torres-Puliza D, Muller-Karger FE, Kramer PA (2008) Regional-scale seagrass habitat mapping in the Wider Caribbean region using Landsat sensors: applications to conservation and ecology. Remote Sens Environ 112:3455–3467
Wang Y (2010) Remote sensing of coastal environments: an overview. In: Wang J (ed) Remote sensing of coastal environments. CRC Press, Boca Raton
Wang Y, Christiano M, Traber M (2010) Mapping salt marshes in Jamaica Bay and terrestrial vegetation in Fire Island National Seashore using QuickBird satellite data. In: Wang J (ed) Remote sensing of coastal environments. CRC Press, Boca Raton
Wang L, Sousa WP, Gong P (2004) Integration of object-based and pixel-based classification for mapping mangroves with IKONOS imagery. Int J Remote Sens 25:5655–5668

Ward DP, Hamilton SK, Jardine TD, Pettit NE, Tews EK, Olley JM, Bunn SE (2012) Assessing the seasonal dynamics of inundation, turbidity and aquatic vegetation in the Australian wet-dry tropics using optical remote sensing. Ecohydrology. doi:10.1002/eco.1270

Waycott M, Duarte CM, Carruthers TJB, Orth RJ, Dennison WC, Olyarnik S, Calladine A, Fourqurean JW, Heck KL, Hughes AR, Kendrick GA, Kenworthy WJ, Short FT, Williams SL (2009) Accelerating loss of seagrasses across the globe threatens coastal ecosystems. Proc Natl Acad Sci 106:12377–12381

Webster TL, Forbes DL, Dickie S, Shreenan R (2004) Using topographic lidar to map flood risk from storm-surge events for Charlottetown, Prince Edward island, Canada. Can J Remote Sens 30:64–76

West GR, Lillycrop WJ, Pope RW (2001) Utilization of airborne LiDAR bathymetry for rapid environmental assessment. Sea Technol 42:10

Williams DJ, Rybicki NB, Lombana AV, O'brien TM, Gomez RB (2003) Preliminary investigation of submerged aquatic vegetation mapping using hyperspectral remote sensing. Environ Monit Assess 81:383–392

Wilson BA, Rashid H (2005) Monitoring the 1997 flood in the Red River Valley using hydrologic regimes and RADARSAT imagery. Can Geogr 49:100–109

Wolter PT, Johnston CA, Niemi GJ (2005) Mapping submerged aquatic vegetation in the US Great Lakes using Quickbird satellite data. Int J Remote Sens 26:5255–5274

Wozencraft JM, Lillycrop WJ (2003) Airborne coastal mapping, past, present, and future. J Coast Res 38:207–215 (special issue)

Wozencraft JM, Millar D (2005) Airborne lidar and integrated technologies for coastal mapping and charting. Marine Technol Soc J 39:27–35

Yang X (2009) Remote sensing and geospatial technologies for coastal ecosystem assessment and management. Springer-Verlag, Berlin

Yang J, Artigas FJ (2010) Mapping salt marsh vegetation by integrating hyperspectral and LiDAR remote sensing. In: Wang J (ed) Remote sensing of coastal environment. CRC Press, Boca Raton

Yang C, Everitt JH, Fletcher RS, Jensen JR, Mausel PW (2009) Mapping black mangrove along the south Texas gulf coast using AISA+ hyperspectral imagery. Photogramm Eng Remote Sens 75:425–436

Young SS, Wang CY (2001) Land-cover change analysis of China using global-scale Pathfinder AVHRR Landcover (PAL) data, 1982–92. Int J Remote Sens 22:1457–1477

Yuan D, Elvidge CD, Lunetta RS (1998) Survey of multispectral methods for land cover change analysis. In: Lunetta RS, Elvidge CD (eds) Remote sensing change detection: environmental monitoring methods and applications. Ann Arbor Press, Ann Arbor, p 21

Zhang X (2010) On the estimation of biomass of submerged vegetation using Landsat thematic mapper (TM) imagery: a case study of the Honghu Lake, PR China. Int J Remote Sens 19:11–20

Zhou G (2010) Coastal 3D change pattern analysis using LiDAR series data. In: Wang Y (ed) Remote sensing of coastal environments. CRC Press, Boca Raton

Chapter 2
Advanced Techniques for Mapping Biophysical Environments on Carbonate Banks Using Laser Airborne Depth Sounding (LADS) and IKONOS Satellite Imagery

Charles W. Finkl, Christopher Makowski, and Heather Vollmer

Abstract Mapping seafloor environments on the continental shelf, over the past several decades, has undergone rapid transitions from early, relatively low-resolution techniques, such as echo sounding in deeper waters and digital aerial photography in shallower waters, to modern advancements like high-density airborne laser bathymetry and multi-spectral satellite imagery that can now detect seafloor reflectance at depths ranging to 50–60 m. Passive imaging systems require clear waters that typically exist on carbonate banks in many regions of the world ocean. Carbonate banks in the south Florida region provide nearly ideal conditions for mapping submarine topography and interpreting geomorphological and biophysical environments. A hierarchical open-ended classification system was developed for both open-ocean and key (low carbonate islands) environments. These classification systems, which are based on cognitive recognition of seafloor features interpreted from LADS and IKONOS imagery, are directly applied in GIS cartography programs to create comprehensive, informative, and interactive products. Examples from the open ocean southeast coast and Marquesa Islands illustrate the applicability and usefulness of advanced remote sensing techniques intercalated with GIS programs and classificatory schema for organizing seafloor typologies. This new technology and its associated classification systems permit major advancements in the detailed mapping of seafloors that have never before been achieved for margins of regional seas.

C.W. Finkl (✉) • C. Makowski
Coastal Education and Research Foundation (CERF), Coconut Creek, FL, USA

Florida Atlantic University, Boca Raton, FL, USA
e-mail: cfinkl@cerf-jcr.com

H. Vollmer
Coastal Education and Research Foundation (CERF), Asheville, NC, USA

C.W. Finkl and C. Makowski (eds.), *Remote Sensing and Modeling: Advances in Coastal and Marine Resources*, Coastal Research Library 9,
DOI 10.1007/978-3-319-06326-3_2,

2.1 Introduction

This chapter considers recent advances in the mapping of seafloor environments on carbonate shelves of southern Florida where water clarity is not an issue because turbidity is very low. These shallow coastal waters thus provide ample opportunity to determine a range of seafloor typologies, within which there are topological variations, using remote sensing techniques. Two different platforms were selected to determine their applicability and appropriateness in several carbonate submarine environments, using two study areas (Fig. 2.1). By reference to other methodologies such as side scan sonar and aerial photography, modern LADS (open ocean mainland coast) and IKONOS (Florida Keys) products are contrasted and compared as optimum imageries for cognitive recognition of seafloor typologies based on LADS bathymetry and IKONOS spectral reflectance. This section includes a brief review of remote sensing of carbonate shelves off distal south Florida and indicates the advantages of coupling classified remotely sensed imagery with GIS.

2.1.1 Remote Sensing of Seafloor Features

Over the past several decades, characterization of seafloor environments has made several important advancements. Early trends were based on side scan sonar imagery and seismic reflection profiling in shallower waters. Shallow water carbonate banks were further characterized with the advent of aerial photography, which continues to be a useful tool today but has limited application in regional (small scale) studies. The wider availability of airborne laser techniques in the past couple of decades provided opportunity for production of high-density bathymetric maps. LADS images, for example, have been used to advantage for small-scale mapping projects on the southeast Florida continental shelf. Although IKONOS satellite images do not provide bathymetric data, they are an excellent means of obtaining seafloor environmental (habitat) data based on processing spectral reflectance to produce near photographic-quality images.

2.1.1.1 Three-Dimensional Hachure Maps

Seafloor topography on the continental shelf was traditionally determined using soundings that were contoured into isobaths. These early bathymetric charts of shelf topography provided rudimentary insight into seafloor morphology, but horizontal positional accuracy and imprecise leadline depth sounding resulted in mapping errors and low-resolution interpretations of morphology. The advent of acoustic remote sensing techniques developed around WWII produced higher resolution maps of seafloor topography. An early landmark achievement in remote sensing during the twentieth century was Marie Tharp's construction of a physiographic

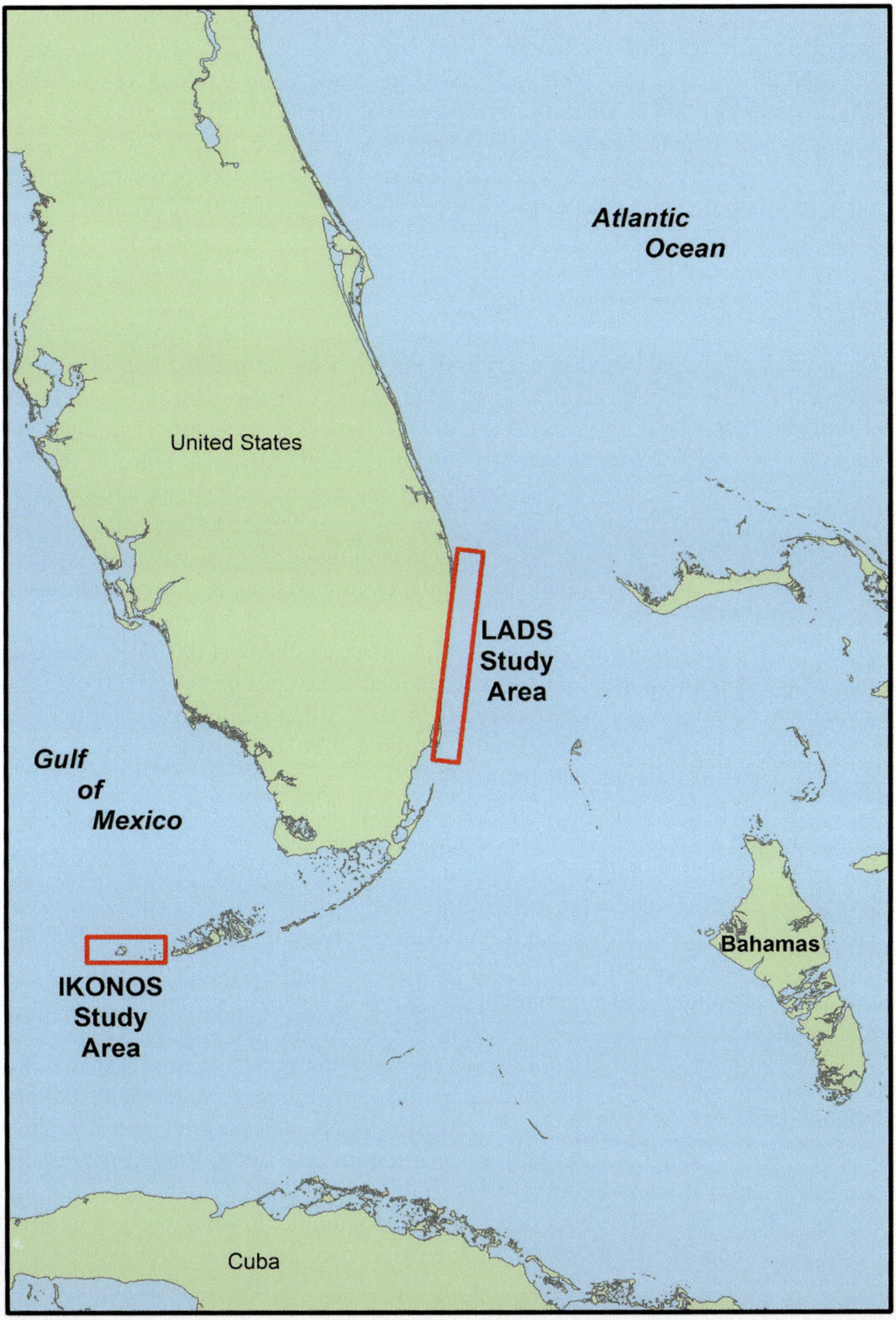

Fig. 2.1 Study areas in tropical and subtropical southern Florida. The LADS (Laser Airborne Depth Sounding) survey covers an area of about 600 km^2 on the southeast open ocean coast where there are shelf-edge coral reefs, carbonate rock reefs, and sediment flats on the narrow continental shelf. The IKONOS survey covers an area of about 422 km^2 on platform coral reefs and carbonate sediment banks in the Florida Keys

map of ocean basins. Working alongside Bruce Heezen, Tharp combined a hachuring cartographic technique with an assumed light source to depict topographic features that were related to the underlying geology (*e.g.*, Barton 2002; Doel et al. 2006; Heezen and Tharp 1965, 1966, 1977; Moody 2007). The first map showing the entire ocean basin was published in 1977 as the *World Ocean Floor* (Barton 2002; Heezen and Tharp 1977), and variations are still used today in modern geographic software applications (*e.g.*, Google™ Earth).

2.1.1.2 Coastal Aerial Photography

The advent of coastal aerial imagery introduced visual records of shallow marine environments, providing greater detail than what could be achieved by acoustic sounding. After WWII, the U.S. Coast and Geodetic Survey (C&GS) worked with the U.S. Army Air Service to acquire coastal aerial photographs. Some of the first coastal aerial imagery projects used oblique photography with single-lens cameras. While occasional nearshore bottom features were detected, these photographs provided very narrow fields of view that covered moderately-sized coastal areas.

By the 1930s and 1940s, multi-lens cameras improved surveying capabilities by allowing the photographer to acquire imagery of the coast at different angles. While vertical stereo-paired images of the coastline were available at this time, extensive sub-bottom feature information was still unobtainable because the effectiveness of aerial photography was, and still currently is, largely dependent on the clarity of the water column, which is generally limited to about 10 m in clear water (Moore 2000; Richards 1980; Thieler and Danforth 1994).

The introduction of color film with appropriate filters in the 1950s visually captured coral reefs, submerged rock outcrops, and sand flat areas. Color photography was further enhanced by Specht et al. (1973) in the 1970s when they developed experimental water-penetrating film by manipulating the blue wavelength region of the spectrum that is transmitted through the water column. Because these aerials provided remarkably clear images of the seafloor, the technology was restricted for military and government use, denying access by the public or research community.

Today, high-resolution digital orthoimagery is the preferred format for coastal aerial photographs. By incorporating visible and infrared wavelengths, digital orthoimagery is widely deployed for interpretation of shorelines and nearshore benthic environments. The digital format of the images, versus analog processing, allows for faster processing turnaround times. An additional advantage of digital aerials is that they can be georeferenced in a geographic information system (GIS) interface. Sheppard et al. (1991) showed that modern aerial photography was effective for mapping shallow marine habitats in the Caribbean. Mumby et al. (1999) and Thieler and Danforth (1994) confirmed the deduction that interpretation of color aerial photography is one of the most effective methods to conduct detailed coastal habitat mapping. Many other researchers lend credence to that postulation (*e.g.*, Anders and Byrnes 1991; Ekeborn and Erkkila 2002;

Gorman et al. 1998; Kenny et al. 2003; Lewis 2002; Moore 2000; Mount 2003; Mumby and Harborne 1999; O'Regan 1996; Ramsey and Laine 1997; Shoshany and Degani 1992; Smith and Rumohr 2005; Smith and Zarillo 1990). That being said, today's coastal aerial photographs are still limited by water clarity and depth.

2.1.1.3 Airborne Laser Imagery (LIDAR and LADS)

One advanced approach to mapping seafloor typology features high-density airborne laser bathymetry (ALB). First developed in the 1960s and 1970s, ALB is a light detection and ranging (LIDAR) technique that uses visible, ultraviolet, and near infrared light to optically sense a contour target through active and passive systems (Brock and Purkis 2009; Guenther et al. 2000; Irish and Lillycrop 1999; Irish et al. 2000). A laser pulse is usually emitted from the underbelly of a low-flying (~200–500 m elevation) aircraft (*e.g.*, helicopter, small plane) and a receiver records two back reflections: one from the water's surface (*i.e.* a passive system reliant upon surface reflectance) and one from the seafloor (*i.e.* an active system reliant upon penetration of the water column) (Guenther et al. 2000; Irish and Lillycrop 1999). In this way, researchers were able to interpret bathymetric configurations in coastal regions from airborne laser reflectance (*e.g.*, Brock and Purkis 2009; Deronde et al. 2008; Finkl et al. 2004, 2005a, b; Gesch 2009; Irish and Lillycrop 1997; Irish et al. 2000; Kempeneers et al. 2009; Klemas 2011a, b, c; Long et al. 2011; Stockdon et al. 2002, 2009; Stoker et al. 2009).

LADS was developed by Australia's Defense, Science, and Technology Organization (DSTO) for the Royal Australian Navy in order to provide an expedited means to survey and chart coastal regions. Flying at an altitude of approximately 500 m with an average speed of 75 m/s, LADS surveys can detect submarine geomorphological features to a depth of 70 m, with a swath width up to 288 m and a sounding horizontal spot density of 6 m. The LADS infrared laser emits a vertical beam pulse that reflects off the sea surface, while a visible green wavelength beam (~532 nm) propagates through the water column to reflect benthic topography. As the infrared pulses provide an initial sea-surface reference, the returning green wavelength pulses are collected and processed by a receiving telescope that contains spectral, spatial, and polarizing filters. The resulting read-out image produces an accurate waveform bottom reflection representation of the benthic topography in relation to the sea-surface reflectance signature. However, as with the SHOALS surveys, the effectiveness of LADS data is limited by turbidity in the water column. Suspended particulate matter, dissolved organic matter, phytoplankton, and dinoflagellate blooms contribute to the scattering and absorption of optical sensors from LIDAR surveys. Only in those regions that have been classified as suitable Case I or Case II coastal waters (*e.g.*, southeast Florida) is the water column visibility clear enough to effectively run these depth-sounding laser surveys (Bukata et al. 1995; Finkl et al. 2004, 2005a, b; Irish and Lillycrop 1997; Klemas 2011a).

2.1.1.4 Satellite Imagery (IKONOS)

As the space program was born in the second half of the twentieth century, so was a new coastal imaging acquisition technology with the use of orbiting satellites. Through the utilization of hyperspectral and multi-spectral sensors, satellites provide a continuous stream of coastal photographs without the logistical hardships of deploying a vessel or aircraft. Instead of an acoustic or light reflectance, satellite sensors create an image-based visual approach to discerning physical and biological bottom features of the ocean floor. Typically, hyperspectral sensor datasets constitute a range of 100–200 spectral bands of relatively narrow bandwidths (5–10 nm). On the other hand, multi-spectral sensor datasets are only composed of a few spectral bands (5–10), but have a relatively large range of bandwidths (70–400 nm). The visual detection of submarine features is dependent upon on the spectral coverage of the spectrometer and the overall spectral resolution (*i.e.* the pixel size of the satellite image covering the earth's surface) of the acquired images. There are many satellite sensors currently in orbit around the Earth today, some with a high spatial resolution (*i.e.* 0.6–4 m) and others with a medium spatial resolution (*i.e.* 4–30 m).

The IKONOS satellite was launched in 1999 and is a good example of a high spatial resolution satellite sensor. Achieving a 0.8 m panachromatic resolution and a 3.2 m multi-spectral resolution, IKONOS uses five spectral bands that include blue, green, red, near infrared, and panachromatic.

2.2 Incorporating Classification Schemes with Advanced Remote Sensing Images

LADS and IKONOS images are used as examples of platform products that can serve as base maps for interpretation of seafloor bathymetric features and environments. The following section summarizes some of the salient steps in the preparation of image attributes that are incorporated into GIS for further analysis. Essential procedures include image enhancement, on-screen digitizing, determination of the range of seafloor features to be mapped, preparation of mapping units in legend format, creation of a hierarchical classification system, and development of new map unit symbolization. Although these are the generic procedures, explanations are separated into development of new hierarchical classification schemes for bathymetric (LADS images) and spectral data (IKONOS images).

2.2.1 Development of a Geomorphological Typology Based on LADS Imagery

Because the bathymetric data is so dense, onscreen and printed products produce patterns and shapes that are identifiable in terms of landform units. Pattern recognition and shape detection (Campbell 1996; Schowengerdt 1997) thus become relevant and important tools for interpreting the LADS bathymetry. Digital image enhancement techniques can be applied using specialized processing modules in programs like Arc GIS Image Analyst®, Idrisi® (Clark University), ERDAS Imagine®, PCI®, Surfer®, *etc.*

Images of practical interest include digital terrain models (DEM) that are generated by data interpolation (Kriging) and grid-generation represented in 3D surfaces by triangular irregular networks (TIN). Fourier analysis is a common mathematical technique for separating an image into its various spatial frequency components. On the basis of a Fourier Transform, it is possible to emphasize certain frequency groups and recombine them into an enhanced image (Campbell 1996). Such filters de-emphasize certain frequencies and pass (emphasize) others. High pass filters emphasize fine detail and edges whereas lowpass filters, which suppress high frequencies, smooth an image and reduce "salt and pepper" noise. Lowpass (mean) filters generalize an image. After an image is enhanced, it is prepared for on-screen digitizing using, for example, a large format smart board (interactive whiteboard). This procedure is possible because morphological units are comprised by combinations of depth, shape, and arrangement of soundings, and shadow patterns. The final digital product is thus compiled in a spatial context that facilitates analysis and computation of selected parameters.

Prior to embarking on image interpretation, the study area should be visually inspected to ascertain the range of features that can be identified (see discussions in Finkl et al. 2004, 2005a, b). A list of features that occur should be compiled to make a comprehensive legend. There are many possibilities for interpretation of features and the orientation depends on the purpose. Because the development of a classification scheme can be an endless task, it is necessary to focus on the purpose of the survey and to rationalize procedures for consistently recognizing features that are identifiable at specific scales of observation. A useful nominal scale of observation for regional LADS bathymetry is about 1:800. Consideration should be the balance between what can be seen, what can be mapped, and what is useful or practical to delineate. The natural spatial heterogeneity of morphological units on the seafloor determines to a large extent what should be mapped. In a sense, then, most natural units are predetermined and they reflect the units that have been mapped and described by other researchers. Table 2.1 is an example of the kind of classification that can be developed from study of LADS imageries. The typology that is presented here shows how seafloor features can be rationally organized and defined. These classificatory units are then merged into mapping units (*cf.* Fig. 2.3a, b).

The LADS high-density bathymetric data sets provide good discrimination of geomorphological units, and this cognitive recognition of various geomorphological units leads to the development of a seafloor typology (*e.g.*, Banks et al. 2007; Finkl 2005; Finkl and Banks 2010). Validation of typologies is achieved by seatruthing that is supported by geophysical surveys (*e.g.*, sidescan sonar and seismic reflection profiling), by geotechnical (*e.g.*, vibracore, jet probe, and grab sample) surveys, and by bottom samples retrieved by divers (*e.g.*, Finkl and Benedet 2005; Finkl and Khalil 2005).

Table 2.1 Typology of morphological features on the southeast Florida continental shelf, based on interpretation of laser airborne depth sounding (LADS) for water depths to 55 m

Province	Subprovince	Comments
A. Sandy (soft) bottom types	1. Shoreface sand flats (−10–25 m depth)	Sand bodies that are shore-attached, between the beachface and offshore rock outcrops
	(a) Sand waves (parabathic)	Shore-parallel waves, large-scale ripple fields
	(b) Smooth seafloor topography	No sand waves with planar bedforms or small-scale ripples
	2. Hummocky (pock-marked) shoreface sands (−20–25 m)	Seafloor surface expression of irregular patterns of low-relief dimples, scour holes
	3. Inner shoreface slope (diabathic channels)	Cross-shore rectilinear channel fields, individual channels up to 120 m wide by 3 m deep
	(a) High relief	>1.5 m with approximate 300 m lateral spacing
	(b) Low relief	<0.5 m with approximate 100 m lateral spacing
	4. Inter-reefal sand flats (north of Biscayne Bay)	Sand bodies, up to 15–20 m thick, between nearshore rock outcrops of the Anastasia
		Formation or reefs of the Florida Reef Tract (FRT)
	5. Intertidal mud flats with mangroves	Shallow-water fine-grained, unconsolidated carbonate accumulations
	6. Banks	Shallow-water backreef flats with skeletal sand overlying limestone and coral
B. Limestone rock[a]	1. Ridge flats (−25–27 m) and depressions (−27–37 m) (ridge and valley, ridge field)	Elongated basins separated by flat-topped ridges, karstified limestone
	2. Fore-basin parabathic ridge system (21–25 m depth)	Shore-parallel ridge crests seaward of basins
	3. Beach ridge plain (lithified ridge systems)	Fossilized ridge and swale topography, northward topographic extension of FRT
	4. Offshore ramp [marine terraces] (−34–37 m)	Terraces seaward of reefs
	a. False crest (top of ramp, −34–37 m)	Seaward inflection of upper slope below the crest to form a lower summit
	b. Shelf break (bottom of ramp, −52–55+ m)	Transition from the continental shelf to the slope
	5. Inshore marine terrace [−1.5–6 m, Anastasia Fm.]	Multiple ridges, partly covered by thin veneer of sand with discernable rock structure

(continued)

Table 2.1 (continued)

Province	Subprovince	Comments
	6. Key (emergent carbonate sand cover over limestone)	Fossilized reef environments, small coral and limestone islands and reefs
C. Channels, paleochannels and related features	1. Structurally controlled meander belt	Structurally controlled meanders entrenched in limestone, nominal sedimentary infilling
	2. Trace channel cuts	Vestige of paleo-valleys, largely buried by sedimentary cover
	3. Infilled valleys	Paleo-valleys filled with sand but crests of valley side slopes clearly visible
	4. Tidal channels	Inlets, drowned paleo-inlets
	5. Ebb-tidal deltas	Sediment accumulations on the seaward side of tidal inlets
D. Florida reef tract (coral-algal reef system)	1. Coral reef	Coral and algal reefs extending from the Dry Tortugas to Martin County
	(a) Barrier (1st, −7–9 m; 2nd, −10–14 m; 3rd, −15–25 m)	Parabathic series of reefs that are near to the shore but separated from it
	(b) Patch	Small isolated reef nearly equant in shape, barrier reef fragments
	(c) Backreef ledge	Shore-facing ledge up to 2 m or more in height
	(d) Backreef rubble slope	Overwashed rubble that accumulates on the backside of a barrier reef
	(e) Forereef rubble slope	Spur and groove topography with coral ridges separated by sand channels
	(f) Platform	Relatively flat-lying bench along the forereef of the FRT
	2. Reef gap (incl. rubble fans)	Break in line of barrier reefs produced by corridors that link inter-reefal troughs with forereef slope on seaward margin of the FRT
	(a) Ramp	Seaward-sloping sedimentary accumulations in reef gaps
	(b) Apron (landward rubble mound)	Arcuate overwash deposits on the backside of barrier reefs that surmount the seaward-most portions of the inter-reefal sand flats
	3. Deepwater reef	Coral reefs occurring below the shelf break in water depths generally greater than 50 m

(continued)

Table 2.1 (continued)

Province	Subprovince	Comments
E. Structural and chemical limestone (karst) bedrock features	1. Karst noye (drowned solution pits, dolines, sink holes)	Drowned limestone terrain
	2. Lineaments, faults, fissures	Linear features characteristic of limestone terrains
	3. Ridge crests	Continuous seabed elevations flanked by side slopes
	4. Trough axis	Approximate center of elongated depression or trough in drowned limestone terrain
F. Continental slope (undifferentiated)		

The classificatory units are based on cognitive interpretation of bottom morphology (bathymetry), depth, exposed and shallowly buried geological structures, and composition of sedimentary materials

[a]Anastasia Fm., Biscayne Aquifer, Tamiami Fm. – Hawthorne Group, Upper Floridan Aquifer System exposed as hardgrounds to form bottom types (Modified from Finkl et al. 2005a)

2.2.2 Geomorphological Symbolization

There are many examples of specialized geomorphological symbols (*e.g.*, Demek 1972; Gardiner and Dackombe 1979; Ollier 1977). Application of geomorphological symbols should focus on detailed geomorphological field mapping and suggestions by Gellert (1988) for coastal mapping. Other sources (*e.g.*, Butler et al. 1986; Elvhage 1980; Gierloff-Emden 1985) may be consulted to select the most useful types of symbols that could be adapted for depiction of shelf environments. Also useful are standards provided by the Federal Geographic Data Committee, Geologic Data Subcommittee (FGDC 2006). The stock symbol set in ArcGIS features topographic symbolization that is partly useful, but additional features are required to adequately depict the range of marine features on the continental shelf.

Additional sources include symbolization from guides to detailed and medium-scale geomorphological mapping (*i.e.* Demek 1972; Demek and Embleton 1976) and various reports of the IGU Commission on Geomorphological Survey and Mapping (viz., Report of the 6th meeting in Canada, 1972; 7th meeting in Brno, 1973; 8th meeting in the USSR, 1974; 9th meeting in CSSR, 1975; 10th meeting in The Netherlands, 1975; 11th meeting in the USSR, 1976; and 12th meeting in Finland, 1977) including legends to the International Geomorphological Map of Europe (Bashenina et al. 1971, 1977). More complete and relevant symbolization can be adapted from Demek (1972) for detailed geomorphological mapping, especially maritime and lacustrine landforms as detailed by Bashenina et al. (1968). Other sources for symbolization and feature names in the marine environment included Milard (1996) and guidelines proposal from terminology for the *Standardization of Undersea Feature Names* (IHB 1989) and lacustrine and marine features (FGDC 2006). Although these works form a basis for mapping, symbolization (*i.e.* symbols, colors, and patterns) can be devised for specific study areas. In the example cited here, geomorphological symbolization was specifically

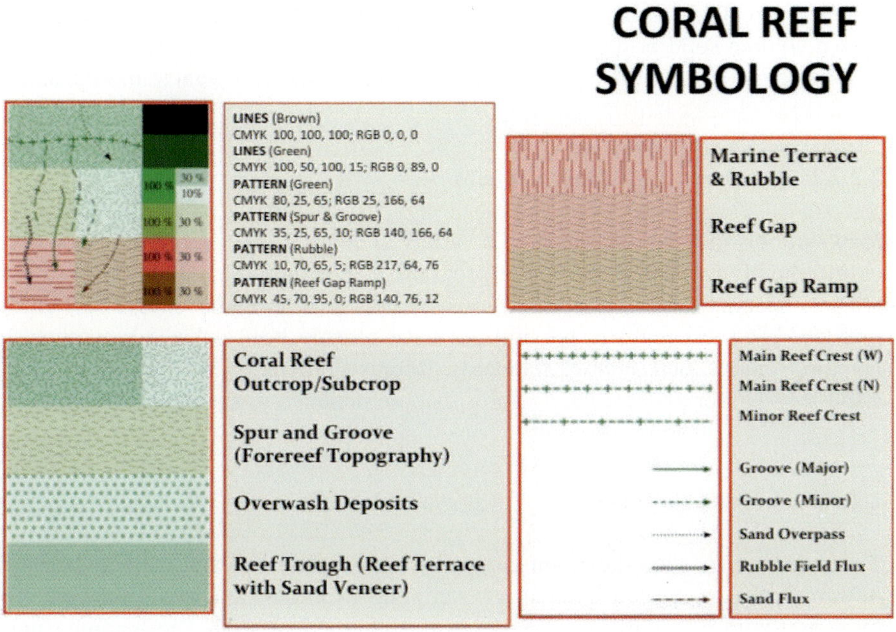

Fig. 2.2 Example of coral reef symbology developed for geomorphological mapping of the seafloor based on cognitive interpretation of LADS imagery. *Lines*, *colors*, *patterns*, and *shading* were assembled in a representative manner to depict the types of features mapped. This kind of symbology was used to construct the map shown in Fig. 2.3a. This symbolization was added as a complimentary layer extension in the standard ESRI ArcGIS® palate (symbol library)

developed for the southeast Florida continental shelf that is broadly characterized by carbonate bedrock exposure (hardgrounds), drowned karst features, barrier and shelf-edge coral reef systems, and unconsolidated carbonate and siliciclastic deposits (see Finkl and Andrews 2008; Finkl et al. 2008). Brief explanations of the legend, symbols, and signs follows.

2.2.2.1 Color Tints and Shades

Maps of seafloor topography interpreted from LADS bathymetric imagery were designed so that symbols were scale independent. The example provided in Fig. 2.2 for coral reefs shows how to build a set of map symbols. The box in the upper left-hand corner of Fig. 2.2 contains the complete symbol set for coral reef mapping units. Line colors are green and brown. The CMYK and RGB notations are supplied so the color can be duplicated. The shade in each tint is denoted by a percentage compared to 100 %. Pattern colors are indicated in the same way and identified in boxes in the lower left-hand corner and upper right-hand corner of Fig. 2.2. The boxes in the lower right-hand corner show line symbols. Within this symbol group, reef crests are denoted by green-colored crosses or crosses with dashes. Solid or dashed green lines with arrowheads indicate forereef grooves. Sand overpass, rubble field flux areas, and sand flux (movement) areas are indicated by brown-colored solid or dashed lines with arrowheads. Color tints, shades, and lines may be combined to produce a wide range of symbols that represent seafloor features.

2.2.2.2 Symbols, Signs, and Ciphers

Map symbols may be derived from a variety of sources (see preceding discussion) or specifically created in a graphics program. The potential combinations of lines, arrows, hachures, and closed polygons (regular and irregular shapes) is almost limitless but most realistic possibilities have already been achieved, as noted in mapping guides, and symbol standards. Representationally, the graphic symbols stand for seafloor geomorphic features that are depicted in simplistic form.

2.2.2.3 Legend Patterns

The map legends should contain all symbology on the maps and include cadastral information, survey area boundaries, sampling locations, and anthropogenic features in addition to the geomorphological units they portray. The mapping units within and between categories collectively combine to produce legend patterns that define similar or dissimilar seafloor features. Cognizance of mapping unit patterns and choice of mapping symbolization normally requires several iterations to achieve compatibility of resulting patterns while maintaining interpretability.

2.2.3 Interpretation and Classification of IKONOS Imagery

IKONOS satellite imagery is increasingly used as a basis for mapping marine habitats (*e.g*., Dial et al. 2003; Finkl and Vollmer 2011; Hochberg et al. 2003; Maeder et al. 2002; Mumby and Edwards 2002; Palandro et al. 2003; Steimle and Finkl 2011). Andrefouet et al. (2003), for example, incorporated multiple IKONOS satellite images of the Arabian Gulf, Indian Ocean, Indo-Pacific, Pacific, and Caribbean biogeographic zones in order to map geomorphologic zones (*e.g*., reef flats, forereef, patch reef, lagoon) and biological communities (*e.g*., seagrass beds, macroalgae coverage, coral overgrowth). Their mapping results help justify the use of IKONOS satellite supplied images as a means to appropriately interpret and classify marine carbonate environments.

Once IKONOS satellite images are enhanced, they can be imported into ArcGIS® ArcMap for onscreen cognitive interpretation of seafloor features and incorporation of classification schemes. Effective interpretation of coastal and benthic marine environments require that the various features are cognitively discriminated by their spectral reflectance characteristics (Hochberg and Atkinson 2000). The fundamental focus of cognitive interpretation is to isolate the portion of upwelling light radiance that penetrates the atmosphere, water column, dissolved

Table 2.2 Typology of morphological features on the Marquesas carbonate bank, based on cognitive interpretation of IKONOS satellite imagery

Bottom type	Material	Geomorphic unit	Type of cover	Coverage
(1) Unconsolidated sediment	(1) Sand	(1) Shoreline intertidal	(1) Live Coral	(1) Continuous
	(2) Mud			(2) Patchy
(2) Coral reef and hardbottom	(1) Spur and groove	(2) Lagoon		(3) Sparse
	(2) Individual or aggregated patch reef		(2) Seagrass	(1) Continuous
		(3) Bank/Shelf		(2) Patchy
	(3) Aggregate reef			(3) Sparse
	(4) Scattered coral/rock in Unconsolidated sediment	(4) Back reef	(3) Macroalgae	(1) Continuous
				(2) Patchy
		(5) Ridge and swale		(3) Sparse
	(5) Pavement		(4) Encrusting/coralline algae	(1) Continuous
	(6) Rock/boulder	(6) Reef crest		(2) Patchy
	(7) Reef rubble			(3) Sparse
	(8) Pavement with sand channels	(7) Forereef	(5) Turf algae	(1) Continuous
		(8) Channel		(2) Patchy
(3) Man-made and terrestrial	(1) Engineering works	(9) Dredged		(3) Sparse
		(10) Vertical wall	(6) Emergent vegetation	(1) Marsh
	(2) Land	(*) Bank/shelf escarpment		(2) Mangrove
			(7) Uncolonized	(1) Uncolonized
(4) Unknown	(1) Unknown	(#) Unknown	(8) Unknown	(1) Unknown

The *color-coded* classificatory units include platform reef environments (*e.g.*, coral reef and hardbottom, unconsolidated sediment), biological overlays, and terrestrial and sub-terrestrial realms. The *numerical codes* are grouped by color, where bottom types and materials go together as a couplet and the other categories are independent

organic material, and turbidity, thereby ultimately, reflecting the specific interpreted features from the seafloor (Dobson and Dustan 2000; Maritorena et al. 1994). However, contrast and brightness of the reflected light can vary amongst different IKONOS images, therefore, color, the main variable in autoclassification algorithms, should be disqualified as the sole conclusive factor for seafloor bottom interpretation. Rather, cognitive interpretation of seafloor environments is contingent upon the ability to collectively distinguish between different color, tone, texture, pattern, and relative spectral reflectance variations.

Just as with the LADS images, local working knowledge of a study area, referred to as contextual editing by Green et al. (2000), becomes a necessary requirement when interpreting seafloor boundaries from IKONOS images. By applying prerequisite knowledge (collateral data) of the geomorphology and biological coverages in tandem with the elements of color, tone, texture, pattern, and relative spectral reflectance, a comprehensive cognitive interpretation of ecological zonations and physical gradients can be accomplished. The process first involves determining the relative brightness, color, and tone of a pixel, or set of pixels, in the IKONOS imagery as light, medium, or dark. The frequency of color and tone variations can then be identified as coarse, medium, or fine texture and pattern. Cognitive knowledge of the study area should also be applied to the repetition and amalgamation of textural patterns, as well as, the relative spectral reflectance of visual cues, to differentiate seafloor features along the various carbonate banks. This is especially important when variable water depths cause seafloor signatures to spectrally appear lighter or darker, which can produce false-positives in autoclassification schemes.

An established scheme of mapping and classificatory units (Tables 2.2 and 2.3) can be applied directly to completed cognitive vector delineation. An individual

Table 2.3 Typology of morphological features on the Marquesas carbonate bank, based on cognitive interpretation of IKONOS satellite imagery

Platform reef environment

- Coral reef and hardbottom
  - Aggregated reef
    - Patchy[a] live coral (23312[b])
    - Sparse live coral (23313)
  - Scattered coral/rock in unconsolidated sediment
    - Patchy live coral (24312)
    - Sparse live coral (24313)
    - Patchy seagrass (24322)
    - Sparse seagrass (24323)
    - Sparse macroalgae (24333)
  - Back reef (rubble)
    - Patchy macroalgae (27432)
    - Sparse macroalgae (27433)
- Unconsolidated sediment
  - Bank/shelf seagrass
    - Continuous (11321)
      - With patchy soft coral (11321, 72[c])
      - With patchy macroalgae and soft coral (11321, 82)
    - Patchy (11322)
      - With continuous macroalgae (bloom) (11322, 11)
      - With patchy macroalgae (bloom) (11322, 12)
      - With sparse macroalgae (bloom) (11322, 13)
      - With sparse epiphytes (11322, 22)
      - With spare epiphytes and sparse lugworms (11322, 33)
      - With continuous epiphytes and macroalgae (11322, 51)
      - With patchy epiphytes and macroalgae (11322, 52)
      - With sparse epiphytes and macroalgae (11322, 53)
      - With continuous lugworm field (11322, 61)
      - With patchy lugworm field (11322, 62)
      - With sparse lugworm field (11322, 61)
      - With continuous soft coral (11322, 71)
      - With patchy soft coral (11322, 72)
      - With sparse soft coral (11322, 73)
      - With patchy macroalgae and soft coral (11322, 82)
      - With continuous macroalgae mats (11322, 91)
      - With patchy macroalgae mats (11322, 92)
      - With sparse macroalgae mats (11322, 93)
    - Sparse (11323)
      - With patchy macroalgae (bloom) (11323, 12)
      - With sparse macroalgae (bloom) (11323, 13)
      - With patchy epiphytes and sparse lugworms (11323, 32)
      - With sparse epiphytes and sparse lugworms (11323, 33)
      - With sparse epiphytes and macroalgae (11323, 53)
      - With patchy lugworm field (11323, 62)
      - With sparse lugworm field (11323, 63)

(continued)

Table 2.3 (continued)

With patchy soft coral (11323, 72)
With sparse soft coral (11323, 73)
With sparse macroalgae and soft coral (11323, 83)
With patchy macroalgae mats (11323, 92)
Lagoon seagrass
Patchy (11222)
With continuous macroalgae (bloom) (11222, 11)
With patchy macroalgae (bloom) (11222, 12)
With sparse macroalgae (bloom) (11222, 13)
With patchy epiphytes (11222, 22)
With sparse epiphytes (11222, 23)
With continuous epiphytes and sparse lugworms (11222, 31)
With patchy epiphytes and sparse lugworms (11222, 32)
With sparse epiphytes and sparse lugworms (11222, 33)
With continuous epiphytes and sparse mangroves (11222, 41)
With patchy epiphytes and sparse mangroves (11222, 42)
With sparse epiphytes and sparse mangroves (11222, 43)
With continuous epiphytes and macroalgae (11222, 51)
With patchy epiphytes and macroalgae (11222, 52)
With sparse epiphytes and macroalgae (11222, 53)
With patchy lugworm field (11222, 62)
With sparse lugworm field (11222, 63)
Sparse (11223)
With patchy epiphytes and sparse lugworms (11223, 32)
With sparse epiphytes and sparse lugworms (11223, 33)
With sparse epiphytes and lugworms (11223, 53)
With patchy lugworm field (11223, 62)
With sparse lugworm field (11223, 63)
Macroalgae
Patchy (11332)
Sparse (11333)
Shoreline intertidal (emergent vegetation)
With continuous mangrove (121621)
With patchy mangrove (121622)
With sparse mangrove (121623)
Saline lake (1118, 11)
Shoreline supertidal
Sand dune (vegetated/unvegetated) (111a, 10)
Upland vegetation (111a, 12)
Channel
With continuous seagrass (11821)
With patchy seagrass (11822)
With sparse seagrass (11823)
Uncolonized
Shoreline intertidal (1117)
Channel (1187)

(continued)

Table 2.3 (continued)

Lagoon (1127)
Bank/shelf (1137)
Unknown
Barrier reef environment
Coral reef and hardbottom
Aggregated reef
With patchy live coral (23312
With sparse live coral (23313)
Individual or aggregate patch reef
With patchy live coral (22312)
With sparse live coral (22313)
Scattered coral/rock in unconsolidated sediment
With patchy seagrass (24322)
With sparse seagrass (24323)
Back reef (reef rubble)
With patchy seagrass (27422)
With sparse seagrass (27423)
Unconsolidated sediment
Seagrass
Patchy (11322)
Sparse (11323)
Uncolonized
Bank/shelf (1137)
Unknown

The classificatory units include platform reef and barrier environments (*e.g.*, coral reef and hardbottom, unconsolidated sediment), biological overlays, and terrestrial and sub-terrestrial realms. The numeric code following a mapping unit is the designator that was entered into the GIS attribute table

[a]Dominant biological coverages: continuous = >90 %, patchy = 50–90 %, sparse = 10–50 %

[b]Mapping units are based on a numeric system that was devised for this project so that each polygon received a coded identifier

[c]Some numeric codes carry a suffix following a comma. The suffix codes are: (1) macroalgae bloom, (2) epiphytes, (3) epiphytes and sparse lugworms, (4) epiphytes and sparse mangroves, (5) epiphytes and macroalgae, (6) lugworm field, (7) soft coral, (8) macroalgae and soft coral, (9) macroalgae mats. The percentage cover is as follows: (1) continuous (>90 %), (2) patchy (50–90 %), and (3) sparse (10–50 %)

thematic layered display, and associated legend, is created for each category heading in the classification scheme, with a specific color being assigned to each classifying interpretation unit. Vector polygons may then be filled with a specific classifying color that corresponds to the cognitive interpretation.

Creation of comprehensive attribute tables relates directly to the classification of cognitive digitized polygons. By doing so, a database of spatial query information, including areal extents of classified units, is compiled for analysis. Attribute tables for given imagers are created in ESRI's ArcGIS® 10.1 ArcMap program. The spatial integrity of attributes can be confirmed through a series of topology checks in ArcMap. When errors are detected, the error inspector table can be used to zoom to specific attributes that spatially broke the parameters set forth in the topological rules layer. Once all the errors have been identified and corrected, the topology is validated and attribute data can be exported for analysis.

2.3 Remote Sensing of Carbonate Banks in Southern Florida

Multiple studies have focused on the classification of seafloor types on southeast Florida carbonate banks (*e.g.*, Banks et al. 2007; DaPrato and Finkl 1994; Finkl 2004; Finkl and Warner 2005; Finkl et al. 2005a, b; Finkl and Vollmer 2011; Lidz 2004; Lidz et al. 1997, 2003, 2006; Madden et al. 2008; Madley et al. 2002; Moyer et al. 2003; Palandro et al. 2005; Rohmann and Monaco 2005; Steimle and Finkl 2011; Walker et al. 2008; Warner 1999; Zieman et al. 1989). One of the earlier modern classification studies was conducted by Duane and Meisburger (1969) in their investigation of geomorphology and unconsolidated sediments offshore Palm Beach and Miami-Dade counties. They surveyed 365 km^2 with seismic reflection interpretation of ridges and sandy areas in water depths ranging from 3 to 33 m. Finkl and Warner (2005), using stereo-paired aerial photographs, mapped submarine morphological features offshore Palm Beach County. Using an acquisition scale of 1:3,900, seafloor features were mapped to an approximate water depth of 15 m (about 500 m seaward of the shoreline).

Using LADS bathymetric data, Finkl et al. (2005a, b) developed a hierarchal classification that defined submarine provinces and subprovinces based on bottom topography, water depth, exposed and shallowly buried geological structures, and composition of sediments. The main provinces included: sedimentary (soft) seafloor units; limestone rock; channels, paleochannels, and related features; Florida Reef Tract and the coral-algal system; structural and chemical limestone (karst) bedrock features; and continental slope. Individual mapping units were then extrapolated and applied to the final maps based on the pre-interpreted submarine provinces.

Lidz et al. (2006) used aerial photomosiacs to map 3,140 km^2 of the Florida Keys National Marine Sanctuary (FKNMS). Instead of specific bottom feature interpretations, Lidz et al. (2006) applied a more general approach to mapping such a large area and derived 19 submarine units. Benthic habitat mapping units with the largest contributions to the overall study area included seagrasses on lime mud (864.70 km^2; 27.5 %), seagrasses on carbonate sand (587.63 km^2; 18.7 %), bare carbonate sand (542.80 km^2; 17.3 %), bare lime mud and/or seagrass-covered muddy carbonate sand (302.87 km^2; 9.6 %), bare Pleistocene oolitic limestone (250.35 km^2; 8.0 %), and senile coral reef (70.19 km^2; 2.2 %). By using this general approach to mapping large areas, Lidz et al. (2006) were able to provide new information on previously undetermined seabed morphologies in the FKNMS.

This type of seafloor mapping was expanded by Madden et al. (2008) with the introduction of a classification scheme that was amenable to mapping carbonate bank ecosystems in shallow water. The Madden et al. (2008) system is based on a rigid hierarchical approach that starts with general coral ecosystem geomorphological structures and ends with specific biological covers. Steimle and Finkl (2011) mapped 1,360 km^2 of Florida Bay using a comprehensive hierarchical classification scheme that they developed to included five physiographic realms, 17 morphodynamic zones, 11 geoforms, 38 landforms, six types of surface sediment cover, and a combination of nine biological covers. They produced a new type of map that amalgamated geological properties and biological communities on a carbonate bank.

2.4 Examples of Seafloor Mapping on Carbonate Banks

The following two examples serve as vignettes of seafloor mapping on carbonate banks based on interpretation of bathymetry (LADS surveys) and environments and habitats (IKONOS surveys). These studies draw on prior mapping efforts in the area but provide new classifications of seafloor typologies that advance application of modern technologies. The first case study deals with a LADS survey along an open ocean coast with shelf-edge coral reefs, shelf hardgrounds (exposure of carbonate bedrock), and sedimentary deposits. The second case study focuses on biophysical environments in the Marquesa Keys and features carbonate bank typologies in the distal Florida Keys. Classification schemes were developed for both areas by cognitive interpretation of bathymetric or spectral data. General hierarchical schema were devised in such a way that they are open ended and can be modified according to local typologies. This procedure avoids compiling massive all-inclusive systems that would need to be uploaded into GIS for further analysis.

2.4.1 LADS Survey of Carbonate Shelf

Bathymetric data, derived from LADS (Laser Airborne Depth Sounding Survey) developed by Tenix LADS Corporation (Mawson Lakes, South Australia), was acquired along the Florida southeast coast in 2001 (Broward County) and 2003 (Palm Beach and Miami-Dade counties). The dataset comprised millions of bathymetric data points along a 160-km coastal segment that extends up to 6 km offshore to cover nearly 600 km^2 of seabed (see Fig. 2.1). The high-density bathymetric datasets provide good discrimination of geomorphological units and this cognitive recognition of various geomorphological units leads to the development of a seafloor typology (Table 2.1). Validation of typologies is achieved by seatruthing that is supported by side scan sonar and sub-bottom profiler geophysical surveys, by geotechnical (vibracore) surveys, and by bottom samples and videos retrieved by divers (Finkl et al. 2005a, b).

Some of the morphological units in the study area originated as terrestrial features (*e.g.*, karst nu) that were subsequently drowned by sea-level rise viz. to become karst noye (drowned karst) of which there is ample evidence throughout the study area in the form of solution pits, dolines, and sink holes. Most other features are, however, marine in origin (*e.g.*, the Florida Reef Tract) except for the coastal channels. The main morphological features occurring in the study area, summarized in Table 2.1, include sandy bottom types, rock hardgrounds (exposed bedrock, usually as karst noye), coral reefs and related features. For the Palm Beach County sector of the overall study area, 269 km^2 were mapped with continental slope, ridge fields, and sand flats respectively making up 15.6 %, 27.4 % and 30.6 % of the total area. Diabathic channel fields comprising 19.2 km^2 accounted for 7.1 % of the survey area. Other units of lesser extent individually accounted for less than 2 % of the survey area except for deepwater reefs, forereef rubble slopes, backreef overwash deposits, and sand waves each of which accounted for about 2 %.

Figure 2.3a is a diptych showing seafloor morphology, based on cognitive interpretation of LADS bathymetry, on the narrow continental shelf offshore the Hillsboro Inlet. The LADS bathymetry extends from the beach to about 55 m depth on the upper reaches of the continental slope. The raw LADS bathymetry (left panel) was deliberately color ramped to emphasize depth relations as demarcated by rock reef and coral reef parabathic (shore-parallel) tracts. Interpretation of seafloor geomorphology is shown in the right panel. The key to the symbols used here is laid out in Fig. 2.3b. The amount of detail provided by LADS bathymetry offers a high level of interpretability for sections of the seafloor that was previously not possible. Study of Fig. 2.3a, b in a GIS format emphasizes the usefulness of airborne remote sensing of the seafloor where applications of the acquired and interpreted data are only beginning to be appreciated.

It is hoped that this example provides insight into the advantages of using ALB data to interpret a range of submarine geomorphological features that can be grouped into general mapping units. The mapping units were determined at a scale of 1:25,000 for the project as a whole, zooming to greater detail as required. It is evident that these mapping units generalize what can be interpreted from the ALB data. Detailed mapping is thus possible and even required in many areas to better ascertain relationship between morphologic features and mapping units. One advantage of having the imagery in GIS format is that scales can be easily manipulated for viewing and analysis.

2.4.2 IKONOS Survey of the Marquesas

Imagery was acquired in 2006 over several months and different frames were used for the same geographic area in order to map different seafloor features that were obscured by cloud cover, waves, glint, turbidity in the water column, or seasonal algal blooms.

Shallow water features were interpreted from pan sharpened and color images but near the seaward margin of the shelf, deepwater features (in about 60 m of water) associated with the FRT could only be interpreted from digitally enhanced images. Image enhancement followed procedures outlined by Finkl and DePrato (1993) and Chauvaud et al. (1998) and was conducted using IDRISI Taiga

Fig. 2.3 (**a**) Diptych showing an uninterpreted enhanced LADS image in the *left panel* and the cognitively interpreted geomorphological units based on bathymetry in the *right panel* for the southeast Florida continental shelf (see Fig. 2.1 for overall survey location). The raw LADS image was enhanced, vertically exaggerated, and bathymetrically colorized for easier interpretation. *Symbols* explained in Fig. 2.2 are used to depict the coral reef systems of the Florida Reef Tract and displayed in GIS format using IDRISI. (**b**) Legend to accompany (**a**), showing mapping units displayed in the right panel of the diptych in (**a**). Although this example of mapping units is incomplete as it only refers to the field of view in (**a**), it demonstrates an ability to interpret LADS bathymetry with symbols specifically developed for depicting submarine topographic units in a GIS format

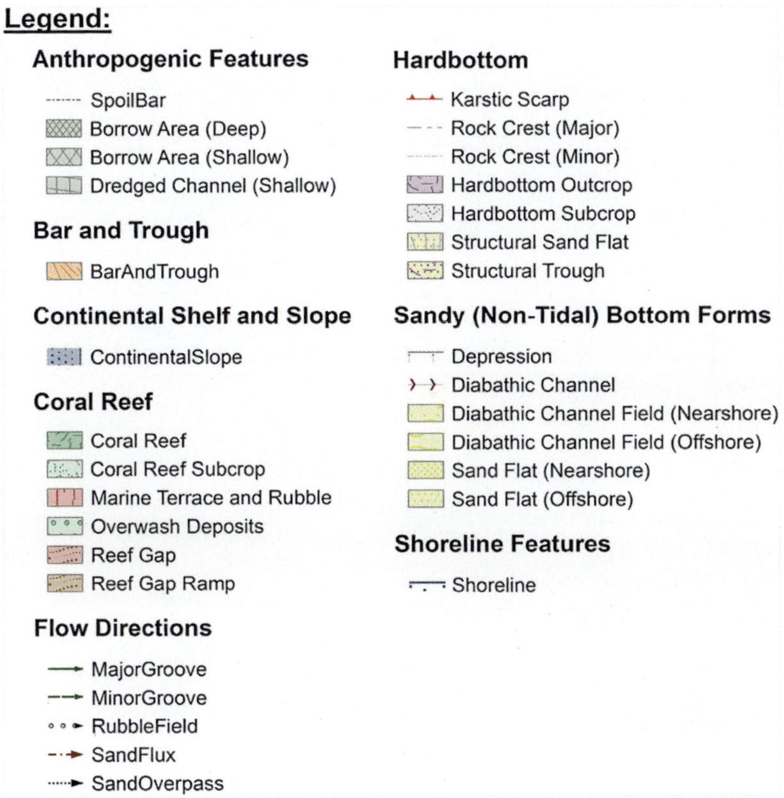

Fig. 2.3 (continued)

integrated GIS and image processing software (Clark University, http://www.clarklabs.org/), which includes nearly 300 modules for the analysis and display of digital spatial information, and involved the following basic steps for enhancing images: (1) import images, (2) determine bands that allow for deepwater features, (3) overlay images (band ratioing), (4) stretch image, and (5) filter image. The linear contrast stretch of the digital data involved the identification of lower and upper bounds from the histogram (minimum and maximum brightness values in the image) and applying a transformation to stretch this range to fill the full spectral range. The data was then filtered using Gaussian and mean filters with filter sizes ranging from 3×3 to 7×7. High-pass and low-pass filters were used to determine the best image quality, which varied from scene to scene.

Figure 2.4 shows the advantages of image enhancement that compares an RGB 8-bit color image (top panel) with a pan-sharpened image (middle panel) with an enhanced image (bottom panel) that shows more detail of deepwater shelf features. Details of the image enhancement procedures are as follows. The IKONOS image was first imported into the software and each band was individually analyzed for its ability to enhance deepwater features. Bands two and three best represented

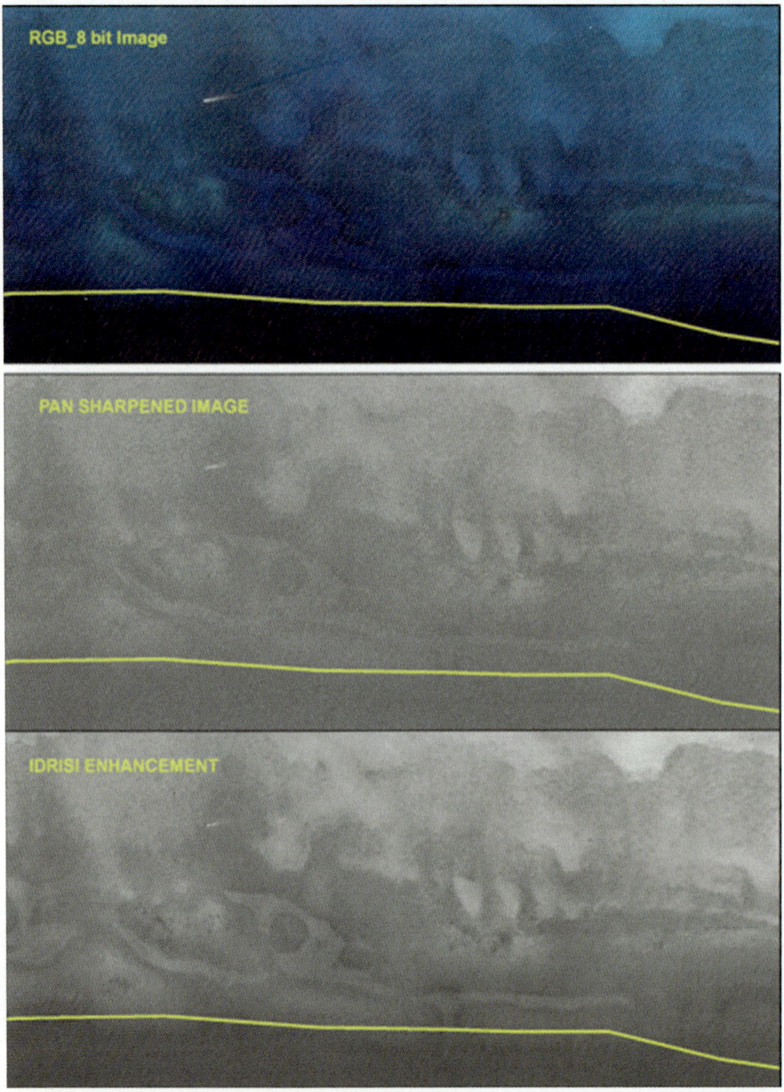

Fig. 2.4 Comparison of image enhancement techniques for deep water coral reef environments that are near the detection limits of the IKONOS radiometer. The spectrally enhanced bottom two images (pan sharpened vs. the IDRISI enhancement) show progressively more detail of bottom features. The yellow line marks the boundary between interpretable and uninterpretable data

deepwater geomorphic features. The overlay module was combined with the maximize option to produce a single image from multiple bands, providing the most detail possible. The maximize option outputs pixels that represent the maximum digital number for those in corresponding positions on the first and second image. A data stretch was then performed on the overlaid image to increase contrast. A linear stretch was found to be the best methodology to further enhance

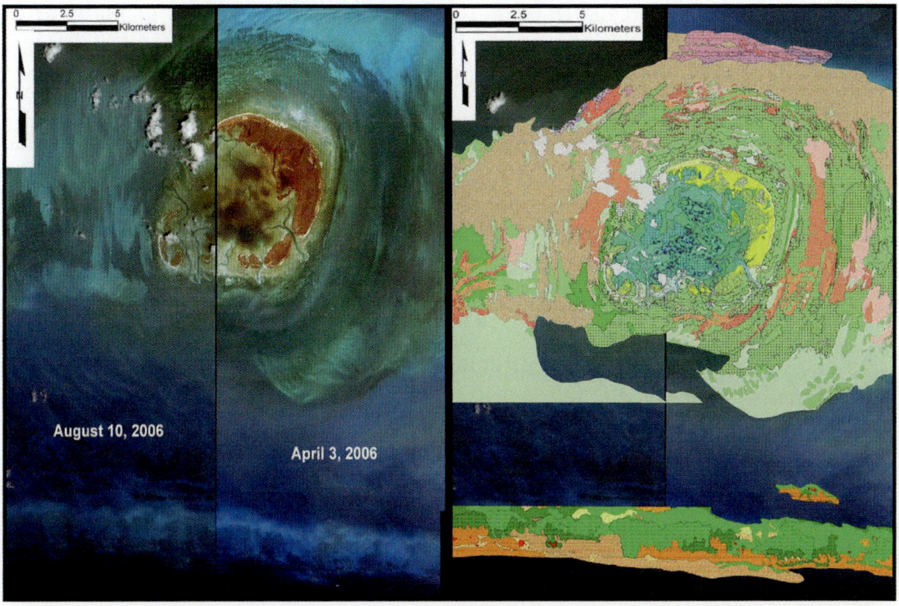

Fig. 2.5 Marquesa Islands (see Fig. 2.1 for overall survey location) diptych showing an uninterpreted enhanced IKONOS image in the *left panel* and cognitively interpreted biophysical units based on spectral reflectance of seafloor barrier and platform coral reef environments in the right panel. The *left panel* is provided for comparison with cognitively interpreted units. Some GIS layers were turned off in the *right panel* for easy comparison of interpreted vs. uninterpreted seafloor. Polygons in this map were compiled from a numerical code for seafloor environments that are summarized in Table 2.1. Application of the code to each polygon produced discrete mapping units, a partial example of which is found in Table 2.2

the underwater features. This stretch linearly scales values between a specified minimum and maximum limit. To further prepare the image for visual display, the stretch was performed in certain instances with 5 % saturation. This saturation concentrates the output values on the less extreme values. The image was then filtered to enhance the deepwater geomorphic features as much as possible. The filtering creates a new image by combining pixel values with immediate neighbors. Gaussian, mean, and median filters were found to be most effective: the Gaussian and mean filters generalized the image whereas median filters removed random noise.

Seafloor features were cognitively interpreted and mapped at a nominal scale of 1:6,000 (Fig. 2.5). They were delineated by relative spectral reflectance, color, textures and patterns of their boundaries as well as contextual inferences from previous studies and *in situ* observations. Visual identification and interpretation of seafloor environments derived from a mosaic of 4 m resolution multi-spectral IKONOS images utilized ESRI's ArcMap Geographic Information System (GIS). Polygon boundaries were digitized in shapefiles georeferenced to the North American 1983 Datum Zone 17 North, Universal Transverse Mercator (UTM) projected coordinate system. Subsequent to initial digital delineations of shallow-marine environments, obscured or irresolute seafloor types were visited in the field for direct visual observations. The *in-situ* observations either confirmed the classified polygon or reclassified incorrect cognitive interpretations.

About 422 km^2 were interpreted with 96 mapping units defined. The Platform Barrier Reef environment had 84 units and the Barrier Reef environment had 12 units. Classification units were defined by a numeric code to allow a "full picture" to the individual units (Table 2.2). These units were defined in a stepwise procedure: the first number is selected from the column for 'Bottom Type,' followed by the associated color units in the second column: 'Material.' Next comes the geomorphologic base, which is independent of the first two columns. The extent of the biological cover is grossly estimated as a percentage of the dominant type of cover. The attribute table was built with a multi-discipline interpretation in mind, by allowing the end user the flexibility of extracting the information of interest, *i.e.*, major biological cover, detailed geological cover, *etc.* Suffixes were added to further interpret the areas biological cover.

A symbology was created for each seafloor feature, providing a range of mapping units. An example of how this was set up is summarized in Table 2.3, which lists the classification but does not show the mapping unit symbolization. The typology breaks down the overall environment into two major provinces: platform reef environment and barrier reef environment. Each is further subdivided into the dominant types of features that can be cognitively interpreted from the IKONOS imagery. Geomorophological and biological units were color-coded by major type, then a pattern was applied to represent minor types and finally a gradation of colors was applied to represent changes in feature densities (frequency of occurrence) within the area. A portion of the mapped area is shown in Fig. 2.4 as an example of the level of detail that can be extracted from the satellite imagery. A legend is not provided because it is so extensive. The figure should be regarded as an illustration of what can be done using the system developed here, as summarized in Tables 2.2 and 2.3.

Interpretation of seafloor environments from IKONOS satellite imagery is possible with major geomorphological features being most easily determined. Subdivision of seafloor environments is, however, complicated by biological covers that can mask underlying geology and geomorphology. Dense seagrass can, for example, look very similar to patch reefs making differentiation problematical (*e.g.*, Mumby et al. 1999). Other complicating factors include variable growth patterns of mangroves, presence of algal blooms, variable water depths and suspension of particulates that can change the appearance of the same features, and hard-bottoms with aggregate corals, sponges, and turf algae. Lugworms (sand worms) (*Arenicola cristata*) that burrow in soft sandy or muddy sediments bring substrate materials to the surface of the seafloor imparting a grayish-blue tone to some lighter colored sediment flats. Where lugworm fields were especially dense, they were identified as discrete mapping units due to tonal differences in the IKONOS imagery. Field inspection is usually required to resolve mapping units that appear similar and this is particularly so for dense circular seagrass beds with sand halos which resemble patch reefs and patchy hardgrounds with variable biological covers.

Development of a hierarchical classification and mapping units was a somewhat complicated process due to complex spatial distribution patterns (Tables 2.2 and 2.3). The Platform Reef Environment was broken down into two main units: coral reefs (aggregate reefs, individual or patch reefs, and scattered corals in unconsolidated sediment) and unconsolidated sediment (uncolonized sand, sand with seagrass, sand with macroalgae, and with turf algae, sand with emergent vegetation, mud with marsh, and mud with mangrove).

The Barrier Reef Environment was likewise broken down into two main mapping units: coral reef and hardbottom (aggregate reef, patch reefs, scattered coral in unconsolidated sediment, and backreef) and unconsolidated sediment (uncolonized or with seagrass and macroalgae). There were different combinations that could be used, but this system seemed to capture salient natural features. Inclusion of biological covers enormously complicated the range and number of mapping units, which were initially and most importantly based on geomorphological parameters. An advantage of presenting the classification and mapping efforts in a GIS framework is that very complicated marine systems can be differentially queried, broken down into component parts, subsumed within overarching units, or graphically displayed to show various types of spatial interrelationships.

2.5 Discussion

Advancements in the interpretation of seafloor biophysical environments on carbonate banks in tropical and subtropical environments, such as in southern Florida, include the use of laser airborne depth sounding (LADS) data and IKONOS satellite imagery and are facilitated by generally clear, shallow waters. Turbidity (due to algal blooms and suspension of particulate matter) and coloration of the water column by tannic acids from surface runoff of the Florida Everglades degrade data quality and under extremely adverse conditions severely limit the acquisition of usable products for interpretation from passive systems. Therefore, airborne laser bathymetric and satellite multispectral data are thus conditioned by optimum environmental conditions that characterize these areas for much of the year. As with any advanced remote sensing platform, there are pros and cons associated with acquisition and interpretation of the data.

Once usable data is acquired under near-optimal conditions, interpretation depends upon the skill set and expertise of the interpreter. Data acquisition is thus one part of the problem and its interpretation is another. Presently there is a plethora of data and a dearth of interpretations. Interpretive skills are very much dependent on formal scientific background or training in such fields as remote sensing, geomorphology, marine geology, and marine biology as well as subfield specializations within these broader disciplines. Some of the pros and cons related to interpretation of airborne high-density bathymetric data and satellite mulitspectral data are discussed in what follows.

2.5.1 Pros and Cons Associated with High Density Airborne Laser Bathymetry (ALB) Data

Numerous pitfalls are associated with the production of geomorphological maps when interpreted from ALB data without optimal geological control. Sub-bottom seismic data helps to define some units, but it may not always be available. The quality of geomorphological units interpreted from the ALB imagery depends on the expertise of researchers. In general, the greater the experience of the researcher (*i.e.* familiarity with geomorphological mapping, landform classification, and terrain analysis in different shelf settings), the more accurate the map. Nevertheless, a great deal of morphological and morphometric information can be acquired by interpretation of high density ALB data represented in three-dimensional digital terrain models, which can be interpreted in terms of bathymorphometric units. This new information provides increased insight into understanding of seafloor features on the continental shelves. Image enhancement is limited by one data band and lack of access to proprietary LADS algorithms that could assist manipulation of the data.

A primary advantage of ALB technology is laser acquisition of sounding data in digital format that provides millions of data points for nearshore seabed topography in a fraction of the time required by conventional surveys. Airborne data acquisition permits rapid day or night survey of large areas that are difficultly accessible. The digital terrain model generated from dense ALB datasets permits variation of pixel size, provides a degree of data separation or overlap, and is amenable to filtering techniques for data enhancement.

The resulting hard copy color maps provide picture-like renditions of the seabed that provide for the first time, accurate depiction of ALB as bathymorphometric images. This latter property is often taken for granted, in spite of the fact that until these bathymetric datasets and associated imagery appeared, we had no good idea of the complexity and continuity of detailed seafloor topography. More than three decades ago, Duane and Meisburger (1969) delineated the approximate positions of reefs, hardgrounds, and sand flats associated with the Florida Reef Tract (FRT). Aerial photography shows nearshore bottom features, but lacks depth information. Satellite imagery also provides limited access to nearshore bottom features, but no previous system of seafloor mapping or image analysis provides the kind of spatial resolution of bottom features over large expanses of the seabed as the newly acquired high dense bathymetric data using ALB systems. Seafloor discrimination on the basis of acoustic classes from sidescan sonar and single- or multibeam bathymetric survey shows a high-level of correlation with interpreted LADS bathymetric classes.

2.5.2 Pros and Cons Associated with Satellite Multispectral Data

Satellite multispectral data provides many advantages when interpreting biophysical environments along carbonate banks. In the tropics and subtropics, marine carbonate banks can be extremely complex to interpret because the sedimentary cover may range from meters thick to only a few centimeters over short distances. Sand waves and dunes are common, but interdunal areas can be nearly sediment-free with the underlying limestone exposed by strong currents. On the reef platform around the Marquesas atoll, for example, there were areas of carbonate sediments that covered the underlying limestone but there were also mosaics of sponges, soft corals and seagrass in areas with thin sediment cover. Even though these areas were dominated by a sedimentary cover, sediment must be restricted to a relatively thin layer (*i.e.* a few centimeters or less) so that species of soft corals and sponges could take advantage of rocky footholds in areas that become exposed by currents. The sedimentary cover in many places is so thin that the underlying rock structure (lineaments, fractures, variations in microtopography due to heterogeneity of the limestone make up) is clearly visible in the IKONOS multispectral imagery. Variations in thickness of the sedimentary cover complicate interpretation of the IKONOS imagery because very small differences on the order of a few centimeters and the velocity of bottom currents determine whether the sediment is uncolonized or is thick enough to support variable densities of seagrass.

Geological and geomorphological features, which form the basis of the mapping units, are greatly expanded upon with the consideration of biological covers that imparted many different tonal combinations in the satellite images. In order to comprehend all of the variations in the IKONOS imagery based on color, texture, tone, pattern, and saturation, it is necessary to construct logical and systematic mapping procedures that first identify bedrock and unconsolidated sediments (*e.g.*, sand, mud), then biophysical features (*e.g.*, coral reefs), and finally various types of biological covers (*e.g.*, live coral, seagrass) The incorporation of all these major biological categories in the IKONOS imagery provides an interpretation advantage that ALB datasets lack. Whereas the ALB allows for a more accurate interpretation of the geomorphology, IKONOS images provide a more complete interpretation of the biological covers growing on the geologic features. Even though the increased number of mapping units in the IKONOS imagery is necessarily large, the mapping effort does not become over-complicated or convoluted. However, there are many challenges associated with the incorporation of biological units when mapping seafloor environments, as well as subtidal and supratidal zones. Beaches on the Marquesas atoll, for example, are composed of macroalgal *Halimeda* platelets and perched on limestone platforms, making it difficult to accurately delineate. Furthermore, in the IKONOS imagery, tonal differences between beaches, intertidal platforms, and lugworm fields can be very subtle and cognitive context becomes the main interpretive guide.

Even though spectral data from IKONOS satellite imagery have been used in multiple benthic interpretive efforts as an ideal platform (*e.g.*, Dial et al. 2003; Finkl and Vollmer 2011; Hochberg et al. 2003; Maeder et al. 2002; Mumby and Edwards 2002; Palandro et al. 2003; Steimle and Finkl 2011), there are some limitations associated with the IKONOS images. IKONOS satellite imagery is typically not freely available to the public; therefore, it must be purchased. In addition to availability and cost, another limitation may include color and tone variations between individual IKONOS scenes, which were acquired at different times. Disparity in the appearance of adjacent scenes making up a composite of the study area occurs due to efforts to minimize cloud cover and variable underwater clarity. While deselection of unusable scenes provides the most effective means for collating a composite image of the study area, noticeable lines of transition from one scene to another are an unwanted result. These variable transitions between scenes eliminate the possibility of using autoclassification algorithms for seafloor interpretation, which would result in false-positives (*i.e.* spurious results) by interpreting the same marine landform unit differently in multiple areas. However, even with these limitations, IKONOS satellite images provide one the most effective remote sensing platforms for the application of newly developed seafloor classification schemes.

2.6 Conclusions

The last four decades has seen remarkable progress in the advancement of remote sensing capabilities. With the advent of technological progress, there has been a corresponding increase in the levels of interpretation and mapping of coastal marine environments. Increasing perfection of LADS and IKONOS technologies now permits a level of image cognition that was heretofore not possible. Airborne digital bathymetry facilitates the mapping of large expanses of seafloor at levels of detail where topographic assemblages can be detailed. Satellite imagery is likewise capable of depicting seafloor environments based on multispectral reflectance. These technological advancements bring together the nature and character of the submarine world into better focus on shallow carbonate banks. Although the two examples discussed here represent advanced technologies that retain advantages and disadvantages, the pros far out weight the cons to produce resilient databases of value to coastal research and resource management. The main problem today is that much more remotely sensed data can be collected than analyzed.

As shown in the LADS survey of the open ocean southeast Florida coast, enhancement of digital imagery created from dense bathymetric data can be used to highlight selected features, detect previously unnoted features, or digitally select certain features from an array of seabed features for specialized study. A good example of application of the new bathymorphometric data is the realization that the Florida Reef Tract contains long continuous troughs between reef systems that have been infilled with sediments. These bathymorphometric maps also show that the classical three-reef system is more complicated than originally perceived.

The example of the Marquesa Islands in the Florida Keys shows that IKONOS satellite imagery provides a suitable platform for mapping coastal (terrestrial) marine (seafloor) environments because the pixel size and number of bands permits differentiation of seafloor environments based on geomorphological units and biological covers. The southern Florida Keys hosts a great diversity of submarine environments in the area containing the only known atoll in the Atlantic Ocean. Interpretation of seafloor features seen in the IKONOS imagery provides an opportunity to develop biogeomorphological classification systems that could be used to characterize shallow water carbonate bank environments in lagoons, sediment flats, hardbottoms, tidal channels, and coral reefs.

References

Anders FJ, Byrnes MR (1991) Accuracy of shoreline change rates as determined from maps and aerial photographs. Shore Beach 59(1):17–26

Andrefouet S, Kramer P, Torres-Pulliza D, Joyce KE, Hochberg EJ, Garza-Perez R, Mumby PJ, Riegl B, Yamano H, White WH, Zubia M, Brock JC, Phinn SR, Naseer A, Hatcher BG, Muller-Karger FE (2003) Multi-site evaluation of IKONOS data for classification of tropical coral reef environments. Remote Sens Environ 88:128–143

Banks KW, Riegl BM, Shinn EA, Piller WE, Dodge RE (2007) Geomorphology of the southeast Florida continental reef tract (Miami-Dade, Broward, and Palm Beach Counties, USA). Coral Reefs 26:617–633

Barton C (2002) Marie Tharp, oceanographic cartographer, and her contributions to the revolution in the Earth sciences. In: The earth inside and out: some major contributions to geology in the twentieth century. The Geological Society of London, Oxford, pp 215–228

Bashenina NV, Gellert J, Joly F, Klimaszewski M, Scholz E (1968) Project of the unified key to the detailed geomorphological map of the world, Folia Geographica (Krakow, Polska Akademia Nauk), Series Geographica-Physica, vol II. Uniwersytetu Jagiellońskiego, Kraków, pp 11–12, 18

Bashenina NV, Blagovolin NS, Demek J, Dumitrashko NV, Ganeshin GS, Gellert JF, Leontyev OK, Mirnova AV, Scholz E (Compilers) (1971) Legend to the international geomorphological map of Europe 1:2,500,000, 5th version. Czechoslovak Academy of Sciences, Institute of Geography, Brno, 30p

Bashenina NV, Blagovolin NS, Castiglioni CB, Demek J, Dumitrashko NV, Gams I, Embleton C, Ganeshin GS, Gellert JF, Leontyev OK, Mirnova AV, Scholz E, Starkel L (1977) Legend to the international geomorphological map of Europe 1:2,500,000. Czechoslovak Academy of Sciences, Institute of Geography, Brno, 23p

Brock JC, Purkis SJ (2009) The emerging role of Lidar remote sensing in coastal research and resource management. In: Brock JC, Purkis SJ (eds) Coastal applications of airborne Lidar, Journal of coastal research, special issue no. 53. Coastal Education and Research Foundation, West Palm Beach, pp 1–5

Bukata RP, Jerome JH, Kondratyev KY, Posdnyakov DV (1995) Optical properties and remote sensing of Inland and Coastal Waters. CRC Press, Boca Raton, p 365

Butler MJA, LeBlanc C, Belbin JA, MacNeill JL (1986) Marine resource mapping: an introductory manual, FAO fisheries technical paper 274. Food and Agriculture Organization of the United Nations, Rome, 256p

Campbell JB (1996) Introduction to remote sensing, 2nd edn. Taylor & Francis, London, 622p

Chauvaud S, Bouchon C, Maniere R (1998) Remote sensing techniques adapted to high resolution mapping of tropical coastal marine ecosystems (coral reefs, seagrass beds and mangrove). Int J Remote Sens 19(18):3625–3639

DaPrato GW, Finkl CW (1994) Applications of a layered classification approach, based on spatial primitives obtained from TM data from Southern Florida, for the analysis of turbid marine waters using fuzzy sets. Int Arch Photogramm Remote Sens 30(4):69–76

Demek J (ed) (1972) Manual of detailed geomorphological mapping. Academia (for the International Geographical Union, Commission on Geomorphological Survey and Mapping), Prague, 344p

Demek J, Embleton C (eds) (1976) Guide to medium-scale geomorphological mapping. Czechoslovakia: International Geographical Union, Brno, Commission on Geomorphological Survey and Mapping (Preliminary edition for the 11th Commission meeting, Kiev, 1976, 23rd International Geographical Union), 339p

Deronde B, Houthuys R, Henriet JP, Van Lancker V (2008) Monitoring of the sediment dynamics along a sandy shoreline by means of airborne hyperspectral remote sensing and LIDAR: a case study in Belgium. Earth Surf Process Landf 33:280–294

Dial G, Bowen H, Gerlach F, Grodecki J, Oleszczuk R (2003) IKONOS satellites, imagery, and products. Remote Sens Environ 88(1–2):23–36

Dobson EL, Dustan P (2000) The use of satellite imagery for detection of shifts in coral reef communities. In: Proceedings, American Society for Photogrammetry and Remote Sensing, Washington, DC

Doel RE, Levin TJ, Marker MK (2006) Extending modern cartography to the ocean depths: military patronage, cold war priorities, and the Heezen-Tharp mapping project, 1952–1959. J Hist Geogr 32(3):605–626

Duane DB, Meisburger EP (1969) Geomorphology and sediments of the nearshore continental shelf Miami to Palm Beach, Florida, CERC technical memorandum 29. U.S. Army Corps of Engineers, Washington, DC, 47p

Ekebom J, Erkkila A (2002) Using aerial photography for identification of marine and coastal habitats under the EU's habitats directive. Aquat Conserv 13(4):287–304

Elvhage C (1980) An experimental series of topogeomorphological maps. Geografiska Annaler Ser A 62:105–111

FGDC (Federal Geographic Data Committee) [prepared for the Federal Geographic Data Committee by the U.S. Geological Survey] (2006) FGDC Digital Cartographic Standard for Geologic Map Symbolization. Federal Geographic Data Committee, Reston, VA. Document number FGDC-STD-013-2006, 290p

Finkl CW (2004) Coastal classification: systematic approaches to consider in the development of a comprehensive scheme. J Coast Res 20(1):166–213

Finkl CW (2005) Nearshore geomorphological mapping. In: Schwartz ML (ed) The encyclopedia of coastal science. Kluwer Academic, Dordrecht, pp 849–865

Finkl CW, Andrews JL (2008) Shelf geomorphology along the southeast Florida Atlantic continental platform: barrier coral reefs, nearshore bedrock, and morphosedimentary features. J Coast Res 24(4):823–849

Finkl CW, Banks KW (2010) Mapping seafloor topography based on interpretation of airborne laser bathymetry: examples from the southeast Florida Atlantic continental shelf. In: Martorino L, Puopolo K (eds) New oceanography research developments: marine chemistry, ocean floor analyses and marine phytoplankton. Nova Science Publishers, Hauppauge, pp 163–187

Finkl CW, Benedet L (2005) Jet probes. In: Schwartz ML (ed) The encyclopedia of coastal science. Kluwer Academic, Dordrecht, pp 707–716

Finkl CW, DaPrato GW (1993) Delineation and distribution of nearshore reefs in subtropical southeast Florida coastal environments using Thematic Mapper imagery. In: MTS '93 conference proceedings (Marine Technology Society annual meeting, Long Beach, CA), pp 90–96

Finkl CW, Khalil SM (2005) Offshore exploration for sand sources: general guidelines and procedural strategies along deltaic coasts. In: Finkl CW, Khalil SM (eds) Saving America's wetland: strategies for restoration of Louisiana's coastal wetlands and barrier islands, Journal of coastal research, special issue no. 44., pp 203–233

Finkl CW, Vollmer H (2011) Interpretation of bottom types from IKONOS satellite images of the southern Key West National Wildlife Refuge, Florida, USA. J Coast Res SI (64). In: Proceedings of the 11th international coastal symposium, pp 731–735

Finkl CW, Warner MT (2005) Morphological features and morphological zones along the inner continental shelf of southeastern Florida: an example of form and process controlled by lithology. J Coast Res SI (42): 79–96

Finkl CW, Benedet L, Andrews JL (2004) Laser Airborne Depth Sounder (LADS): a new bathymetric survey technique in the service of coastal engineering environmental studies, and coastal zone management. In: Proceedings of the 17th annual national conference on Beach Preservation Technology (February 11–13, 2004, Lake Buena Vista, FL). Florida Shore and Beach Preservation Association, CD-ROM, Tallahassee, 15p

Finkl CW, Benedet L, Andrews JL (2005a) Interpretation of seabed geomorphology based on spatial analysis of high-density airborne laser bathymetry. J Coast Res 21(3):501–514

Finkl CW, Benedet L, Andrews JL (2005b) Submarine geomorphology of the continental shelf off southeast Florida based on interpretation of airborne laser bathymetry. J Coast Res 21(6):1178–1190

Finkl CW, Becerra JE, Achatz V, Andrews JL (2008) Geomorphological mapping along the upper southeast Florida Atlantic continental platform; I: mapping units, symbolization, and GIS presentation of interpreted seafloor topography. J Coast Res 24(6):1388–1417

Gardiner V, Dackombe RV (1979) Geomorphological field manual. Allen & Unwin, London, 254p

Gellert JF (1988) Applied geomorphological survey and mapping on coasts. Zeitschrift fur Geomorphologie, N.F., Supplement Band 68, 223p

Gesch DB (2009) Analysis of lidar elevation data for improved identification and delineation of lands vulnerable to sea-level rise. J Coast Res SI (53): 49–58

Gierloff-Emden HG (1985) Documentation and presentation of the relief of the sea on maps. In: Fuchs K, Soffel H (eds) Numerical data and functional relationships in science and technology. Springer, Dordrecht. (Landolt-Bornstein, Group V Geophysics. Lehrbuch d. Allgemeinen Geographie, vol 5, Subvolume B [Relief of the Seafloor)]), pp 267–271

Gorman L, Morang A, Larson R (1998) Monitoring the coastal environment; Part IV: mapping, shoreline changes, and bathymetric analysis. J Coast Res 14(1):61–92

Green EP, Mumby PJ, Edwards AJ, Clark CD (2000) Remote sensing handbook for tropical coastal management. UNESCO Publishing, Paris, p 316

Guenther GC, Cunningham AG, LaRocque PE, Reid DJ (2000) Meeting the accuracy challenge in airborne lidar bathymetry. In: Proceedings of EARSeL-SIG-workshop LIDAR, Dresden/FRG, 16–17 June 2000, 27p

Heezen BC, Tharp M (1965) Tectonic fabric of the Atlantic and Indian Oceans and continental drift. Philos Trans R Soc Lond Ser A Math Phys Sci 258(1088):90–106

Heezen BC, Tharp M (1966) Physiography of the Indian Ocean. Philos Trans R Soc Lond Ser A Math Phys Sci 259(1099):137–149

Heezen BC, Tharp M (1977) World ocean floor. U.S. Navy, Washington, DC

Hochberg EJ, Atkinson MJ (2000) Spectral discrimination of coral reef benthic communities. Coral Reefs 19:164–171

Hochberg EJ, Andrefouet S, Tyler MR (2003) Sea surface correction of high spatial resolution Ikonos images to improve bottom mapping in near-shore environments. IEEE Trans Geosci Remote Sens 41(7):1724–1729

IHB (International Hydrographic Bureau) (1989) Standardization of undersea feature names. International Hydrographic Organization and the Intergovernmental Oceanographic Commission, Monaco, 28p

Irish JL, Lillycrop WJ (1997) Monitoring new pass, Florida with high density lidar bathymetry. J Coast Res 13(4):1130–1140

Irish JL, Lillycrop WJ (1999) Scanning laser mapping of the coastal zone: the SHOALS system. ISPRS J Photogramm Remote Sens 54(2–3):123–129

Irish JL, McClung JK, Lillycrop WJ (2000) Airborne lidar bathymetry: the SHOALS system. Int Navig Assoc PIANC Bull 103:43–53

Kempeneers P, Deronde B, Provoost S, Houthuys R (2009) Synergy of airborne digital camera and Lidar data to map coastal dune vegetation. J Coast Res SI(53): 73–82

Kenny AJ, Cato I, Desprez M, Fader G, Schuttenhelm RTE, Side J (2003) An overview of seabed-mapping technologies in the context of marine habitat classification. ICES J Mar Sci 60:411–418

Klemas V (2011a) Remote sensing techniques for studying coastal ecosystems: an overview. J Coast Res 27(1):2–17

Klemas V (2011b) Remote sensing of wetlands: case studies comparing practical techniques. J Coast Res 27(3):418–427

Klemas V (2011c) Beach profiling and Lidar bathymetry: an overview with case studies. J Coast Res 27(6):1019–1028

Lewis JB (2002) Evidence from aerial photography of structural loss of coral reefs at Barbados, West Indies. Coral Reefs 21:49–56

Lidz BH (2004) Coral reef complexes at an atypical windward platform margin: late quaternary, southeast Florida. Geol Soc Am Bull 116:974–988

Lidz BH, Shinn EA, Hine AC, Locker SD (1997) Contrasts within an outlier-reef system: evidence for differential quaternary evolution, south Florida windward margin, U.S.A. J Coast Res 13(3):711–731
Lidz BH, Reich CD, Shinn EA (2003) Regional quaternary submarine geomorphology in the Florida keys. Bull Geol Soc Am 115:845–866
Lidz BH, Reich CD, Peterson RL, Shinn EA (2006) New maps, new information: coral reefs of the Florida keys. J Coast Res 22(2):260–282
Long TM, Angelo J, Weishampel JF (2011) LiDAR-derived measures of hurricane- and restoration-generated beach morphodynamics in relation to sea turtle nesting behaviour. Int J Remote Sens 32(1):231–241
Madden C, Goodin K, Allee B, Finkbeiner M, Bamford D (2008) Coastal and marine ecological classification standard. NOAA and NatureServe, Arlington, 77p
Madley KA, Sargent B, Sargent FJ (2002) Development of a System for Classification of Habitats in Estuarine and Marine Environments (SCHEME) for Florida. Unpublished report to the U.S. Environmental Protection Agency, Gulf of Mexico Program (Grant Assistance Agreement MX-97408100). Florida Marine Research Institute, Florida Fish and Wildlife Conservation Commission, St. Petersburg, 43p
Maeder J, Narumalani S, Rundquist DC, Perk RL, Schalles J, Hutchins K, Keck J (2002) Classifying and mapping general coral-reef structure using IKONOS data. Photogramm Eng Remote Sens 68(12):1297–1305
Maritorena S, Morel A, Gentili B (1994) Diffuse reflectance of oceanic shallow waters: influence of water depth and bottom albedo. Limnol Oceanogr 39(7):1689–1703
Milard F (1996) Symboles et Abre'viations figurant sur les cartes marines françaises. Service Hydrgoraphique et Oce'anographique de la Marine (SHOM), Brest, 102p (Based on Chart Specifications of the IHO. XIIth International Hydrographic conference 1982 in Monaco)
Moody R (2007) Beyond plate tectonics: 'plate' dynamics. Infinite Energy 74:1–13
Moore LJ (2000) Shoreline mapping techniques. J Coast Res 16(1):111–124
Mount R (2003) The application of digital aerial photography to shallow water seabed mapping and monitoring: how deep can you see? In: Proceedings of coastal GIS 2003, University of Wollongong, Australia, 7–8 June 2003, 10p
Moyer RP, Riegl B, Banks K, Dodge RE (2003) Spatial patterns and ecology of benthic communities on a high-latitude south Florida (Broward County, USA) reef system. Coral Reefs 22:447–464
Mumby PJ, Edwards AJ (2002) Mapping marine environments with IKONOS imagery: enhanced spatial resolution can deliver greater thematic accuracy. Remote Sens Environ 82:248–257
Mumby PJ, Harborne AR (1999) Development of a systematic classification scheme of marine habitats to facilitate regional management and mapping of Caribbean coral reefs. Biol Conserv 88:155–163
Mumby PJ, Green EP, Edwards AJ, Clark CD (1999) The cost-effectiveness of remote sensing for tropical coastal resources assessment and management. J Environ Manag 55(3):157–166
O'Regan PR (1996) The use of contemporary information technologies for coastal research and management: a review. J Coast Res 12(1):192–204
Ollier CD (1977) Terrain classification: methods, applications and principles. In: Hails JR (ed) Applied geomorphology. Elsevier, Amsterdam, pp 277–316
Palandro D, Andrefouet S, Dustan P, Muller-Karger FE (2003) Change detection in coral reef communities using IKONOS satellite sensor imagery and historical aerial photographs. Int J Remote Sens 24(4):873–878
Palandro D, Thoms K, Kristen K, Muller-Karger F, Greely T (2005) Satellite remote sensing of coral reefs: by learning about coral reefs, students gain an understanding of ecosystems and how cutting-edge technology can be used to study ecological change. Sci Teach 72(6), 51p
Ramsey EW III, Laine SC (1997) Comparison of landsat thematic mapper and high resolution photography to identify change in complex coastal wetlands. J Coast Res 13(2):281–292

Richards DG (1980) Water-penetration aerial photography. Int J Naut Archaeol Underw Explor 9(4):331–337

Rohmann SO, Monaco ME (2005) Mapping southern Florida's shallow-water coral ecosystem: an implementation plan, NOAA Technical Memorandum NOS NCCOS 19. Center for Coastal Monitoring and Assessment, Silver Spring, 45p

Schowengerdt RA (1997) Remote sensing models and methods for image processing, 2nd edn. Academic, San Diego, 560p

Sheppard CR, Matheson K, Bythel JC, Murphy P, Myers CB, Blake B (1991) Habitat mapping in the Caribbean for management and conservation: use and assessment of aerial photography. Aquat Conserv 5:277–298

Shoshany M, Degani A (1992) Shoreline detection by digital image processing of aerial photography. J Coast Res 8(1):29–34

Smith CJ, Rumohr H (2005) Imaging techniques. In: Eleftheriou A, McIntyre A (eds) Methods for the study of marine benthos. Blackwell Science, Oxford, pp 87–111

Smith GL, Zarillo GA (1990) Calculating long-term shoreline recession rates using aerial photographic and beach profiling techniques. J Coast Res 6(1):111–120

Specht MR, Needler D, Fritz NL (1973) New color film for water photography penetration. Photogramm Eng 39:359–369

Steimle JT, Finkl CW (2011) Interpretation of seafloor topologies based on IKONOS satellite imagery of a shallow-marine carbonate platform: Florida Bay to the Florida Reef Tract. J Coast Res SI 64. In: Proceedings of the 11th international coastal symposium, pp 825–830

Stockdon HF, Sallenger AH, List JH, Holman RA (2002) Estimation of shoreline position and change using airborne topographic lidar data. J Coast Res 18(3):502–513

Stockdon H, Doran KS, Sallenger AH Jr (2009) Extraction of lidar-based dune-crest elevations for use in examining the vulnerability of beaches to inundation during hurricanes. J Coast Res SI (53): 59–65

Stoker JM, Tyler DJ, Turnipseed DP, Van Wilson K Jr, Oimoen MJ (2009) Integrating disparate lidar datasets for a regional storm tide inundation analysis of Hurricane Katrina. J Coast Res SI (53): 1–13

Thieler ER, Danforth WW (1994) Historical shoreline mapping (I): improving techniques and reducing positioning errors. J Coast Res 10(3):549–563

Walker BK, Riegl B, Dodge RE (2008) Mapping coral reef habitats in southeast Florida using a combined technique approach. J Coast Res 24(5):1138–1150

Warner MT (1999) Analysis of coastal morphodynamic zones based on detailed mapping in Palm Beach County. Master's thesis, Florida Atlantic University, Boca Raton, 110p

Zieman JC, Fourqurean JW, Iverson RL (1989) Distribution, abundance and productivity of seagrasses and macroalgae in Florida Bay. Bull Mar Sci 44(1):292–311

Chapter 3
Terrestrial Laser Scanner Surveying in Coastal Settings

Michael A. O'Neal

Abstract Over the last decade, there has been a proliferation of commercially available tripod-mounted Terrestrial Laser Scanner (TLS) systems that use the phase difference or the time-of-flight of emitted pulses of light to rapidly acquire high-density topographic and surface reflectance data. These TLS systems have been well received in the Earth science community because of their ability to collect 10^2–10^5 measurements per second of azimuth, zenith, distance, intensity, and surface color data at distances ranging from 10^0 to 10^3 m. A TLS instrument's portability, ease of use, and rate of data collection opens the possibility of collecting detailed topographic data at sites where such surveys may not have been possible before. The application of TLS data to current Earth sciences research has allowed us to better understand of the character, timing, rates, and spatial scales of different processes that have been difficult, if not impossible, to evaluate using traditional survey techniques. However, the successes achieved with TLS systems in certain projects may result in unrealistic expectations regarding the density and quality of data that can reasonably be achieved in different settings. This chapter seeks to orient potential or experienced TLS users to its applicability in above-water coastal settings. An emphasis is placed on providing insight into both the variety of current research using TLS data, as well as the compromises in spatial resolution that necessarily arise from field conditions and survey design.

M.A. O'Neal (✉)
Department of Geological Sciences, University of Delaware, Newark, USA
e-mail: oneal@udel.edu

C.W. Finkl and C. Makowski (eds.), *Remote Sensing and Modeling: Advances in Coastal and Marine Resources*, Coastal Research Library 9,
DOI 10.1007/978-3-319-06326-3_3,

3.1 Introduction

Tripod-mounted Terrestrial Laser Scanners (TLS) have been well received by the coastal research community because of their ability to capture detailed topography in dynamic or difficult to survey areas. Their recent proliferation is largely due to the limited technical expertise required for operation and the user-friendly interfaces and software for automated data collection and processing. The high-resolution topographic and surface-reflectance data generated by TLS systems have been used to reevaluate many long held assumptions, while also opening new avenues of research that were difficult, if not impossible, to pursue using traditional survey techniques. However, developing a final product that is appropriate for the intended research purpose first requires an understanding of the limitations of the TLS system and an awareness of the errors that arise throughout the survey and data processing. Accounting for such errors through a well-planned survey design optimizes both the quality and usefulness of the data. Any successful TLS-dependent study must seek out a compromise among: (1) the smallest features that are to be captured in the scans; (2) the minimum number of survey locations required to observe features of interest; (3) the environmental conditions that affect the reflectance data collected by the instrument; and (4) the practical limits of the final models produced.

3.2 The TLS Instrument

TLS systems analyze the phase difference or the time-of-flight of emitted pulses of light to acquire high-density topographic and surface reflectance data. The basic goal of the use of such systems is to rapidly collect dense datasets of millimeter- to centimeter-accurate topographic coordinates (i.e., on the order of 10^2–10^5 points per second) at distances between 10^0 and 10^3 m. The resulting datasets from TLS systems contain the three-dimensional coordinates from the surface reflections along any vector; collectively, these data form a point cloud (Fig. 3.1). Many TLS systems also return other information about the target surfaces, such as color and an estimate of reflection intensity. There are many different types of

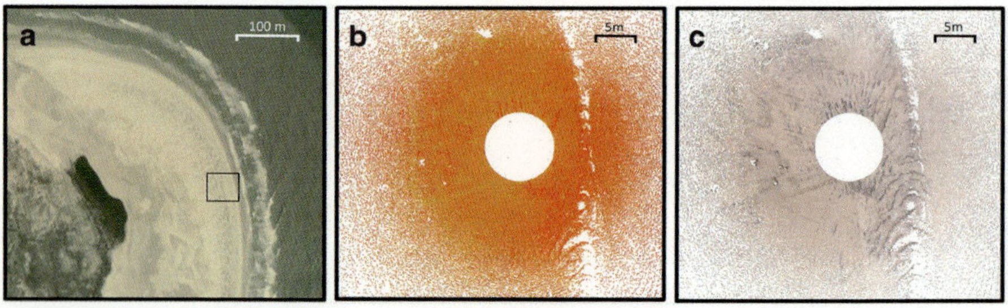

Fig. 3.1 (**a**) A aerial image of the northern end of Cedar Island, VA showing a TLS survey area for images (**b**) and (**c**) (*inset box*), (**b**) a point cloud of intensity data from a TLS survey, and (**c**) a point cloud with surface color data

commercially available TLS systems that vary by intended use, with tradeoffs between accuracy, the maximum distance measured, and the rate of data collection (see Petrie and Toth (2009) for a technical overview of the advantages and limitations of many currently available commercial TLS systems).

Developing a survey with an awareness of the limitations and errors inherent to the instrument will ultimately improve the quality of the data retrieved. Errors inherent to the instrument vary between instrument manufacturers, and are largely dependent on characteristics of the emitted and return pulses (i.e., pulse width, energy, noise, and detector sensitivity) and the limitations and/or calibration of internal moving parts. These errors are largely out of the control of the user, but must be accounted statistically according to manufacturer-reported performance measurements under laboratory conditions. Naturally, specifications of manufacturer accuracy can be difficult to translate into the variety of operational and environmental conditions presented in the field (Ussyshkin and Boba 2008). Reshetyuk (2009) provides an exhaustive overview on the statistical evaluations required for field calibration of terrestrial laser scanners.

3.3 Field Surveys and Scanner Setup

The laser's interaction with the natural environment may make collected data problematic to merge or georeference, difficult to analyze, or simply unreliable. In field settings, the reflective properties of surveyed surfaces (albedo) and atmospheric conditions affect the path of the laser beam and its intensity upon return to the detector. Many instruments allow for some level of atmospheric correction at the time of the survey. However, certain characteristics of the scanned surface such as roughness, reflectivity, and color cause scattering that impart additional position errors when measuring the natural environment (e.g., Bohler et al. 2003; Lichti and Gordon 2004). Even over topographically homogenous coastal terrains, subtle differences in the reflective properties of surface materials as a result of color, grain size, and lithology, or even secondary characteristics such as detritus or water content, may add to these positional errors at increased distances and angles of incidence (Kaasalainen et al. 2008, 2011; Luzi et al. 2009).

Geometry between the laser and the target surface introduces incidence angle errors that can be controlled, to a certain degree, by the user via careful site selection and survey design. Laser scan measurements that are perpendicular to a surface yield the most accurate reflectance data. At large incidence angles, the laser footprint becomes elongated, yielding distance and the beam intensity measurements that are necessarily less likely to represent a point rather than a summation of a broad surface (Schaer et al. 2007; Ussyshkin et al. 2009; Soudarissanane et al. 2011). At incidence angles above 60°, positional errors dominate the data (Soudarissanane et al. 2009). However, limiting the incidence angle is not always practical in field settings and field-testing the effects of positional error may be necessary before analyzing the data. Many researchers seek to minimize incidence angles by increasing the elevation of the scanner, by mounting the instrument atop very large or telescoping tripods, atop a stabilized vehicle, or from scaffolding

Fig. 3.2 An example of elevating a TLS using a vehicle stabilized with jacks

(Brown and Hugenholtz 2013; Hobbs et al. 2010; Pietro et al. 2008) (Fig. 3.2). Note that because the range of scan angles is fixed on most units, elevating the instrument does increase the area immediately under and around the instrument that cannot be surveyed. Because the unscanned area needs to be filled from another vantage point, this method may increase the number of survey sites and scans required to obtain a dense point cloud of the survey area.

TLS instrument placement must also seek to minimize the effects of long scan distances. Effective field measurement distances for most instruments are no more than a few hundred meters, typically far less than the maximum laboratory-tested distances of many units. Beyond a certain distance, positional errors of the far field landscape will exceed instrument standards such that no or noisy data will be returned. Additional consideration must also be given to potential effects on far field measurements introduced by the angular geometry of the instrument during data collection. Most instruments rotate at user-prescribed incremental angles about a vertical axis from a fixed mechanical platform. Substantially greater numbers of observations will be collected from positions closer to the instrument than in the far field area, i.e., tens of thousands vs. a single, or no observation (Pietro et al. 2008). Also, the linear nature of each beam pass creates an angular gap in the integrated point cloud that increases with distance from the instrument (Fig. 3.3). Despite higher data densities near the instrument, it is the maximum size of the far field spacing that dictates the smallest size of the landscape feature that can be measured in a useful way by TLS data (Soudarissanane et al. 2011). Instrument software is now designed to help the user collect a regular spacing across all distances,

Fig. 3.3 An *aerial view* of a point cloud data dataset showing the increasing gap between scan lines in farfield areas

prescribing specific angular steps to collect the highest possible density of data, though there are actual physical limitations of the instrument that limit how thorough such processes can be. Software algorithms also filter data by design, throwing away some of the denser data in favor of collecting a more uniform point cloud, which may not be preferable for all scientific applications.

In coastal settings, attempting to survey complex forms and/or large areal extents (i.e., dunes, beaches, and sea cliffs) necessarily requires a more involved survey design. Line-of-sight issues, or 'shadowing' of features with respect to the scanner location, limit the applicability of a single position scan for most studies. Positioning the TLS at regular geographically or topographically controlled intervals throughout the study area may improve areal coverage and additionally help to minimize the effects of increased incidence angles and distance-related issues (Fig. 3.4). For small areas on the order of 10^0–10^1 m^2, larger scan angles are easier to avoid so that dense datasets with millimeter-accurate data are possible (e.g., van Gaalen et al. 2011). At larger geographic scales, each additional scan substantially increases the survey time, physical effort required, and also far field positional errors.

A field strategy for a multi-scan survey must anticipate the subsequent data merging process by including surfaces that can be used as a reference between adjacent point clouds. Surveys of the same area from different locations typically use manufacturer-provided reflective targets. These objects can be pinpointed within a survey by automated processes built into the instrument software (Fig. 3.5). Targets must be thoughtfully placed within scanning distance of neighboring survey stations. The targets then become an embedded part of each scan, their position subject to the same incidence-angle and distance-related errors that affect other surfaces. Additionally, because target positions are calculated by the manufacturer's software using a statistical fit, they themselves carry some error.

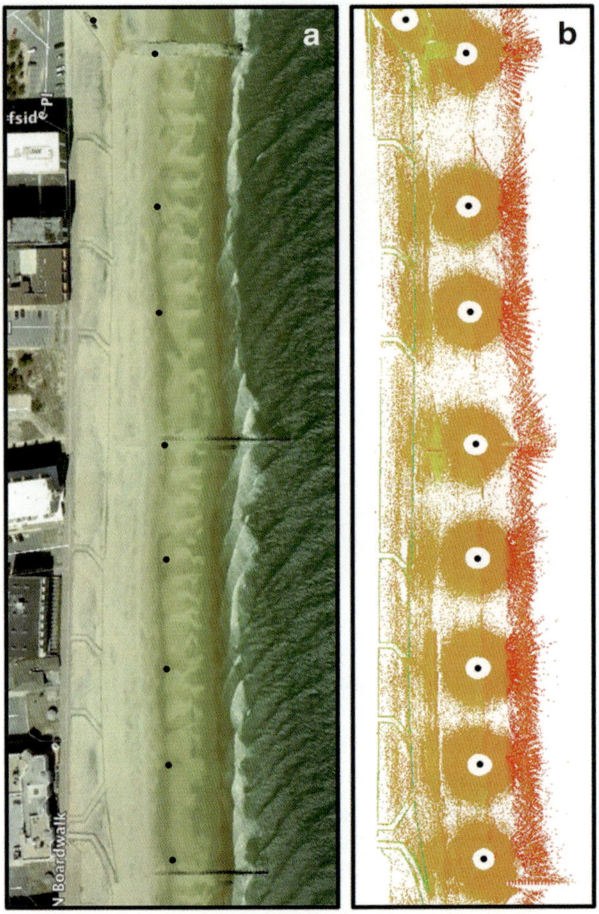

Fig. 3.4 (**a**) An *aerial view* of a TLS survey area along the northern end of Rehoboth Beach, DE. (**b**) *Aerial view* of TLS point clouds of the beach in image (a), showing nine separate survey sites (instrument location represented by *black dot*) (Note the increasing gap between scan lines, as well as the drop in data density, with distance from the instrument)

Post-scan statistical analyses of all data in adjacent point clouds can provide a greater understanding of quality-of-fit and improve the merging outcome.

When scanning larger geographic extents, relying on relativistic reference targets between adjacent scans often results in cumulative errors, manifested in final surface models as irregular geometries and overall slopes that do not well represent the actual scanned area (Pietro et al. 2008; Hobbs et al. 2010; Olsen et al. 2010). Scanning permanent benchmarks embedded in concrete by the surveyor may improve merging statistics, but also may be difficult to install and/or maintain in dynamic coastal settings. Many successful larger-scale studies rely on geodetic-quality GPS coordinates for objects in each scan (i.e., survey targets or stationary features of the built environment), so that each scan can be georeferenced independently. Differential GPS systems may be used to this end, although real-time kinematic GPS surveying does not require a permanent benchmark (e.g., Hobbs et al. 2010).

Fig. 3.5 (**a**) A *shaded* relief image from a TLS scan of a boulder surface and (**b**) a point cloud of the same area in (a) depicting scans completed in four cardinal directions as four different colors

3.4 Data Merging and Modeling

Merging and georeferencing of point clouds that include well-placed reference points can be relatively easy. Two or more scans can be aligned precisely by matching reference points common to both scans, statistically aligning overlapping point clouds (Olsen et al. 2010), or by providing GPS coordinates for control points in each point cloud (Fig. 3.6). Despite the method used to merge point clouds, the statistical fit of control points and/or georeferencing adds to the error budget beyond the internal errors of the instrument.

By design, the TLS system captures many unwanted or irrelevant features in the survey domain (Fig. 3.7). There is no single filtering method that is appropriate for all data sets, and researchers commonly apply a combination of several techniques to specific areas depending on the types of information to be omitted. Manual filtering is possible using a variety of software applications. However, the time involved in visually manipulating the large number of points typical of TLS surveys limits the usefulness of this approach. Therefore, a variety of automated data reduction techniques are often necessary (e.g., Zhang et al. 2003). Most digital terrain models can be derived by filtering non-ground features (objects or data artifacts) from the desired ground surface using slope-elevation relationships, or by analyzing simple geometric characteristics of points, or groups of points, in relation to neighbors (e.g., Pietro et al. 2008; Roggero 2001; Guarnieria et al. 2009).

Fig. 3.6 Images depicting GPS surveys of hard structures (*left*) and TLS spherical targets (*right*) used to georeference point could collected for these features

Fig. 3.7 An image depicting many of the non-terrain objects that may be included in TLS surveys in populated regions with point cloud examples of a (**a**) fence, (**b**) beach umbrella, and (**c**) a person

Most studies will require additional, research-specific algorithms and strategies for the classification and interpretation of their unique point clouds. For example, coastal settings may require analysis of intensity data to evaluate or remove landscape features like vegetation (e.g., Guarnieria et al. 2009).

The orientation and topographic complexity of the surveyed landscape will have a profound influence on the type of model produced. Although many analyses may be able to directly utilize point cloud data, many researchers will require the conversion of point cloud data to grid (raster) or triangular irregular network

(TIN; vector) models for analysis. A grid provides a 2-D matrix of values for use in a scalable GIS infrastructure, suitable for evaluating surface changes; a disadvantage of grid models is that each cell contains an estimated value, interpolated from the original data. A TIN retains the original point cloud locations and does not distinguish between plan and profile views, a particularly useful attribute in analyzing complex vertically oriented features like coastal cliffs (e.g. Hobbs et al. 2010; Young et al. 2010). However, TINs can be computationally intense to analyze. Cloud-to-cloud statistical comparisons are also possible, but come with inherent statistical errors that result from the imperfect overlap of points between surveys (Girardeau-Montaut et al. 2005). Regardless of whether the data are modeled as a point cloud, a grid, or a TIN, it is critical that the researcher understand and report errors accumulated in arriving at the final model. Cumulative errors should also be included in any statistical evaluations of landscape change using these data (e.g. Hobbs et al. 2010; Pietro et al. 2008; Young et al. 2010).

3.5 Applications of TLS in Coastal Settings

With a TLS instrument and a well-planned surveying strategy, a researcher can observe surface characteristics of landscapes that range in area from just a few to tens of thousands of square meters. Data obtained at either of these scales can provide insight into active landscape-forming processes and/or surface mapping. When the spatial scale of TLS data coincides with other elevation datasets, like those from GPS or airborne lidar (ALS), TLS-based digital terrain models can be used for calibration, orthorectification, and error analysis (e.g., TLS/ALS comparisons of coastal cliff retreat presented by Young et al. 2010 and GPS comparisons by Coveney et al. 2010).

The simplest and most obvious product from a TLS survey is digital terrain model (DTM). Even as an emerging technology, researchers quickly observed the usefulness of TLS data in developing more accurate estimates of area and volume as compared to traditional survey techniques or even airborne LIDAR (Girardeau-Montaut et al. 2005). An important value of TLS instruments lies in the ability to perform repeated surveys in settings that were traditionally difficult to monitor (Young et al. 2010). TLS-derived DTMs have been used to provide grain-scale surface roughness data for atmosphere-surface models (Hugenholtz et al. 2013), to determine the relative ages and the rates and types of surface processes at decimeter scales (Nield et al. 2011; van Gaalen et al. 2011), and to better improve large-scale models used to classifying coastal flooding and disaster management strategies (Mastronuzzi and Pignatelli 2011; Pignatelli et al. 2010). When the topographic data from a TLS survey are accompanied by intensity measurements, individual points can be segregated and/or classified based on spectral characteristics, indicating lithological changes in a scanned feature (Hobbs et al. 2010) or the moisture content of ephemeral landforms (Nield et al. 2011). Because intensity values measured by the TLS are an estimate, they require a correction based on standardized lab or field target if they are to be quantitatively (Pfeifer et al. 2008).

Simple topographic models studied over time can provide new insights or improved understanding regarding the timing and rates of changes in above-water coastal settings. TLS surveys prove especially useful when a time series of centimeter-scale data are required from settings like marshes and mud flats, which have been difficult to monitor using traditional survey techniques (Thiebes et al. 2013). Many TLS surveys that focus on geomorphic changes over time not only provide more accurate and detailed data then has been previously available, but are also able to rapidly repeat surveys so as to take advantage of changes that occur over a single tidal cycle (i.e., strip mobility related to moisture content (Nield et al. 2011)), multiple tidal cycles (i.e., beach cusp morphodynamics (van Gaalen et al. 2011), or event and seasonal forcings on beach geometry (e.g., Pietro et al. 2008; Theuerkauf and Rodriguez 2012). TLS surveys have tackled problems ranging in scope from the mobility of a single sandstrip (Nield et al. 2011) to large-scale geomorphic analyses of how the number, height, and orientation of dunes change from season to season (Montreuil et al. 2013).

All of the aforementioned uses of TLS data can be visualized and evaluated from either an aerial or planview form; however, a key usefulness of TLS over airborne systems is in its ability to capture dense datasets of steep coastal topography. One of the primary areas where TLS has improved our knowledge of coastal process has in the study of coastal cliffs. The vertical nature of these areas makes detailed airborne datasets difficult to obtain and costly to repeat. Likewise, traditional survey methods are simply too difficult to apply in rugged vertical terrain. TLS surveys in these settings have substantially increased our understanding of the timing, rates, and spatial scales of cliff retreat and mass wasting processes (e.g., Hobbs et al. 2010; Lim et al. 2010; Poulton et al. 2006; Rosser et al. 2005; Young et al. 2010), and have also helped elucidate human dimensions of these problems in terms of threats to property (Olsen et al. 2009). Together, the body of TLS-based survey research in this area has illuminated coastal mass wasting and retreat to reveal an interconnected process, in which each fall is part of a continuum of change.

3.6 Conclusions

TLS proves to be a cost-effective survey tool for collecting high-density topographic and reflective data in above-water coastal settings. Terrestrial laser scanning in coastal areas continues to open exciting avenues of research that were difficult, if not impossible, using traditional survey techniques. To date, the primary limitations of TLS surveys stem from the need for better survey and data analysis techniques, a key focus of this manuscript. The current array of TLS-based research covers aspects of surface characteristics and mapping, surface dynamics and landcover changes, and validation of other surface measurements and techniques, just to name a few. The ability to complete repeated surveys without the great expense associated with airborne or other mobile systems provides an opportunity to evaluate, or re-evaluate, many long-held assumptions and poorly understood stochastic and/or unpredictable processes. Both the density and areal extent of data collected make TLS technology superior to traditional techniques, even though we have not amassed enough records to compare with long-term maps and aerial imagery. The interpretative value of TLS is even more robust when combined with readily available geospatial or geophysical (Lim et al. 2010).

References

Boehler W, Bordas V, Marbs A (2003) Investigating laser scanner accuracy. In: IAPRS (ed) Proceedings in the CIPA 2003 XVIII international symposium, vol XXXIV(5/C15), Institute for Spatial Information and Surveying Technology, Antalya, Turkey, pp 696–701

Brown OW, Hugenholtz CH (2013) Quantifying the effects of terrestrial laser scanner settings and survey configuration on land surface roughness measurement. Geosphere 9(2):367–377

Coveney S, Stewart Fotheringham A, Charlton M, McCarthy T (2010) Dual-scale validation of a medium-resolution coastal DEM with terrestrial LiDAR DSM and GPS. Comput Geosci 36(4):489–499

Girardeau-Montaut D, Roux M, Marc R, Thibault G (2005) Change detection on points cloud data acquired with a ground laser scanner. Int Arch Photogramm Remote Sens Spat Inf Sci 36(Part 3):W19

Guarnieria A, Vettorea A, Pirottia F, Maranib M (2009) Filtering of TLS point clouds for the generation of DTM in salt-marsh areas. In: Bretar F, Pierrot-Deseilligny M, Vosselman G (eds) Laser scanning 2009, IAPRS, vol XXXVIII, Part 3/W8, Paris, France, 1–2 Sept 2009

Hobbs PRN, Gibson A, Jones L, Pennington C, Jenkins G, Pearson S, Freeborough K (2010) Monitoring coastal change using terrestrial LiDAR. Geol Soc Lond Spec Publ 345(1):117–127

Hugenholtz CH, Brown OW, Barchyn TE (2013) Estimating aerodynamic roughness (Z_0) from terrestrial laser scanning point cloud data over un-vegetated surfaces. Aeolian Res 10:161–169

Kaasalainen S, Kukko A, Lindroos T, Litkey P, Kaartinen H, Hyyppa J, Ahokas E (2008) Brightness measurements and calibration with airborne and terrestrial laser scanners. Geosci Remote Sens IEEE Trans 46(2):528–534

Kaasalainen S, Jaakkola A, Kaasalainen M, Krooks A, Kukko A (2011) Analysis of incidence angle and distance effects on terrestrial laser scanner intensity: search for correction methods. Remote Sens 3(10):2207–2221

Lichti DD, Gordon SJ (2004) error propagation in directly georeferenced terrestrial laser scanner point clouds for cultural heritage recording. In: Proceedings of FIG working week, Athens, 22–27 May 2004. http://www.fig.net/pub/athens/

Lim M, Rosser NJ, Allison RJ, Petley DN (2010) Erosional processes in the hard rock coastal cliffs at Staithes, North Yorkshire. Geomorphology 114(1):12–21

Luzi G, Noferini L, Mecatti D, Macaluso G, Pieraccini M, Atzeni C, Schaffhauser A, Fromm R, Nagler T (2009) Using a ground – based SAR interferometer and a terrestrial laser scanner to monitor a snow – covered slope: results from an experimental data collection in Tyrol (Austria). IEEE Trans Geosci Remote Sens 47(2):382–393

Mastronuzzi G, Pignatelli C (2011) Determination of tsunami inundation model using terrestrial laser scanner techniques. The tsunami threat—research and technology, pp 219–236

Montreuil AL, Bullard J, Chandler J (2013) Detecting seasonal variations in embryo dune morphology using a terrestrial laser scanner. J Coast Res Spec Issue 65:1313–1318

Nield JM, Wiggs GF, Squirrell RS (2011) Aeolian sand strip mobility and protodune development on a drying beach: examining surface moisture and surface roughness patterns measured by terrestrial laser scanning. Earth Surf Process Landforms 36(4):513–522

Olsen MJ, Johnstone E, Driscoll N, Ashford SA, Kuester F (2009) Terrestrial laser scanning of extended cliff sections in dynamic environments: parameter analysis. J Surv Eng 135(4):161–169

Olsen MJ, Johnstone E, Kuester F, Driscoll N, Ashford SA (2010) New automated point-cloud alignment for ground-based light detection and ranging data of long coastal sections. J Surv Eng 137(1):14–25

Petrie G, Toth CK (2009) Terrestrial laser scanners. In: Topographic laser ranging and scanning principles and processing. CRC Press, Boca Raton, pp 87–128

Pfeifer N, Höfle B, Briese C, Rutzinger M, Haring A (2008) Analysis of the backscattered energy in terrestrial laser scanning data. In: Proceedings in the XXIth ISPRS Congress, silk road for information from imagery, vol 37, p B5

Pietro LS, O'Neal MA, Puleo JA (2008) Developing terrestrial-LIDAR-based digital elevation models for monitoring beach nourishment performance. J Coast Res 24(6):1555–1564

Pignatelli C, Piscitelli A, Damato B, Mastronuzzi G (2010) Estimation of the value of Manning's coefficient using terrestrial laser scanner techniques for the assessment of flooding by extreme waves. Zeitschrift für Geomorphologie Supplementary Issues 54(3):317–336

Poulton CV, Lee J, Hobbs P, Jones L, Hall M (2006) Preliminary investigation into monitoring coastal erosion using terrestrial laser scanning: case study at Happisburgh, Norfolk. Bull Geol Soc Norfolk 56, 45–64

Reshetyuk Y (2009) Self-calibration and direct georeferencing in terrestrial laser scanning. PhD dissertation, Umeå University

Roggero M (2001) Airborne laser scanning: clustering in raw data. IAPRS 34(Part 3/W4):227–232

Rosser NJ, Petley DN, Lim M, Dunning SA, Allison RJ (2005) Terrestrial laser scanning for monitoring the process of hard rock coastal cliff erosion. Q J Eng Geol Hydrogeol 38(4):363–375

Schaer P, Skaloud J, Landtwing S, Legat K (2007) Accuracy estimation for laser point cloud including scanning geometry. In: Proceedings of 5th international symposium on Mobile Mapping Technology (MMT2007), Padua, Italy

Soudarissanane S, Lindenbergh R, Menenti M, Teunissen P (2009) Incidence angle influence on the quality of terrestrial laser scanning points. XXXVIII(Part 3, W8):183–188

Soudarissanane S, Lindenbergh R, Menenti M, Teunissen P (2011) Scanning geometry: influencing factor on the quality of terrestrial laser scanning points. ISPRS J Photogramm Remote Sens 66(4):389–399

Thiebes B, Wang J, Bai S, Li J (2013) Terrestrial laserscanning of tidal flats—a case study in Jiangsu Province, China. J Coast Conserv 17(4):813–823

Theuerkauf EJ, Rodriguez AB (2012) Impacts of transect location and variations in along-beach morphology on measuring volume change. J Coast Res 28(3):707–718

Ussyshkin RV, Boba M (2008) Performance characterization of a mobile lidar system: expected and unexpected variables. In: ASPRS conference proceedings, Portland, 27 Apr–2 May 2008 (on CDROM)

Ussyshkin V, Boba M, Sitar M (2009) Performance characterization of an airborne lidar system: bridging system specifications and expected performance. Int Arch Photogramm Remote Sens Spat Inf Sci 37:177–182

van Gaalen JF, Kruse SE, Coco G, Collins L, Doering T (2011) Observations of beach cusp evolution at Melbourne Beach, Florida, USA. Geomorphology 129(1):131–140

Young AP, Olsen MJ, Driscoll N, Flick RE, Guitarrez R, Guza RT, Johnstone E, Kuester F (2010) Comparison of airborne and terrestrial lidar estimates of seacliff erosion. Photogramm Eng Remote Sens 76(4):421–427

Zhang KQ, Chen SC, Whitman D, Shyu ML, Yan JH, Zhang CC (2003) A progressive morphological filter for removing nonground measurements from airborne lidar data. IEEE Trans Geosci Remote Sens 41(4):872–882

Chapter 4
Advances in Applied Remote Sensing to Coastal Environments Using Free Satellite Imagery

Cristina Lira and Rui Taborda

Abstract Remote sensing emerges as a very effective technique to capture the dynamics of the coastal system, as it provides a holist view of the system at a wide range of spatial and temporal scales. However, the recent thrive of these systems has led to a broad variety of sensors and data, which can difficult the choice of the optimal sensor for a practical application. In mesoscale coastal environment applications and considering the universe of the solutions, Landsat program arises as a good compromise between spectral, radiometric, spatial and temporal resolutions, combined with free data access, supported by an efficient data sharing platform. The capabilities of the Landsat program was recently extended with the launch of a new satellite – Landsat 8, with improved radiometric and spectral resolution, opening the door to new studies. This work describes the applicability of these images in four case studies that demonstrates the potentialities of the Landsat program in what concerns: (1) time coverage – long-term evolution of an ephemeral ebb delta island; (2) frequency of coverage – seasonal evolution of a short-lived beach; (3) radiometric resolution – shoreline detection and extraction; (4) spectral resolution – bathymetric data retrieval.

4.1 Introduction

The coastal zone is one of the most dynamic areas on Earth, with changes occurring at a wide range of time and spatial scales. This fact, makes systematic mapping of this type of regions a challenge. Thus, the understanding of coastal dynamics

C. Lira (✉)
Instituto Dom Luiz, Centro de Geologia, Geology Department, Faculdade de Ciências da Universidade de Lisboa, Lisbon, Portugal
e-mail: fclira@fc.ul.pt

R. Taborda
Instituto Dom Luiz, Geology Department, Faculdade de Ciências da Universidade de Lisboa, Lisbon, Portugal

C.W. Finkl and C. Makowski (eds.), *Remote Sensing and Modeling: Advances in Coastal and Marine Resources*, Coastal Research Library 9,
DOI 10.1007/978-3-319-06326-3_4, © Springer International Publishing Switzerland 2014

requires a larger time and spatial scale approach, which time-limited and local studies are incapable of providing. Existing studies, based only on time-limited in situ data collections fall out of date very quickly, due to the constant adjustment to natural forces and human modifications. Therefore, new scientific approaches are needed to improve the understanding of coastal morphology in order to obtain correct evolution models whose application is capable of support management decisions and risk assessment.

From the numerous techniques available to acquire data in coastal systems remote sensing stands out as it can (1) provide a holistic view of the system as the acquired data covers a large area, potentially allowing the observation and understanding of the entire variety of the coastal phenomena; (2) provide data at different temporal scales from hourly to decadal; (3) provide an historical retrospective view of the Earth surface and processes; (4) represents an additional data source, complementing field surveys and in situ measurements.

From the large set of remote sensors currently available, this study focus on the analysis of satellite sensors whose characteristics makes their data suitable to use in mesoscale coastal environment studies, aiming at demonstrating their potentialities and recent advances. This demonstration is supported with several application examples that addresses a range of relevant issues found in several coastal environments. The study targets mesoscale research (i.e. the time scale span months to decades and the spatial scale hundreds of meters to tens of kilometers) since it's increasingly directed towards providing information related to management problems (Horn 2002).

4.2 Remote Sensors: An Overview

From the beginning of remote sensing, about 150 years ago, many different remote sensing systems have been developed to detect and measure energy patterns from different portions of the electromagnetic spectrum (Yang 2009). Nowadays, there is a wide range of remote sensors available and choosing the one that is best for our work can be a challenging task.

Characteristics often used to describe and compare these systems can be grouped into four different types of resolution: spatial, spectral, radiometric, and temporal (Klemas 2011). Spatial requirements include ground resolution (minimum mapping unit) and coverage (swath-width); spectral resolution refers to the number, location and width of the spectral bands; radiometric resolution concerns the pixel depth (i.e. the intensity resolution of each spectral band and the sensor's ability to distinguish between two objects of similar reflectance) and temporal resolution is linked with the frequency of coverage (i.e. hourly, daily, seasonal) and coverage to the extent of the acquisition program (start and end of the period covered for a particular sensor).

With the advent of new technologies, satellites are beginning to overcome the problem of spatial resolutions allowing, simultaneously a good space coverage and a very high ground resolution (<1 m). Additionally, and because there are satellites in space from the 1970s they can also provide a good time-interval coverage.

Table 4.1 Remote sensing constraints in coastal studies

	Open ocean	Estuaries	Beach
Spatial resolution	1–10 km	20–200 m	1–30 m
Coverage area	2,000 × 2,000 km	200 × 200 km	200 × 200 km
Frequency of coverage	1–6 days	0.5 days–decadal	0.5 days–decadal
Radiometric resolution	10–12 bits	8–12 bits	8–12 bits
Spectral resolution	Multispectral/ hyperspectral	Multispetral/ hyperspectral	Multispectral/ hyperspectral

Adapted from Klemas (2009)

Nevertheless, the choice of a specific sensor continues to be a compromise between their main characteristics applied to a specific study, thus the user should familiarize first with the strengths and weaknesses of the available sensors in combination with the desired application. The first step in this process should be the precise identification of the characteristics to extract from the imagery in order to identify the range of sensors that can provide that information. Additionally, sensor choice should depend not only the target spatial, temporal, spectral and radiometric resolutions, but also account for more practical aspects like availability and cost.

Most coastal environments show great variations over larger extents, spatial complexity and temporal variability, which means that these type of environments require high spatial, spectral, radiometric and temporal resolution (Table 4.1, adapted from Klemas 2009). The principle coastal zone related sensors on satellite platforms and their main characteristics and applications are summarized in Table 4.2 (adapted from Klemas 2009, 2011; Mather 2004).

Radar imagery is primarily used in oceanography, imaging the sea surface with the capability to map swell, internal waves, oil slicks, determine sea surface winds and currents, sea level and wave height, etc. (Ikeda and Dobson 1995; Martin 2004), but it lacks the information contained in the visible (VIS) and infra-red (IR) regions of the electromagnetic spectrum. Nevertheless, radar can penetrate fog and clouds, making it valuable in areas where cloud cover persists, particularly, in emergency applications (storms, hurricanes).

The use of very high spatial resolution (VIR) sensors will improve the detail of the analysis, but this kind of image products are available only by commercial companies (high cost) and because they display great detail (ground resolution), they present lower area coverage, lacking the global spatial vision that sometimes coastal studies mandate. Thus, to keep costs practical, large coastal areas should be analyzed using medium to high resolution satellite sensors (15–30 m) and only small and critical areas mapped with airborne or VIR satellite sensors.

4.2.1 The Landsat Program

With over 40 years of collecting spectral information from Earth's surface and creating an historical archive unmatched in quality, detail, coverage, and length,

Table 4.2 Principal coastal zone related sensors and their main characteristics

		Spectral band		SR	RR	Cycle	Swath width			
Satellite	Sensor	Number	(μm)	(m)	bits	(days)	(km)	Temporal coverage	Availability	Applications
Landsat 1, 2, 3	MSS	4 – Green	0.5–0.6	60[a]	8	18	185	1972–1983	F	LULC
		5 – Red	0.6–0.7	60[a]	8	18	185			CD
		6 – NIR	0.7–0.8	60[a]	8	18	185			V
		7 – NIR	0.8–1.1	60[a]	8	18	185			CA
Landsat 4, 5	TM	1 – Blue	0.45–0.52	30	8	16	185	1982–2012	F	
		2 – Green	0.52–0.60	30	8	16	185			
		3 – Red	0.63–0.69	30	8	16	185			
		4 – NIR	0.76–0.90	30	8	16	185			
		5 – NIR	1.55–1.75	30	8	16	185			
		6 – Thermal IR	10.40–1.50	30	8	16	185			
		7 – MIR	2.08–2.35	120	8	16	185			
Landsat 7	ETM+	1 – Blue	0.45–0.52	30	8	16	185	1999 – now	F	
		2 – Green	0.52–0.60	30	8	16	185			
		3 – Red	0.63–0.69	30	8	16	185			
		4 – NIR	0.77–0.90	30	8	16	185			
		5 – NIR	1.55–1.75	30	8	16	185			
		6 – Thermal IR	10.40–12.50	60	8	16	185			
		7 – MIR	2.08–2.35	30	8	16	185			
		8 – PAN	0.52–0.90	15	8	16	185			
Landsat 8	OLI	1 – Coastal aerosol	0.43–0.45	30	16	16	185	2013 – now	F	
		2 – Blue	0.45–0.51	30	16	16	185			
		3 – Green	0.53–0.59	30	16	16	185			
		4 – Red	0.64–0.67	30	16	16	185			
		5 – NIR	0.85–0.88	30	16	16	185			
		6 – SWIR 1	1.57–1.65	30	16	16	185			

		7 – SWIR 2	2.11–2.29	30	16	16	185			
		8 – PAN	0.50–0.68	15	16	16	185			
		9 – Cirrus	1.36–1.38	30	16	16	185			
	TIRS	10-Thermal IR	10.60–11.19	100[b]	16	16	185			
		11-Thermal IR	11.50–12.51	100[b]	16	16	185			
SPOT 1, 2, 3	HVR	1 – Green	0.50–0.59	20	8	26	60	1986–2009	C/A	LULC
		2 – Red	0.61–0.68	20	8	26	60			CD, CA
		3 – NIR	0.79–0.89	20	8	26	60			
		PAN	0.50–0.73	10	8	26	60			
SPOT 4	HRVIR	B1 – Green	0.50–0.59	20	8	5	60	1998 – now	C	LULC
	VGT	B2 – Red	0.61–0.68	20	8	5	60			CD, CA
		B3 – NIR	0.78–0.89	20	8	5	60			V, DEM
		B4 – SWIR	1.58–1.75	20	8	5	60			B
		PAN	0.61–0.68	10	8	5	60			
SPOT 5	HRG	PAN	0.48–0.71	2.5–5	8	2–3	60	2002 – now	C	LULC
		B1 – Green	0.50–0.59	10	8	2–3	60			CD, CA
		B2 – Red	0.61–0.68	10	8	2–3	60			V, DEM
		B3 – NIR	0.78–0.89	10	8	2–3	60			B
		B4 – SWIR	1.58–1.75	20	8	2–3	60			
IKONOS		1 – Blue	0.45–0.52	3.2	11	<3	11.3	1999 – now	C	CA, B
		2 – Green	0.52–0.60	3.2	11	<3				V, DEM
		3 – Red	0.63–0.69	3.2	11	<3				
		4 – NIR	0.76–0.90	3.2	11	<3				
		5 – PAN	0.45–0.90	0.82	11	<3				
NOAA	AVHRR	1 – Red	0.58–0.68	1,100	10	2/day	2,400	1978 – now		V, Cir
	LAC	2 – NIR	0.725–1.1	1,100	10	2/day	2,400			
		3 – SWIR	3.55–3.93	1,100	10	2/day	2,400			
		4 – MIR	10.5–11.3	1,100	10	2/day	2,400			
		5 – Thermal IR	11.5–12.5	1,100	10	2/day	2,400			

(continued)

Table 4.2 (continued)

		Spectral band		SR	RR	Cycle	Swath width			
Satellite	Sensor	Number	(µm)	(m)	bits	(days)	(km)	Temporal coverage	Availability	Applications
Orbview 2	SeaWiFS	1 – Violet	404–422	1,100	10	1	2,800	1997 – now	C/F	OC
		2 – Blue	433–453	1,100	10	1	2,800			
		3 – Blue-Green	480–500	1,100	10	1	2,800			
		4 – Blue-Green	500–520	1,100	10	1	2,800			
		5 – Green	545–565	1,100	10	1	2,800			
		6 – Red	660–680	1,100	10	1	2,800			
		7 – NIR	745–785	1,100	10	1	2,800			
		8 – NIR	845–885	1,100	10	1	2,800			
QuicBird		1 – Blue	450–520	2.44	11	1–3.5	16.5	2001 – now	C	LULC
		2 – Green	520–60	2.44	11	1–3.5	16.5			CD, CA
		3 – Red	630–690	2.44	11	1–3.5	16.5			V, B
		4 – NIR	760–890	2.44	11	1–3.5	16.5			
		5 – PAN	450–900	0.61	11	1–3.5	16.5			
GeoEye		1 – Blue	450–510	1.65	11	2.1–8.3	15.2	2008 – now	C	LULC
		2 – Green	510–580	1.65	11	2.1–8.3	15.2			CD, CA
		3 – Red	655–690	1.65	11	2.1–8.3	15.2			V, B
		4 – NIR	780–920	1.65	11	2.1–8.3	15.2			
		5 – PAN	450–800	0.41	11	2.1–8.3	15.2			
WorldView 2		1 – Coastal Blue	400–450	2	11	1.1–2.7	16.4	2008 – now	C	LULC
		2 – Blue	450–510	2	11	1.1–2.7	16.4			CD, CA
		3 – Green	510–580	2	11	1.1–2.7	16.4			V, B, DEM
		4 – Yellow	585–625	2	11	1.1–2.7	16.4			
		5 – Red	630–690	2	11	1.1–2.7	16.4			

		6 – NIR	705–745	2	11	1.1–2.7	16.4			
		7 – NIR	770–1,040	2	11	1.1–2.7	16.4			
		8 – PAN	450–800	0.5	11	1.1–2.7	16.4			
ASTER	VNIR	1 – Green	0.52–0.60	15	8	16	60	1999 – now	F[c]	LULC
		2 – Red	0.63–0.69	15	8	16	60			CD
Terra		3 – NIR	0.76–0.86	15	8	16	60			V
	SWIR	4	1.600–1.700	30	8	16	60			B
		5	2.145–2.185	30	8	16	60			CA
		6	2.185–2.225	30	8	16	60			
		7	2.235–2.285	30	8	16	60			
		8	2.295–2.365	30	8	16	60			
		9	2.360–2.430	30	8	16	60			
	TIR	10	8.125–8.475	90	12	16	60			
		11	8.475–8.825	90	12	16	60			
		12	8.925–9.275	90	12	16	60			
		13	10.25–10.95	90	12	16	60			
		14	10.95–11.65	90	12	16	60			
ASAR	ENVISAT	1	C – band	30		<3	500	2002–2012	C/A	Cir, W
		2	4 pol							
RADARSAT –1, 2	1	C – band	10–100	16	26	100	1995–now	C	MS, I, LULC	

Adapted from Klemas (2009), Klemas (2011), and Mather (2004)

A Available under project submission, *B* Bathymetry, *C* Commercial, *CA* Coastal applications dynamics, *CD* Change detection, *Cir* Circulation, *DEM* Digital elevation models, *F* Available on-line for free download, *I* Ice, *LULC* Land use land cover, *MIR* Middle Infrared, *MS* Marine surveillance, *NIR* Near Infrared, *OC* Ocean Color, *PAN* Panchromatic, *RR* Radiometric resolution, *SR* Spatial Resolution, *SWIR* Shortwave Infrared, *V* Vegetation, *W* Waves

[a]Original MSS pixel size was 79 × 57 m; production systems now resample the data to 60 m

[b]Resampled to 30 m to match multispectral bands

[c]ASTER level-1B data products over the U.S. and Territories from the NASA ASTER mission

Landsat program has offered the world with continuous and consistent monitoring of critically important global resources from which coastal subjects are no exception. Additionally, as of December 2009 all Landsat data in the USGS archive became free[1] to the public, which has, since then, stimulated the use of multispectral imagery in coastal applications.

Besides the fact that Landsat data is freely available online, it is used in coastal studies because it is systematically collected (high temporal frequency), provides large footprints (allowing meso to macro spatial scale studies), has the largest satellite data available historic archive (highest temporal coverage) and provides information on the VIS and IR ranges of the electromagnetic spectrum, where the major coastal information is registered.

Landsat images have been used in different type of studies: in coastal management the National Oceanic and Atmospheric Administration Coastal Change Analysis Program (NOAA) uses Landsat 5 to provide coastal managers with consistent and reliable source of information; shoreline evolution was successfully analyzed from Landsat images (Yu et al. 2011; Ahmad and Lakhan 2012; Huang et al. 2012; Mukhopadhyay 2012); Klemas (2001) uses this imagery to detect coastal environmental indicators; Ryu et al. (2008) derive intertidal morphology and Pe'eri et al. (2012) states that multispectral satellite remote sensing, such as Landsat, is capable of characterize bathymetry information in clear shallow-water.

Nevertheless, Landsat is not free from limitations. The medium spatial resolution can hinder more local studies or extraction of very small characteristics and the use of VIR images can became necessary when studying small objects. Additionally, Landsat images are also dependent upon cloud coverage limiting the use of valuable information on clear sky days or low cloud coverage.

4.3 The Landsat Operational History

Landsat 1 was launched on July 23, 1972 and known, at that time, as the Earth Resources Technology Satellite (ERTS) and operated until January 1978, outliving its design life by 5 years. It was the first Earth-observing satellite to be launched with the express intent to study and monitor our planet's landmasses (NASA). It carried two instruments: a camera system built by the Radio Corporation of America (RCA) called the Return Beam Vidicon (RBV), and the Multispectral Scanner (MSS) (Short et al. 1976).

Landsat 2 was launched on January 22, 1975, and removed from operation on February 25, 1982, due to yam control problems. Still considered an experimental project and operated by NASA, Landsat 2 carried the same sensors as its predecessor: the RBV and the MSS. Landsat 3 was launched on March 5, 1978 and putted on

[1] As of October 1, 2008, all Landsat 7 data became free to the public. In December 2009, all Landsat data in the USGS archive followed suit.

standby mode on March, 1983, being decommissioned on September 7, 1983. Landsat 3 carried the same sensors as its predecessor, but RBV instrument had an improved 38 m ground resolution and used two RCA cameras which both imaged in one broad spectral band (green to near-infrared; 0.505–0.750 μm) instead of three separate bands (green, red, infrared) like its predecessors (Mika 1997; USGS).

Launched on July 16, 1982, Landsat 4 was significantly different than the former and did not carry the RBV instrument. In addition to the MSS instrument, Landsat 4 (and Landsat 5) carried a sensor with improved spectral and spatial resolution, known as the Thematic Mapper (TM). This instrument had seven spectral bands, collecting data from the blue, green, red, near-infrared, mid-infrared (two bands) and thermal infrared portions of the electromagnetic spectrum. Within a year of its launch Landsat 4 experienced problems and the data downlink capability was lost completely in 1993 (USGS).

Carrying the same payload, the MSS and the TM instruments as its predecessor, the Landsat 5 was launched on March 1, 1984, which was designed and built at the same time as Landsat 4. In 1987, its TDRSS transmitter (Ku-band) failed making downlinking data acquired outside the range of U.S. ground receiving antennas impossible. The MSS instrument was turned off in August of 1995 and in November 2011, the TM instrument stopped acquiring images due to a rapidly degrading electronic component. A few months later, engineers turned the MSS instrument back on, and implemented new capabilities to ingest the raw instrument data at the ground station. USGS announced, on December 21, 2012, that Landsat 5 would be decommissioned. Nevertheless, Landsat 5 outlived its 3-year design life, delivering high-quality and global data for 28 years and 10 months, which officially set a new Guinness World Records title for "Longest-operating Earth observation satellite."

During the life span of the Landsat 5, the Landsat program experienced a privatization era, during which Landsat coverage standards declined and many observations from 1984 to 1999 were missed because there was no obvious and immediate buyer (Williamson 1997; Goward and Williams 1997). On July 1, 2001, operational control was officially returned to the federal government, which also relinquished their commercial right to Landsat data.

On April 15, 1999, Landsat 7 was successfully launched carrying the Enhanced Thematic Mapper Plus (ETM+) Earth observing instrument, which replicates the capabilities of the highly successful TM instruments. This new sensor included additional features that make it a more versatile and efficient instrument for global change studies, land cover monitoring and assessment, and large area mapping than its design forebears: a panchromatic band with 15 m spatial resolution, an on-board, full aperture, 5 % absolute radiometric calibration, a thermal IR channel with 60 m spatial resolution and an on-board data recorder (USGS). Landsat 7 experienced a failure of its scan line corrector (SLC) on May 2003, and since then Landsat 7 have been acquiring images with SLC-off mode, but the ETM + still acquires approximately 75 % of the data for any given scene and images continued to be highly used.

4.4 The New Landsat 8

This year, the Landsat program has given us another step forward, with the launch of Landsat 8 sensor, with new functionalities. Landsat 8 was launched on February 11, 2013, with a payload consisted of two science instruments – the Operational Land Imager (OLI) and the Thermal Infrared Sensor (TIRS). These two sensors provide seasonal coverage of the global landmass and surrounding areas at a spatial resolution of 30 m (visible, NIR, SWIR); 100 m (thermal); and 15 m (panchromatic), every 16 days. These new instruments represent an evolutionary advance in technology. OLI is a push-broom sensor with a four-mirror telescope and 12-bit quantization, collects data for visible, near infrared, and short wave infrared spectral bands as well as a panchromatic band. It has a 5-year design lifetime (USGS). The comparison between OLI spectral bands to Landsat 7's ETM + bands can be seen in Table 4.1. OLI provides two new spectral bands, one tailored especially for detecting cirrus clouds and the other for coastal zone observations. TIRS collects data for two more narrow spectral bands in the thermal region formerly covered by one wide spectral band on Landsat 4–7, with a spatial resolution of 100 m and also a 12-bit quantization. Landsat 8 is required to return 400 scenes per day to the USGS data archive (150 more than Landsat 7), increasing the probability of capturing cloud-free scenes for the global landmass, scene size is 185-km-cross-track-by-180-km-along-track and a cartographic accuracy of 12 m or better (including compensation for terrain effects).

4.5 Application Examples

The following application examples, applied in four different sites (Fig. 4.1) intend to address four main characteristics of the Landsat program, highlighting its potentialities in coastal environment uses: (1) time coverage; (2) frequency of coverage; (3) radiometric resolution and (4) spectral resolution. Figure 4.1 displays the location of the four study sites: A – Long-term evolution of the Bugio ephemeral ebb delta island; B – Seasonal evolution of the Prainha transient beach; D – Shoreline detection and extraction along the Costa da Caparica's beaches and D – Bathymetric data retrieval on Sado ebb delta.

4.5.1 Shoreline Detection and Extraction

In the geological context, processes of erosion, transport and deposition involve the removal, transport and deposition of a quantity of material which, from the viewpoint of the shoreline concept implies considering the shore as a three-dimensional surface whose shape is constantly changing. In this context the shoreline is set with

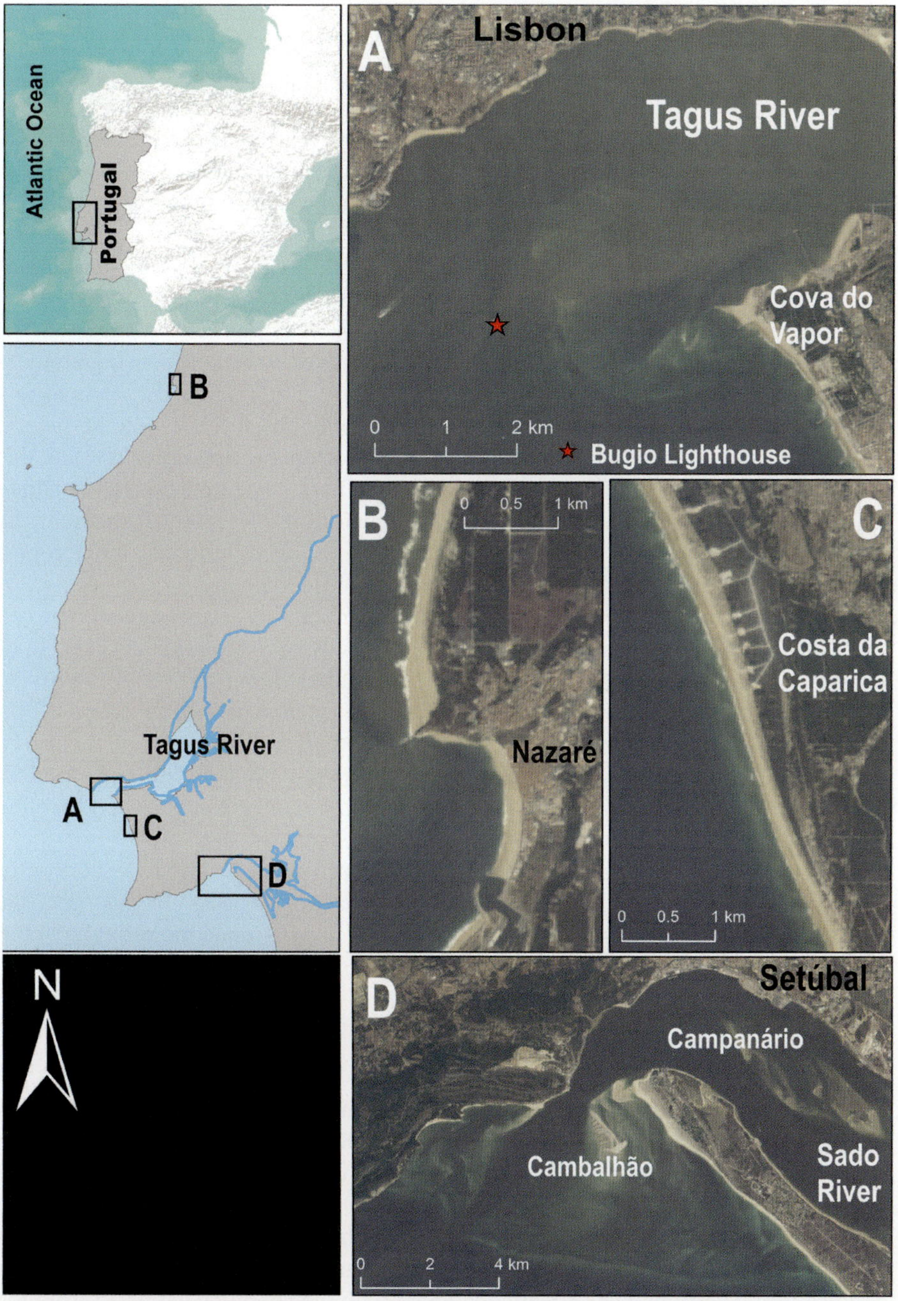

Fig. 4.1 Location of the study sites. (**a**) Long-term evolution of the Bugio ephemeral ebb delta island; (**b**) Seasonal evolution of the Prainha transient beach; (**c**) Shoreline detection and extraction and (**d**) Bathymetric data retrieval on Sado ebb delta. Basemap: 2013-09-08 Landsat 8 image

a principal objective: to characterize the mobility of the sand in space through time. Even considering a static sandy coastline, the shoreline, regarded as the water/land interface, changes because the water level is constantly changing. However, in what concerns context of coastal evolution this concept should necessarily be independent of water level variations. Due to the extremely dynamic nature of the concept, its definition has rely on several criteria. The adopted criteria should be the one that best lists and quantifies the beach horizontal/vertical relations in the sense that it is both robust and reproducible. Boak and Turner (2005) lists a variety of shoreline criteria, which depending of the data source has different capabilities and limitations. According to Parker (2003), the best shoreline definition is the one least susceptible to variability that is not related to actual physical changes in the shoreline itself as, for example, the dune toe. However, in remote sensing applications the HWL (high water level) mark is most commonly adopted criteria as is a feature very simple to identify (sometimes is the only possible) despite the several drawbacks widely acknowledged. For example, Morton and Speed (1998) state this is not a morphological feature, but instead an ephemeral "line in the sand", sensitive to short-term fluctuations in wave and tide conditions. Recent advances in the radiometric resolution of satellite imagery eases the recognition of more robust proxies that depends on differences in the radiometric intensity.

Radiometric resolution is a measure of a sensor's ability to distinguish between two objects of similar reflectance, and higher radiometric resolutions is capable of improving the distinctions between reflectance values. Landsat 8 sensors provide improved signal-to-noise (SNR) radiometric performance quantized over a 12-bit dynamic range, which turns into 4,096 potential grey levels in an image compared with only 256 grey levels in previous versions. Improved signal to noise performance is capable of improving the characterization of land cover, such as shoreline detection using the vegetation line.

This application example try to exploit the improved radiometric resolution of the Landsat images to support the shoreline extraction process. In this case, the shoreline is based on the seaward vegetation line, used as dune toe proxy.

4.5.1.1 Methodology

The shoreline was extracted from two satellite images: a Landsat 5 image dated of 2010-07-30 and a Landsat 8 image of 2013-09-08 along a sandy shoreline stretch (Fig. 4.1c). Detection and extraction was made manually by digitizing the visible vegetation line in ArcGIS 10.1 following Ford (2013). A consistent scale was used and the definition of the vegetation boundary highlighted with the false color composition of 742 bands (Landsat 5) and 753 (Landsat 8), joint with the true color composition. Shoreline results were compared with the shoreline extracted from the 2010 orthophotomaps with 0.5 m of spatial resolution (baseline) of the same area using the same proxy. The two satellite derived shorelines were compared with the baseline using the Digital Shoreline Analysis System (DSAS) (Thieler et al. 2009).

Table 4.3 Mean distance and root mean square error (RMSE) between the two Landsat images and the baseline

Landsat sensor	Date	Mean distance (m)	RMSE (m)
TM	2010-07-30	9.44	12.47
OLI	2013-09-08	5.98	7.27

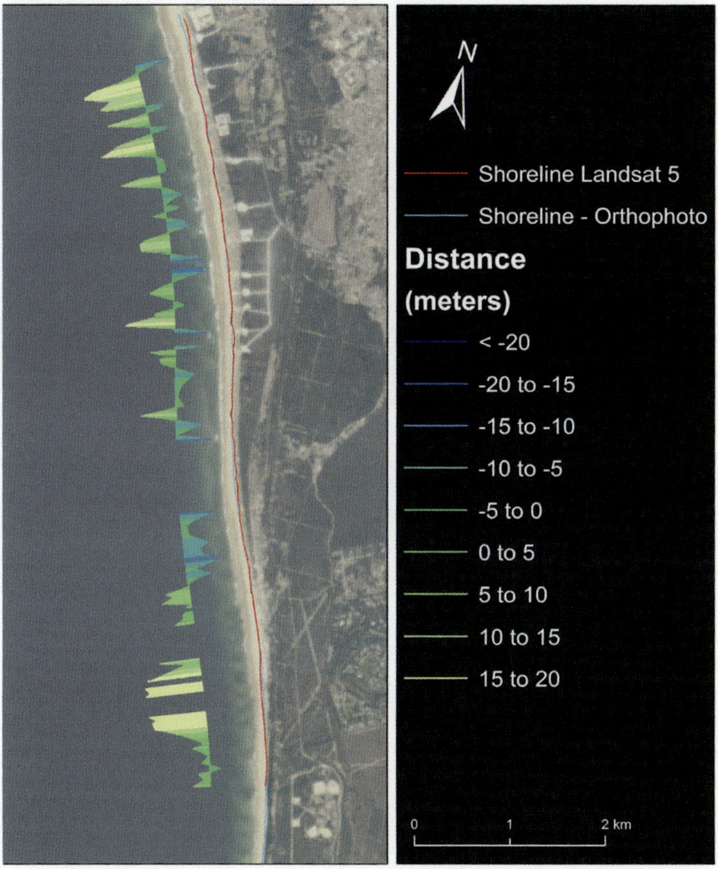

Fig. 4.2 Deviation along the shoreline extracted from the Landsat 5 (2010-07-30)

4.5.1.2 Results

Table 4.3 presents the results of the comparison between lines identified and extracted from satellite imagery and a baseline digitized over a high resolution orthophotomap. The Landsat 5 had a mean distance error of 9.44 m with 12.47 m of RMSE; and Landsat 8 was able to extract the same line with a mean distance error of 5.98 m and 7.27 m of RMSE. Spatial differences alongshore are represented in Figs. 4.2 and 4.3. Results show that Landsat 8 enhanced radiometric resolution can

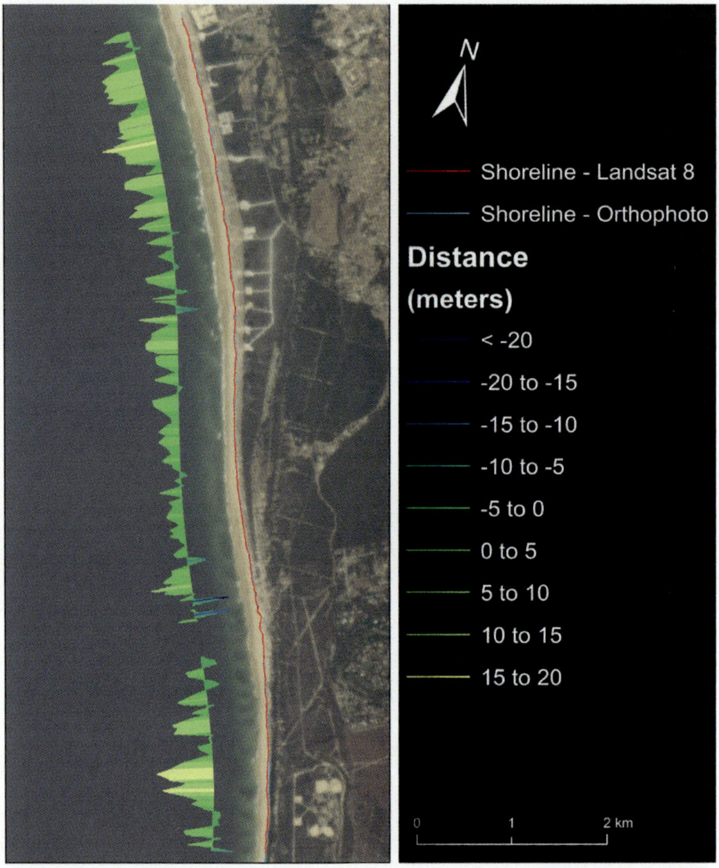

Fig. 4.3 Deviation along the shoreline extracted from the Landsat 8 (2013-09-08)

help to improve the shoreline detection procedure, making it useful in shoreline evolution studies of extremely dynamic and large coastline environments.

4.5.2 Long-Term Evolution of an Ephemeral Ebb Delta Island

River mouths are coastal regions with particularly complex morphodynamic characteristics. Knowledge of the temporal and spatial evolution of these spaces is fundamental to the understanding of their dynamics and to define the sediment budget of the adjacent coastal sectors. A notable element of this dynamic is associated with the development of emerged sand bodies, with extremely variable configuration in space and time that often exhibit an ephemeral nature. Such case is the Bugio Island, located in the Tagus ebb delta (Fig. 4.1a). The evolution of this island has exhibited a pattern where long (pluri-annual) stabilization periods alternate with sudden changes in its configuration, in a clear link to the adjacent coastal area. The main objective of this application example was to study the evolution of the configuration and location of the emerged section of the Bugio sandbank through the use of satellite images from Landsat program over the past 40 years. The long time coverage of the Landsat program makes it the ideal tool to study the decadal evolution of these dynamic areas, especially in cases where other data sources are not available. This is the case of the Tagus ebb delta where the limited information available makes Landsat images an invaluable asset.

4.5.2.1 Methodology

The outline of Bugio Island was defined by the land-water interface extracted from the different Landsat images using the ratio between two bands: b2/b4, for TM and ETM + images and b1/b4 and b3/b6 for the MSS imagery (Alesheikh et al. 2007; Gens 2010). Results drove to a successful mapping of the shoreline; root mean square error (RMSE), computed from the automatic extraction of the Bugio Lighthouse, was estimated in 20.6 m which corresponds to a subpixel precision for all spatial resolutions used (Table 4.4). This data was complement with an aerial photography dated from 1958.

4.5.2.2 Results

Aerial photographs from 1958 show that the Bugio Island was located ~1 km NE of the Bugio lighthouse and displayed an elongated horseshoe-shape. Results of the analysis of satellite images are summarized in Figs. 4.4, 4.5 and Table 4.4. Between 1973 and 1984 the position of Bugio Island did not change significantly but exhibited relevant changes in shape, which altered from a double islet configuration to a single trapezoidal shape and a u-shape with a bay opened to the estuary. In 1987, Bugio Island enlarges and display an arcuate configuration, with the convex side pointing seaward. Afterwards, the island suffers a counterclockwise rotation and progressively stretched towards Cova do Vapor. From 1993 onwards, the area of the island was progressively reduced until its disappearance in 2002. The results show very irregular evolution and dynamics of Bugio Island, with periods of relative stability alternating with episodes in which the variations are extremely fast (Fig. 4.4). This example shows the potential of using the full temporal range of the Landsat program, allowing to know the past in a striving to understand the future.

Table 4.4 Landsat images used is the study, date, time and respective tide level

Landsat sensor	Date	Time (UTC)	Resolution (m)	Tide (MSL) (m)	Landsat sensor	Date	Time (UTC)	Resolution (m)	Tide (MSL) (m)
MSS	1973-03-29	10:50:40	60	0.44	TM	1992-07-12	10:37:46	30	0.57
MSS	1975-08-01	10:37:53	60	0.41	TM	1993-07-31	10:38:15	30	0.57
MSS	1977-04-21	10:15:00	60	−1.12	TM	1994-07-02	10:33:14	30	0.67
MSS	1978-11-01	10:22:22	60	−0.67	TM	1995-07-21	10:19:18	30	0.79
TM	1984-05-03	10:41:43	30	−0.88	TM	1996-08-08	10:30:24	30	0.86
TM	1987-07-31	10:40:07	30	−0.69	TM	1997-07-26	10:45:35	30	−0.09
TM	1988-12-16	10:46:23	30	0.22	TM	1998-08-30	10:53:00	30	0.12
TM	1989-03-14	10:43:00	30	−0.02	TM	1999-07-08	11:07:17	30	0.71
TM	1990-12-22	10:39:28	30	−0.79	ETM+	2000-11-15	11:04:51	30	−1.22
TM	1991-08-11	10:37:08	30	0.95	ETM+	2001-05-26	11:04:19	30	−1.04

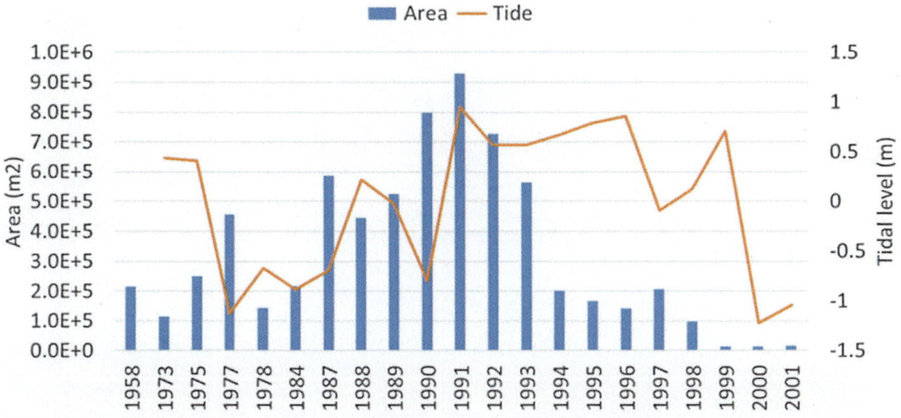

Fig. 4.4 Evolution of the Bugio Island area over the period 1958–2001 and the respectively tidal level of each Landsat image used

4.5.3 Seasonal Evolution of a Transient Beach

Seasonal beach changes can be connected to both cross-shore and longshore sediment transport processes: while the former relates to the transference of sand along the beach profile, reflecting the natural adjustment of the beach profile to the seasonal changes in wave energy, the latter is connected to modifications in planform geometry, usually in relation to seasonal changes in wave direction.

The detection of these changes, depends not only on an adequate temporal coverage but also on the relation between the magnitude of the changes and spatial resolution of the satellite imagery. For the large majority of cases this inhibits the application of this methodology to the evaluation of cross-shore changes. On the other hand, planform changes, usually related with beach rotation or sediment headland bypassing, can attain a magnitude compatible with satellite resolution making this technique suitable for the study of these processes.

The case study used to demonstrate satellite imagery potential to the study of these processes concerns to a transient beach in relation to headland sediment bypassing of Nazaré headland, the Prainha beach (Figs. 4.1b and 4.6). This beach usually grows in summer, in connection to the low energetic, short period, northerly waves, that promotes sediment bypassing and fades out in early autumn with the onset of higher energetic waves.

4.5.3.1 Methodology

The shoreline was extracted using the water/land boundary using the Red (Band4)/NIR (band5) ratio of the Landsat 8 imagery. Line elevation was estimated from the tide level at the time of acquisition. Results were validated using three fieldwork topographic surveys dated from 2013-09-19, 2013-10-09 to 2013-10-13, more or less concomitant with three satellite images (Fig. 4.7).

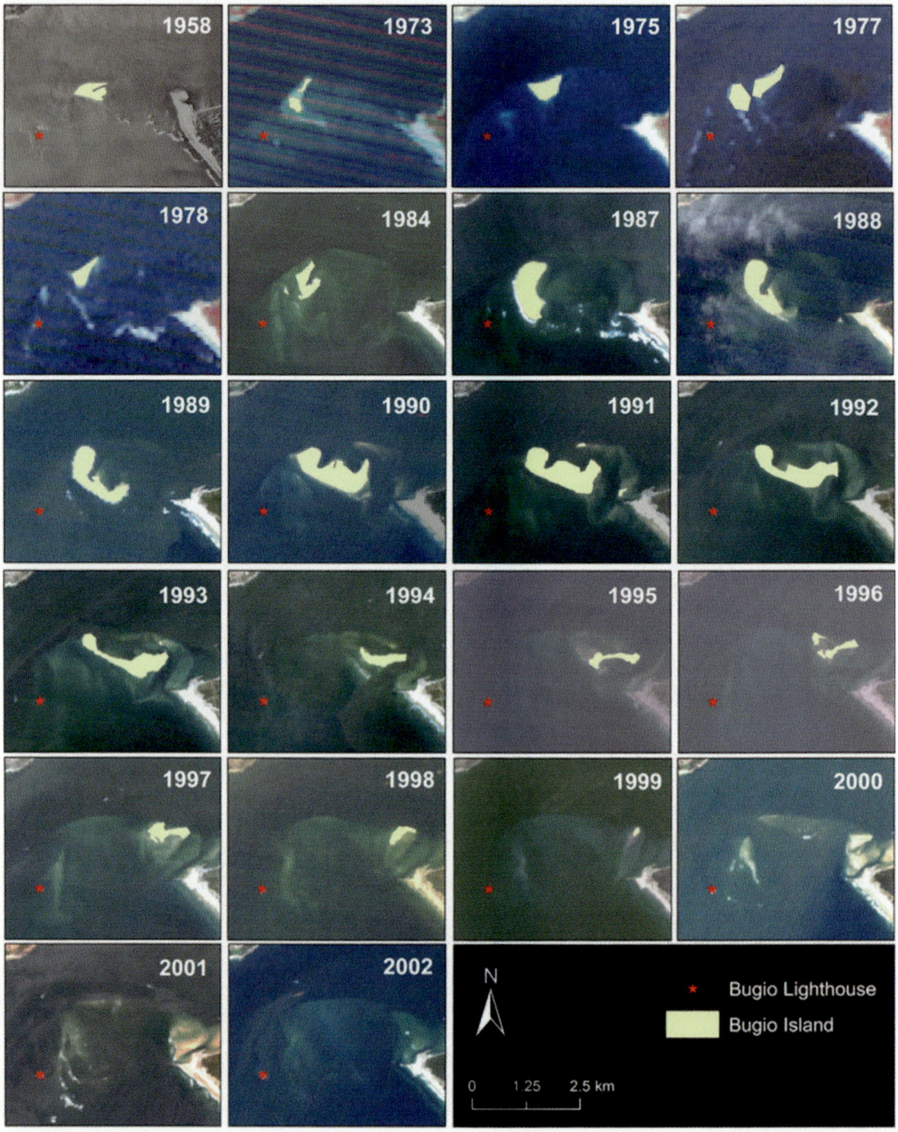

Fig. 4.5 Evolution of the Bugio Island between the years 1958 and 2002. Base map images: 1958 – aerial photograph; 1973–1978 – Landsat MSS 754 false color images, 1984–1999 – Landsat TM true color images; 2000–2002 – Landsat ETM + true color images

4.5.3.2 Results

The short-term evolution of Prainha beach is displayed on Fig. 4.7. The comparison of the shoreline extracted from satellite with field data showed a reasonable match, validating the applied methodology.

Fig. 4.6 Ephemeral beach development. *Left*: no ephemeral beach south of the Nazaré headland (Source: Esri basemap, 2011-03-18). *Right*: beach development south of the Nazaré headland visible on the Landsat 8 image of 2013-09-24

The beach started to develop on July, 2013 and has reached its maximum extent by the end of September, with ~1 km in length and ~100 m width. After that, and until mid October, the beach migrated eastwards with a noticeble retreat on the western section while preserving more less the area and shape. By the end of October, the beach receded and became significantly shorter, with a maximum width of ~50 m. From this point onwards the beach fades-out progressively until it will eventually vanishes (Fig. 4.6 – right).

The 16 days Landsat frequency of coverage appears to be effective in the attempt to study and monitor these type of seasonal phenomena, allowing the mapping of shoreline evolution and configuration, effectively complementing fieldwork experiments and data.

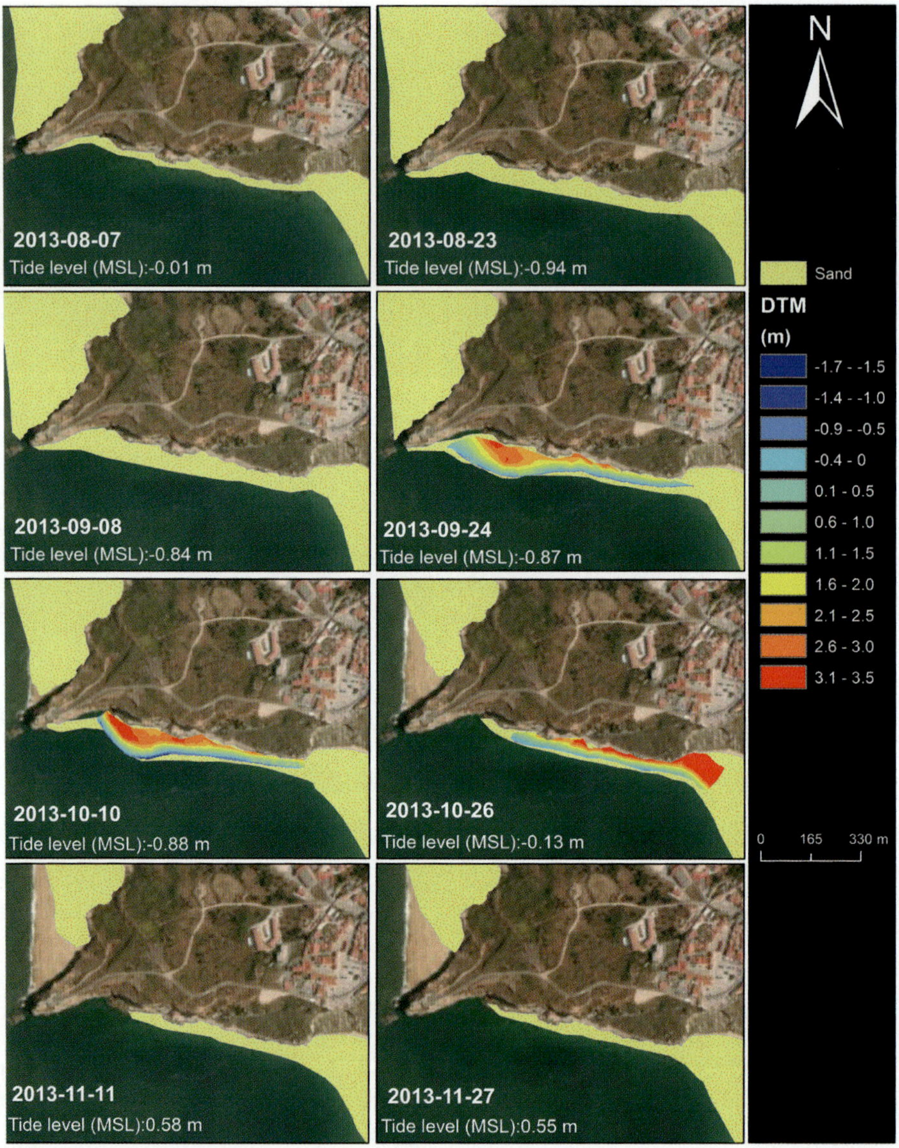

Fig. 4.7 Evolution of the ephemeral beach south of the Nazaré headland. The sand feature extracted by satellite imagery and DTM is the field validation

4.5.4 Bathymetric Data Retrieval

Airborne hyperspectral sensors are the best ones to derive bathymetric data, reducing the problem of spectral coverage, but they are still inappropriate for monitoring dynamical processes (Durand et al. 2000). Although passive optical systems are limited in water depth penetration and constrained by water turbidity (Stumpf and Holderied 2003), the use of this type of data might be the only viable way to characterize extensive shallow-water areas, extremely dynamic regions or areas not covered by other, more accurate, methods.

The Landsat 8 coastal aerosol band (Band 1: 0.4333–0.4530 μm) has two objectives: (1) estimate the concentration of aerosols in the atmosphere, which may be used to refine the atmospheric correction procedures such as dark object subtraction and (2) provide new studies of the coastal and inland waters. Relative to the MODIS and the SeaWiFS sensors, the Landsat 8 allows for better imaging of these shallow water areas due to its superior spatial and radiometric resolutions (Table 4.1).

The following application example intends to demonstrate the potential of this new spectral band to derive and map shallow-water bathymetry.

4.5.4.1 Methodology

The bathymetry was derived using the bottom albedo[2]-independent bathymetry algorithm developed by Stumpf and Holderied (2003). This algorithm is based on the hypothesis that when the sea floor is covered by bright sand, the albedo is the same when the bottom is at the same depth. When the water is clear, this algorithm is capable of extracting the bathymetry up 25 m water depths; however results are intended to offer a general overview of the bathymetry and to highlighted relative changes and should not be used for navigational purposes. The pre-processing stage should include sun de-glinting (Hochberg et al. 2003; Hedley et al. 2005), because water usually presents variable sun glint across images, as a result of the ocean swell, causing widely variable incident angles for the reflection of the Sun's radiation, which may mask bottom features. Additionally, water should be as clear as possible, so that turbidity doesn't mask bottom albedo.

In this work, the bathymetry of Sado river ebb-delta (Fig. 4.1d) was derived from two Landsat images (Table 4.5) using the R-G-B-NIR bands. The difference in both images was the representation of the blue radiation range: Landsat 7 image used the ETM + Band 1 while the Landsat 8 image used the OLI Band 1 (coastal aerosol band).

4.5.4.2 Results

The shallow water bathymetry over the outer section of the Sado estuary derived from the Landsat imagery can be observed in Figs. 4.9 and 4.10. Results obtained from both images exhibits an overall pattern that closely matches the information portrayed in the nautical chart (Fig. 4.8). In fact, all the major morphological features, as the main channel, the ebb-tide delta and the Campanário tidal flats

[2] Albedo – the proportion of the incident light or radiation that is reflected by a surface.

Table 4.5 Landsat images used to derive bathymetry, date, time and respective tide level

Landsat sensor	Date	Time (UTC)	Tide (MSL) (m)
ETM+	2008-10-20	11:03:16	−0.60
OLI	2013-09-08	11:16:32	−0.84

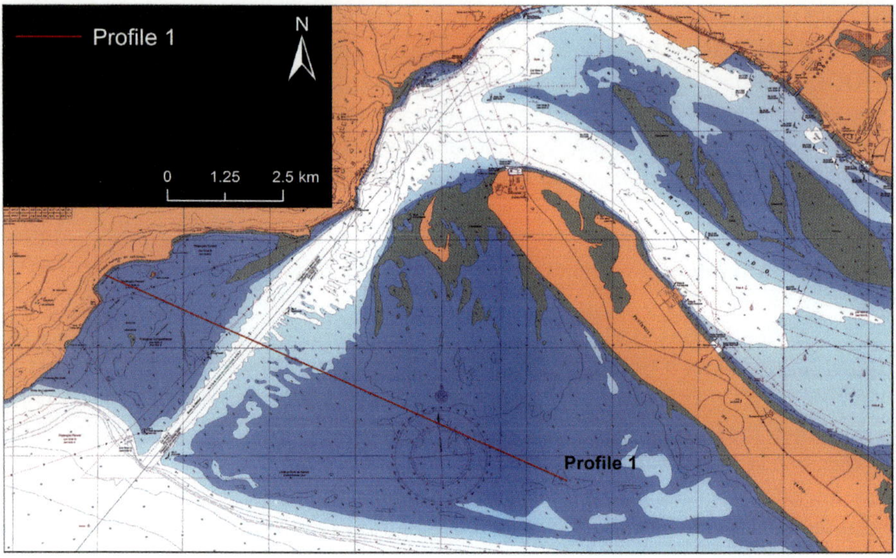

Fig. 4.8 Nautical Chart (November 2008) and localization of profile 1

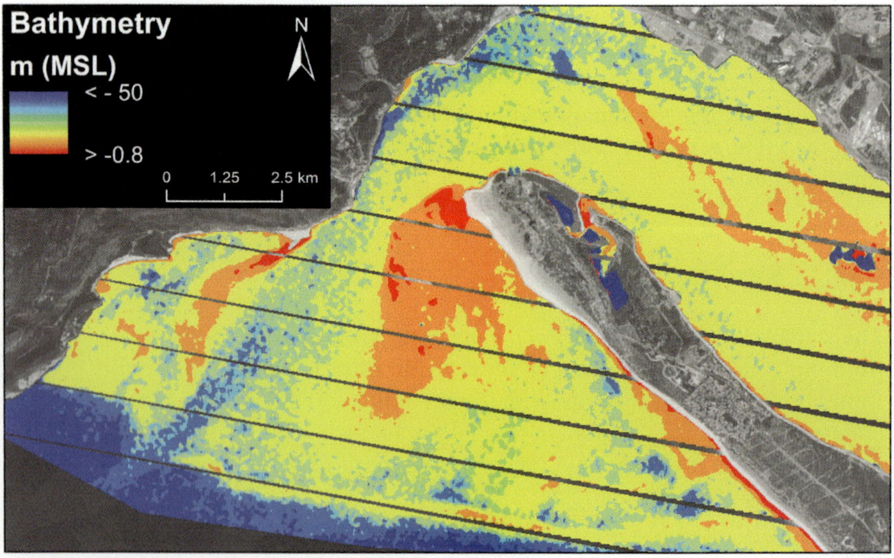

Fig. 4.9 Bathymetric derived data from 2008-10-20 Landsat 7 image

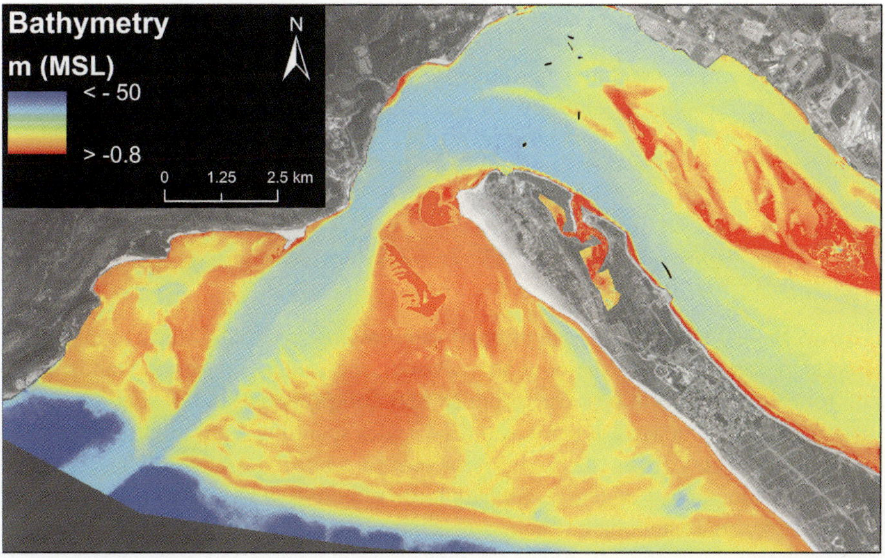

Fig. 4.10 Bathymetric derived data from 2013-09-08 Landsat 8 image

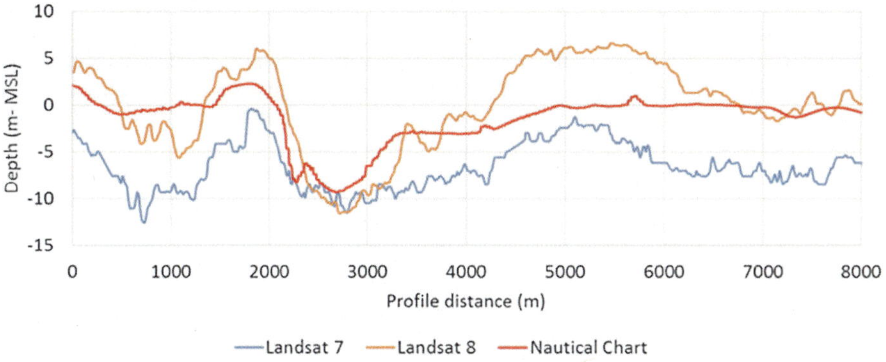

Fig. 4.11 Bathymetry derived from Landsat 7 (2008-10-20) and Landsat 8 (2013-09-08) in comparison with the nautical chart

display a very similar configuration. When comparing Landsat 7 with the Landsat 8 derived bathymetry, is possible to see that the former is clearly noisier, while Landsat 8 enables an impressive representation of the bathymetry over the area, with a detail that makes possible the definition of the bedform geometry over the entire ebb delta.

The larger bathymetric differences between images are observed over the Cambalhão shoal since this is area subject to significant morphological changes over time; to a large extent these discrepancies can be justified by the differences in the date of acquisition. A test profile (Profile 1 – Fig. 4.8) over the ebb delta that cross the main channel was performed in the three data sources. Results (Fig. 4.11) show a reasonable agreement between satellites derived data and the bathymetric chart, which is better for the Landsat 8 image. The weak resolution of the bathymetric chart precludes a more detailed comparison but is sufficient to demonstrate the huge potentiality of the Landsat 8 images to derive bathymetric data.

4.6 Conclusion

This work synthetizes the application of satellite imagery to the study of mesoscale coastal processes. From the available sensors, the Landsat program stands out as the tool that offers the best compromise between potential and availability to the understanding of coastal dynamics at time scales spanning from months to decades and spatial scales from hundreds of meters to tens of kilometers. The 16 days Landsat frequency of coverage appears to be effective in the attempt to study and monitor seasonal phenomena such as beach rotation and sediment bypassing, while the temporal coverage (40 years) of this program allows the characterization of decadal shoreline changes. The recent launch of Landsat 8, with improved spectral and radiometric resolutions, further enhances its capabilities in what concerns shoreline detection and shallow water bathymetry extraction.

Acknowledgments Landsat images from the years 1991 to 1998 are data provided by the European Space Agency, in the frame of ESA Project – ID 14512 – An Integrated Approach to Shoreline Evolution: Application to West Portuguese Coast. The remaining Landsat images are available at the USGS website (http://earthexplorer.usgs.gov/).

The first author is supported by a Postdoc grant funded by the Fundação para a Ciência e Tecnologia – FCT (grant # SFRH/BPD/81800/2011). The authors are also supported by the Beach to Canyon project – Beach to Canyon Head Sedimentary Processes (# PTDC/MAR/114674/2009).

The authors would also like to thanks Ana Bastos for the vectorization of the nautical chart and Mónica Ribeira for the review.

References

Ahmad SR, Lakhan VC (2012) GIS-based analysis and modeling of coastline advance and retreat along the coast of Guyana. Mar Geod 35:1–15. doi:10.1080/01490419.2011.637851

Alesheikh A, Ghorbanali A, Nouri N (2007) Coastline change detection using remote sensing. Int J Environ Sci Technol 4:61–66

Boak EH, Turner IL (2005) Shoreline definition and detection: a review. J Coast Res 14:688–703

Durand D, Bijaoui J, Cauneau F (2000) Optical remote sensing of shallow-water environmental parameters: a feasibility study. Remote Sens Environ 73(2):152–161. doi:10.1016/S0034-4257(00)00090-0

Ford M (2013) Shoreline changes interpreted from multi-temporal aerial photographs and high resolution satellite images: Wotje Atoll, Marshall Islands. Remote Sens Environ 135:130–140. doi:10.1016/j.rse.2013.03.027

Gens R (2010) Remote sensing of coastlines: detection, extraction and monitoring. Int J Remote Sens 31:1819–1836. doi:10.1080/01431160902926673

Goward SN, Williams DL (1997) Landsat and earth systems science: development of terrestrial monitoring. Photogramm Eng Remote Sens 63(7):887–900

Hedley JD, Harborne AR, Mumby PJ (2005) Simple and robust removal of sun glint for mapping shallow-water benthos. Int J Remote Sens 26(10):2107–2112. doi:10.1080/01431160500034086

Hochberg EJ, Andrefouet S, Tyler MR (2003) Sea surface correction of high spatial resolution Ikonos images to improve bottom mapping in near-shore environments. IEEE Trans Geosci Rem Sens 41(7):1724–1729

Horn DP (2002) Mesoscale beach processes. Prog Phys Geogr 26:271–289. doi:10.1191/0309133302pp336pr

Huang H, Liu Y, Qiu Z (2012) Morphodynamic evolution of the Xiaoqing River mouth: a Huanghe River-derived mixed energy estuary. Chin J Oceanol Limnol 30:889–904. doi:10.1007/s00343-012-1295-4%201

Ikeda M, Dobson FW (1995) Oceanographic applications of remote sensing. CRC Press, New York

Klemas V (2001) Remote sensing of landscape-level coastal environmental indicators. Environ Manage 27:47–57. doi:10.1007/s002670010133

Klemas V (2009) Sensors and techniques for observing coastal ecosystems. In: Yang X (ed) Remote sensing and geospatial technologies for coastal ecosystem assessment and management, Lecture notes in geoinformation and cartography. Springer, Berlin/Heidelberg, 561 p. doi:10.1007/978-3-540-88183-4 1

Klemas V (2011) Remote sensing techniques for studying coastal ecosystems: an overview. J Coast Res 27:2–17. doi:10.2112/JCOASTRES-D-10-00103.1

Martin S (2004) An introduction to remote sensing. Cambridge University Press, Cambridge

Mather PM (2004) Computer processing of remotely-sensed images: an introduction. Wiley, Chichester

Mika AM (1997) Three decades of Landsat instruments. Photogramm Eng Remote Sens 63(7):839–852

Morton RA, Speed FM (1998) Evaluation of shorelines and legal boundaries controlled by water levels on sandy beaches. J Coast Res 14:1373–1384

Mukhopadhyay A (2012) Automatic shoreline detection and future prediction: a case study on Puri Coast, Bay of Bengal, India. Eur J Remote Sens 45:201–213. doi:10.5721/EuJRS20124519

NASA. Landsat 7 science data user's handbook. Available online: http://ltpwww.gsfc.nasa.gov/IAS/handbook/handbook_toc.html

NOAA. Coastal change analysis program regional land cover. Available online: http://www.csc.noaa.gov/digitalcoast/data/ccapregional

Parker BB (2003) The difficulties in measuring a consistently defined shoreline – the problem of vertical referencing. J Coast Res Spec Issue 38:44–56

Pe'eri S, Azuike C, Alexander L, Parrish C, Armstrong A (2012) Beyond the chart: the use of satellite remote sensing for assessing chart adequacy and completeness information. In: Proceedings of the 2012 Canadian hydrographic conference: the Arctic, old challenges, new approaches, Niagara Falls, Canada, 15–17 May

Ryu JH, Kim CH, Lee YK, Won JS, Chun SS, Lee S (2008) Detecting the intertidal morphologic change using satellite data. Estuar Coast Shelf Sci 78:623–632. doi:10.1016/j.ecss.2008.01.020

Short NM, Lowan PD, Freden SC, Finch WA (1976) Mission to earth: Landsat views the world, NASA special publication 360. NASA Science and Technical Information Office, Washington, DC

Stumpf RP, Holderied K (2003) Determination of water depth with high-resolution satellite imagery over variable bottom types. Limnol Oceanogr 48(1):547–556

Thieler ER, Himmelstoss EA, Zichichi JL, Ergul A (2009) The Digital Shoreline Analysis System (DSAS) version 4.0: an ArcGIS extension for calculating shoreline change. U.S. Geological Survey, Reston. Open Report 2008–1278

USGS. Landsat Mission Timeline. Available online: http://landsat.usgs.gov/about_mission_history.php

Williamson RA (1997) The Landsat legacy: remote sensing policy and the development of commercial remote sensing. Photogramm Eng Remote Sens 63(7):877–885

Yang X (2009) Remote sensing, geospatial technologies and coastal ecosystems. In: Yang X (ed) Remote sensing and geospatial technologies for coastal ecosystem assessment and management. Lecture notes in geoinformation and cartography. Springer, Berlin/Heidelberg, p 561. doi:10.1007/978-3-540-88183-4%201

Yu K, Hu C, Muller-Karger F (2011) Shoreline changes in west-central Florida between 1987 and 2008 from Landsat observations. Int J Remote Sens 32(23):8299–8313. doi:10.1080/01431161.2010.535045

Chapter 5
Remote Sensing and Modeling of Coral Reef Resilience

Anders Knudby, Simon J. Pittman, Joseph Maina, and Gwilym Rowlands

Abstract A new paradigm has emerged for management of coral reefs in an era of changing climate – managing for resilience. A fundamental need for such management to be effective is our ability to measure and map coral reef resilience. We review the resilience concept and factors that may make a coral reef more or less resilient to climate-driven impacts, and focus on recent advances in a trio of technologies – remote sensing, spatial distribution modeling, and ecosystem simulation – that promise to improve our ability to quantify coral reef resilience across reefs. Remote sensing allows direct mapping of several ecosystem variables that influence reef resilience, including coral and algal cover, as well as a range of coral reef stressors, as exemplified by three case studies. Spatial distribution modeling allows exploitation of statistical relationships between mappable environmental variables and factors that influence resilience but which cannot be mapped directly, such as herbivore biomass. Ecosystem simulation modeling allows predictions to be

A. Knudby (✉)
Department of Geography, Simon Fraser University, Burnaby, BC, Canada
e-mail: aknudby@sfu.ca

S.J. Pittman
Biogeography Branch, National Oceanic and Atmospheric Administration, Silver Spring, MD, USA

Centre for Marine and Coastal Policy Research, Marine Institute, Plymouth University, Plymouth, UK

J. Maina
ARC Centre of Excellence for Environmental Decisions, The University of Queensland, St Lucia, QLD 4072, Australia

Wildlife Conservation Society, 2300 Southern Boulevard, Bronx, NY 10460, United States of America

G. Rowlands
Oceanographic Center, Nova Southeastern University, Fort Lauderdale, FL, USA

Khaled bin Sultan Living Oceans Foundation, Hyattsville, MD, USA

C.W. Finkl and C. Makowski (eds.), *Remote Sensing and Modeling: Advances in Coastal and Marine Resources*, Coastal Research Library 9,
DOI 10.1007/978-3-319-06326-3_5,

made for the trajectories of reef ecosystems, given their initial state, interactions between ecosystem components, and a realistic current and future disturbance regime. Together, these technologies have the potential to allow production of coral reef resilience maps. We conclude with a fourth case study that illustrates integration of resilience maps into a multi-objective decision support framework. Implementation of the managing for resilience paradigm is still in its infancy, but the rapidly advancing technologies reviewed here can provide the resilience maps needed for its successful operationalization.

5.1 Introduction

Global climate change is widely recognized as a major threat to tropical coral reefs, primarily due to an increase in the frequency and magnitude of thermal stress events leading to coral mortality from bleaching and disease outbreaks (Eakin et al. 2010). Climate change is also predicted to increase the frequency of severe storms and the acidity of ocean water that can cause physical damage to reef structure and reduce coral growth rates (Hoegh-Guldberg et al. 2007; van Hooidonk et al. 2014; Manzello et al. 2013). These broad-scale threats interact with more local stressors, such as exposure to materials from land that reduce water quality and fishing that removes functionally important species from the ecosystem (Hughes and Connell 1999; Wilson et al. 2006; Ban et al. 2014), to degrade reefs. The ecological consequence for coral reefs from multiple stressors operating at multiple scales will depend on the structural and functional attributes of the coral reef ecosystem, as well as the type, magnitude and duration of stress (Carilli et al. 2009; McClanahan et al. 2014). In many locations, interacting stressors have undermined the resilience of coral-dominated communities, resulting in a phase-shift to a less desirable algal-dominated community, with impaired provisioning of ecosystem goods and services (Moberg and Rönnbäck 2003; Hughes et al. 2010; Pratchett et al. 2014).

Although the functional integrity of coral reefs is likely to be impacted more where multiple severe stressors exist, the complex spatio-temporal heterogeneity of environmental systems causes coral reefs and associated organisms to exhibit a spatially complex stress response (Mumby and Steneck 2008; Elmqvist et al. 2003). This presents a major challenge for management of coral reef ecosystems where the identification of areas with different levels of resilience to stressors is required to prioritize actions and design place-based conservation strategies (Game et al. 2008; McClanahan et al. 2009; McLeod et al. 2008).

A new *resilience* paradigm has thus emerged for coral reef management, with conservation objectives that aim to enhance or sustain the resilience of coral reefs to the range of stressors they face (Nyström et al. 2008; McClanahan et al. 2012). A major challenge for conservation ecology has thus become the provision of information on resilience at spatial and temporal scales relevant to management. Typically, indicators of resilience are measured through in-water surveys, yet management domains are geographically broad and therefore in-water surveys alone are of limited utility and prohibitively expensive for agencies to conduct. Rapid and cost-effective techniques are therefore required to provide detailed and ecologically relevant spatial information on coral reef resilience across broad and structurally complex geographical areas. Advances in spatial technologies such as remote sensing and spatial modeling show great potential to address this challenge (Maina et al. 2008; Rowlands et al. 2012; Knudby et al. 2013a). For example, ecological studies have demonstrated that the diversity and abundance of herbivores (fishes and invertebrates) are important indicators of coral reef health through their function in controlling algal biomass (Burkepile and Hay 2008; Cheal et al. 2010). Models estimate that increased herbivory by parrotfish after protection from fishing may increase reef resilience sixfold (Mumby et al. 2013b). While little is known about the geography of herbivory and its implication for predicting resilience, spatial modeling has shown that it is possible to map herbivore species distributions, biomass, and the functional richness of herbivores across the seascape (Pittman and Brown 2011; Knudby et al. 2013a; Pittman and Knudby 2014). Research is now urgently required to determine the utility of mapping indicators of resilience as a functionally meaningful spatial proxy for reef resilience. To support science-based decision making in marine conservation, this chapter presents a review of how remote sensing and spatial distribution modeling can be used to map resilience indicators and how such maps can be integrated through simulation modeling to provide spatially explicit estimates of coral reef resilience. Finally we outline an approach for integrating such information through a spatial decision support framework that uses local expert knowledge, remote sensing data and in-water surveys to identify and prioritize coral reef ecosystems for conservation action.

5.2 Coral Reef Ecosystem Resilience and Indicators of Resilience

The concept of resilience and stability in ecology are not new. Holling (1973) defined resilience as "*a measure of the persistence of systems and their ability to absorb change and disturbance and still maintain the same relationships between populations or state variables*". Stability was defined as "*the ability of a system to return to an equilibrium state after a temporary disturbance*" with a more rapid return indicative of greater stability. Holling (1996) differentiated between *ecological resilience* and *engineering resilience*, two aspects of a system's stability that have different consequences for evaluating, understanding, and managing complexity and change. *Engineering resilience* assumes that the system in question has a single stable state that it will return to in the absence of disturbance, and is typically quantified as the magnitude of deviation from, and speed of return to, the stable state following a disturbance. *Ecological resilience*, on the other hand, assumes that multiple stable states exist, each bounded by a domain of attraction. It is thus considered the amount of disturbance the system can be exposed to without moving beyond its current domain of attraction and transitioning to another stable state (Holling 1996). Other authors have also considered ecosystem characteristics such as durability and robustness (Dawson et al. 2010). Clarification and debate on the use of terminology for both conceptualization and operationalization of resilience is dealt with elsewhere in the literature (Gunderson 2000; Levin and Lubchenco 2008; Mumby et al. 2014).

The spatial and temporal scales at which the system is considered influences our understanding of stability and resilience (Holling 1992). Although the relevant spatial and temporal scales are rarely specifically defined, it is broadly accepted that multiple stable states exist for coral reef ecosystems (Knowlton 1992; Mumby et al. 2013a, but see Dudgeon et al. 2010). Typically these include a desired coral-dominated state that, given a combination of press and pulse disturbances, can be replaced by an undesired macroalgae-dominated state. Ecological resilience has thus been broadly adopted as the relevant resilience concept by the coral reef community, however, it is difficult or impossible to measure in the absence of observed transitions between stable states, and thus not practical as a basis for resilience assessment or management. Engineering resilience, on the other hand, can be assessed by focusing on its two components, often termed "resistance" (to disturbance) and "recovery" (from disturbance). In order to manage for resilience, factors that influence resistance and recovery must be identified and protected. A recent survey of expert opinion and scientific evidence (McClanahan et al. 2012) identified 11 principal factors that influence the resistance and/or recovery of coral reef ecosystems to climate-driven disturbances – and are also feasible to assess from local field observations at relatively fine spatial scales. These include aspects of the coral fauna (presence of stress-resistant coral species, diversity of coral species, high levels of coral recruitment, and absence of coral disease) and competition for space (low presence of macroalgae), as well as moderators of competition (herbivore biomass), the physical environment (high annual temperature variability, low nutrient and sediment levels), and direct human impacts (physical impacts and fishing pressure). These 11 factors can thus act as resilience indicators and may function as a list of mapping targets that combined have the potential to characterize the resilience of a coral reef ecosystem to climate-driven disturbance. For coral reefs, resilience has also been linked to connectivity through consideration of the movement of organisms across the broader seascape within which coral reefs are embedded (Mumby and Hastings 2008; Olds et al. 2012; Melbourne-Thomas et al. 2011). This landscape ecology approach is still in its infancy, but offers a pragmatic and ecologically rational way that remote sensing and mapping can be used to identify coral reefs that vary in their connectivity to neighboring habitat types (Pittman and Olds 2014).

5.3 What Remote Sensing Can Map on Coral Reefs

Remote sensing from water-, air- and space-borne platforms is now a core tool for the mapping, monitoring and management of coral reef ecosystems (Mumby et al. 2004a; Knudby et al. 2007). Sensor technology and the analytical tools for

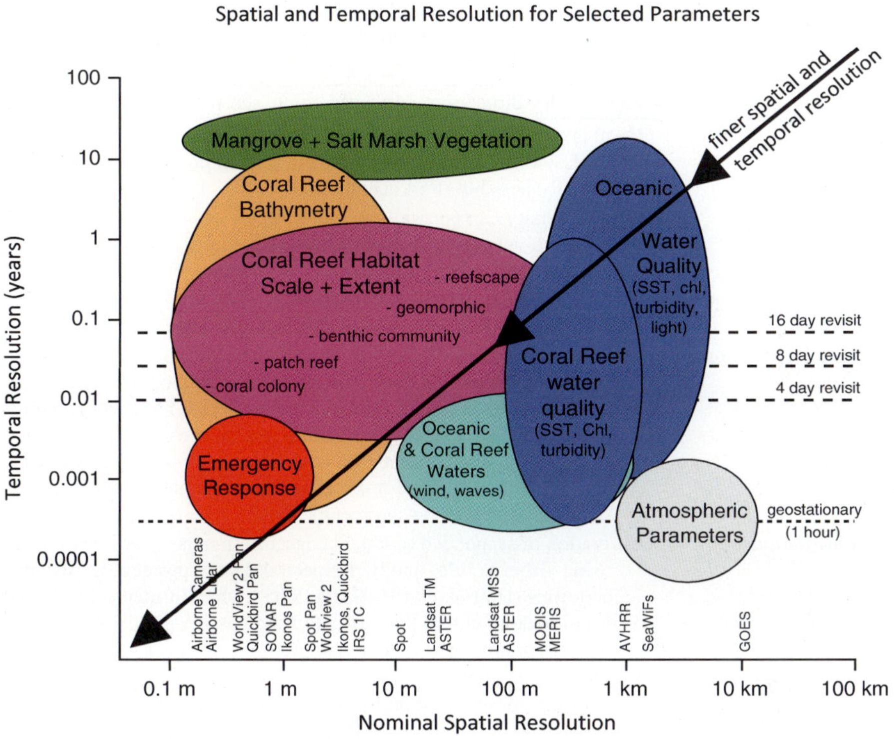

Fig. 5.1 Spatial and temporal scales for mapping and monitoring coral reefs and their environment (Reproduced with permission from Jupiter et al. (2013))

interpreting marine remote sensing data have advanced significantly in the past decade, and now provide reliable, repeatable and cost-effective quantitative assessments of habitat distributions and conditions over spatially extensive areas (Goodman et al. 2013). Remote sensing can provide a synoptic view of coral reef ecosystems by mapping a wide range of biological and physical variables from the water column to the seafloor, as well as on adjacent terrestrial areas that are required to characterize coral reef resilience, all at a wide range of spatial and temporal scales (Fig. 5.1). The diversity of information that can now be provided by remote sensing using passive sensors such as aerial photography, multispectral and hyperspectral sensors, or active sensors such as acoustic and lidar instruments, thus allows complex scientific and management-related questions to be addressed. Mapping the 11 resilience indicators identified by McClanahan et al. (2012), however, remains a significant challenge. Here we describe the capabilities of acoustic and optical sensors for mapping characteristics of the seafloor, the waters above, and the seascape context that is relevant to understanding and mapping reef resilience (Table 5.1).

Table 5.1 List of the 11 resilience indicators identified by McClanahan et al. (2012), and their potential for begin mapped either directly using remote sensing or indirectly through spatial distribution modeling

Resilience indicator	Potential for direct mapping with remote sensing
Stress-tolerant coral taxa	Coral taxa can generally not be mapped with remote sensing (Hochberg et al. 2003). Local mapping of distinct and prominent taxa is occasionally possible (Purkis et al. 2006)
Coral diversity	Coral diversity can generally not be mapped with remote sensing, although general measures of habitat diversity can be derived (Mumby 2001; Harborne et al. 2006; LeDrew et al. 2004)
Historical temperature variability	Historical SST variability can be directly derived from the long record of global SST data products (McClanahan et al. 2007)
Nutrients (pollution)	Derivation of near-surface concentrations of chlorophyll-a, CDOM and CDM is possible for Case I waters using ocean colour data, but is unlikely to be effective for coral reefs where RTM inversion algorithms must be used to derive nutrient concentrations (Giardino et al. 2007). The combination of airborne lidar and hyperspectral data is also promising for direct observations of nutrient concentrations on coral reefs (Aitken et al. 2010)
Sedimentation	Derivation of suspended sediment concentration has been demonstrated using airborne lidar and hyperspectral data (Epps et al. 2010) and acoustics (Hoitink and Hoekstra 2005), while limited progress has been made to derive this variable from passive optical data in coral reef environments (Ouillon et al. 2008)
Herbivore biomass	Several case studies show that the biomass of herbivorous fish can be mapped from lidar or multispectral data (Purkis et al. 2008; Pittman et al. 2009)
Physical human impacts	While natural physical impacts on reefs can be derived from models and historical observations, anthropogenic physical impacts on coral reefs cannot generally be mapped with remote sensing
Coral disease	Coral disease can generally not be mapped with remote sensing. However, remotely sensed SST can be used to predicted disease outbreak risk (Heron et al. 2010)
Macroalgae	Macroalgal cover can be mapped either with spectral unmixing techniques (Goodman and Ustin 2007) or using benthic cover classification techniques (Mumby et al. 1997; Roelfsema et al. 2013)
Coral recruitment	Coral recruitment can generally not be mapped with remote sensing. However, factors that influence coral recruitment and post-settlement survival, and which can be mapped with remote sensing, include suitable settlement substrate (Mumby et al. 2004b), suspended sediment concentration (Hoitink and Hoekstra 2005) and others
Fishing pressure	While fishing pressure cannot be directly mapped using remote sensing, proxy variables such as distance to settlements, port infrastructure or markets may provide quantification of relative fishing pressure between locations (Rowlands et al. 2012)

5.3.1 Mapping the Seafloor

Both the three-dimensional structure of the seafloor and the biological composition of benthic coral reef habitats can be mapped accurately and at high spatial resolution using active sensors, including ship-based acoustic systems (single-beam, side-scan, multi-beam) and airborne lidar, as well as passive optical (multispectral, hyperspectral) remote sensing. The basic principle employed in acoustic systems is that a transmitter emits a sound pulse and measures the time until its reflection off the seafloor is registered by a receiver. Using an estimation of the speed of sound in water (Chen and Millero 1977), the two-way travel time of the pulse (from the emitter to the seafloor, then back to the receiver) is then converted to a one-way vertical distance that equals water depth (Riegl and Guarin 2013). Airborne lidar operates using a similar principle. A short laser pulse is emitted, and the time until its reflection off both the sea surface and the seafloor is measured. The two-way travel times are then converted to one-way vertical distances, and the difference between the distances to the sea surface and the seafloor equals water depth (Purkis and Brock 2013). Given the positional accuracy obtainable with kinematic GPS and the timing accuracy of laser pulse emission and return registration, airborne lidar can achieve typical positional accuracies of seafloor points of 15 cm vertically and 1 m horizontally (Purkis and Brock 2013). The energy contained in the laser pulse is lost through refraction and backscattering at the sea surface, as well as absorption and scattering at the seafloor and in the water column, thus limiting the depth to which a signal reflected off the seafloor can reliably be detected to 60 m even in exceptionally clear water, and much less in more turbid conditions. This stands in contrast to the energy in the sound wave emitted by acoustic instruments, the attenuation of which is largely determined by frequency while being relatively insensitive to turbidity, and which can easily reach the seafloor at any depth relevant to studies of coral reefs (Tolstoy and Clay 1966). Typical positional accuracies of seafloor observations from multi-beam acoustic instruments are in the order of a few centimeters vertically and ~0.5 m horizontally (Ernstsen et al. 2006).

As an alternative to active sensors, the use of multi- or hyperspectral data to derive maps of bathymetry also has a long history (Lyzenga 1978; Jupp 1988; Bierwirth et al. 1993; Stumpf et al. 2003; Dekker et al. 2011). The most commonly used empirical methods rely on field observations for local calibration of model coefficients, and apply simplifying assumptions to constrain what is an inherently underdetermined problem (Lee et al. 1998). More recently, methods that do not require coincident field data for calibration have been developed for hyperspectral remote sensing data. Pioneered by Lee et al. (1998, 1999), these methods use radiative transfer models (RTM) to simulate the above-water spectral reflectance that would be observed in atmospherically corrected satellite data, given a set of inputs that include (or can be used to derive) water depth, water optical properties, and seafloor spectral reflectance (Mobley et al. 2005). The use of these methods with multispectral data is a promising but little explored approach to bathymetry mapping (Hedley et al. 2012), with typical vertical accuracies of ~1–2 m in optically shallow waters (Dekker et al. 2011). Although bathymetry per se is not an important factor determining coral reef resilience, it may influence several of the identified resilience indicators such as coral diversity and recruitment through its influence on local current patterns. Thus, bathymetry and its derivatives can serve as useful spatial proxies for a range of variables that cannot be reliably mapped (Pittman and Knudby 2014). For example, measures of seafloor structural complexity, derived from seafloor terrain models, have repeatedly been shown to function as a key predictor of several reef resilience indicators such as herbivore biomass and coral abundance (Pittman et al. 2009; Knudby et al. 2010a, b; Pittman and Brown 2011).

Although acoustic and lidar-based methods are also rapidly developing to provide more detailed information on benthic cover (Park et al. 2010; Foster et al. 2013; Pittman et al. 2013), derivation of information on coral reef biota primarily relies on passive optical remote sensing, from which reef geomorphological zonation (Smith et al. 1975; Andréfouët and Guzman 2005; Purkis et al. 2010) and benthic cover types (Ahmad and Neil 1994; Green et al. 1996; Mumby et al. 1997; Newman et al. 2007; Phinn et al. 2012) can be mapped. The number of benthic cover types that can reliably be distinguished using passive optical remote sensing methods depends on the platform and the sensor type, the depth and optical properties of the water, as well as the inherent spectral separability of the benthic cover types in question. Collectively research in this field suggests that high sensor spatial resolution (Andréfouët et al. 2003; Mumby and Edwards 2002), high sensor spectral resolution (Capolsini et al. 2003), and the presence of one or more bands operating in the 400–500 nm (blue) spectrum (Hedley et al. 2012), in addition to suitable environmental conditions (limited specular reflection off the sea surface, clear and shallow water) are important for production of detailed and accurate map products. Notable recent methodological developments have included object-based (Roelfsema et al. 2013) and semi-automated (Suzuki et al. 2001) delineation of geomorphology, a shift from per-pixel to object-based classification of benthic habitat (Leon and Woodroffe 2011; Phinn et al. 2012; Roelfsema et al. 2013), and multi-image approaches to improve map accuracy (Knudby et al. 2014). For example, to overcome limitations of any one technique, Costa and Battista (2013) developed a novel, semi-automated object- and pixel-based technique to map coral reefs in the Caribbean from multibeam echo sounder imagery. They produced maps with high accuracy (74–97 %) for geomorphological types, detailed biological cover types and live coral cover.

Due to the widespread use of fractional live coral cover as an indicator of reef health, methods have also been developed to map this biological variable (Hochberg et al. 2003; Goodman and Ustin 2007; Joyce et al. 2013; Mumby et al. 2004b). These methods typically rely on spectral derivative or unmixing approaches applied to airborne hyperspectral data, and are most successful in clear shallow water with few spectrally similar non-coral benthic cover types (Dekker et al. 2011). Individual coral species cannot generally be distinguished with remote sensing (Hochberg and Atkinson 2000; Hochberg et al. 2003), except in rare circumstances (Purkis et al. 2006). As a result, coral diversity or the presence of stress-resistant corals can also not be directly inferred from remote sensing data. Although it has been demonstrated that corals affected by disease have distinct spectral reflectance characteristics when measured in-situ (Anderson et al. 2013), coral disease is also unlikely to be detectable with existing remote sensing instruments, as is coral recruitment. As such, remote sensing is not capable of directly mapping any of the four aspects of the coral fauna identified as important resilience indicators. Estimates of macroalgal cover, on the other hand, can be derived from the same spectral unmixing approaches used to map live coral cover (Goodman and Ustin 2007; Lee et al. 1999; Hedley 2013).

5.3.2 *Mapping the Water Column*

In addition to mapping the seafloor, remote sensing can be used to characterize the physical environment surrounding a reef ecosystem. Applications of satellite remote sensing have been demonstrated for mapping surface layer concentrations of chlorophyll (Morel and Prieur 1977; Moses et al. 2009), coloured dissolved organic matter (CDOM) (Morel and Gentili 2009) and suspended sediment (Globcolour 2008), as well as sea surface temperature (SST) (Maina et al. 2008; McClanahan et al. 2007; Strong et al. 2000). SST has routinely been mapped from satellite data since the early 1970s, with a globally consistent temperature records available from 1981 (Casey et al. 2010). The accuracy of temperature estimates has improved through time with development of increasingly sophisticated sensors (e.g. MODIS, AATSR). The long global SST data record allows remote sensing to directly map the historical temperature variability of coral reef sites, one of the factors identified as important for coral reef resilience. Surface layer chlorophyll concentration was first mapped on a global scale by the CZCS sensor (1978–1986), and has been continuously mapped since first operation of the SeaWiFS sensor (1997–2010), with additional products available from the MERIS (2002–2012) and MODIS (2000–present) sensors. Derivation of near-surface chlorophyll concentration is based on observations of ocean colour, with algorithms for chlorophyll concentration retrieval differing between sensors, and as a result of regional optimization (O'Reilly et al. 2000; Curran and Dash 2005). Ocean colour is similarly the basis for operational algorithms used to derive concentrations of CDOM (Morel and Gentili 2009) or CDM (which includes CDOM as well as coloured detrital materials) (Siegel et al. 2005), as well as suspended sediment (Globcolour 2008). While ocean colour and SST algorithms are now routinely applied to data from several satellite sensors, producing freely available data with daily near-global coverage at 4 km spatial resolution (NASA 2014), it is important to note the algorithms employed were developed and calibrated for Case I waters (Jerlov 1968), oceanic waters where ocean colour is negligibly influenced by terrigenous particles, and where the spectral radiance recorded by satellite sensors is not influenced by seafloor reflection. Application is thus questionable for coral reef environments (Ouillon et al. 2008), where RTM inversion algorithms hold the greatest promise for mapping water optical quality (and constituents) from space (Giardino et al. 2007). Combinations of airborne lidar and hyperspectral data have also shown great promise for direct derivation of water quality parameters including chlorophyll (Aitken et al. 2010) and suspended sediment concentrations (Epps et al. 2010), the latter of which can also routinely be mapped with acoustic methods (Bunt et al. 1999; Hoitink and Hoekstra 2005).

5.3.3 Mapping and Modeling the Seascape Context

The ability of a coral reef to recover after disturbance and the rate and trajectory of that recovery will be influenced by its connections with the surrounding seascape (Mumby and Hastings 2008; Jones et al. 2009). Connectivity is critical for replenishment of corals, fishes and other species that form coral reef ecosystems, but can also be linked to the spread of pollution, invasive species and diseases (Hughes et al. 2010). In coral reef ecology, connectivity is usually studied with particle dispersal models where the focus is on identifying the distribution and movement pathways of larval transport. Progress in larval connectivity modeling has been thoroughly reviewed in the existing literature (Jones et al. 2009). Here we focus on how remote sensing data can help to identify coral reefs that are structurally connected to other benthic habitat types by the movement of post-settlement fishes.

The knowledge of ecological connectivity between coral reefs and neighboring habitat types such as seagrasses and mangroves is not new (Parrish 1989), yet it is only recently that this inter-habitat connectivity has been linked to reef resilience (Mumby and Hastings 2008; Olds et al. 2013). Underwater observations of the spatial patterns of fish distributions on coral reefs near and far from 'nursery seascapes', such as mangroves and seagrasses, have highlighted the importance of structural connectivity (Grober-Dunsmore et al. 2009; Nagelkerken et al. 2012). For example, in Belize, Mumby et al. (2004c) found that mangroves strongly influence the community structure of fish on neighbouring coral reefs and boosted biomass of some reef-associated fish species. Subsequent simulations indicated that enhanced herbivory by parrotfishes on deeper reefs near mangroves resulted in greater coral recovery from intense hurricanes, whereas reefs without ecosystem connectivity had lower capacity for recovery (Mumby and Hastings 2008). In the Bahamas, high levels of parrotfish herbivory led to a twofold increase in coral recruitment (Mumby et al. 2007a, b). Thus, for some reefs, connectivity to mangroves infers greater resilience to disturbance. A similar mangrove-enhanced trophic cascade that reduced algal cover and enhanced coral recruitment and reef resilience was identified in eastern Australia (Olds et al. 2013).

In landscape ecology, structural connectivity is measured by the spatial arrangement of patches in the seascape using a wide range of spatial pattern metrics from simple proximity measures (i.e. nearest neighbor distances) to more complex graph theoretic approaches (Calabrese and Fagan 2004). Although rarely applied to coral reef ecosystems, these metrics can be readily applied to benthic habitat maps to calculate a spatial proxy for functional connectivity (Grober-Dunsmore et al. 2009; Wedding et al. 2011). Very little is currently known about the relevance of seascape pattern for reef resilience or the ecological consequences of movements on populations and ecosystem patterns and processes. The landscape ecology approach, with its focus on spatially-explicit pattern-process analysis, offers great potential to support ecologically effective strategies for restoring and optimizing connectivity. For example, when marine reserves are placed to protect well connected habitats then the rate of recovery will likely be enhanced (Olds et al. 2013). Landscape ecology concepts and tools together with remote sensing data are set to make major contributions to the new study of spatial resilience (Nyström and Folke 2001; Cumming 2011a). Spatial resilience focuses on the importance of location, connectivity, and context for resilience, based on the idea that spatial variation in patterns and processes at different scales both impacts and is impacted by local system resilience (Cumming 2011b).

5.3.4 Mapping Threats and Stressors

Natural physical impacts on the reef environment include long-term wave exposure as well as extreme events such as hurricanes. Long-term wave exposure can be derived at coarse resolution (~25 km) directly from wind wave models (Tolman and Alves 2005) or extracted from climatological reanalyses (Caires and Sterl 2005), while hurricane exposure can be depicted statistically from historical hurricane data (Edwards et al. 2011). However, the anthropogenic physical impacts identified as an important factor influencing reef resilience (McClanahan et al. 2012) include damage from reef trampling and diving, ship groundings and coral mining, all of which are difficult or impossible to map directly using remote sensing. While these may correlate with proxies such as the distance to human settlements, data from interviews or field observations are more likely to provide spatial information on anthropogenic physical impacts. Similarly, fishing pressure, another important resilience indicator, cannot be mapped directly using remote sensing, although proxies such as distance to settlements, port infrastructure or markets may be used to provide quantification of relative fishing pressure between different sites (Rowlands et al. 2012).

5.4 Direct Monitoring

An alternative to mapping resilience indicators is to map and monitor relevant reef state variables such as the fractional cover of corals and macroalgae through a period before, during and after a major disturbance, to assess the state changes caused by the disturbance (resistance), as well as the time taken to reach the pre-disturbance state (recovery), thus providing a direct measure of engineering resilience to the specific disturbance event. In-situ hyperspectral measurements have shown that significant differences in spectral reflectance exist between corals and macroalgae (Holden and LeDrew 1998; Myers et al. 1999; Hochberg and Atkinson 2000), and studies employing airborne hyperspectral data have shown the potential to utilize these differences for monitoring. However, the limited availability of hyperspectral data means that any operational monitoring of live coral and macroalgal cover must rely on multispectral satellite data, at least until the launch of EnMAP expected in 2015. Encouragingly, a few studies using time series of Landsat TM/ETM+ data (Dustan et al. 2002; Phinney et al. 2002; Palandro et al. 2003) have demonstrated correlations between measures derived from multi-spectral satellite data and transitions from coral- to algal-dominated reef benthos over a ~15 year period for two reef sites in Florida. Although no remote sensing-based coral reef monitoring system exists and successful replication of these studies elsewhere is needed, these studies suggest that monitoring of live coral and macroalgal cover may be possible both retrospectively and operationally. The potential of this approach will be further improved by availability of data from sensors with improved spatial and spectral resolutions such as Landsat 8 OLI and Sentinel-2 MSI (launch expected 2014).

5.5 Spatial Distribution Modeling

For resilience indicators that cannot be mapped directly, spatial distribution modeling can be used to predict distributions across geographical space by extrapolating the statistical relationship between remotely sensed environmental variables and georeferenced field observations (Franklin 2009; Pittman and Brown 2011; Knudby et al. 2013a). Spatial distribution modeling relies on a two-step process. In the first step, a predictive model is calibrated based on an observed statistical relationship between the resilience indicator in question and one or more environmental variables with known spatial distribution. For example, several studies have demonstrated statistical relationships between fish herbivore biomass and environmental variables such as geomorphologic zone (Friedlander and Parrish 1998), seafloor structural complexity (Pittman et al. 2009), and habitat heterogeneity (Purkis et al. 2008). Model calibration requires georeferenced field observations of the resilience indicator in question, as well as maps of the environmental variables thought to structure its spatial distribution. In the second step, the predictive model is applied to the environmental variable maps to produce per-pixel predictions of the resilience indicator (Franklin 2009; Knudby et al. 2013a). Spatial distribution modeling has been extensively used to map distributions of species and habitats in both terrestrial (Guisan and Thuiller 2005; Elith and Leathwick 2009) and marine environments (Cheung et al. 2008; Pittman et al. 2009; Knudby et al. 2013b). Even when maps of the relevant environmental variables are not available or possible to derive from remote sensing, spatial interpolation/extrapolation methods can be used to predict the spatial distribution of resilience indicators (Knudby et al. 2013a; Mumby et al. 2013b). Spatial predictive modeling has also been used to forecast changes to the quality of fish habitat due to declines in the topographic complexity of coral reef terrains (Pittman et al. 2011). Such scenario modeling can be used to investigate the potential impact of multiple stressors such as thermal stress, ocean acidification, storms and land-based sources of pollution that can adversely influence the structural complexity of coral reefs through coral mortality, mechanical breakage, reduced growth and survival and erosion (Graham and Nash 2013). These questions address crucial knowledge gaps in our understanding of climate impacts on coral reef fish, fisheries and coastal livelihoods in a changing world (Wilson et al. 2010). Linkages between remotely sensed measures of reef complexity and the structure and function of coral reefs may provide a rapid and cost-effective way to assess the spatial complexity of reef resilience and predict impacts to a wide range of goods and services provided by coral reefs (Pratchett et al. 2014).

5.6 Case Studies

With a combination of remote sensing and spatial distribution modeling, it is possible to map biological, physical and human factors known to influence coral reef resilience, albeit with varying and often unknown degrees of accuracy. In addition to factors related to resilience (resistance and recovery), remote sensing has successfully been used to map the exposure to climate-driven disturbance experienced at different reef sites. Few examples of such resilience-related mapping exists for tropical coral reefs, here we present three case studies that represent the leading edge of this new frontier in marine spatial ecology.

5.6.1 High-Resolution Mapping of Selected Resilience Indicators in Fiji

In their study of the traditional fisheries management area of Knudby et al. (2013a) used spatial predictive modeling to map four of the resilience indicators listed in Table 5.1: stress-tolerant coral taxa, coral diversity, herbivore biomass, and coral recruitment (quantified as the density of juvenile corals), as well as herbivore functional group richness and the live cover of corals and coralline algae, both of which may also influence reef resilience. The range of resilience indicators mapped in this study was limited by available field data, and focused on indicators that could not be mapped with existing methods but were thought to exhibit significant small-scale variability within the study area, a >260 km^2 complex reef system. Georeferenced field observations of the resilience indicators were derived from the Wildlife Conservation Society's reef monitoring program, from which data for 66–72 sites were available depending on resilience indicator. IKONOS and QuickBird satellite images were used to derive maps of bathymetry, geomorphology and reef benthos (Knudby et al. 2011), which in turn were used to map a coral cover index, seafloor structural complexity and habitat richness, all of which were calculated at spatial scales ranging from the smallest possible (individual 16 m^2 pixels) to those incorporating large parts of the neighbouring landscape (>3 km^2). Spatial layers quantifying distances to land, nearest seagrass bed, and nearest mangrove stand were also derived, as was a layer describing the conservation status of different areas (unprotected, tabu, no-take reserve). Two types of spatial predictive models were then used to produce maps of each resilience indicator. Regression modeling was conducted with random forest (Breiman 2001), a non-parametric tree-based ensemble classifier that predicts per-pixel values for each resilience indicator based on their statistical relationship with the

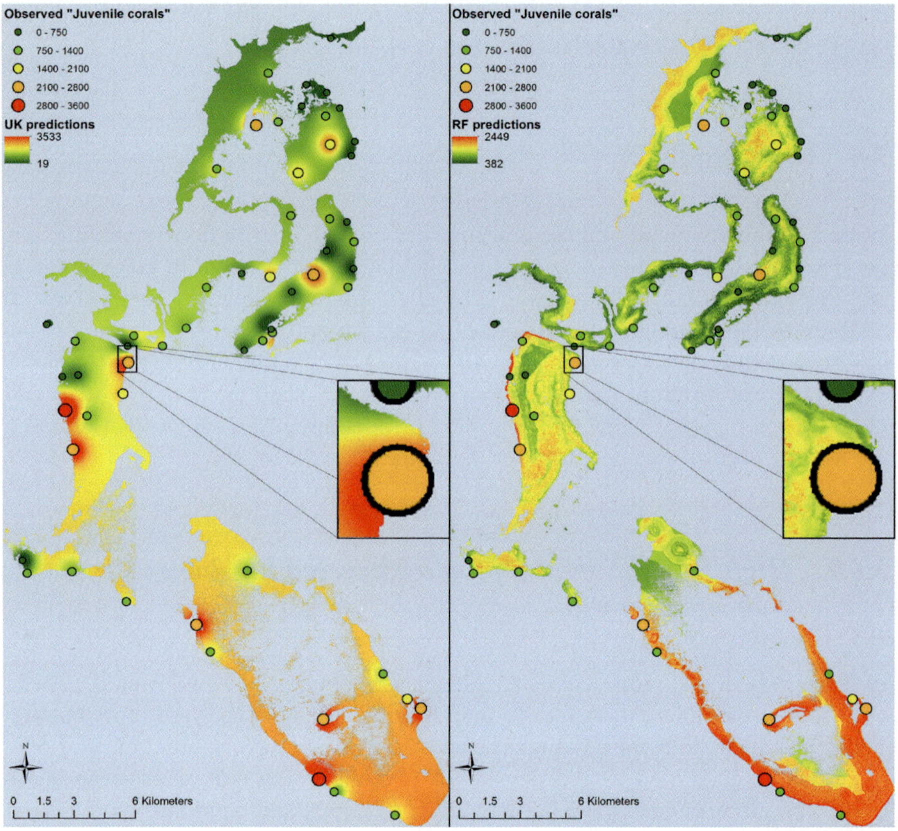

Fig. 5.2 Predictions of the spatial distribution of juvenile coral density across the Kubulau traditional fisheries management area, Fiji (Knudby et al. 2013a). *Points* indicate field observations while the coloured background illustrates the model predictions. Universal Kriging (*UK*) predictions are shown on the *left*, Random Forest (*RF*) predictions are shown on the *right*

spatial data layers. Spatial interpolation/extrapolation was conducted using universal kriging (Krige 1951), a geostatistical method that does not utilize the spatial data layers but rather fits a deterministic trend to the large-scale spatial variation in the value of each resilience indicator, and then uses locally optimized spatial interpolation of the residuals to account for small-scale spatial variations (Goovaerts 1997).

The results from Knudby et al. (2013a) suggested that, based on the data available, the spatial distribution of stress-tolerant coral taxa and herbivore biomass could be reasonably well predicted using the random forest model, while the density of juvenile corals could only be poorly predicted using the universal kriging model and coral diversity was essentially unpredictable. An example output, the predicted spatial distribution of juvenile coral density from each of the two models, is provided in Fig. 5.2. Numerous factors contribute to limited predictability of field-measured resilience indicators in this study, including mismatch between the spatial scale of field sites (belt transects, point intercept transect and quadrats, all covering different areas) and that of the satellite data (4 m for IKONOS, 2.4 m for QuickBird), imperfect georeferencing of both field sites and satellite data, limited sample distribution for both resilience indicators and environmental predictors, and, importantly, inability to incorporate information on past disturbance history, direct human impacts from fishing and other reef use, the influence of source populations and ocean and tidal currents on coral and fish recruitment, and the direct influence of wave exposure on post-settlement coral survival and growth. Some of these limiting factors can be easily addressed in future studies, while others will require substantial effort in data collection and analysis.

In addition, the results illustrated that the choice of modeling method, beyond producing different estimates of prediction error, also produced markedly different mapped predictions for each resilience indicator. As would be expected, predictions based on the random forest model tightly matched distributions of influential environmental predictors and thus changed quickly across steep environmental gradient such as reef edges, while predictions based on universal kriging varied more smoothly through space but were also better able to account for spatial clustering of high or low resilience indicator values when these were unrelated to the reef environment. The relative strengths of the two modeling methods suggest a potential for error reduction with hybrid models such as regression kriging (Hengl et al. 2004).

5.6.2 High-Resolution Mapping of a Coral Reef Resilience Index in Saudi Arabia

Focusing on coral reefs in the Saudi Arabian Red Sea, Rowlands et al. (2012) mapped aspects of coral reef resilience using a geographic information system (GIS) approach. A novel metric termed the *Remote Sensing Resilience Index* (*RSRI*) was developed that quantified and mapped important factors influencing reef resilience in the Saudi Arabian Red Sea (Fig. 5.3). *RSRI* was calculated at 1 km^2 ground resolution across a geographic extent of 20,000 km^2. The spatial resolution and extent are appropriate for addressing both local and regional management concerns. *RSRI* maps three key spatial gradients: (1) human use gradients; (2) physical gradients; and (3) biological and sediment gradients. These were considered within two classes of resilience indicators: (1) those which positively affect the reef community, termed *landscape indicators* and incorporated into the Coral Resistance Index (*CRI*); and (2) indicators that negatively impact reef communities, termed *stress indicators* and incorporated into the Coral Stress Index (*CSI*) (see Box 5.1).

Meter-scale habitat and water depth mapping, satellite derived SST, night time imagery, and spatial modelling were all used to develop map-based indices of

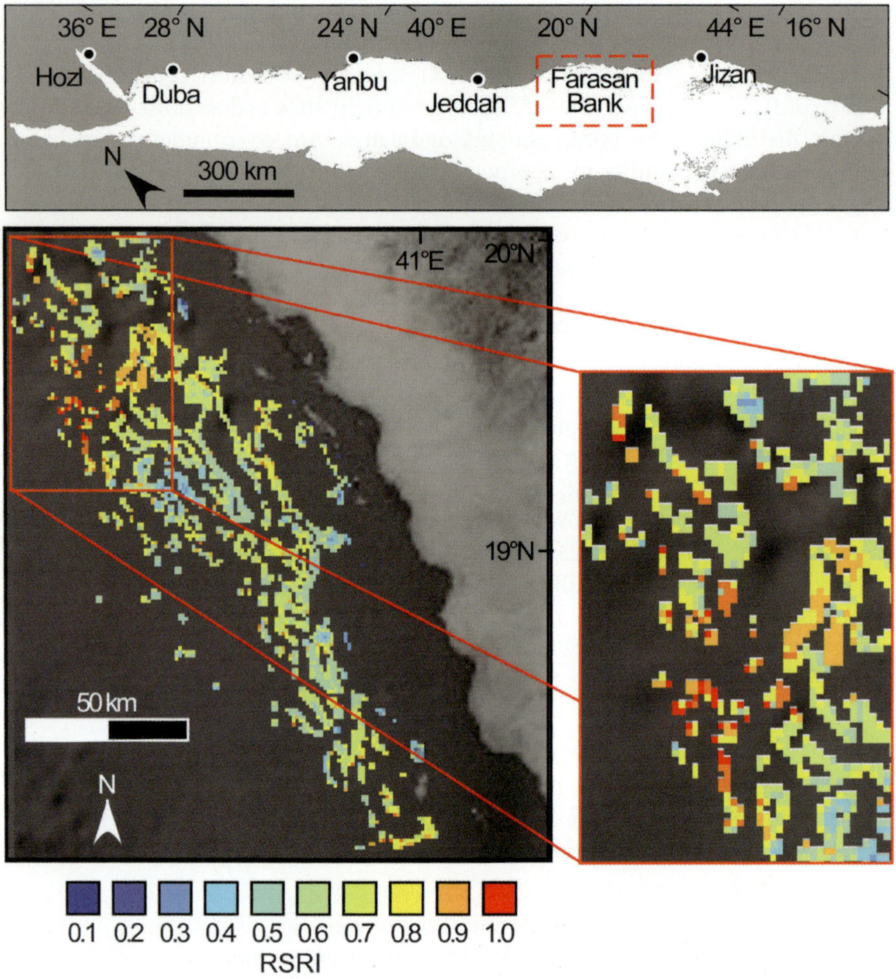

Fig. 5.3 The remote sensed resilience index (*RSRI*) mapped for the outer Farasan Banks in the Saudi Arabian Red Sea. The inset image to the right highlights a region with many high RSRI grids (Adapted from Rowlands et al. 2012)

important resilience indicators. Data sources included both freely available satellite data (NASA, Google Earth), and archived and tasked QuickBird satellite (DigitalGlobe Inc.) imagery funded by the Khaled bin Sultan Living Oceans Foundation. In the original formulation of *RSRI* three landscape indicators were considered: (1) live coral abundance; (2) framework abundance; and (3) water depth variability, and three stress indicators were considered: (1) fishing; (2) industrial development; and (3) temperature stress. Subsequent analysis by Rowlands (2013) used a more refined approach, using maps of water depth and wave exposure to incorporate the vulnerability of sites to coral bleaching events and modelling the distribution of high coral sites with greater certainty.

Box 5.1. Equations used to calculate the Remote Sensing Resilience Index (RSRI)

The remote sensed resilience index (*RSRI*), is calculated according to Eq. 5.1 to be:

$$RSRI = \frac{CRI_n - CSI_n}{\max(CRI_n - CSI_n)} \tag{5.1}$$

Positive influence is provided within the coral resistance index (*CRI*) shown in Eq. 5.2, while negative influence is provided by the coral stress index (*CSI*) shown in Eq. 5.3. All indices are normalized to a 0–1 scale.

$$CRI_n = \frac{((Lw_1 \cdot LF_1) + \ldots .(Lw_n \cdot LF_{\mathrm{n}}))}{\max((Lw_1 \cdot LF_1) + \ldots .(Lw_n \cdot LF_n))} \tag{5.2}$$

$$CSI_n = \frac{((Sw_I \cdot SF_1) + \ldots .(Sw_n \cdot SF_n))}{\max((Sw_1 \cdot SF_1) + \ldots .(Sw_n \cdot SF_n))} \tag{5.3}$$

LF and *SF* represent the map-based landscape and stress resilience indicators. *Lw* and *Sw* represent weighting factors, attributed to each resilience indicator and adjusted to model a range of resilience scenarios.

By adjusting the weight attributed to each input index, a number of resilience scenarios are considered for Saudi Arabia. Noting that resilient reefs can act as both repositories and also sources of reef organisms enabling recovery after a chance mortality event (Nyström and Folke 2001; Pinsky et al. 2012), several scenarios were modeled to explore the role of local feedback. Changing the relative importance of input indices, and calculating with or without local feedback produced nonlinear responses in *RSRI*, emphasizing the need to understand spatial dynamics when assessing spatial resilience.

The study showed that seascape morphology exerts a strong controlling role on reef resilience through the abundance of habitat, topography, and exposure to anthropogenic and physical stress. Understanding of the processes creating morphology in the Saudi Arabian Red Sea and elsewhere will therefore be helpful in targeting research and management towards the most resilient reef sites.

5.6.3 *Mapping Exposure of Coral Reefs to Climate-Driven Environmental Stress*

In addition to spatial information on the factors that enable coral reefs to resist and recover from climate-driven impacts, managing for coral reef resilience requires a better understanding of the magnitude and spatial distribution of threats, their interactions, and respective roles they play in impacting coral reefs. Threats maps and ecosystem vulnerability studies in historical and future time scales and at multiple spatial scales have therefore become an important research focus (Bruno et al. 2007; Maina et al. 2008, 2011; Selig et al. 2010; Burke et al. 2011; Gove et al. 2013; van Hooidonk et al. 2013, 2014). Often these regional and global scale models utilize modelled and remotely sensed data as inputs, highlighting the increasingly important role of remote sensing in decision support. Among these, multivariate stress models (Maina et al. 2008, 2011) have used a relatively more complicated multivariate approach including fuzzy logic techniques of integrating multiple variables and stress proxies derived from remote sensing data (Logan et al. 2014). Fuzzy logic provides a platform for translating the absolute values of environmental variables into indices that represent the exposure of coral reefs to the respective threats. Such procedures are informed by the perceived role of the environmental variable on coral health, as supported by empirical findings or expert knowledge. Fuzzy logic models are also easy to apply as they allow for use of fuzzy operators, which are useful in synthesizing the partial exposures into an overall exposure or vulnerability metric. Moreover, this modeling framework is relatively flexible for modifications to accommodate any lurking variables and new science.

In their global analysis, Maina et al. defined three broad categories of environmental stressors. These included: (1) radiation stress indicators which are considered to be the primary climatic drivers of coral reef exposure (temperature, UV light and doldrums); (2) stress-reinforcing indicators (sedimentation and eutrophication) which have been shown to reinforce radiation stressors; and (3) stress-reducing indicators (temperature variability and tidal amplitude) which act to reduce or balance thermal stress. Following these definitions, a systems analytical approach and fuzzy logic functions were used to represent the interactions among radiation stress variables, stress reinforcing variables and stress reducing variables. For each of these broad stress categories, constituent variables were derived from various sensors within the ocean realm.

SST time series from 1982 to 2010, produced using data from NOAA's Advanced Very High Resolution Radiometer (AVHRR) sensors, were used to produce global maps of mean absolute temperature, mean maximum temperature, and temperature variability. Thermal stress metrics, including time series of weekly SST anomalies (WSSTAs), defined as the weekly averaged temperature in excess of 1 °C or more above that week's long term average value, and thermal stress anomalies (TSAs), defined as the temperature excess of 1 °C or more above the climatologically warmest week of the year were computed. Cumulative estimates of TSAs and WSSTAs were calculated from annual averages in the 27 year time series. Further, for each year, a maximum duration (in weeks) that WSSTA and TSA were greater than or equal to 1 °C were computed and averaged over 27 years. These two cumulative metrics, the mean annual cumulative and mean yearly maximum duration, represent the characteristic magnitude and duration of the anomalies at a given location, which are important predictors of coral stress (Bruno et al. 2007; Selig et al. 2010). UV-erythemal (biologically damaging) irradiance at the Earth's surface was mapped in a 1 by 1.25° grid using data from 1996 to 2001 from the total ozone mapping spectrometer (TOMS) (Herman et al. 1999; Vasilkov et al. 2001). Doldrums (i.e. wind conditions with a daily mean of less than 3 ms^{-1}) were computed using daily averaged wind speeds (2000–2009) and the averaged 10-year mean monthly wind speeds (1995–2004). To estimate the magnitude and consistency of wind regimes in a given location, a doldrums metric was computed by taking the annual average maximum number of days that wind speeds were greater than 3 ms^{-1} over 10 years (2000–2009) and multiplying this by the 10-year mean monthly average. To represent tidal conditions, modelled tidal data (Le Provost et al. 1998) was used where tidal ranges were computed as the long term averaged difference between the weekly maximum and minimum simulated tidal heights.

As outlined earlier, concentrations of nutrients and suspended sediment are operationally mapped from ocean colour data, but algorithms are calibrated for Case I waters and fail in turbid coastal waters (Morel and Bélanger 2006) as well as in shallow areas where sunlight reflected off the seafloor increases the reflectance of the water/seafloor system, leading to overestimations of suspended particles (Boss and Zaneveld 2003). Given these problems, until special algorithms that take into account the complexity in coral reef areas (e.g. Ouillon et al. 2008) are developed and incorporated in the standard processing chains of the current ocean color satellites, the usefulness of ocean color data for coral reef applications will remain limited (Boss and Zaneveld 2003; Mumby et al. 2004a). To overcome this, Maina et al. (2011) analyzed the global ocean color data such that appropriate algorithms were applied in the respective water types. Further they applied a depth flag of 30 m to remove shallow water before extrapolating values from nearby deep-water pixels into the shallow areas. While this product avoided the problems associated with high reflectivity of complex coastal areas, optical properties estimates in highly turbid coastal areas such as river mouths were underestimated (Gove et al. 2013).

Based on the above data sets, modeling of coral reef stress exposure involved two key steps. First, summaries of environmental layers comprised within each of the three categories of stressors were converted into partial exposures using fuzzy logic equations. Second, exposure layers for each of the three categories were synthesized using fuzzy-logic operators. The resulting multivariate exposure model provided estimates of coral reef exposure to climate stress worldwide where coral reefs are found.

The multivariate stress model by Maina et al. (2011) found high within-region variability, and spatial heterogeneity of exposure to radiation and both reinforcing and reducing stress categories. Furthermore, the magnitudes of radiation and reinforcing stressors were analyzed in order to facilitate appropriate management decisions. For instance, while reduction of climate-driven stress is impossible in the short term, management can act to reduce reinforcing stress on corals through a pollution reduction strategy. Findings from the analysis of the exposure model advanced coral threat mapping by providing evidence of spatial differences in exposure to multiple stressors, highlighting potential utility of the spatial adaptive management approach to coral reef conservation, and developing a frame work for exposure analysis applicable at local scales for implementation purposes. While the model has found application in social vulnerability analyses Cinner et al. (2013), its implementation for local scale coral reef management is yet to be realized. Ideally, application of the model at the local scale would require downscaling where in situ measurements of the constituent variables in the model would be used. Furthermore, other useful information not currently incorporated in the model, for example habitat typologies and socio-economic indicators, would improve the models performance and its utility in aiding management decisions.

5.7 Spatially Explicit Resilience Modeling

Informative as they may be, maps that quantify either exposure to stress and disturbance or factors that influence resilience do not by themselves allow quantification of coral reef resilience. Engineering resilience can be quantified as the magnitude of deviation from, and speed of return to, the stable state following a disturbance, while ecological resilience can be quantified as the amount of disturbance the system can be exposed to without moving beyond its current domain of attraction and transitioning to another stable state (Holling 1996). To properly quantify resilience we thus need mechanistic models that allow simulation of how key ecosystem state variables (e.g. coral and algae cover) through interaction with other components of their ecosystems respond to both chronic stress and periodic disturbances through time (Mumby and Hastings 2008). Parameters used in such models should at a minimum include the critical state variables, factors they interact strongly with and which therefore are important for determination of resilience, and realistic stress and disturbance regimes that the ecosystems can be subjected to in model simulations.

An early example of a coral reef ecosystem simulation model was developed by McClanahan (1995). Although not explicitly developed to study coral reef resilience, the model quantifies a set of core ecosystem state variables and their interactions using an energy-circuit diagram (Odum 1983). Model coefficients that quantify development of individual ecosystem components and interactions between different components are derived from field observations. Using fixed time steps the model can be run until negligible change in state variables is seen and a *stable state* has emerged, using either a business-as-usual scenario or one of several simplified management scenarios (e.g. allowing fishing activity to remove all piscivores, or both piscivores and herbivores). Although results were demonstrated by McClanahan (1995) to broadly match field observations from a number of Kenyan reefs, the model suffers from several limitations, including fixed instantiation (the starting values of all state variables), no inclusion of large-scale disturbance events, no inclusion of information on recruitment, and a lack of spatiality that prevents modeling of connectivity.

A more recent model developed by Mumby et al. (2007a, b) similarly allows simulation of states and interactions between core ecosystem variables, focusing on coral-algal dynamics. Allowing variable instantiation this model illustrates how, in the absence of an external disturbance regime, the initial state of the system determines whether it will gradually develop toward a coral-dominated or an algal-dominated stable state. The model also allows introduction of a stochastic disturbance regime (e.g. hurricanes causing 20 % coral mortality occurring on average every 20 years) as well as chronic stresses (e.g. annual removal of a certain amount of fish herbivore biomass) to examine the effect of pulse disturbances and chronic stresses on coral-algal dynamics. The use of a stochastic disturbance regime additionally allows multiple model runs with identical instantiations to produce different results, which in turn enabled Mumby et al. to define a useful measure of coral reef resilience – the probability that the reef would be in a coral-dominated state after a given time period. This model was recently applied to the Belize Barrier Reef System (Mumby et al. 2013b), with a spatially explicit disturbance regime defined on a 16 km^2 cell-by-cell basis using information on hurricane tracks from 1909 to 2008 and predictions of future sea surface temperatures derived from two climate scenarios. Reef state variables were derived from interpolation of field observations. This spatially explicit model instantiation and disturbance regime definition allowed exploration of differences in resilience between different parts of the studied reef system, and also allowed spatially explicit evaluation of potential management interventions (e.g. protection of parrotfishes was shown to be important for reef resilience in parts of central Belize, but less so in southern Belize where reef resilience was already high).

Another model, developed by Melbourne-Thomas et al. (2011) specifically as a decision-support tool, provides a core of ecosystem variables and interactions similar to the previously discussed models but is inherently spatial in nature and also specifically incorporates connectivity by larval transport (though not by movement of adults). Larval dispersal is simulated using lower-resolution hydrodynamic and particle-tracking modules and includes basic larval behavior (i.e. coral, fish and urchin larvae behave differently). Once arrived at a sink reef location, larval settlement is limited by availability of settlement habitat and mortality of recruits is modeled differently than adult mortality during the first year after settlement. A simulation applied to the Meso-American Reef System, using the Millennium Coral Reef Mapping Project (IMaRS 2004) to define reef locations and spatially distributed field observations of reef state (Garcia-Salgado et al. 2006) to define initial values of state variables, showed that the model realistically captures ecosystem dynamics and reproduces known historical trajectories of state variables. This simulation illustrates the current state-of-the-art in resilience mapping and modeling, relying on a combination of spatially explicit data on state variables and disturbance regime parameters, and modeling of ecosystem dynamics through time to assess likely future ecosystem states, including the likelihood that part of or all of the reef ecosystem will flip to an alternate stable state.

5.8 Management Applications

In many parts of the world resource managers are operating with limited funds, and therefore require cost effective, reliable, spatially-explicit and easily interpretable information to help prioritize coral reef sites for risk assessments and conservation actions. Managers may be required to develop a conservation investment portfolio that prioritizes management actions to coral reefs of highest conservation status or exhibiting greatest resilience to stressors, or they may wish to spread investments over a wider range of risks that include reefs with low resilience but high importance to local industry. Reefs with lower resilience may have potential to rapidly regenerate to their former status if targeted action is taken to reduce stressors i.e. reducing runoff through watershed restoration. The great problem for current coral reef conservation is to operationalize our understanding of ecosystem resilience and apply it for management. Here we illustrate one approach to addressing this challenge with a decision-support framework for prioritizing coral reef units for conservation action, currently under development in the U.S. Virgin Islands (USVI).

Coral reefs across the USVI vary geographically in their diversity, structure, resilience and economic and aesthetic importance to people. To prioritize actions, effective management requires knowledge of the locations of the ecologically and economically most important coral reefs, their resilience to stressors and expectations for rapid recovery when management action is taken to reduce stressors. Interdisciplinary research is underway in the USVI as a collaboration between NOAA, The Nature Conservancy, University of the Virgin Islands, USVI government, local professional dive industry (dive operators and scientific divers) and other local partner agencies to develop a decision-support tool that integrates environmental data (remote sensing, ocean models and in-water biological surveys) with local expert knowledge from underwater observations, maps of stressors, and predictions of the locations and attributes of resilient reefs. To quantify structural connectivity, NOAA's benthic habitat maps are being used to rank reefs based on their proximity to seagrasses and mangroves. In addition, sites of ecological and biological significance are identified using spatial predictive models of fish species richness and known locations of fish spawning aggregations. Sites with high biodiversity have been associated with high topographic complexity. For unsurveyed areas, predictive modeling used bathymetry data as a proxy by classify the seafloor terrain into high, medium and low topographic complexity. These patterns will be combined with locations identified as important to local professional SCUBA divers (scientific and commercial dive industry).

The decision support framework (Fig. 5.4) will identify and rank the most important reefs for conservation concern in USVI. By evaluating resilience metrics and incorporating a resilience index into a comprehensive and transparent spatial matrix of site importance this project will objectively rank sites into several categories that will identify: Class A Reefs) the most important and best examples of coral reefs in the USVI; Class B reefs) coral reefs that have high potential to become Class A reefs with strategic conservation actions; Class C reefs) coral reefs

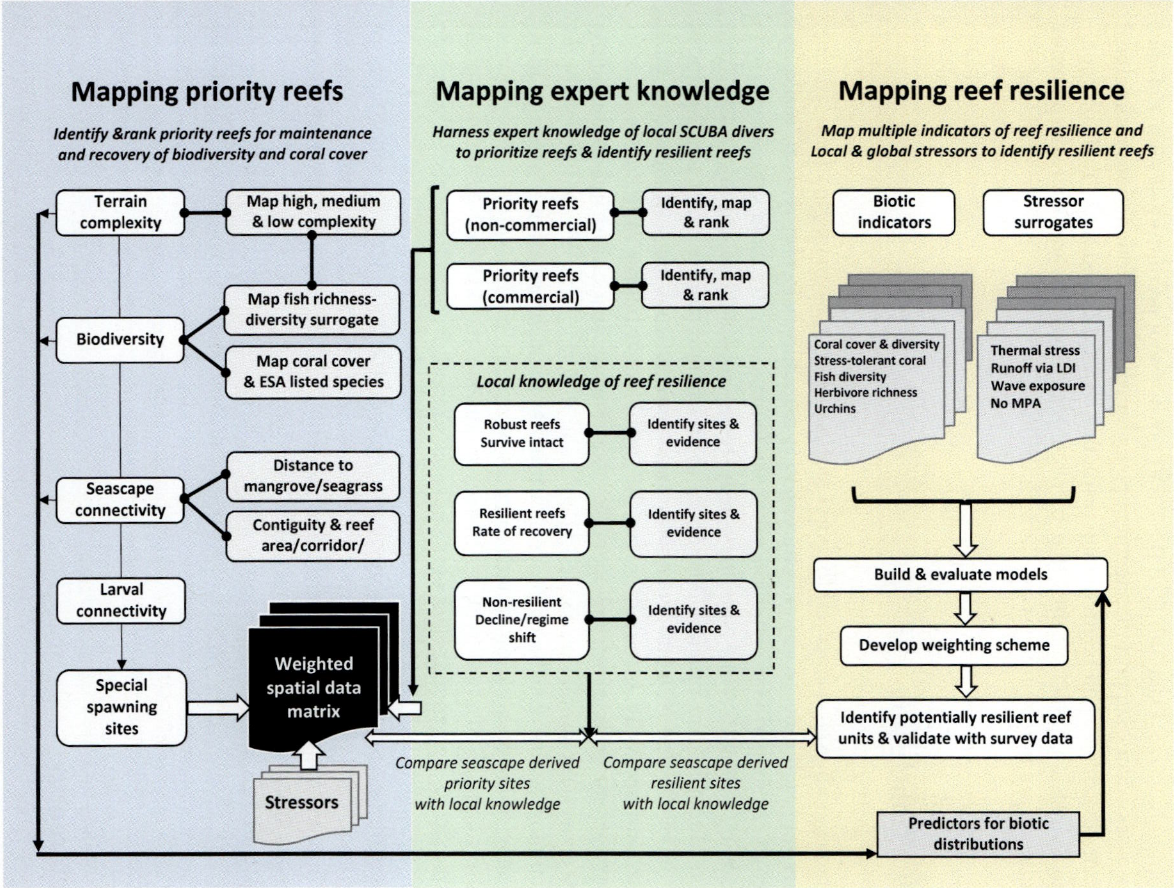

Fig. 5.4 Schematic illustrating the decision-support framework being developed to support resilience-focused coral reef management in USVI

that have some importance, but have low potential for recovery even with considerable conservation effort.

The framework is being designed to address multiple management objectives with a greater understanding of the spatial heterogeneity of reef resilience being a central objective. Data will feed the USVI government permitting system, enhance risk assessments, support management plans for MPAs, lay foundations for regional ocean governance and begin to map resiliency to identify and rank "reefs of hope". This synoptic product ensures that management processes and strategic planning decisions are guided by best available information.

5.9 Conclusion

A new paradigm has emerged for coral reef management – managing for resilience. This approach aims to sustain and enhance the ability of coral reefs to resist and recover from periodic disturbances made increasingly frequent and severe by a changing climate, while simultaneously being subjected to chronic stress from more direct human impacts. Managing for resilience requires, first and foremost, an understanding of what contributes to or detracts from the resilience of a coral reef, and secondly information on the spatial distribution of resilience across management areas of varying size. Remote sensing, spatial distribution modeling and ecosystem simulation modeling combine to form a trio of rapidly developing technologies that can be employed to provide such information. Although in its infancy, this combination of technologies is already being explored for regional-scale mapping of coral reef resilience, producing the kind of spatially explicit information demanded. Through decision-support frameworks this information can be translated into management action. Maps of reef resilience will show improvement as: (1) research sheds new light on the influence of a wider range of environmental factors on aspects of coral reef resilience; (2) remote sensing technology improves and distribution modeling matures; 3) ecosystem simulation models become increasingly adept at incorporating a wider range of species or functional groups as well as information on larval and adult connectivity, seascape context, disease and invasive species. Improved maps of resilience will enable better management on coral reefs.

References

Ahmad W, Neil D (1994) An evaluation of Landsat Thematic Mapper (TM) digital data for discriminating coral reef zonation: Heron Reef (GBR). Int J Remote Sens 15:2583–2597

Aitken J, Ramnath V, Feygels V, Mathur A, Kim M, Park JY, Tuell G (2010) Prelude to CZMIL: seafloor imaging and classification results achieved with CHARTS and the Rapid Environmental Assessment (REA) Processor. Proc SPIE 7695:76950S-1. doi:10.1117/12.851915

Anderson DA, Armstrong RA, Weil E (2013) Hyperspectral sensing of disease stress in the Caribbean reef-building coral, *Orbicella faveolata* – perspectives for the field of coral disease monitoring. PLoS One 8(12):e81478. doi:10.1371/journal.pone.0081478

Andréfouët S, Guzman HM (2005) Coral reef distribution, status and geomorphology-biodiversity relationship in Kuna Yala (San Blas) archipelago, Caribbean Panama. Coral Reefs 24:31–42

Andréfouët S, Kramer P, Torres-Pulliza D, Joyce KE, Hochberg EJ, Garza-Perez R, Mumby PJ, Riegl B, Yamano H, White WH (2003) Multi-site evaluation of IKONOS data for classification of tropical coral reef environments. Remote Sens Environ 88(1–2):128–143

Ban SS, Graham NA, Connolly SR (2014) Evidence for multiple stressor interactions and effects on coral reefs. Glob Chang Biol, pp 681–697 doi:10.1111/gcb.12453

Bierwirth PN, Lee TJ, Burne RV (1993) Shallow sea-floor reflectance and water depth derived by unmixing multispectral imagery. Photogramm Eng Remote Sens 59:331–338

Boss E, Zaneveld JRV (2003) The effect of bottom substrate on inherent optical properties: evidence of biogeochemical processes. Limnol Oceanogr 48:346–354

Breiman L (2001) Random forests. Mach Learn 45:5–32

Bruno JF, Selig ER, Casey KS, Page CA, Willis BL et al (2007) Thermal stress and coral cover as drivers of coral disease outbreaks. PLoS Biol 5:e124

Burke LK, Reytar MS, Spalding M, Perry A (2011) Reefs at risk revisited. World Resources Institute, Washington, DC

Bunt JAC, Larcombe P, Jago CF (1999) Quantifying the response of optical backscatter devices and transmissometers to variations in suspended particulate matter. Cont Shelf Res 19:1199–1220

Burkepile DE, Hay ME (2008) Herbivore species richness and feeding complementarity affect community structure and function on a coral reef. Proc Natl Acad Sci 105(42):16201–16206

Caires S, Sterl A (2005) A new nonparametric method to correct model data: application to significant wave height from the ERA-40 re-analysis. J Atmos Oceanic Tech 22:443–459

Calabrese JM, Fagan WF (2004) A comparison-shopper's guide to connectivity metrics. Front Ecol Environ 2:529–536

Capolsini P, Andréfouët S, Rion C, Payri C (2003) A comparison of Landsat ETM+, SPOT HRV, IKONOS, ASTER, and airborne MASTER data for coral reef habitat mapping in South Pacific islands. Can J Remote Sens 29(2):187–200

Carilli JE, Norris RD, Black BA, Walsh SM, McField M (2009) Local stressors reduce coral resilience to bleaching. PLoS One 4(7):e6324. doi:10.1371/journal.pone.0006324

Casey KS, Brandon TB, Cornillon P, Evans R (2010) The past, present and future of the AVHRR Pathfinder SST Program. In: Barale V, Gower JFR, Alberotanza L (eds) Oceanography from space: revisited. Springer, London. Springer Dordrecht Heidelberg London New York doi:10.1007/978-90-481-8681-5_16

Cheal AJ, MacNeil MA, Cripps E, Emslie MJ, Jonker M, Schaffelke B, Sweatman H (2010) Coral–macroalgal phase shifts or reef resilience: links with diversity and functional roles of herbivorous fishes on the Great Barrier Reef. Coral Reefs 29(4):1005–1015

Chen C-T, Millero FJ (1977) Speed of sound in seawater at high pressures. J Acoust Soc Am 62:1129–1135

Cheung WWL, Lam VWY, Pauly D (2008) Modelling present and climate-shifted distribution of marine fishes and invertebrates, Fisheries Centre research reports 16(3). The Fisheries Centre, University of British Columbia, Vancouver

Cinner JE, Huchery C, Darling ES, Humphries AT, Graham NAJ, Hicks CC, Marshall N, McClanahan TR (2013) Evaluating social and ecological vulnerability of coral reef fisheries to climate change. PLoS One 8(9):e74321. doi:10.1371/journal.pone.0074321

Costa BM, Battista TA (2013) The semi-automated classification of acoustic imagery for characterizing coral reef ecosystems. Int J Remote Sens 34(18):6389–6422

Cumming GS (2011a) Spatial resilience in social-ecological systems. Springer, London

Cumming GS (2011b) Spatial resilience: integrating landscape ecology, resilience, and sustainability. Landsc Ecol 26(7):899–909

Curran PJ, Dash J (2005) Algorithm theoretical basis document, ATBD 2.22, chlorophyll index. University of Southampton. Available at: https://earth.esa.int/instruments/meris/atbd/atbd_2_22.pdf

Dawson TP, Rounsevell MD, Kluvánková-Oravská T, Chobotová V, Stirling A (2010) Dynamic properties of complex adaptive ecosystems: implications for the sustainability of service provision. Biodivers Conserv 19(10):2843–2853

Dekker AG, Phinn SR, Anstee J, Bissett P, Brando V, Casey B, Fearns P, Hedley J, Klonowski W, Lee ZP, Lynch M, Lyons M, Mobley CD, Roelfsema C (2011) Intercomparison of shallow water bathymetry, hydro-optics, and benthos mapping techniques in Australian and Caribbean coastal environments. Limnol Oceanogr Methods 9:396–425

Dudgeon SR, Aronson RB, Bruno JF, Precht WF (2010) Phase shifts and stable states on coral reefs. Mar Ecol Prog Ser 413:201–216

Dustan P, Dobson E, Nelson G (2002) Landsat Thematic Mapper: detection of shifts in community composition of coral reefs. Conserv Biol 15:892–902

Eakin CM et al (2010) Caribbean corals in crisis: record thermal stress, bleaching, and mortality in 2005. PLoS One 5(11):e13969

Edwards HJ, Elliott IA, Eakin CM, Irikawa A, Madin SJ, McField M, Morgan JA, van Woesik R, Mumby PJ (2011) How much time can herbivore protection buy for coral reefs under realistic regimes of hurricanes and coral bleaching? Glob Chang Biol 17:2033–2048

Elith J, Leathwick JR (2009) Species distribution models: ecological explanation and prediction across space and time. Annu Rev Ecol Evol Syst 40:677–697. doi:10.1146/annurev.ecolsys.110308.120159

Elmqvist T, Folke C, Nyström M, Peterson G, Bengtsson J, Walker B, Norberg J (2003) Response diversity, ecosystem change, and resilience. Front Ecol Environ 1(9):488–494

Epps S, Lohrenz S, Tuell G, Barbor K (2010) Development of a Suspended Particulate Matter (SPM) algorithm for the Coastal Zone Mapping and Imaging Lidar (CZMIL). Proc SPIE 7695:769514. doi:10.1117/12.851901

Ernstsen VB, Noormets R, Hebbeln D, Bartholomä A, Flemming BW (2006) Precision of high-resolution multibeam echo sounding coupled with high-accuracy positioning in a shallow water coastal environment. Geo-Mar Lett 26:141–149

Foster G, Gleason A, Costa B, Battista T, Taylor C (2013) Acoustic applications. In: Goodman JA, Purkis SJ, Phinn RP (eds) Coral reef remote sensing. Springer, Dordrecht

Franklin J (2009) Mapping species distributions. Cambridge University Press, Cambridge, UK

Friedlander AM, Parrish JD (1998) Habitat characteristics affecting fish assemblages on a Hawaiian coral reef. J Exp Mar Biol Ecol 224:1–30

Game ET, Watts ME, Wooldridge S, Possingham HP (2008) Planning for persistence in marine reserves: a question of catastrophic importance. Ecol Appl 18(3):670–680

Garcia-Salgado MA, Camarena TL, Vasquez MG, Gold B, Galland GG, Nava M, Alarcon GD, Ceja VM (2006) Baseline of the status of the Mesoamerican barrier reef systems: results of synoptic monitoring from 2004 and 2005, vol 1. Project for the conservation and sustainable use of the Meso-American Barrier Reef System (MBRS). Project Coordinating Unit, Belize City

Giardino C, Brando V, Dekker A, Strömbeck N, Candiani G (2007) Assessment of water quality in Lake Garda (Italy) using Hyperion. Remote Sens Environ 109(2):183–195

Globcolour (2008) Global ocean colour for carbon cycle research. Observation de la Terre – Environnement (ACRI-ST), Sophia Antipolis. doi:10.1594/PANGAEA.695832

Goodman JA, Ustin SL (2007) Classification of benthic composition in a coral reef environment using spectral unmixing. J Appl Remote Sens 1:011501. doi:10.1117/1.2815907

Goodman JA, Purkis SJ, Phinn SR (eds) (2013) Coral reef remote sensing. Springer, Dordrecht

Goovaerts P (1997) Geostatistics for natural resource evaluation. Oxford University Press, Oxford, p 496

Gove JM, Williams GJ, McManus MA, Heron SF, Sandin SA, Vetter OJ, Foley DG (2013) Quantifying climatological ranges and anomalies for Pacific coral reef ecosystems. PLoS One 8:e61974

Graham NAJ, Nash KL (2013) The importance of structural complexity in coral reef ecosystems. Coral Reefs 32(2):1–12

Green E, Mumby P, Edwards A, Clark C (1996) A review of remote sensing for the assessment and management of tropical coastal resources. Coast Manag 24:1–40

Grober-Dunsmore R, Pittman SJ, Caldow C, Kendall MA, Fraser T (2009) Chapter 14: A landscape ecology approach for the study of ecological connectivity across tropical marine seascapes. In: Nagelkerken I (ed) Ecological connectivity among tropical coastal ecosystems. Springer, Dordrecht, pp 493–530

Guisan A, Thuiller W (2005) Predicting species distribution: offering more than simple habitat models. Ecol Lett 8:993–1009

Gunderson L (2000) Resilience in theory and practice. Ann Rev Ecol Syst 31:425–439

Harborne AR, Mumby PJ, Zychaluk K, Hedley JD, Blackwell PG (2006) Modeling the beta diversity of coral reefs. Ecology 87:2871–2881

Hedley J (2013) Hyperspectral applications. In: Goodman JA, Purkis SJ, Phinn RP (eds) Coral reef remote sensing. Springer, Dordrecht

Hedley J, Roelfsema C, Koetz B, Phinn S (2012) Capability of the Sentinel 2 mission for tropical coral reef mapping and coral bleaching detection. Remote Sens Environ 120:145–155

Hengl T, Heuvelink G, Stein A (2004) A generic framework for spatial prediction of soil variables based on regression-kriging. Geoderma 120:75–93

Heron SF, Willis BL, Skirving WJ, Eakin CM, Page CA, Miller IR (2010) Summer hot snaps and winter conditions: modelling white syndrome outbreaks on Great Barrier reef corals. PLoS One. doi:10.1371/journal.pone.0012210

Herman JR, Krotkov N, Celarier E, Larko D, Labow G (1999) Distribution of UV radiation at the Earth's surface from TOMS measured UV-backscattered radiances. J Geophys Res 104:12059–12076

Hochberg EJ, Atkinson MJ (2000) Spectral discrimination of coral reef benthic communities. Coral Reefs 19:164–171

Hochberg EJ, Atkinson MJ, Andréfouët S (2003) Spectral reflectance of coral reef bottom-types worldwide and implications for coral reef remote sensing. Remote Sens Environ 85:159–173

Hoegh-Guldberg O et al (2007) Coral reefs under rapid climate change and ocean acidification. Science 318(5857):1737–1742

Hoitink AJF, Hoekstra P (2005) Observations of suspended sediment from ADCP and OBS measurements in a mud-dominated environment. Coast Eng 52(2):103–118

Holden H, LeDrew EF (1998) Spectral discrimination of healthy and non-healthy corals based on cluster analysis, principal components analysis and derivative spectroscopy. Remote Sens Environ 65:217–224

Holling CS (1973) Resilience and stability of ecological systems. Annu Rev Ecol Syst 4:1–23

Holling CS (1992) Cross-scale morphology, geometry, and dynamics of ecosystems. Ecol Monogr 62(4):447–502

Holling CS (1996) Engineering resilience vs. ecological resilience. In: Schulze PC (ed) Engineering within ecological constraints. National Academy Press, Washington, DC, pp 31–43

Hughes TP, Connell JH (1999) Multiple stressors on coral reefs: a long-term perspective. Limnol Oceanogr 44(3):932–940

Hughes TP, Graham NA, Jackson JB, Mumby PJ, Steneck RS (2010) Rising to the challenge of sustaining coral reef resilience. Trends Ecol Evol 25(11):633–642

IMaRS (2004) Millennium Coral Reef Mapping Project. University of South Florida, Institute for Marine Remote Sensing, St. Petersburg

Jerlov NG (1968) Optical oceanography, Elsevier oceanography series 5. Elsevier, New York, 194 p

Jones GP, Almany GR, Russ GR, Sale PF, Steneck RS, Van Oppen MJH, Willis BL (2009) Larval retention and connectivity among populations of corals and reef fishes: history, advances and challenges. Coral Reefs 28(2):307–325

Joyce K, Phinn S, Roelfsema C (2013) Live coral cover index testing and application with hyperspectral airborne image data. Remote Sens 5(11):6116–6137

Jupiter S, Roelfsema CM, Phinn SR (2013) Science and management. In: Goodman JA, Purkis SJ, Phinn RP (eds) Coral reef remote sensing. Springer, Dordrecht

Jupp DLB (1988) Background and extensions to depth of penetration (DOP) mapping in shallow coastal waters. In: Proceedings of the international symposium on remote sensing of the coastal zone, Gold Coast, pp IV.2.1–IV.2.19

Knowlton N (1992) Thresholds and multiple stable states in coral reef community dynamics. Am Zool 32(6):674–682

Knudby A, LeDrew E, Newman C (2007) Progress in the use of remote sensing for coral reef biodiversity studies. Prog Phys Geogr 31:421–434

Knudby A, LeDrew EF, Brenning A (2010a) Predictive mapping of reef fish species richness, diversity and biomass in Zanzibar using IKONOS imagery and machine-learning techniques. Remote Sens Environ 114:1230–1241

Knudby A, Brenning A, LeDrew E (2010b) New approaches to modelling fish-habitat relationships. Ecol Mod 221:503–511

Knudby A, Roelfsema C, Lyons M, Phinn S, Jupiter S (2011) Mapping fish community variables by Integrating field and satellite data, object-based image analysis and modeling in a traditional Fijian Fisheries management area. Remote Sens 3:460–483. doi:10.3390/rs3030460

Knudby A, Jupiter S, Roelfsema C, Lyons M, Phinn S (2013a) Mapping coral reef resilience indicators using field and remotely sensed data. Remote Sens 5(3):1311–1334

Knudby A, Kenchington E, Murillo FJ (2013b) Modeling the distribution of Geodia sponges and sponge grounds in the Northwest Atlantic. PLoS One 8(12):e82306. doi:10.1371/journal.pone.0082306

Knudby A, Nordlund LM, Palmqvist G, Wikström K, Koliji A, Lindborg R, Gullström M (2014) Using multiple Landsat scenes in an ensemble classifier reduces classification error in a stable nearshore environment. Int J Appl Earth Obs Geoinform 28:90–101

Krige DG (1951) A statistical approach to some mine valuations and allied problems at the Witwatersrand. University of Witwatersrand, Johannesburg

Le Provost C, Lyard F, Molines JM, Genco ML, Rabilloud F (1998) A hydrodynamic ocean tide model improved by assimilating a satellite altimeter-derived data set. Geophys Res Lett 103:5513–5529

LeDrew EF, Holden H, Wulder MA, Derksen C, Newman C (2004) A spatial statistical operator applied to multidate satellite imagery for identification of coral reef stress. Remote Sens Environ 91:271–279

Lee Z, Carder KL, Mobley CD, Steward RG, Patch JS (1998) Hyperspectral remote sensing for shallow waters. I. A semianalytical model. Appl Optics 37(27):6329–6338

Lee Z, Carder KL, Mobley CD, Steward RG, Patch JS (1999) Hyperspectral remote sensing for shallow waters: 2. Deriving bottom depths and water properties by optimization. Appl Optics 38:3831–3843

Leon J, Woodroffe CD (2011) Improving the synoptic mapping of coral reef geomorphology using object-based image analysis. Int J Geogr Inf Sci 25:949–969

Levin SA, Lubchenco J (2008) Resilience, robustness, and marine ecosystem-based management. Bioscience 58(1):27–32

Logan CA, Dunne JP, Eakin CM, Donner SD (2014) Incorporating adaptive responses into future projections of coral bleaching. Glob Chang Biol 20:125–139

Lyzenga DR (1978) Passive remote-sensing techniques for mapping water depth and bottom features. Appl Optics 17:379–383

Maina J, Venus V, McClanahan TR, Ateweberhan M (2008) Modelling susceptibility of coral reefs to environmental stress using remote sensing data and GIS models in the western Indian Ocean. Ecol Model 212:180–199

Maina J, McClanahan TR, Venus V, Ateweberhan M, Madin J (2011) Global gradients of coral exposure to environmental stresses and implications for local management. PLoS One 6:e23064

Manzello D, Enochs I, Musielewicz S, Carlton R, Gledhill D (2013) Tropical cyclones cause CaCO3 undersaturation of coral reef seawater in a high-CO2 world. J Geophys Res Oceans 118(10):5312–5321

McClanahan T (1995) A coral reef ecosystem-fisheries model: impacts of fishing intensity and catch selection on reef structure and processes. Ecol Model 80:1–19

McClanahan TR, Ateweberhan M, Sebastian CR, Graham NAJ, Wilson SK, Bruggemann JH, Guillaume MMM (2007) Predictability of coral bleaching from synoptic satellite and in situ temperature observations. Coral Reefs 26:695–701

McClanahan TR, Cinner JE, Graham NAJ, Daw TM, Maina J, Stead SM, Wamukota A, Brown K, Venus V, Polunin NVC (2009) Identifying reefs of hope and hopeful actions: contextualizing environmental, ecological, and social parameters to respond effectively to climate change. Conserv Biol 23(3):662–671

McClanahan TR, Donner SD, Maynard JA, MacNeil MA, Graham NAJ et al (2012) Prioritizing key resilience indicators to support coral reef management in a changing climate. PLoS One 7(8):e42884. doi:10.1371/journal.pone.0042884

McClanahan TR, Graham NA, Darling ES (2014) Coral reefs in a crystal ball: predicting the future from the vulnerability of corals and reef fishes to multiple stressors. Curr Opin Environ Sustain 7:59–64

McLeod E, Salm R, Green A, Almany J (2008) Designing marine protected area networks to address the impacts of climate change. Front Ecol Environ 7(7):362–370

Melbourne-Thomas J, Johnson CR, Fung T, Seymour RN, Cherubin LM, Arias-Gonzalez JE, Fulton EA (2011) Regional-scale scenario modeling for coral reefs: a decision support tool to inform management of a complex system. Ecol Appl 21(4):1380–1398

Moberg F, Rönnbäck P (2003) Ecosystem services of the tropical seascape: interactions, substitutions and restoration. Ocean Coast Man 46(1):27–46

Mobley CD, Sundman LK, Davis CO, Bowles JH, Downes TV, Leathers RA, Montes MJ, Bissett WP, Kohler DDR, Reid RP, Louchard EM, Gleason A (2005) Interpretation of hyperspectral remote-sensing imagery by spectrum matching and look-up tables. Appl Optics 44(17):3576–3592

Morel A, Bélanger S (2006) Improved detection of turbid waters from ocean colour sensors information. Remote Sens Environ 102:237–249

Morel A, Gentili B (2009) A simple band ratio technique to quantify the coloured dissolved and detrital organic material from ocean colour remotely sensed data. Remote Sens Environ 113:998–1011

Morel A, Prieur L (1977) Analysis of variations in ocean color. Limnol Oceanogr 22:709–722

Moses WJ, Gitelson AA, Berdnikov S, Povazhnyy V (2009) Estimation of chlorophyll-a concentration in case II waters using MODIS and MERIS data-successes and challenges. Environ Res Lett 4:045005

Mumby P (2001) Beta and habitat diversity in marine systems: a new approach to measurement, scaling and interpretation. Oecologia 128:274–280

Mumby PJ, Edwards AJ (2002) Mapping marine environments with IKONOS imagery: enhanced spatial resolution can deliver greater thematic accuracy. Remote Sens Environ 82(2–3):248–257

Mumby PJ, Hastings A (2008) The impact of ecosystem connectivity on coral reef resilience. J Appl Ecol 45(3):854–862

Mumby PJ, Steneck RS (2008) Coral reef management and conservation in light of rapidly evolving ecological paradigms. Trends Ecol Evol 23(10):555–563

Mumby P, Green E, Edwards A, Clark C (1997) Coral reef habitat-mapping: how much detail can remote sensing provide? Mar Biol 130:193–202

Mumby PJ, Skirving W, Strong AE, Hardy JT, LeDrew EF, Hochberg EJ, Stumpf RP, David LT (2004a) Remote sensing of coral reefs and their physical environment. Mar Pollut Bull 48(3–4):219–228

Mumby PJ, Hedley JD, Chisholm JRM, Clark CD, Ripley H, Jaubert J (2004b) The cover of living and dead corals from airborne remote sensing. Coral Reefs 23:171–183

Mumby PJ, Edwards AJ, Arias-Gonzalez JE, Lindeman KC, Blackwell PG, Gall A, Gorczynska MI, Harborne AR, Pescod CL, Renken H, Wabnitz CCC, Llewellyn G (2004c) Mangroves enhance the biomass of coral reef fish communities in the Caribbean. Nature 427:533–536

Mumby PJ, Harborne AR, Williams J, Kappel CV, Brumbaugh DR, Micheli F, Holmes KE, Dahlgren CP, Paris CB, Blackwell PG (2007a) Trophic cascade facilitates coral recruitment in a marine reserve. Proc Natl Acad Sci U S A 104:8362–8367

Mumby PJ, Hastings A, Edwards HJ (2007b) Thresholds and the resilience of Caribbean coral reefs. Nature 450:98–101. doi:10.1038/nature06252

Mumby PJ, Steneck RS, Hastings A (2013a) Evidence for and against the existence of alternate attractors on coral reefs. Oikos 122:481–491

Mumby PJ, Wolff NH, Bozec YM, Chollett I, Halloran P (2013b) Operationalizing the resilience of coral reefs in an era of climate change. Conserv Lett 1–12. doi: 10.1111/conl.12047

Mumby PJ, Chollett I, Bozec Y-M, Wolf NH (2014) Ecological resilience, robustness and vulnerability: how do these concepts benefit ecosystem management? Curr Opin Environ Sustain 7:22–27. doi:10.1016/j.cosust.2013.11.021

Myers MR, Hardy JT, Mazel CH, Dustan P (1999) Optical spectra and pigmentation of Caribbean reef corals and macroalgae. Coral Reefs 18:179–186

NASA (2014) OceanColor Web. http://oceancolor.gsfc.nasa.gov/. Accessed 29 Jan 2014

Nagelkerken I, Sheaves M, Baker R, Connolly RM (2012) The seascape nursery: a novel spatial approach to identify and manage nurseries for coastal marine fauna. Fish and Fisheries doi:10.1111/faf.12057

Newman C, Knudby A, LeDrew E (2007) Assessing the effect of management zonation on live coral cover using multi-date IKONOS satellite imagery. J Appl Remote Sens 1:011504

Nyström M, Folke C (2001) Spatial resilience of coral reefs. Ecosystems 4(5):406–417

Nyström M, Graham NAJ, Lokrantz J, Norström AV (2008) Capturing the cornerstones of coral reef resilience: linking theory to practice. Coral Reefs 27(4):795–809

O'Reilly JE, 24 Co-authors (2000) In: Hooker SB, Firestone ER (eds) SeaWiFS postlaunch calibration and validation analyses, Part 3, NASA technical memorandum, 2000-206892, vol 11. NASA Goddard Space Flight Center, Greenbelt, Maryland 49 pp

Odum HT (1983) Systems ecology: an introduction. Wiley, New York

Olds AD, Pitt KA, Maxwell PS, Connolly RM (2012) Synergistic effects of reserves and connectivity on ecological resilience. J Appl Ecol 49(6):1195–1203

Olds AD, Albert S, Maxwell PS (2013) Mangrove-reef connectivity promotes the effectiveness of marine reserves across the western Pacific. Global Ecol Biogeogr 22(9):1040–1049

Ouillon S, Douillet P, Petrenko A, Neveux J, Dupouy C, Froidefond J, Andréfouët S, Munoz-Caravaca A (2008) Optical algorithms at satellite wavelengths for total suspended matter in tropical coastal waters. Sensors 8:4165–4185

Parrish JD (1989) Fish communities of interacting shallow-water habitats in tropical oceanic regions. Marine Ecol Progr Ser 58:143–160

Palandro D, Andréfouët S, Muller-Karger FE, Dustan P, Hu C, Hallock P (2003) Detection of changes in coral reef communities using Landsat 5/TM and Landsat 7/ETM+ data. Can J Remote Sens 29:201–209

Park JY, Ramnath V, Feygels V, Kim M, Mathur A, Aitken J, Tuell G (2010) Active-passive data fusion algorithms for seafloor imaging and classification from CZMIL data. Proc SPIE 7695:769515. doi:10.1117/12.851991

Phinn SR, Roelfsema CM, Mumby PJ (2012) Multi-scale, object-based image analysis for mapping geomorphic and ecological zones on coral reefs. Int J Remote Sens 33:3768–3797

Phinney J, Muller-Karger F, Dustan P (2002) Using remote sensing to reassess the mass mortality of Diadema antillarum 1983–1984. Conserv Biol 15:885–891

Pinsky ML, Palumbi SR, Andréfouët S, Purkis SJ (2012) Open and closed seascapes: where does habitat patchiness create populations with high fractions of self-recruitment? Ecol Appl 22 (4):1257–1267

Pittman SJ, Brown KA (2011) Multi-scale approach for predicting fish species distributions across coral reef seascapes. PLoS One 6(5):e20583
Pittman SJ, Knudby A (2014) Predictive mapping of coral reef fish species and communities. In: Bortone SA (ed) Interrelationships between coral reefs and fisheries. CRC Press, Boca Raton, FL, USA
Pittman SJ, Olds AD (2014) Shifting perceptions from patch to seascapes in reef fish ecology. In: Mora C (ed) Ecology of fishes on coral reefs, The functioning of an ecosystem in a changing world, Cambridge University Press, Cambridge, UK
Pittman SJ, Costa BM, Battista TA (2009) Using lidar bathymetry and boosted regression trees to predict the diversity and abundance of fish and corals. J Coast Res 53:27–38
Pittman SJ, Costa B, Jeffrey CF, Caldow C (2011) Importance of seascape complexity for resilient fish habitat and sustainable fisheries. In: Proceedings of the Gulf and Caribbean Fisheries Institute, 63, 420–426. Gulf and Caribbean Fisheries Institute, c/o Harbor Branch Oceanographic Institution Inc., Fort Pierce
Pittman SJ, Costa B, Wedding LM (2013) LiDAR applications. In: Coral reef remote sensing. Springer, Dordrecht, pp 145–174
Pratchett MS, Hoey AS, Wilson SK (2014) Reef degradation and the loss of critical ecosystem goods and services provided by coral reef fishes. Curr Opin Environ Sustain 7:37–43
Purkis SJ, Brock JC (2013) Lidar overview. In: Goodman JA, Purkis SJ, Phinn RP (eds) Coral reef remote sensing. Springer, Dordrecht
Purkis SJ, Myint SW, Riegl BM (2006) Enhanced detection of the coral Acropora cervicornis from satellite imagery using a textural operator. Remote Sens Environ 101(1):82–94
Purkis SJ, Graham NAJ, Riegl BM (2008) Predictability of reef fish diversity and abundance using remote sensing data in Diego Garcia (Chagos Archipelago). Coral Reefs 27:167–178
Purkis SJ, Rowlands GP, Riegl BM, Renaud PG (2010) The paradox of tropical karst morphology in the coral reefs of the arid Middle East. Geology 38(3):227–230
Riegl B, Guarin H (2013) Acoustic methods overview. In: Goodman JA, Purkis SJ, Phinn RP (eds) Coral reef remote sensing. Springer, Dordrecht
Roelfsema C, Phinn S, Jupiter S, Comley J, Albert S (2013) Mapping coral reefs at reef to reef-system scales, 10s–1000s km(2), using object-based image analysis. Int J Remote Sens 34 (18):6367–6388
Rowlands G (2013) Remote sensing the diversity, distribution and resilience of coral reef environments. Ph.D. thesis, Nova Southeastern University, Fort Lauderdale
Rowlands G, Purkis S, Riegl B, Metsamaa L, Bruckner A, Renaud P (2012) Satellite imaging coral reef resilience at regional scale. A case-study from Saudi Arabia. Mar Pollut Bullet 64(6):1222–1237
Selig ER, Casey KS, Bruno JF (2010) New insights into global patterns of ocean temperature anomalies: implications for coral reef health and management. Glob Ecol Biogeogr 19:397–411
Siegel DA, Maritorena S, Nelson NB, Behrenfeld MJ (2005) Coloured dissolved organic matter and its influence on the satellite-based characterization of the ocean biosphere. Geophys Res Lett 32:L20605. doi:10.1029/2005GL024310
Smith VE, Rogers RH, Reed LE (1975) Automated mapping and inventory of Great Barrier Reef zonation with Landsat. Oceans 7:775–780
Strong AE, Kearns EJ, Gjovig KK (2000) Sea surface temperature signals from satellites – an update. Geophys Res Lett 27(11):1667–1670
Stumpf RP, Holderied K, Sinclair M (2003) Determination of water depth with high-resolution satellite imagery over variable bottom types. Limnol Oceanogr 48(1):547–556
Suzuki H, Matsakis P, Andréfouët S, Desachy J (2001) Satellite image classification using expert structural knowledge: a method based on fuzzy partition computation and simulated annealing. In: Proceedings of the annual conference of the international association for mathematical geology, Cancun, 6–12 Sept 2001
Tolman HL, Alves JHGM (2005) Numerical modeling of wind waves generated by tropical cyclones using moving grids. Ocean Model 9:305–323

Tolstoy I, Clay CS (1966) Ocean acoustics. Theory sand experiment in underwater sound. McGraw Hill, New York, p 293

van Hooidonk R, Maynard JA, Planes S (2013) Temporary refugia for coral reefs in a warming world. Nat Clim Chang 3:508–511

van Hooidonk R, Maynard JA, Manzello D, Planes S (2014) Opposite latitudinal gradients in projected ocean acidification and bleaching impacts on coral reefs. Glob Chang Biol 20:103–112

Vasilkov A, Krotkov N, Herman J, McClain C, Arrigo K, Robinson W (2001) Global mapping of underwater UV irradiances and DNA-weighted exposures using total ozone mapping spectrometer and sea-viewing wide field-of-view sensor data products. J Geophys Res 106:27205–27219

Wedding LM, Lepczyk CA, Pittman SJ, Friedlander AM, Jorgensen S (2011) Quantifying seascape structure: extending terrestrial spatial pattern metrics to the marine realm. Mar Ecol Prog Ser 427:219–232

Wilson SK, Graham NA, Pratchett MS, Jones GP, Polunin NV (2006) Multiple disturbances and the global degradation of coral reefs: are reef fishes at risk or resilient? Glob Chang Biol 12(11):2220–2234

Wilson SK, Adjeroud M, Bellwood DR, Berumen ML, Booth D, Bozec YM, Syms C (2010) Crucial knowledge gaps in current understanding of climate change impacts on coral reef fishes. J Exp Biol 213(6):894–900

Chapter 6
An Assessment of Physiographic Habitats, Geomorphology and Evolution of Chilika Lagoon (Odisha, India) Using Geospatial Technology

Ashis Kr. Paul, Sk. Majharul Islam, and Subrata Jana

Abstract The coastal lagoon 'Chilika' along the Bay of Bengal shore (Odisha state, India) support a productive wetland ecosystem that influences the livelihood of local people by providing large scale fishing activity within the lagoon water body, pasturing in the lagoon fringe marshes by rearing livestock and promoting ecotourism in the islands and spits of the lagoon system using nature as magnet of attractions. The lagoon is shore parallel elongated water body enclosed with barrier spits and narrow tidal entrances, and extended in between Mahanadi deltaic distributary channels to the north east and Rishikulya River to the southwest. The average water spread area of Chilika is 1,165 km^2 during rainy season and gradually it shrinkages to 906 km^2 during summer months with average depth ranges from 1.70 to 3.70 m.

Physiographically, the lagoon habitats are categorized as (i) Daya-Bhargavi deltaic flats with extensive fresh water wetlands, (ii) brackish water marshes and tidal flats, (iii) islands with marshes and transition vegetation, (iv) islands with mangrove swamps, (v) fresh water weeds of lagoon fringes, (vi) sand spits and dunes, (vii) and lagoon water bodies.

All these habitats are showing significant variations related to geological, sedimentological, hydrological, climatic and ecological factors. Northeastern parts of the lagoon water body have been swallowed by deltaic sedimentation; barrier back shores, islands and other lagoon fringes have shrunk as the result of bordering accretion and encroachment of marshes, weeds and swamps; however, southwestern parts of the lagoon are narrowed, and in some places near-segmented by growth of spits and coalescence of headlands and islands. Opening of tidal inlets have been

A.Kr. Paul (✉) • Sk.M. Islam • S. Jana
Department of Geography and Environment Management, Vidyasagar University, Midnapore 721102, West Bengal, India
e-mail: akpaul_geo2007@yahoo.co.in

C.W. Finkl and C. Makowski (eds.), *Remote Sensing and Modeling: Advances in Coastal and Marine Resources*, Coastal Research Library 9,
DOI 10.1007/978-3-319-06326-3_6,

shifted from one place to other place with time across the enclosed barriers to achieve the balance between water levels of the lagoon and open marine shelf, and as a result of interaction of fresh water from rivers with salt water from the sea during floods and cyclones in the region.

Historically, the habitats of Chilika lagoon provided favorable conditions for maritime trade and commerce, as well as modern fishing related livelihood support system to the local people; but dynamic landscape changes of the lagoon with sedimentation have produced significant effects on human systems by changing salinity regime and eutrophication of enclosed water body. The paper highlights the gradual changes of coastal lagoon systems with variety of habitats influenced by sedimentation, fresh water inputs, local sea levels, tidal mixing and dynamic wetland geomorphology.

Chilika habitats, their geomorphologic changes, ecologic responses, and evolution of the lagoon itself are studied in the present paper using geospatial technology with temporal image data, as well as with available archaeological remains of past maritime activities and existing dating records for better management option and conservation of degraded habitats.

6.1 Introduction

The coastal lagoon of Chilika (Odisha state, India) is subject to variable rates and types of geomorphological change and create a variety of habitats which are colonized by marsh plants, mangroves, other coastal plants of sand strand, weeds and algae. Chilika lagoon section of the coastal region is subject to low tidal range (below 2 m), a hot and humid climate with monsoonal rainfall, frequent landfall of Bay of Bengal cyclones, and high rates of sediment discharge of Daya and Bhargavi rivers particularly in the monsoon months. These four factors with open marine environment adjacent to the lagoon are largely responsible for the nature of geomorphic change which occurs within the lagoon system. Rishikulya river mouth to the extreme southwest of Chilika also contribute a high rate of sand size sediment discharge which takes part in the deposition of modern barrier spits dominated by north ward long shore drift.

Types of plants and weeds are distributed throughout the area in a variety of physiographic habitats. It is the objective of this paper (i) to identify and assess the wetland habitats of the coastal lagoon, (ii) to discuss about the spatial variation of wetland vegetations, (iii) to examine the geomorphic change in response to various processes, and (iv) to explore the lagoonal deposits associated with sea level movements in the Holocene and Late-Holocene phases (Fig. 6.1).

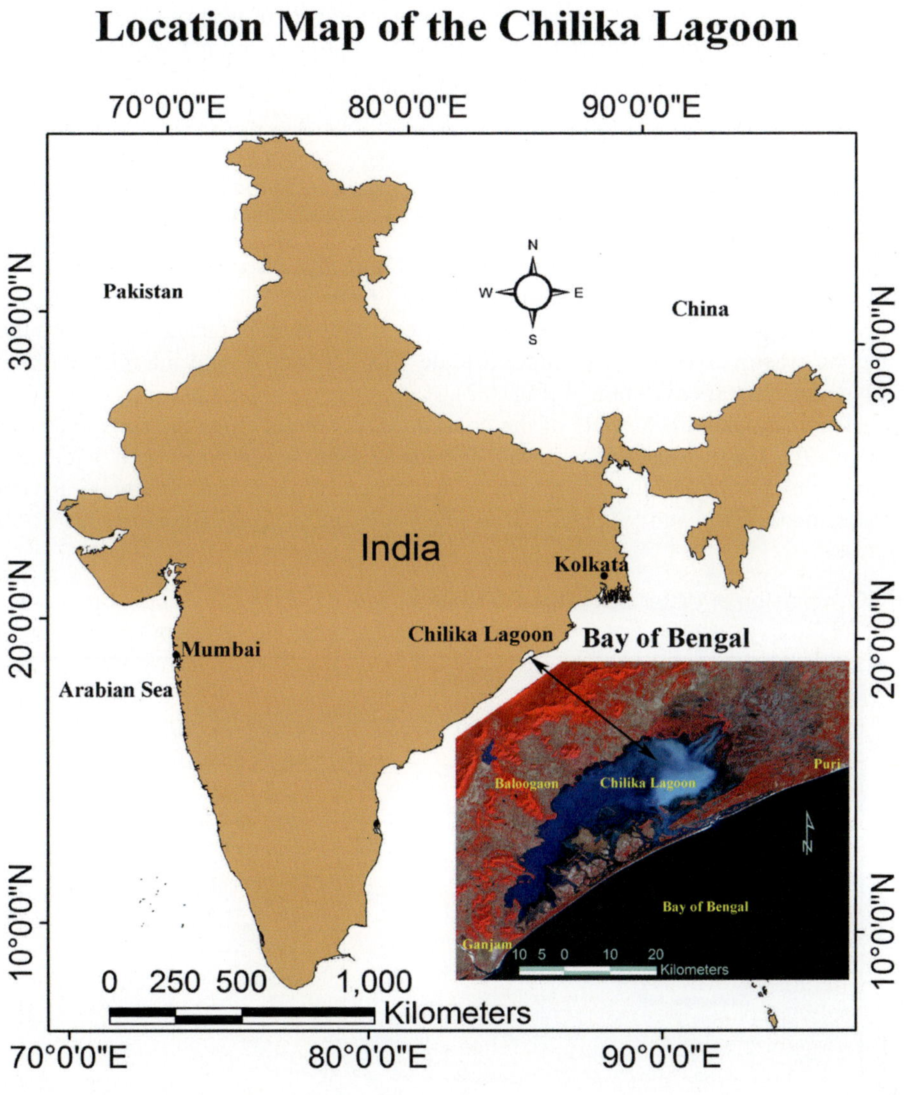

Fig. 6.1 Location map of Chilika lagoon and its adjacent areas (Odisha, India)

6.2 Methods of Study

Stratigraphic sections from the field stations; waves, winds and tides related data from INCOIS and CDA; Landsat OLI, ETM$^+$, TM, MSS, ASTER and QuickBird images from NASA; historical records of geo-archeological remains from existing literatures; total station survey of sample areas of Sea mouth of modern spits; SOI toposheets and Geological maps of the study area; textural analysis of sediments collected from Chilika lagoon habitats; vegetation identification from different physiographic habitats are considered for the purpose of present study in Chilika lagoon and adjacent areas.

6.3 Geomorphology of Chilika Lagoon

The largest brackish water lagoon is physiographically associated with four or five diverse landform units at or along the Bay of Bengal coast. Geomorphological diversity of these landform units can indicate the direction of changes by which the lagoon was evolved over a temporal scale with interactive and interplaying processes of coastal environment (Fig. 6.2).

The major landform units of the coastal lagoon include: (a) the spits and tidal inlets, (b) the lagoon fringe areas, (c) the lagoon water body with wide and elongated basin, (d) the distributaries fan lobes of Daya and Bhargavi Rivers (Mahanadi delta), and (e) open marine shelf adjacent to spits. The minor landform

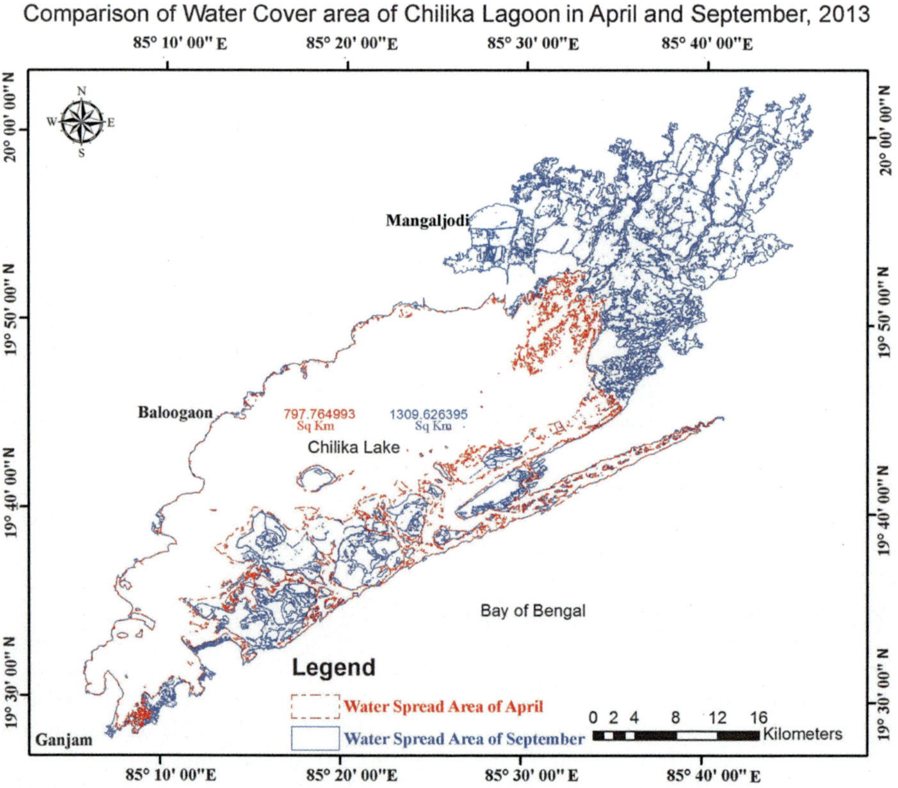

Fig. 6.2 Seasonal water covers areas and wetland types of Chilika lagoon

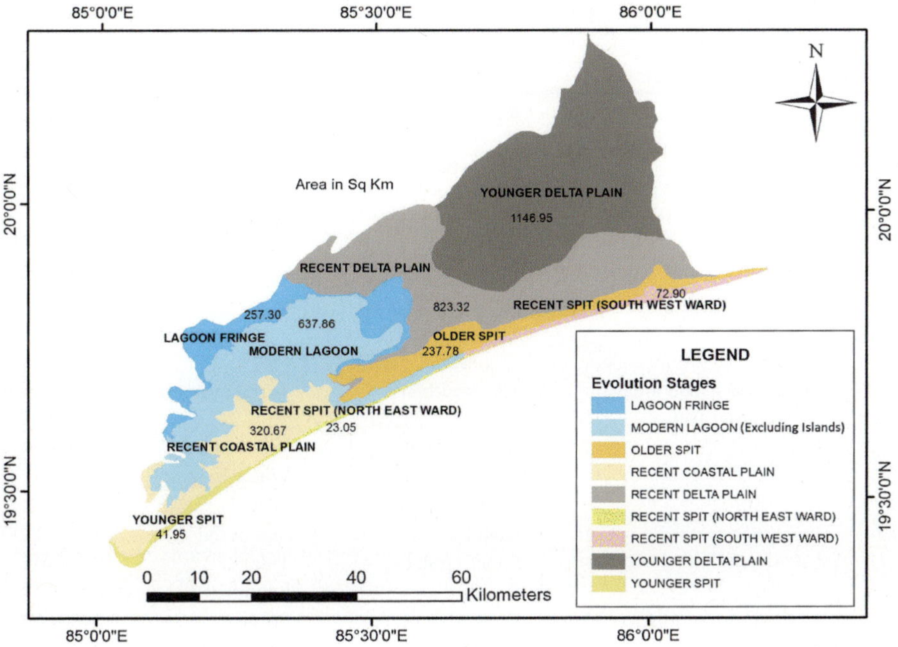

Fig. 6.3 Geomorphic divisions of Chilika lagoon and its fringe area

units, associated processes and possible age of their formation are estimated in the following ways (Fig. 6.3, Table 6.1).

1. There are two major groups of sand spit as barrier against the open marine environment and sheltered lagoonal environment parallel to the present day shoreline of Bay of Bengal. The landward older spits (Satapada-Puri section) are originated from Mahanadi delta side and extended towards south of the lagoon shores as viewed from the images and S.O.I. toposheets of the present study area. The length, width and elevation of these spits indicate the high volume of sediment input capacity of Mahanadi distributary Rivers (Devi River and Chandrabhaga River) into the coastal zone at different sediment abundant phases and dominant long shore drift for shaping the spits from the north east direction. Chandrabhaga River was abandoned at the later stage and Devi River discharged its sediment plumes into the sea in north east ward direction as evidenced from drainage features of Mahanadi delta, and for which the supply of sediment was reduced into the landward spits with south ward weakened longshore drift in the region. The sea ward younger spits are however, developed and extended from both sides with seasonal change of longshore drifts (from north west in winter and from south west in summer months), and supply of sediments from Rishikulya River and Devi River in the modern period. In width the younger spits range from 300 to 500 m and they are backed by beach ridges and sand dunes in most of the places except in south

Table 6.1 Landform assemblages of Chilika coastal lagoon

Sl. No.	Major landform units	Minor landform units	Associated processes	Possible ages
1.	The spits and tidal inlets	Sea front spits	Waves, long shore currents	30–300 YBP
		Landward spits	Waves, long shore currents, tides	1850–1300 YBP
		Spit front beach	Waves	2200–2100 YBP
		Spit back dunes	Winds	
		Tidal inlets	Tides	
		Back spit islands, flats, bars, marshes and channel pans	Wind waves, winds and tides	
2.	The lagoon fringe areas or lacustrine fringes	Terraced banks	Surface runoff	3750–1850 YBP
		Emerged flats	Winds, runoff	4300–3100 YBP
		Marshes and weeds	Water logging	
		Tidal flats	tides	
3.	The lagoon water body with wide and elongated basin	Islands	Sub-aerial	4000–3500 YBP
		Shoaling flats	Tidal	
		Basins	Lagoonal	
4.	Distributary fan lobes of Daya and Bhargavi Rivers	Delta plains	Fluvial	5000–4000 YBP
		Marshes	Fresh water	
		Channels	Fluvial	
		Water bodies	Flood water	
		Tidal flats	Tidal	
5.	Open marine shelf adjacent to spits	Shallow shelf	Marine	8000–6000 YBP
		Shelf break	Marine	

central shores adjacent to 'New mouth' (new opening of tidal inlet). The spits are flat in form and ranging from 2 to 3 m in elevation and liable to flooding in high magnitude storm events in the Bay of Bengal coast. As the balance of coastal lagoon is significantly affected due to reduced supply of sediments and fresh water discharges by Daya and Bhargavi rivers in the near past the younger spits are breached in two another places and existing inlet is also shifted towards north east. Because of such dynamic behavior of the younger spits a spit back dune body of 20 m elevation and 270 m width is washed out within 2.5 years. Around the month of July, 2012 another mouth or inlet is also opened across the spit towards north of existing one. Ample of evidences of geomorphic signatures prove the wave reworking as well as over wash activities of the low lying spits in the raised water levels (period of storms, monsoon months, H.A.T. phases) and for which the spit front beaches are scarped and spit back dunes are cliffed in many places (Fig. 6.4, Table 6.2, Figs. 6.5 and 6.6).

2. The lacustrine fringe areas of Chilika lagoon is dominated by fresh water wetlands on the landward sides (particularly near river mouths of Daya, Bhargavi, Rishikulya and others) and brackish water wetlands on the barrier back margins parallel to the present shoreline of Bay of Bengal. More over the seasonal shrinkage of lagoon body (in the drier months) also produce a sub-aerial

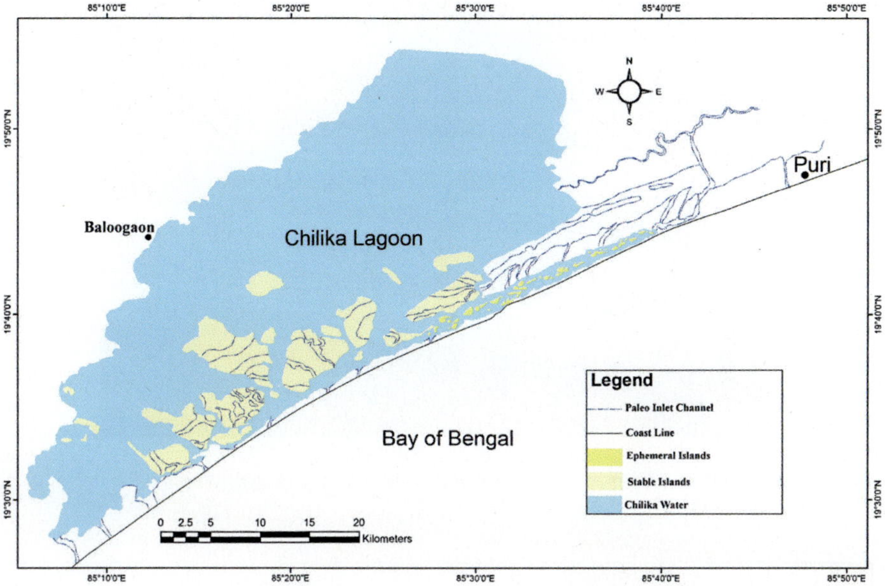

Fig. 6.4 Palaeo channels along the spits and islands of Chilika lagoon

Table 6.2 Tidal inlet opening and variation of tidal prisms of Chilika lagoon

Year	Date	No. of opening	Tidal prisms
1972	07 Nov.	03	93,312,000 m^3
1980	18 Jan.	02	62,208,000 m^3
1990	28 Nov.	01	62,208,000 m^3
2000	12 Sept.	01	38,880,000 m^3
2000	15 Nov.	02	54,532,000 m^3
2004	02 Nov.	01	62,208,000 m^3
2009	13 Sept.	02	93,312,000 m^3
2013	26 Apr.	01	202,176,000 m^3

exposure of an extensive areas of temporal wetlands dominated by lagoon filled sediments. The lagoon water body of Chilika covers about 1,309.63 km^2 areas in the monsoon month (September), and it reduces to 797.77 km^2 in the month of April (Non-monsoon month).

The average water spread area of Chilika is 790 km^2 according to the Remote Sensing data of 1996, though according to Government of Odisha record, the area covered by the lake is around 1,055 km^2 which swells to 1,165 km^2 during the rainy season and gradually shrinks to 906 km^2 during summer. The total area of islands inside the lake is 223 km^2. The average depth of the lake varies from 1.73 to 3.70 m during rainy seasons and 0.93 to 2.60 m during summer.

About 259 km^2 areas of seasonal wetland emerge at the fringe of the lagoon in winter and summer months every year. Diversified wetland types are visible at the lagoon fringe with characteristic soils and vegetations. Peaty soils, sandy soils, clayey soils, loamy soils and humus rich organic soils are found in the back

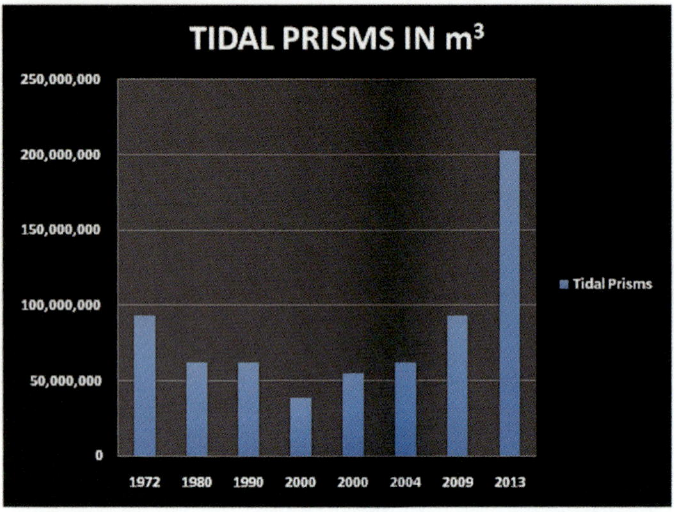

Fig. 6.5 Variation of tidal prisms in a temporal scale at Chilika lagoon through the opening and closing of tidal inlets

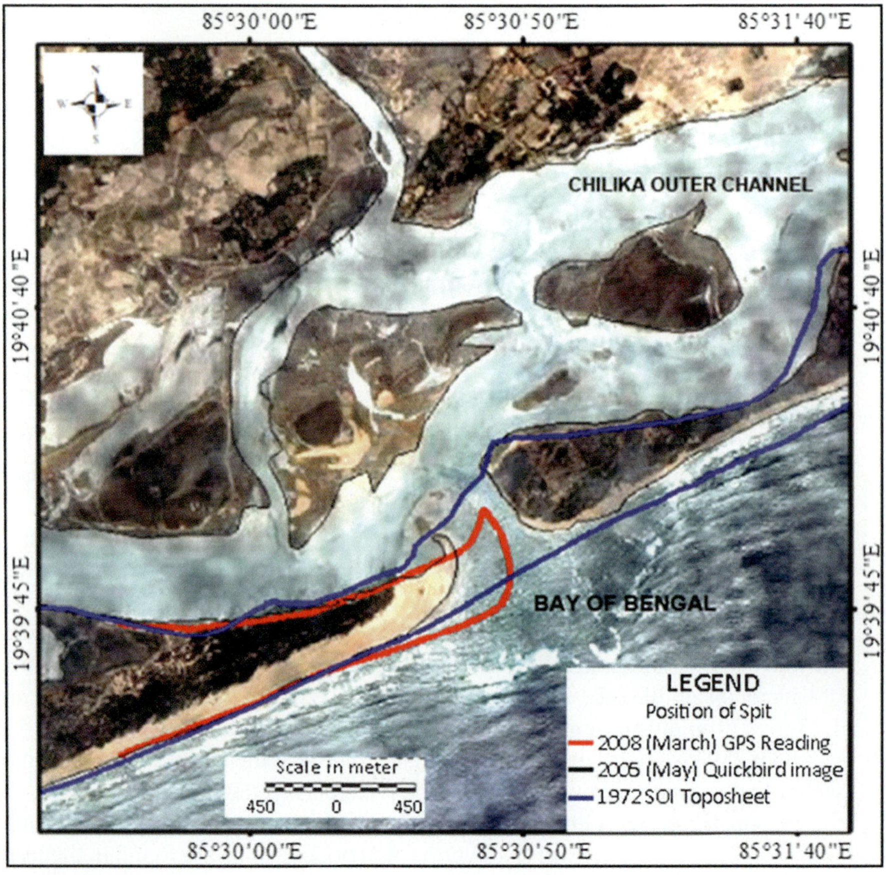

Fig. 6.6 Barrier spit extension, hooking, breaching and retreat along the present day shoreline of Chilika lagoon

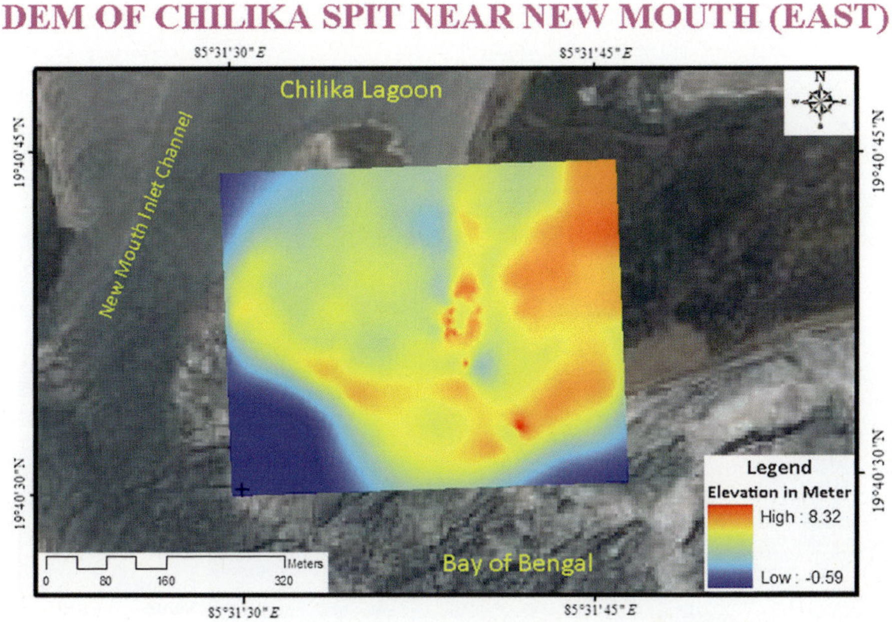

Fig. 6.7 Nature of spit surface morphometry of Chilika near sea mouth with total station survey

spit lagoon fringes of Chilika. In most cases, the lagoon fill sediments of underlying areas are blanketed by fluvial, fluvio-marine, marine and windblown sediments at the lagoon fringes. A number of soil pits excavated at the lagoon fringe proves the above facts at Chilika (Figs. 6.7, 6.8, and 6.9).

3. The lagoon water body with wide and elongated basin occupies a larger part of embayed coast in this part of eastern India. Balance of the lagoon is achieved between fluvial discharge and sediment input, evaporation, precipitation, tidal link with prisms and sea level change.

   The lagoon was extended further towards northeast along the shoreline up to Puri even in the recent past in elongated form. Ample of geomorphic evidences prove the earlier extension of the lagoon which is partially filled up with deltaic sediments of Mahanadi distributaries (Daya R., Bhargavi R. and Nuna R.) to the north east and reshaped in the present form. The northeastern part of the lagoon is also relatively shallower in compare to the central and southwestern parts due to seasonal monsoon deposition of alluviums transported by Mahanadi distributary rivers in this part annually as lagoon filled sediments. Depth of the lagoon ranges from 0.50 to 1.80 m in northeastern shores and it is ranging from 2.40 to 3.50 m in the southwestern part. Northeastern part of the lagoon is affected by concentration of weeds due to shallowness of lagoon bed, seasonal wash deposition from nearby agricultural fields, buffering of phosphorus, nitrogen and sulphars, and presence of maximum fish firms with modern shrimp culture ponds.

   As long as the parts of spits, bars and associated sand dunes are unstable or dynamic they may act as the major source of sediment supply into the lagoon bed in the form of washover deposition, windblown deposition, and flood delta

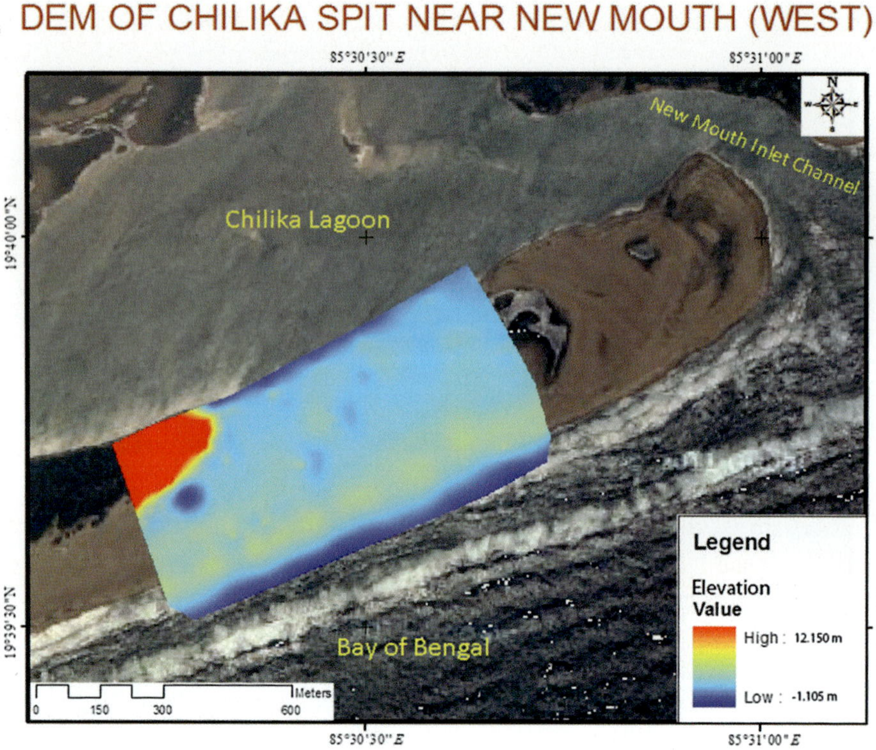

Fig. 6.8 Nature of spit surface morphometry of Chilika towards southwest section (total station survey)

deposition guided by inlet channels. The inlet to the lagoon is the product of interaction between waves, currents, tides and sediment bypassing. Gradually, the inlet is shifting towards the northeast close to the old mouth due to net sediment transport from southwest to northeast by longshore current at present. The lagoon is gradually decreasing in depth due to siltation process. The tidal fluctuation is recorded as 0.20 m in non-monsoon months and only 1.00 m in monsoon months in the lagoon that is driven by the occurrences of open marine Bay of Bengal tides ranging from 0.90 to 2.40 m in front of Chilika sand spit.

The lagoon was formed in between 3,700 years and 5,000 years before present (YBP) when the shoreline was located towards west in the Pleistocene era. Later the shallow shelf of the embayed basin was enclosed by sand spit extending from the mouth of Rishikulya River to the mouth of Kushabhadra River of Mahanadi delta. The northeastern extension of the lagoon was filled by Daya and Bhargavi river system in the post Hirakud situation (a dam across Mahanadi at the off take of Mahanadi delta) that accelerated the sediment flow into the lagoon, causing high rate of siltation (Das and Jena 2008).

Several palaeo stands in the form of indented terraces and topographic highs or ridges with formation of concretions and compact sediments are available off

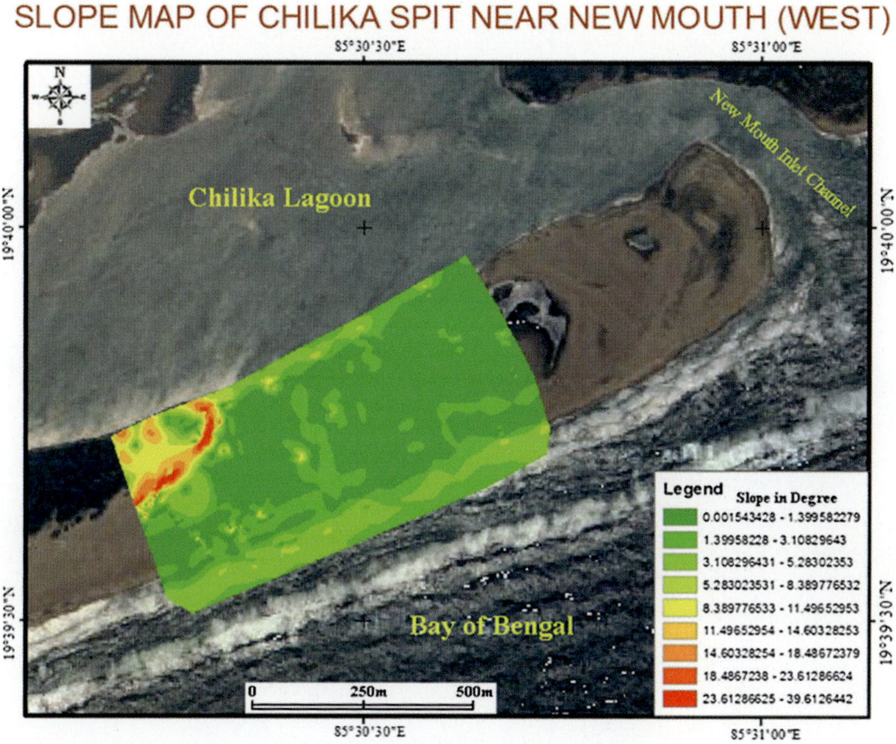

Fig. 6.9 Various slope aspects of Chilika barrier section measured by total station survey. Amount of slope varies from 0-39.61° in the section

Chilika offshores. These are termed as evidence of Holocene sea level still stands off Odisha coast by geologists (Bhattacharya et al. 1994). With the subsequent transgression of the sea the younger sediments are overlain by underlying compact sediments. Such underlying sediments with carbonate samples from −17 m depth off Chilika-Gopalpur-Nizampatnam areas are dated at 8200 + 120 YBP (Rao et al. 1990) which is very significant in this respect of sedimentation rate along the shallow shelf. If it is the beginning of the subsequent sedimentation around the shallow shelf and Chilika embayment, the overlying sediment column might have been deposited at the rate of 4 cm per 100 years. Furthermore, it can be extended as an estimated rate of sediment deposition for the case of Chilika lagoon till today.

As Chilika lagoon or embayment is older of 5,000 years, the upper most sediment depositional column might be approximately 2.00 m in thickness. Gradually by subsequent transgression and regression processes the lagoon bed is filled up and lagoon margins are modified with sedimentation (Table 6.3).

4. Today the distriburary sedment fan lobes of Daya and Bhargavi rivers with palaeo channels can be identified from the satellite images of northern Chilika lagoon. The sediment plumes of both the rivers are extended into Chilika lagoon in the monsoon months, and such evidence proves the time, duration and

Table 6.3 Cumulative frequency of sediment grain size distribution of the inner island of Chilika lagoon

Samples	1.70 mm	1 mm	850 μm	500 μm	425 μm	250 μm	150 μm	125 μm	75 μm	38 μm	<38 μm
**T1	0.052	0.185	0.543	4.613	11.743	29.700	89.089	92.536	98.705	99.831	100
**T2	0.060	0.309	1.098	7.239	17.967	48.627	93.668	95.698	99.396	99.961	100
**T3	0.041	0.218	0.773	5.361	14.610	44.582	93.467	96.207	99.490	99.970	100
**S1	0.050	0.143	0.548	4.482	14.096	51.291	93.407	95.801	99.298	99.928	100
**S2	0.0734	0.305	1.203	9.396	20.777	46.943	91.628	93.946	98.250	99.901	100
**S3	0.056	0.320	1.215	13.915	36.278	66.839	95.849	97.233	99.425	99.960	100
**D1	0.387	1.389	3.119	11.680	22.214	40.870	90.072	92.812	97.981	99.917	100
**D2	0.165	0.682	2.186	12.641	28.053	54.062	92.784	94.629	98.254	99.872	100
**D3	0.3945	1.3468	3.365	13.681	24.867	49.082	91.947	93.788	98.000	99.887	100

***T* Crestal part of the island, *S* Middle slope of the island margin, *D* Basal slope of the island (from Chilika lagoon); 1, 2, 3 represents vertical sampling layers

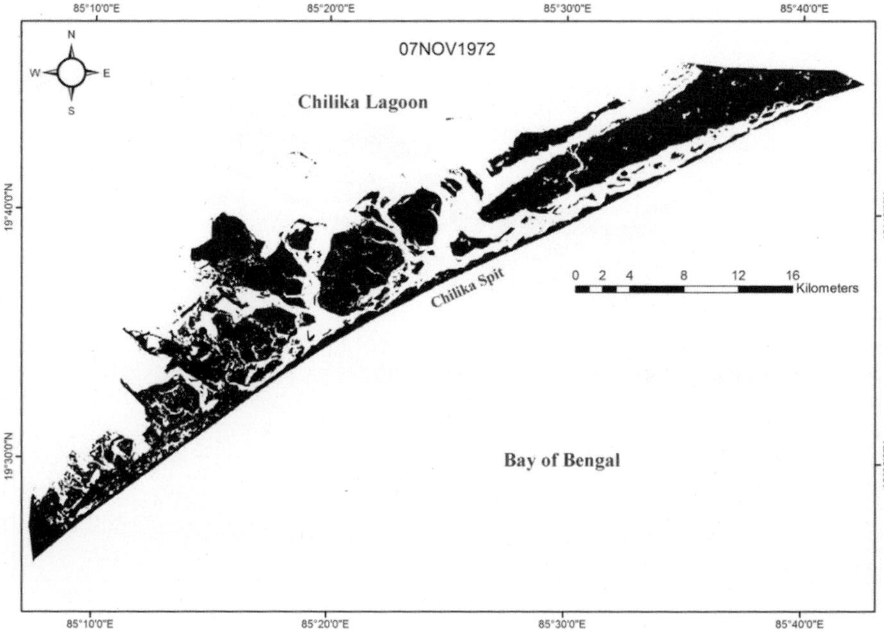

Fig. 6.10 Section of Chilika spits and islands (November, 1972) with tidal water ways and lagoon water bodies

direction of sediment focie of distributary channels as regular phenomenon for lagoon sedimentation in the region. This part of delta fan lobe surfaces can be categorized as older, younger and recent on the basis of their surface characters, elevation, drainage features and remains of historical ports along the palaeo channels close to lagoon fringes. The entire area is now characterized by location of several wetlands, dense pastures, delta plain surface with channels, and dots of islands. The sea front older spit (Satapada-Puri-Konark ridge) was acting as barrier against the distributary channels of Mahanadi delta system (Daya and Bhargavi rivers) in the historical period. Former extension of the lagoon in this part is practically filled up with rapid advancement of sediment fan lobes after the formation of Satapada-Puri spit.

5. The open marine shelf adjacent to Chilika spit is only 40 km wide up to 200 m bathymetric contour and then marked by a significant shelf break in-between 200 and 1,200 m with 1,000 m vertical fall of underwater marine surface. This shelf break might be the part of palaeo shore of late Pleistocene period. The bathymetric chart of Chilika offshore also shows that the shelf break contours are jagged by two canyon like valleys which were probably the mouth of older rivers in submerged condition. A borehole litholog of the shallow marine shelf is prepared by the marine wing of G.S.I. in the EEZ sea bed map off Chilika lake. The depositional facies of sediments are dated as Pliocene to Recent, Miocene and Eocene ages up to 2,000 m depth of sediment column in the shallow offshores (Figs. 6.10, 6.11, 6.12, 6.13, 6.14, 6.15, 6.16, and 6.17).

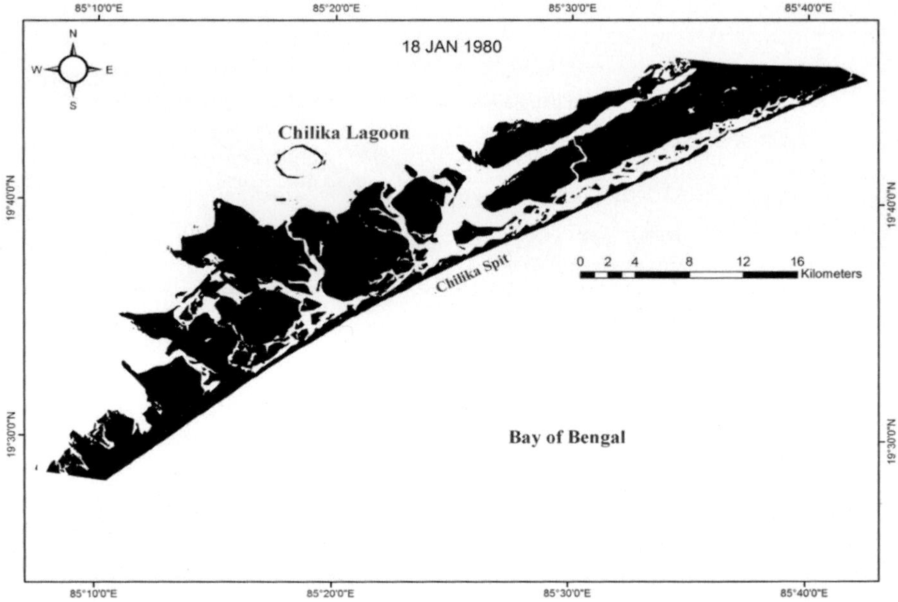

Fig. 6.11 Position of Chilika Old Mouth towards northeast (January, 1980) and associated tidal flat deposits

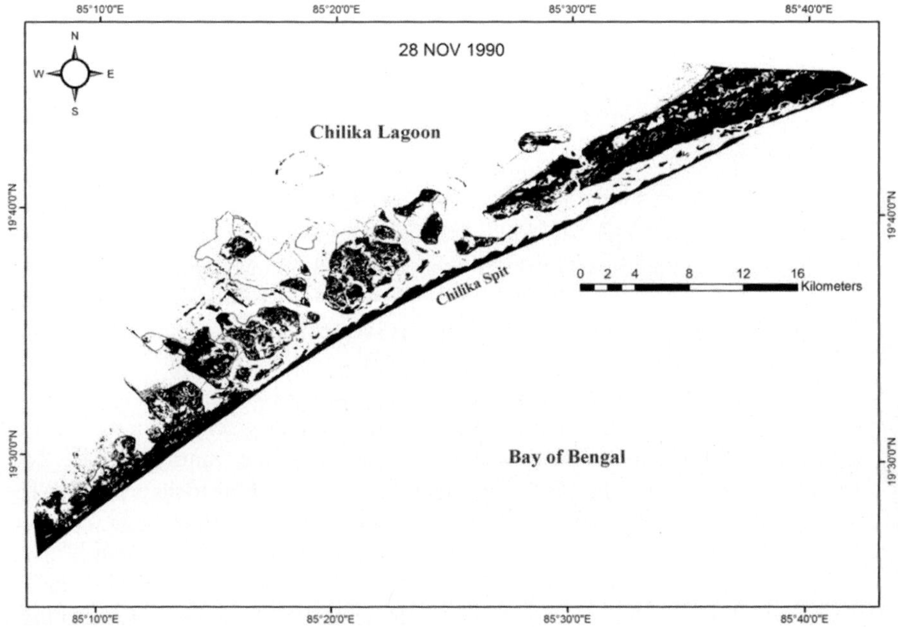

Fig. 6.12 Shifting of Old Mouth towards northeast (November, 1990) with tidal water ways and less vegetation covers

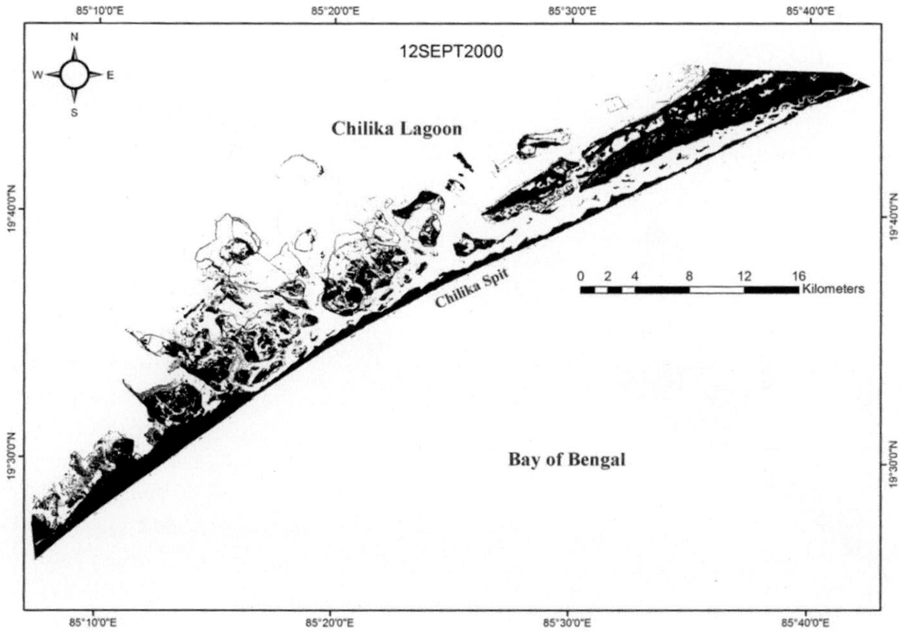

Fig. 6.13 Chilika spits and islands of monsoon month (September, 2000)

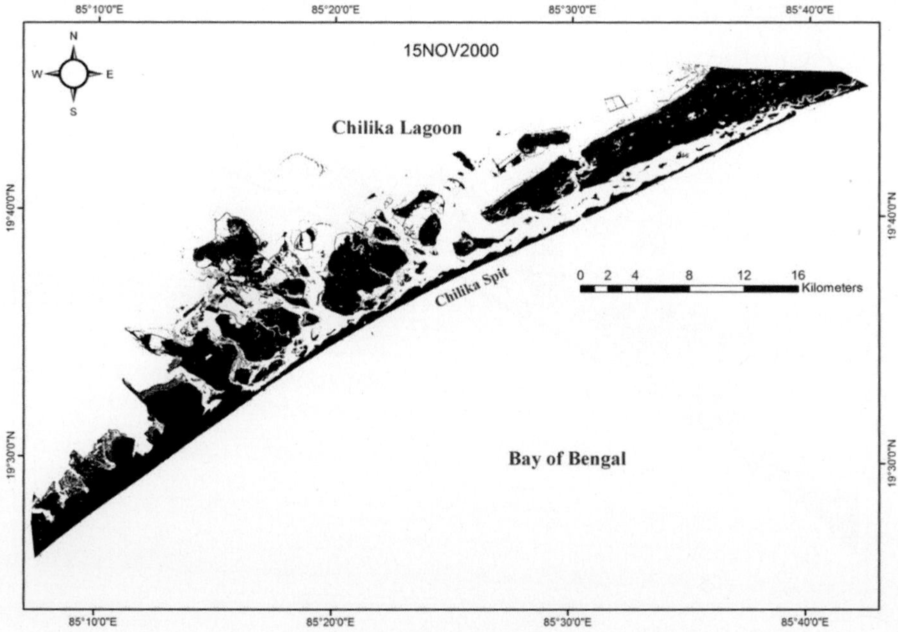

Fig. 6.14 Spit breaching and isolation in November, 2000

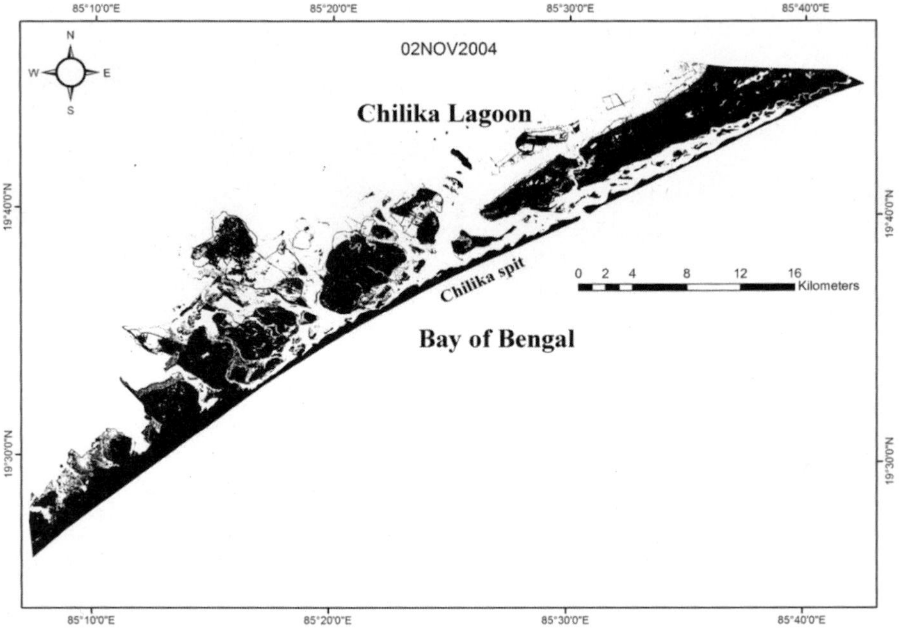

Fig. 6.15 Opening and widening of Chilika New Mouth in November, 2004

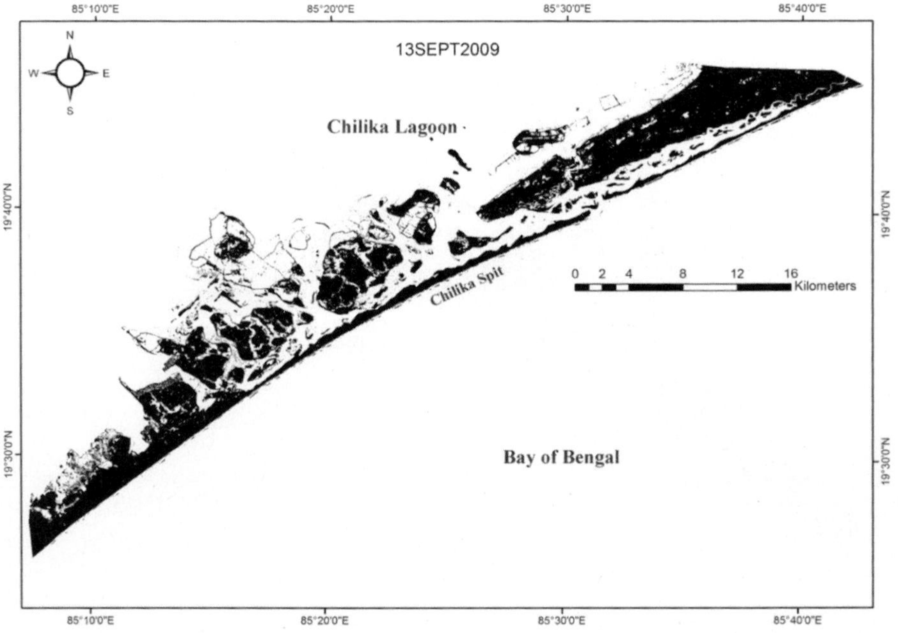

Fig. 6.16 Opening and fragmentation of Chilika spits with islands along the tidal water ways (September, 2009)

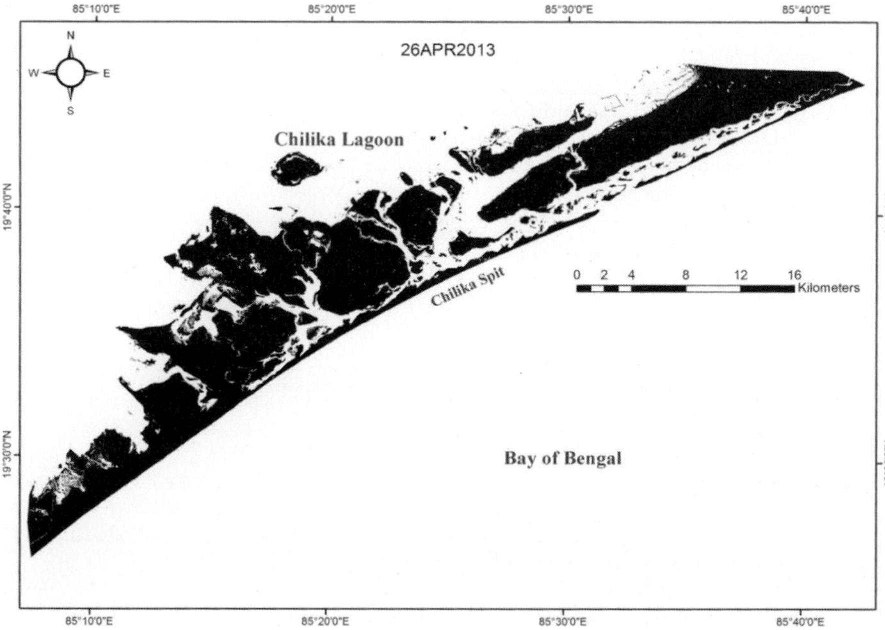

Fig. 6.17 Large tidal flats of Chilika with wide range of vegetation covers (April, 2013)

6.4 Historical Environment of Chilika Lagoon System

Several ancient ports and associated settlements were existing in and around Chilika lagoon in the historical past. Ancient sea going traders used the ports in the region when the water cover area of the lagoon was extensive and favourable for navigation. The remains of past settlements are also explored from different layers particularly at the higher grounds of Chilika lagoon fringes. Present decay of the lagoon is caused by the gradual siltation and dynamic geomorphic processes of marine, fluvial and lagoonal activities over a time scale. Chronology of the development and decay of Chilika lagoon system can be estimated with available historical records of the region (Table 6.4).

6.5 Wetland Types (Chilika Lagoon System)

As water fluctuates in Chilika in dry and wet seasons, extensive wetland types are visible from the images along the fringe areas of the lagoon under different hydrogeomorphic settings. Wetlands of the fringe areas are modified by vertical accretion of sediments caused by the occurrences of storms and river floods at an interval.

Table 6.4 Chronology of the development and the decay of Chilika lagoon system

Sl. no.	Geological and historical exploration sites	Existing records
1	The inner shelf off Odisha off Chilika provides signatures of palaeo stands of the transgressive sea during the Holocene Terraces and topographic highs Sand ridges Concretions and compact sediment	8200 ± 150 YBP (Malini et al. 1993)
2	Excavations at Golbai in north east of Chilika lagoon, and in 'Kanas' Stone anchors along the coast of Chilika Lagoon Handmade pottery Monolithic stone pillars on the shore of the lake with a lamp on the top Bone and polished stone tools Copper tools and ceramics	4300–3100 YBP, Neolithic and Chalcolithic period (Sinha 2000)
3	A shell of Ostrea Virginiana' from the south western edge of the spit has been dated by radiocarbon method	3750 ± 200 YBP (Subrahmanyam et al. 2006)
4	Palur port as 'Paloura', the point of departure for ships bound for south east Asian Countries	AD 150 or 1850 YBP Ptolemy, the Greek Geographer (McCrindle)
5	Decline of 'Manikpatana' port Indegenous and foreign pottery Varieties of seramics Artifacts and coins	Historical evidence from 200–100 BC or 2200–2100 YBP (Pradhan et al. 2000)
6	"Che-li-ta-lo port" Hiuen Tsang visited Odisha Through this port	Activities of the port During Evidenced from 600 to 638 AD or (1400–1338) 1600 to 1338 YBP (Watters 1973)
7	'Kanas' port in the north east of the Chilika lagoon on the bank of Nuna River Ship anchors Stone memoirs	Historical evidence of seventh Century AD or 1300 YBP Archaeological Exploration (Sinha 2000)
8	Jagannath temple at Puri constructed Heritage structure	(Historical evidence from 1000 to 1100 AD or 1000 to 900 YBP) (Sinha 2000)
9	Konark sun temple constructed Heritage structure	(Historical evidence from 1250 or 750YBP) (Sinha 2000)
10	Excavations at Palur and adjacent areas yielded Fine red ware with flaring rims and bowls Local imitation of northern black polished ware (Scattered in the midst of sand dunes) Pottery	Historical evidence of twelve to fourteen Century AD or 800-600 YBP (Basa and Behera 2000)
11	Barrier spit at Chilika lake is dated by OSL method Optically stimulated luminescence	40 years at the top of the spits 300 years at the bottom (Murray and Mohanty 2006)
12	Garh Krushnaprasad Heritage structure at the Chilika Island	Historical evidence of 1800 AD or 200–250 YBP (CDA 2009)

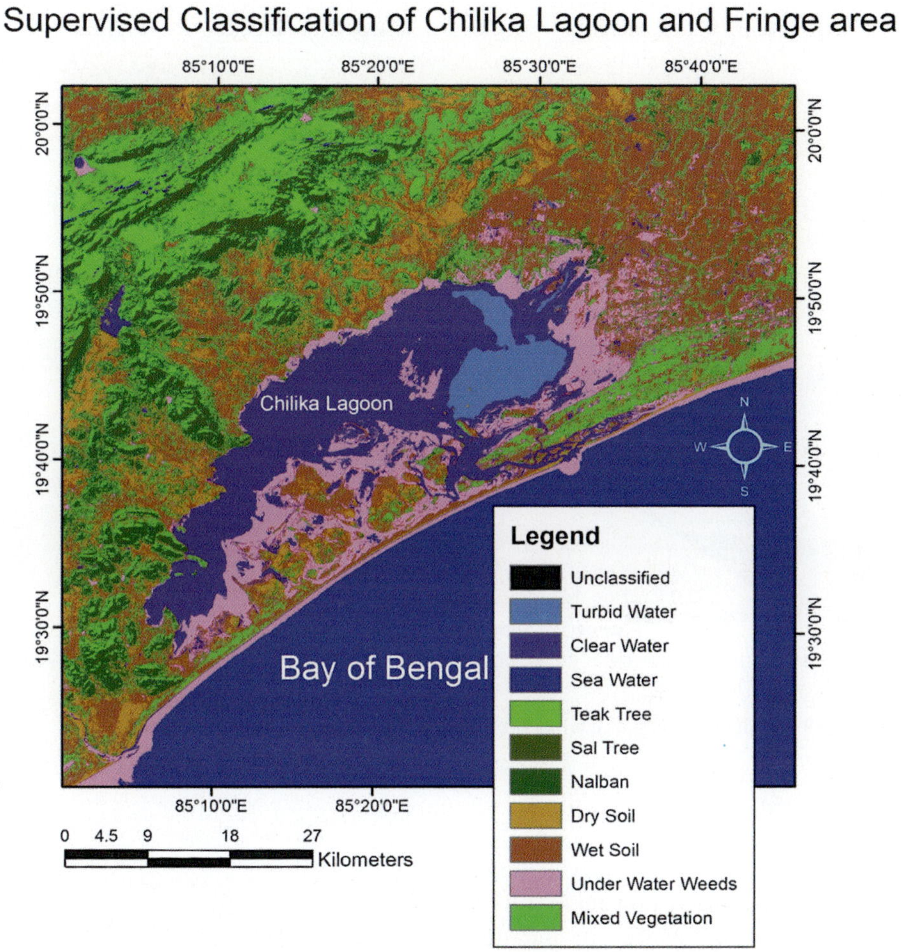

Fig. 6.18 Land covers map in and adjacent areas of the Chilika lagoon, Odisha, India

The diversity of lagoon fringe wetlands is subject to such dynamic depositional environment. Following types of wetlands are identified with the image analysis and field truth verifications (Fig. 6.18).

6.6 Chilika Lagoon Fringe Wetland Types at the Physiographic Settings

(I) **Lagoon fringe at the back spit**:

(a) Halophytic grassland or salt marshes
(b) Extensive bogs
(c) Emerged flat with cover of grasses

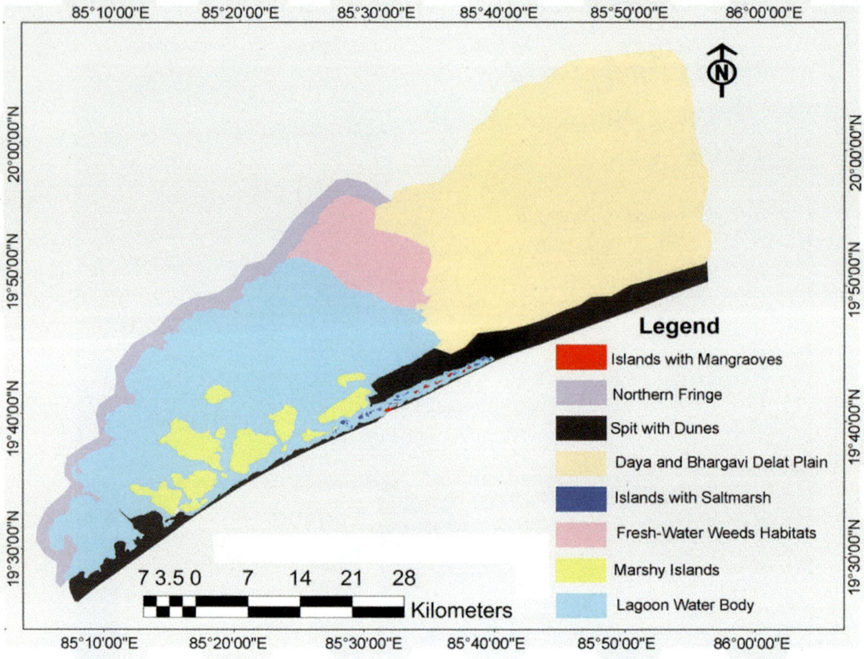

Fig. 6.19 Habitat mapping of Chilika lagoon and its fringe area

(d) Algae encrusted salt flats
(e) Algae encrusted mud flats
(f) Older channel banks with tall grasses
(g) Marginal sand flats
(h) Dots of Islands with interlinked linear sand ridges
(i) Series of protective pond water bodies
(j) Tide pools and mangroves
(k) Lagoon bed muds with oyster banks

(II) **Lagoon fringe at the North-East**:

(a) Algae encrusted mud flats
(b) Fresh water marshes and pastures
(c) Channel and channel fringe marshes
(d) Island with shrubs and woodlands
(e) Pool of water bodies with weeds

(III) **Lagoon fringe at the landward sides (West and Southwest)**:

(a) Tidal flats with algae mat and weeds
(b) Emerged flats
(c) Islands with scrubs and woodlands
(d) Shallow sub-tidal water bodies with thicker weeds (Fig. 6.19)

6.7 Storm Generated Geomorphic Changes

Chilika spit and inner islands are affected by the landfall of 'Phyllin' cyclone in 12th October of 2013. During the storm the velocity of wind was recorded as 240 km per hour at an approaching angle of waves from 50° to 217° with 7 m wave heights at the period of landfall. Such high energy wave environment with strong long shore currents and cross shore currents has produced a significant effect on geomorphology of shorefront position of Chilika spits and lagoon inside adjacent to the shoreline. Entire storm affected areas are visited to record the geomorphic changes produced by 'Phyllin' cyclone in the Chilika system (Table 6.5, Figs. 6.20 and 6.21, Plates 6.1 and 6.2).

6.8 Conclusions

The purpose of this paper has been to discuss the physiographic habitats, geomorphology, and evolution of Chilika lagoon in a humid coastal region of Bay of Bengal (Odisha, India). Following conclusions have been emerged after the present study.

Table 6.5 Storm generated geomorphic changes recorded in Chilika lagoon after 'Phyllin' cyclone

Place	Location type	Types of change	Recovery phase
Rishikulya Mouth (Purnabandh)	River Mouth with good buffer	Spit extension Spit breaching	Partial recovery
Southern spit (Brahmandeo)	Barrier spit with good buffer	Over wash deposits Dune cliffing Beach narrowing (Steeper)	Partial recovery
Rambhartiya spit (P.F.)	Barrier spit with good buffer	Dune cliffing (Recent) Over wash breaching Beach scarping	Partial recovery
Sea mouth of open spit	Low-lying spit with open sea	Inlet migration Spit breaches Over wash deposits	Incomplete recovery
Mirzapur	Lagoon fringe	Tide pools, pans surrounded by sand ridges Overwash sand fans	Incomplete recovery
Manikpatna	Lagoon fringe	Overwash sands Bank erosion Exposure of early settlement remains	Incomplete recovery

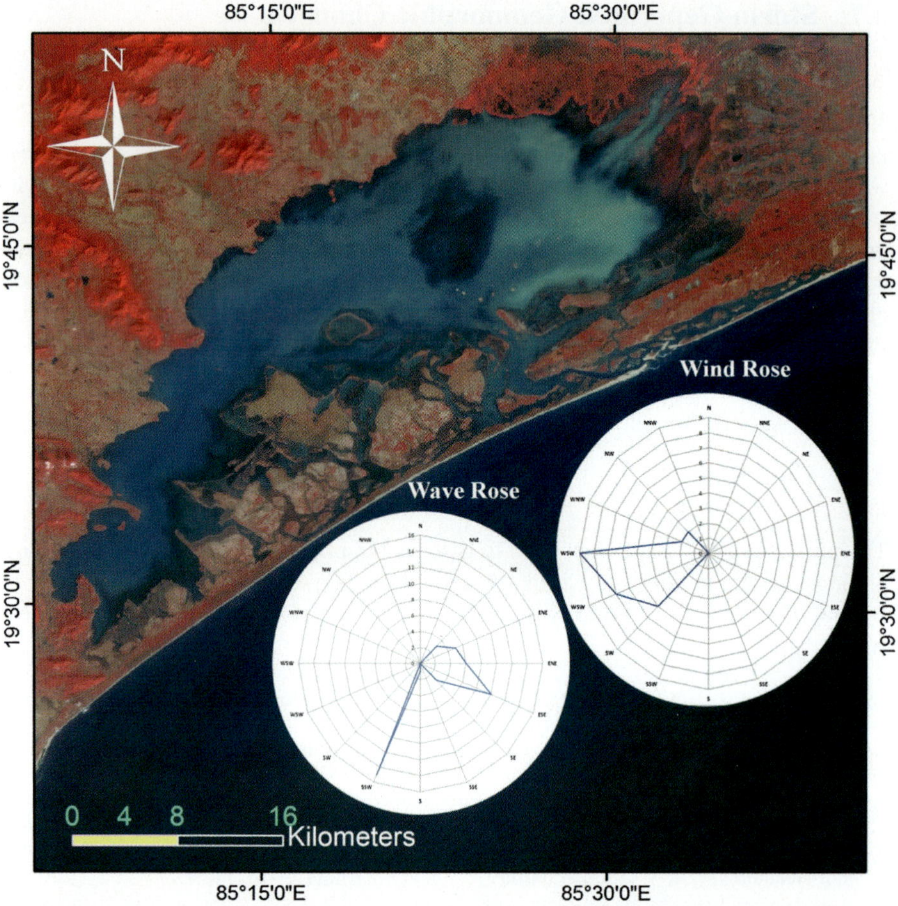

Fig. 6.20 Wind and wave direction during the landfall of cyclone 'Phyllin' in Puri-Chilika coast at Odisha

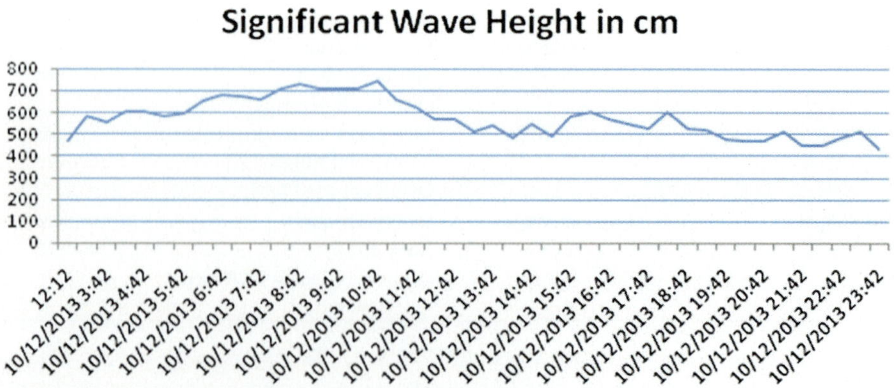

Fig. 6.21 The significant wave heights during and after landfall of 'Phyllin' cyclone around Chilika lagoon (2013)

Plate 6.1 Storm affected sedimentary environment, Chilika lagoon and adjacent areas along the Bay of Bengal shoreline (evidences from the field): *1* Older dunes are cliffed by storm surge impact at Gopalpur on sea along the southwestern shores (to the south of Chilika lagoon); *2* Exposure of a high stand in the lagoon fringe island (Brahmandeo); *3* Sedimentary facies of over wash deposits behind the younger spit of Chilika (southwest section of the lagoon system); *4* Barrier spit is eroded and cliffed at Rambhatiya protected forest. Over wash deposit in the stratigraphical section is significant; *5* Storm signature over the lagoonal facies at the interior lagoon fringe of Chilika; *6* Marsh deposition with peaty soil at the stratigraphical section of lagoon fringe area (landward side of Balugaon); *7* Palaeo beach ridge section of oxidized sands are blanketed by lagoonal and tidal depositions at the interior lagoon fringe near Mirzapur; *8* Colony of mangroves in the emerged island along the tidal water ways of Chilika adjacent to sea mouth; *9* Wash over fan lobe deposits over the lagoonal facies at the fringe area near Manikpatna; *10* Lagoonal facies over the barrier spit sands nearby sea mouth; *11* Layers of old high stands are eroded and exposed by wave erosion during storms and Manikpatna; *12* Ancient pottery remains of early settlements at the section of high stands near Manikpatna

Plate 6.2 Morphilogical features produced by storms and water level fluctuations, Chilika lagoon areas (evidences from the field): *13* Algae mat at the pan-shaped basin arrested mangrove seeds drifted from other islands dominated by mangroves; *14* Mangrove regeneration at the adverse condition of New Mouth spit towards the lagoon; *15* Over wash sand sheets at the lagoon fringe developed after the storm break in Chilika; *16* Extensive shoals of sand ridges developed at the

1. **Physiographic habitats**:
   This study has recognized that the entire areas of Chilika lagoon are characterized by seasonal fluctuations of water level, low tidal range, high monsoon freshets of Daya and Bhargavi river systems into the lagoon, exchange of sea waters through the existing inlets, and marine influences along the barrier spits and sea mouths. The lagoon is eutrophicated due to heavy nutrient loading under such environment. Diversity of physiography has influenced the nature of habitat evolution in this environment. The patterns of vegetation change can be recognized in different physiographic units of the lagoon system. A fringe of mangroves along tidal water ways, salt marsh grasses along the tidal flats, various weeds and long standing grasses along the sheltered lagoon fringes, a thick algae mat on the emerged flats and shallow water substratum with oyster beds, fresh water weeds and other marsh grasses along the inner fringe of lagoon, and other coastal plants along the sand dunes and spits involve vertical accumulation of sediments over a long period of at least since the deterioration of ancient ports (1000–2100 YBP) in and around Chilika lagoon.

   Tidal water ways between Old mouth and Sea mouth (present day) involved the occupance by plants of mud flats and sand flats which are expanding sea ward of back barrier region. Local variations in the rate of vertical accretion and periodic interruptions to the sedimentation process caused by cut and fill, wave erosion, and storm over wash, have influenced the present day vegetation cover on tidal flats.
2. **Geomorphological features and lagoon evolution**:
   There are several high strands in the form of palaeo shorelines of Chilika lagoon along the inner part of spits and islands, southwest ward hill slope (e.g. Ghantashila hills, Palur hills), along the lake margin terraces towards west and also along the northeast ward sand dunes parallel to the back spit. The features indicate the drowning and inundation of Daya and Bhargavi estuaries as well as southwest ward embayment surrounded by hills during the Holocene transgressive phase. The back water was extended towards northeast

Plate 6.2 (continued) inner part of lagoon in the form of flood delta deposits after the storm; *17* Steeper basal slope of Ghantashila Hill with lagoon water level indicates the drowning effect of Holocene transgressive sea in the region; *18* Seasonal water level marks on the vertical hill slope of Ghantashila near Rambha (Chilika lagoon towards southwest corner); *19* Over flowing impact of storm surge water across the dune barrier close to Rishikulya mouth; *20* Retreat of barrier spit dune by storm erosion at Rambhatiya near sea mouth; *21* Winnowing action of back barrier spit by high water level at the time of storm surge; *22* Carbonate rich cemented sands at the island crest (below 5m in elevation) near the inner part of lagoon (indicative of high sea level into the lagoon in the past); *23* Ancient temple excavated from the sediment filled older spit of inner Chilika (Bhabakundaleswar); *24* Younger sand lobe advanced towards the marsh deposits at Mirzapur; *25* Salt water inundated areas of Ganjam coast near Chilika after the storm break of 'Phyllin'; *26* Location of fragmented high stands at the inner part of a Chilika island; *27* Modern buoya anchored over the lagoon water level for monitoring informations of lagoon by Chilika Development Authority (CDA) parallel to the shoreline, and later on by regressive phase of the sea a series of spits develop along shoreline from northeast to southwest direction, and finally a set of spit also develops along the shoreline from the Rishikulya river mouth to northeast direction. During this period the lagoon deposits modify the Chilika basin with tidal, fluvial, coastal marine, and wind storm processes.

Modern barriers are affected by storm generated washover process and moved over the lagoonal facies of Chilika. Flood deltaic deposits also through tidal inlets have created advancing sand ridges or shoals into the lagoon water bodies during repeated storms. Location of several abandoned tidal inlets and lagoon ward encroached islands behind the barrier spits proves the closing of inlets by modern barrier advancement and repeated wash over fan lobes deposit of barrier islands by storms. Modern tidal flats are located on the islands and lagoon fringes. The northeastern parts of lagoon are mostly filled up by deltaic sedimentary fan lobes of Daya and Bhargavi rivers over which many abandoned channels and wetlands are located at present.

References

Basa KK, Behera KS (2000) Maritime archaeology of Orissa. In: Basa K, Mohanty P (eds) Archaeology of Orissa, vol 1, Pratibha Prakashan, Delhi, pp 574–577

Bhattacharya S, Sen SK, Acharyya A (1994) Structural setting of the Chilka Lake granulite–migmatite–anorthosite suite with emphasis on the time relation of charnockites. Precamb Res 66:393–409

CDA (2009) History of Chilika. Chilika Lake Development Authority's website, Odisha. Retrieved 15 December, 2013

Das BP, Jena J (2008) Impact of Mahanadi Basin development on eco-hydrology of Chilika. In: Sengupta M, Dalwani R (eds) Proceedings of Taal2007: the 12th world lake conference, Jaipur, pp 697–702

Malini BH, Rao KS, Rao KN (1993) Evolution and dynamic of Chilika Lake. American Society of Civil Engineers. Biennial symposia: 8th, comprehensive coastal and ocean management: the management of coastal lagoons and enclosed bays; Conference proceeding paper, pp 257–268

Murray AS, Mohanty M (2006) Luminescence dating of the barrier spit at Chilika lake, Orissa, India. Radiat Prot Dosim Oxford J 119(1–4):442–445

Pradhan D, Mohanty P, Mishra J (2000) Manikapatna: an ancient and medieval port on the coast of Orissa. In: Basa KK, Mohanty P (eds) Archaeology of Orissa, vol 1, Pratibha Prakashan, Delhi, pp 473–493

Rao PS, Rao GK, Durgaprasada Rao NVN, Swamy ASR (1990) Sedimentation and sea level variations in Nizampatnam Bay, East Coast of India. Ind J Marine Sci 19:261–264

Sinha BK (2000) Golabai: a protohistoric site on the coast of Orissa. In: Basa KK, Mohanty P (eds) Archaeology of Orissa, vol 1. Pratibha Prakashan, Delhi, pp 322–355. ISBN 81-7702-011-0

Subrahmanyam V, Subrahmanyam AS, Murthy KSR, Murty GPS, Sarma KVLNS, Rani PS, Anuradha A (2006) Precambrian mega lineaments across the Indian sub-continent – preliminary evidence from offshore magnetic data. Curr Sci 90:578–580

Watters T (1973) On Yuan Chawang's travels in India, New Delhi, vol 2, pp 193–195

Chapter 7
Foreshore Applications of *X*-band Radar

G.M. Jahid Hasan and Satoshi Takewaka

Abstract The shallow waters of a nearshore region are dynamic and often hostile. Prediction in this region is usually difficult probably by our limited understanding of the physics or by availability of accurate field data. It is a challenge for traditional in situ instruments to provide these inputs with the appropriate temporal or spatial density at a reasonable cost. Remote sensing provides an attractive alternative. An *X*-band nautical radar system was employed for this study to examine alongshore propagation of low frequency run-up motion around the research pier HORS in Hasaki beach, Japan. Analyses on radar echo images were done to estimate longshore distribution of shoreline positions and inter-tidal foreshore profile using time-averaged images. Spatio-temporal variation of water fronts were digitized manually from cross-shore time-stack images. Run-up heights were then estimated from the digitized water fronts with the help of foreshore slope. Run-up variations under dissipative condition were parameterized with surf similarity parameter. Low frequency variances in the run-up motion were observed, which were traveling in the longshore direction. Longshore structures of this motion were examined and compared with different wave incidences during two typhoon events in the Pacific Ocean. Estimates of morpho-dynamic parameters during passage of different storms were analyzed and are explained in this chapter to demonstrate the potential of radar measurement in capturing essential characteristics of foreshore dynamics.

G.M.J. Hasan (✉)
Department of Civil and Environmental Engineering, Shahjalal University of Science and Technology, Sylhet 3114, Bangladesh

Department of Civil Engineering, Military Institute of Science and Technology, Dhaka 1216, Bangladesh
e-mail: jahid01@yahoo.com

S. Takewaka
Department of Engineering Mechanics and Energy, Graduate School of Systems and Information Engineering, University of Tsukuba, Tsukuba, Ibaraki 305-8573, Japan
e-mail: takewaka@kz.tsukuba.ac.jp

C.W. Finkl and C. Makowski (eds.), *Remote Sensing and Modeling: Advances in Coastal and Marine Resources*, Coastal Research Library 9,
DOI 10.1007/978-3-319-06326-3_7,

7.1 Introduction

The nearshore is the narrow strip of the ocean that borders the continents. It can be dynamically defined as the coastal region that is significantly affected by gravity waves, and spans from inland reach to an offshore depth of O (10 m) beyond which bathymetric change is no longer due to wave motions. The nearshore is the energetic region of the coastal environment where ocean waves shoal and interact with local morphology. The nearshore is a difficult domain to sample and understand. As waves approach an open shoreline from deep water, they undergo several changes. Breaking waves in the surf zone are often intense and wave-driven currents can be strong, which makes work in the surf zone dangerous to both people as well as instruments. Changes in bathymetry can occur on time scales as short as hours due to presence of large storms. Sandy beds can undergo substantial erosion or accretion within short periods, so traditional bottom-mounted sensors are often rapidly scoured out or buried in.

Data are especially difficult to collect in nearshore region during periods of large waves or strong currents. However, these conditions often represent the most energetic periods for hydrodynamics, morpho-dynamics and are the times of greatest scientific interest. The harsh and challenging conditions make long-duration in situ observations in the surf zone problematic but suggest the potential benefits of remote sensing approaches. Although most people associate remote sensing with satellites, a number of other solutions are available, from airborne sensors on manned and unmanned platforms to shore-based sensors mounted on lighthouses, towers or bluffs. All have the advantage that sensors can be installed away from harsh marine conditions; can often have direct access to the power grid, storage, and the internet; and can usually observe a large spatial extent over long durations at a low cost.

The nearshore can be sampled by a full suite of both active and passive remote sensors (cameras, radars, lidars, etc.) using a range of platforms (fixed, flying, floating, and orbiting) and operating across the visible, infrared (IR), microwave, and radio bands of the electromagnetic (EM) spectrum in addition to the in situ sensors. However, for nearshore oceanographic applications, fixed optical cameras and X-band radars are used the most frequently nowadays.

7.1.1 Requirements of Nearshore Monitoring System

The dynamics of the nearshore are driven primarily by ocean wave energy that has been generated elsewhere and propagated into the nearshore domain. Unlike the deeper ocean, the dissipation of nearshore energy occurs in a narrow strip due to depth limitation. These processes generally drive nearshore currents and circulation as well as the sediment transport that creates nearshore sand bars and complex bottom morphology.

The goal of nearshore science is to understand, characterize, and predict the evolving waves, currents, and bathymetry over any nearshore region for which observations are available. Solutions have inevitably consisted of numerical models, which have shown increasing skill when fed with accurate bathymetry and wave data (e.g., Ruessink et al. 2001). For all but a few well-instrumented field sites, prediction accuracy is limited by data availability, particularly for bathymetry data, gathering of which usually requires expensive sampling methods and can be done only occasionally. Thus, data assimilation methods must be used to merge available observations with models to yield dynamically consistent estimates of flow variables and bathymetries (e.g., Wilson et al. 2010).

The requirements for successful sampling of the nearshore are governed by the time and space scales of variability within the system. Visually, the most obvious timescale is associated with the *O* (10 s) periods of surface waves. However, the nearshore spectrum often includes significant motions at longer scales such as infra-gravity waves (30–300 s periods) driven by wave groups (Herbers et al. 1995), very low-frequency motions (102–103 s periods) arising from current instabilities (Oltman-Shay et al. 1989), and longer timescales associated with system modulation by tides (Thornton and Kim 1993), and extreme events such as storms and hurricanes as well. Sand bars and bottom profile shapes evolve on timescales as short as days and force corresponding changes in the hydrodynamics. The 1–10 Hz sampling capability of most existing in situ sensors is adequate for high-frequency needs, but the requirement for extended sampling under difficult conditions usually leads to sensor degradation and excessive expense.

In the spatial domain, the nearshore is a region of high inhomogeneity. Surf zone motions and morphology are rarely uniform alongshore and so must also be sampled in the alongshore direction. Wave-breaking processes generally drive nearshore currents with similar spatial variability and induce sediment transport that creates nearshore sand bars, complex bottom morphologies, and eroding or accreting beaches. Wave motions, currents, and bathymetry all vary strongly over 10–1,000 m cross-shore scales. Proper sampling of this wide range of spatial scales would require a large array of in situ point sensors. In contrast, most remote sensors operate in an imaging mode that is optimized for such spatial sampling needs.

7.1.2 Merits and Demerits Between Different Sensors

Traditional in situ instruments such as buoys or bottom mounted pressure sensors can record precisely the variations of water surface elevations, velocities, sediment concentration, bottom elevation etc. with high sampling capability. These instruments are a powerful tool for monitoring nearshore regions. However, the measurements are expensive and limited in spatial extent. They must withstand wave forces, ocean currents, marine growth, and salt water immersion. In situ point sensors must be constantly monitored and regularly maintained to ensure their successful operation. Moreover, they require minimum depths to operate properly and are limited in study of nearshore processes. Further, during energetic sea states it is not easy to maintain data collection. Remote sensing techniques provide a feasible alternative because they allow sampling over large spatial extents (meters to kilometers) and temporal scales (seconds to years).

Unlike in situ point sensors, remote sensing techniques provide a feasible alternative since they allow sampling over large spatial extents and temporal scales. Every digitized pixel in the remotely sensed area can be thought of as representing a "virtual" wave gauge. Methods of nearshore remote sensing include: Space-borne (satellite) radar, Synthetic Aperture Radar (SAR), High Frequency (HF) radar, *X*-band radar and Video Camera.

Space-borne radar and SAR remote sensing systems are limited by their ability to provide long term temporal information due to their lack of time on scene. Shore-based remote sensing systems, such as video and radar remote sensing systems can be set up at any coastal sites and offer much longer times on scene. Shore-based remote sensing systems avoid the problems associated with deployment and maintenance of in situ point sensors in the marine environment. Logistically there is minimal adjustment to the equipment for each individual deployment as the system is designed to function in a variety of environments. Deployment of shore-based remote sensing systems can typically be made in harsh weather without worry of damage to equipment or personnel.

Aerial photography is one of the earliest basic forms of remote sensing used to image wave fields over an area, and recently video cameras (e.g., Stockdon and Holman 2000; Plant and Holman 1997; etc.) or nautical radars (e.g., Bell 1999; Borge et al. 2004; Hasan and Takewaka 2007a, b etc.) have been employed for temporal coastal imaging supported by digital technologies and becoming popular for continuous monitoring.

For shore-based systems, e.g., a High-Frequency (HF) radar or *X*-band radar, the operation cost is relatively low, especially for the *X*-band radar if the required study area is within a radius of 3–5 km. HF radars are mainly used for large area currents and wave measurements (Gurgel et al. 1999). The principal for a HF and a *X*-band radar are same, all use Bragg effect to detect the scatter waves, but the selected wavelength are different. A HF radar selects large EM wavelength (with the frequency range from 3 to 30 MHz) to produce Bragg scatter waves with the gravity waves itself, but a *X*-band radar uses short EM waves to interact with ripples on top of the gravity waves. The operation range of HF radars is large (normally on the order of 50 km, and can be extended to 200 km with an antenna array) but with a relatively low resolution.

Video cameras can monitor sea surface patterns with high temporal resolution and can provide color images, which enable to detect wave breaking, suspension of foams, sediment concentrations etc. However, the spatial coverage is limited with a single camera, so deployment of multiple cameras is necessary to gain wide coverage. In addition, the use of video cameras is limited to daylight hours and fair atmospheric conditions.

Remote sensors often have the opposite problem of data starvation, i.e., too much data. For example, a single video camera can easily deliver 35 MB S^{-1}, and five cameras are commonly needed to span the full 180° field of view. Twelve hours of daylight would then yield 7.2 TB. Finally, remote sensing data are often surprisingly noisy. The presence of fog, low wind, or rain also contributes to data degradation.

Table 7.1 Radar classifications (after Raemer 1996)

Band	Frequency	Wavelength
HF	3–30 MHz	10–100 m
VHF	30–300 MHz	1–10 m
UHF	300 MHz–1 GHz	30 cm–1 m
L-band	1–2 GHz	15–30 cm
S-band	2–4 GHz	7.5–15 cm
C-band	4–8 GHz	3.75–7.5 cm
X-band	8–12 GHz	2.5–3.75 cm
Ku-band	12–18 GHz	1.67–2.5 cm
K-band	18–27 GHz	1.11–1.67 cm
Ka-band	27–40 GHz	7.5 mm–1.11 cm
V-band	40–75 GHz	4–7.5 mm
W-band	75–110 GHz	2.7–4 mm
mm-band	110–300 GHz	1–2.7 mm

Some of the limitations of video imaging can overcome using *X*-band nautical radar. *X*-band nautical radar is an imaging tool that is capable of tracking movements of wave crests over large spatial regions and is becoming popular for coastal studies nowadays. The advantage in using an *X*-band radar system is its ability to monitor the coastal processes remotely and continuously, under calm conditions as well as stormy. Radar image sequences with large coverage (extending up to several kilometers) offer a unique opportunity to study individual waves and wave fields in space and time. The drawbacks of radars are their relatively low sampling rate and difficulties in detection of sediment suspensions and wave breaking states from the echo backscatter.

7.1.3 Basics of Radar Remote Sensing

Radar systems are classified by the wavelength of their emitted frequency. Marine radars are typically *X*- or *S*-band. Table 7.1 lists radar system classifications based on frequency and wavelength. Radar backscatter is produced from the interaction of electromagnetic (EM) waves and the rough surface of the ocean. The backscatter of the ocean surface is often called "Sea Clutter". Electromagnetic energy is backscattered from the ocean surface by two general methods. Specular reflection is mirror-like reflection that is most predominant for small angles of incidence. The incidence angle is the angle that the incident ray makes with the normal to the surface. The grazing angle is complement of the incidence angle. In specular reflection the EM energy is reflected directly off the surface of the water. Bragg scattering is resonant reflection that occurs when EM energy interacts with waves that have a similar wavelength as the transmitted electromagnetic waves. Small scale capillary waves on the order of 1–3 cm provide the source for Bragg resonant scattering of marine *X*-band radars. Thus, the presence of wind is required to create these small scale waves in order for *X*-band radars to image ocean waves. The intensity of the backscattered energy is most affected by the magnitude of the wind velocity, the incidence angle of the emitted EM energy, and the azimuth angle of the radar relative to the crest of the wave. Bragg scattering is strongest when the radar azimuth angle is aligned with the wave direction.

7.1.4 Swash Motion Studies

Swash is the time varying position of the shoreward edge of water on a beach (e.g., Guza and Thornton 1985), and swash zone is the boundary between the inner surf zone and the sub-aerial beach (e.g., Ruggiero et al. 2004). Swash zone possesses scientific importance since significant amount of total surf zone sediment transport occurs within this region (Osborne and Rooker 1999). Swash zone hydrodynamics play a critical role in design and maintenance of shore protection structures as it is responsible for changing beach morphology by wave induced erosion or deposition (Ruggiero et al. 2001; Sallenger 2000). Despite this importance, there is still considerable debate as to how swash is related to environmental parameters such as the local beach slope or incident wave characteristics etc. Recent literature on swash zone processes (e.g., Elfrink and Baldock 2002; Butt and Russell 2000) highlights the lack of sufficient knowledge within this region.

Wave run-up is expressed as the set of discrete vertical elevations of seawater, measured on the foreshore from still water level (SWL) consisting of two components: a super elevation of mean water level, commonly known as wave setup, and vertical fluctuations about that mean. Many researchers investigated run-up dynamics on intermediate to reflective beaches (e.g., Holman and Sallenger 1985; Holman 1986; Holland and Holman 1999), low energy and mildly dissipative beaches (e.g., Guza and Thornton 1982; Raubenheimer et al. 1995; Raubenheimer and Guza 1996) and highly dissipative beaches (e.g., Ruessink et al. 1998; Ruggiero et al. 2004; Holman and Boyen 1984; Hasan and Takewaka 2009). Most of the studies parameterized wave run-up with environmental conditions and examined its relation with surf similarity parameter. However, there is still uncertainty on longshore propagation of the run-ups, particularly for the low frequency motions.

Miche (1951) hypothesized from laboratory observations that swash motion due to monochromatic waves generally saturated, and shoreline oscillation depends on wave frequency and beach slope. The saturation condition implies swash amplitude does not increase with increasing offshore wave height. Frequency components that are not dissipated and survive in the surf zone are reflected at the shoreline, and are relatively low frequency and have been termed as infra-gravity energy. Several studies have examined relative roles of the infra-gravity band run-up (generally $f < 0.05$ Hz) and wind wave or incident band run-up ($f > 0.05$ Hz) for particular field sites (e.g., Ruessink et al. 1998; Ruggiero et al. 2004; Hasan and Takewaka 2009 etc.). On natural beaches, run-ups at incident frequencies are typically saturated, while run-ups at infra-gravity frequencies are unsaturated and thus increases with offshore wave height (e.g., Guza and Thornton 1982; Holman and Sallenger 1985; Holland et al. 1995 etc.) and also depends on beach morphology (e.g., Hasan and Takewaka 2009). Extensive field evidence supports that energy at infra-gravity frequencies often constitute a large portion of total energy observed at the shoreline during storm events (Holman 1981; Guza and Thornton 1982; Holman and Sallenger 1985; Raubenheimer and Guza 1996).

Holland and Holman (1999) examined wavenumber-frequency structure of swash variations over a range of environmental conditions using video technique with a 250 m longshore coverage. Frequency spectra of their analyses showed energy concentration at the infra-gravity band and wavenumber spectra were dominated by energy at low wavenumbers. Takewaka and Nishimura (2005) used *X*-band radar to analyze qualitatively longshore structure of swash motion during an energetic sea state. Recently, Hasan and Takewaka (2009) also analyzed wave run-up using *X*-band radar images during an energetic sea condition. They could successful in tracing swash variations and its propagation in the longshore direction.

7.1.5 Objectives of the Chapter

Foreshore can be defined as the part of the shore, which is wet due to the varying tide and wave run-up under normal conditions. When ocean waves approach the shore, majority of the wave energy is dissipated across the surf zone through wave breaking. However, a portion of the energy is converted to potential energy in the form of run-up on the foreshore (Hunt 1959). This wave run-up is important to coastal engineers, nearshore oceanographers or coastal planners since run-up motions deliver most of the energy which is responsible for foreshore morphology such as dune and beach erosion (Ruggiero et al. 2001; Sallenger 2000) or beach cusp development (Inman and Guza 1982) etc. Hence, understanding the magnitude, longshore variability or propagation of wave run-up is critical to maintain the foreshore and its adjacent areas properly.

The deployment of a radar system is particularly advantageous over other measurement techniques for foreshore monitoring i.e. spatial analyses of run-up motions. Temporal and spatial variation of water front for a region can be observed with a single radar system. Using wave gauges to understand the spatial behavior of run-up will be extremely hard, laborious and expensive. Traces of water front and its motion can also be captured using video cameras with high temporal resolution; however, as mentioned above, at least several cameras are required in order to gain similar spatial coverage of a radar system.

An *X*-band radar system was operated at the research pier HORS (Hasaki Oceanographical Research Station, Fig. 7.1) at Hasaki beach, Japan for continuous nearshore monitoring since 2002. The present chapter analyzes radar echo images to estimate temporal and spatial variation of wave run-up, its spectral response, and dependence with foreshore slope during different storm events in 2005 over an extent of approximately 3.0 km. Radar echo images were collected for several hours during the passage of two typhoon events. Analyses on the echo images were done to evaluate longshore distribution of shoreline positions and inter-tidal foreshore slopes. Characteristics of the low frequency motions, its propagation and relation with different incident wave fields were analyzed and explained.

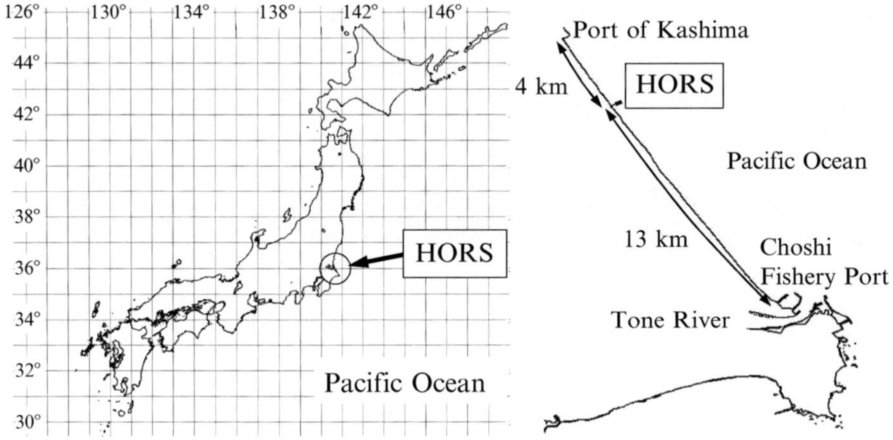

Fig. 7.1 Location of the Hasaki Oceanographical Research Station (HORS) in Japan

Section 7.1 explains the importance of nearshore monitoring, merits and demerits between different sensors used for nearshore monitoring and objectives of this chapter. Setup of the study with incident wave field can be found in Sect. 7.2. Estimation of inter-tidal bathymetry and foreshore slopes using time-averaged radar images are explained in Sects. 7.3 and 7.4. Section 7.5 describes the estimation procedure of wave run-up with their validation, and formulate run-up with foreshore slope and surf similarity parameter. Concluding remarks based on the radar remote sensing are depicted in Sect. 7.6.

7.2 Setup of the Study

7.2.1 Research Pier HORS

X-band radar measurements were conducted at the research pier HORS of the Port and Airport Research Institute (PARI), located in Hasaki, Japan. Main facilities of the station are a 400 m long pier extending into the Pacific Ocean equipped with different in-situ measuring instruments and a research building located nearly 110 m back from the mean shoreline position. HORS is on an almost straight sandy coast approximately 17 km long with the Hasaki Fishery Port at the south end and Kashima Port at the north end of the coast (Fig. 7.1). The pier is located approximately 4 km away from Kashima Port. Water surface elevations are measured routinely with several wave gauges mounted on the pier and bottom profiles along the pier are surveyed on weekdays.

Moreover, airborne ultrasonic wave gauges were deployed in the swash zone with a close spacing along the pier for a few years (Mizuguchi and Seki 2007).

Fig. 7.2 Aerial view of the Hasaki observation pier (*left photo*) and radar antenna (length ~2.8 m) on the roof of the research building (*right photo*). Length of the pier is approximately 400 m

Some gauges were installed well onshore around mean shoreline position to cope with large run-up during storm events. Wave gauges in the foreshore were placed with a small spacing (approximately 3 m). Wave gauges placed close to the shoreline are capable of measuring the elevation of the beach when it is bare during the swash-down phase as well as the water surface elevation when it is inundated during the swash-up phase. Thus the wave gauge system is capable of detecting the water front of the swash motion, and continuous measurements of both waves and beach profiles in the swash zone are possible.

7.2.2 Radar System and Echo Image

The radar system employed in this study is a conventional marine *X*-band radar, usually installed on fishing or recreational boats (Takewaka 2005; Hasan and Takewaka 2007a). The radar antenna (approximately 2.8 m in length, Fig. 7.2) is installed on the roof of the research building, which is 17 m above mean sea level. The figure also gives an aerial view of the HORS pier and its surrounding area. The antenna rotates with a period of approximately 2.5 s and transmits with a beam width of 0.8° horizontally and 25.0° vertically. The echo signals from the sea surface, generally called sea clutter, are captured with a specially designed A/D board installed on a computer. The backscatter echo signals are converted to gray images for further analyses.

Figure 7.3 shows the coordinate system used in this study. A sample radar image with reduced brightness is shown as a background in this figure. The *x*-axis

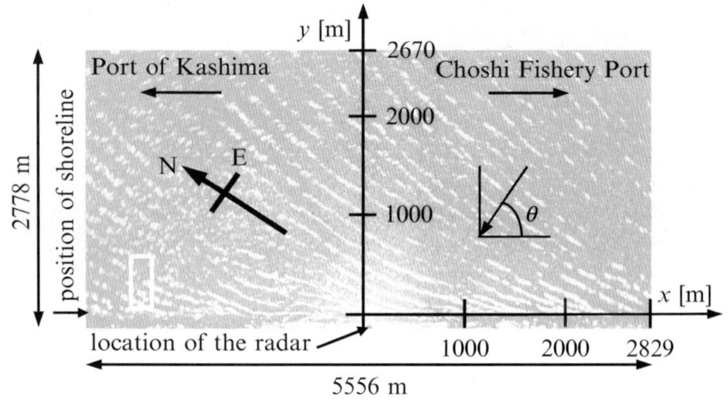

Fig. 7.3 Coordinate system. Location of the radar is at $(x, y) = (51, -108)$

corresponds to the longshore extent and positive towards the Choshi Fishery Port and the y-axis coinciding with the pier and oriented in the offshore direction. The figure also indicates the wave propagation direction θ, with 90° being normal to the shore.

Measurements of the sea state using nautical radars are based on the backscatter of electromagnetic pulse from the sea surface. The echo signals from X-band radar were used to generate images with 1,024 pixels in horizontal and 512 pixels in vertical. The horizontal and vertical extents of an image correspond physically to a longshore extent of 5.6 km (≈3.0 nautical miles) and a cross-shore extent of 2.8 km (≈1.5 nautical miles), respectively, and each pixel corresponds to a square of 5.4 m. The gray images have pixel intensities between 0 (no backscatter) and 255 (highest signal).

The radar images were updated at every 2 s and its processing has been discussed in the author's previous work (Takewaka 2005; Hasan and Takewaka 2007a) in details. Figure 7.4 shows typical radar images captured during two storm events in 2005 having different incident wave field. The slanted white lines are oblique wave crests and the vertical white strip in the middle of the figure close to the bottom edge is the pier. Propagation of waves from different directions is observed from a sequence of radar images. The radar is located at the midpoint of the bottom edge of the diagram indicated by a black dot. Incoming waves mainly from a single direction is observed in the upper panel of Fig. 7.4, whereas waves propagation from multiple directions are shown in the lower panel of the figure.

7.2.3 *Meteorological and Sea State Conditions*

Typhoon BANYAN (Typhoon #7, T7 in 2005) traveled through the western edge of the Pacific Ocean from 21st to 31st of July. Also, typhoon SAOLA (Typhoon #17, T17 in 2005) followed almost similar track in the Pacific Ocean and traveled from 20th to 26th of Sept. Tracks of both the typhoons can be obtained from the authors'

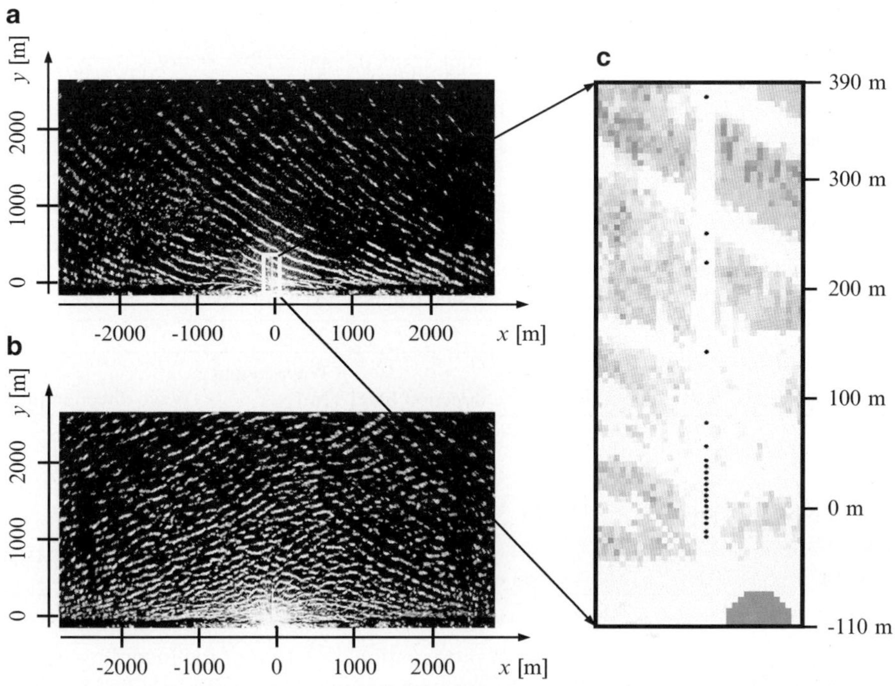

Fig. 7.4 Typical radar echo image during the period of (**a**) typhoon BANYAN (typhoon # 7, T7, $H_s = 3.69$ m and $T_s = 12.1$ s) and (**b**) typhoon SAOLA (typhoon # 17, T17, $H_s = 4.87$ m and $T_s = 9.6$ s) in 2005. Close up view of an echo image showing wave gauge locations (*black dots*) along the pier (**c**)

previous studies (Hasan and Takewaka 2007a, b). Radar echoes were collected continuously for several hours during the passage of the typhoon events and used for these analyses.

Atmospheric pressure is recorded by the Japanese Meteorological Agency (JMA) at the Choshi Fishery Port. Analyzing periods for both the storm events were selected, when the lowest atmospheric pressure was closest to the HORS, which is 22nd hour of Jul 26 during typhoon BANYAN and 14th hour of Sept 25 during typhoon SAOLA. Wind speed and its direction are also collected from JMA. Wind was found weak during 2005/Jul/26/22 h, whereas strong north-eastern wind (~23.5 m/s) was blowing over this region during 2005/Sept/25/14 h which was induced by the typhoon SAOLA.

Offshore waves were measured by the Nationwide Ocean Wave Information Network for Ports and Harbors (NOWPHAS, http://infosv2.pari.go.jp/bsh/ky-skb/kaisho/eng/marine_home_e.htm) station near the Kashima Port. Table 7.2 summarizes the sea state conditions during the analyzing periods obtained from NOWPHAS record. The local datum level (DL) defined at HORS used in this study which is 0.687 m below the datum level of Tokyo Peil (TP) usually used in Japan.

Table 7.2 Sea state conditions during the analyzing hours estimated by NOWPHAS (at Kashima Port)

Parameters	2005/Jul/26/22 h	2005/Sept/25/14 h
Significant wave height H_s	3.69 m	4.87 m
Significant wave period T_s	12.1 s	9.6 s
Significant wave direction θ_s	33°	89°
Tide level	1.21 m	1.39 m

Table 7.3 Estimated wave heights and directions for different frequency bins by NOWPHAS during typhoon BANYAN (2005/Jul/26/22 h) and typhoon SAOLA (2005/Sept/25/14 h)

	2005/Jul/26/22 h		2005/Sept/25/14 h	
Frequency [Hz]	Wave height [m]	Wave direction [deg]	Wave height [m]	Wave direction [deg]
<0.03125	0.31	Not estimated	0.92	Not estimated
0.04 ~ 0.0625	1.73	39	0.95	60
0.0704 ~ 0.0935	2.86	38	2.55	88
0.102 ~ 0.125	1.23	47	2.5	110
0.133 ~ 0.2325	1.34	51	2.54	123
>0.244	0.80	Not estimated	1.10	Not estimated

Sequences of radar images exhibit propagation direction of the incoming waves. Single dominant oblique direction was observed from the wave field during typhoon BANYAN (2005/Jul/26/22 h), whereas incoming waves from several directions were observed from the wave field generated by typhoon SAOLA (2005/Sept/25/14 h). Wave heights and wave directions of different frequencies for the analyzing hours are summarized in Table 7.3 which was processed by NOWPHAS at Kashima Port. Propagations of different frequency waves from different directions were observed during 2005/Sept/25/14 h which can be observed from the table, whereas almost single incoming waves were found during 2005/Jul/26/22 h.

7.2.4 Spectral Characteristics of Incident Wave Fields

Offshore wave field during the period of typhoon BANYAN (2005/Jul/26/22 h) were analyzed by Hasan and Takewaka (2007a) using *X*-band radar images. Temporal variations of echo signals were analyzed with the Fast Fourier Transform (FFT) to examine the dominant frequency wave. A sequence of images of 256.0 s (128 images) was used in the analysis and the peak frequency corresponding to the highest spectral density was determined. Spectrum along a cross-shore line $y = 1{,}000$ m is shown in the upper panel of Fig. 7.5a which is normalized and alongshore averaged. Peak frequency f_p of 0.066 Hz (~15.1 s) is apparent in the spectrum. The lower panel of Fig. 7.5a shows the results of the FFT analyses along the cross-shore displayed as an image view. Spectra were prepared in the same manner as described above. The brightness of the image corresponds to higher spectral strength. The figure depicts the

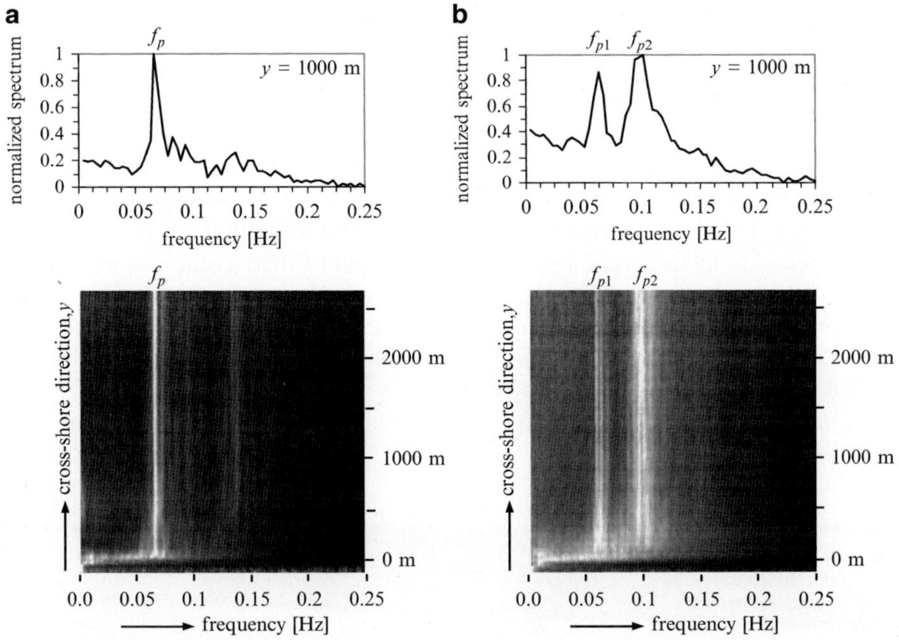

Fig. 7.5 Frequency spectra of the incident wave field for the period of (**a**) typhoon BANYAN (2005/Jul/26/22 h) and (**b**) typhoon SAOLA (2005/Sept/25/14 h) (From Hasan and Takewaka 2007a, b). Brighter intensities in the lower images correspond to higher spectral strength

cross-shore uniformity of the peak frequency f_p 0.066 Hz, which shows dominance of a single wave component during this period.

Temporal variations of radar echo signals were also analyzed with the FFT for the period of typhoon SAOLA (2005/Sept/25/14 h) by Hasan and Takewaka (2007b). A sequence of images of 512.0 s (256 images) was used in the analysis and peak frequencies corresponding to higher spectral strength were determined. Spatially averaged spectra at a cross-shore coordinate $y = 1{,}000$ m is shown in the upper panel of Fig. 7.5b which was processed similarly as described above. Two distinct peaks f_{p1} and f_{p2} are observed in the spectrum. The higher frequency component ($f_{p2} = 0.095$ Hz) is predominant over the other one ($f_{p1} = 0.0625$ Hz). Lower panel of Fig. 7.5b shows the results of the FFT analyses displayed as an image view. The figure shows almost uniformity of the spectral structure in the cross-shore direction. Energetic waves at higher frequency range (0.133 ~ 0.2325 Hz) is observed by NOWPHAS as shown in Table 7.3, which is not resolved in the spectra estimated using radar images. Since this frequency range is close to the Nyquist frequency of radar measurement (0.25 Hz for this analysis), the radar system failed to catch these waves.

In the followings, longshore structures of the run-up motion were analyzed and compared between different incident wave fields, namely 22 h of 2005/Jul/26 as a wave field with dominant single wave incidence and 14 h of 2005/Sept/25 as a wave field with multiple wave incidences.

7.3 Implication of Averaged Images

7.3.1 Inter-tidal Bathymetry

Inter-tidal foreshore profiles (i.e., bathymetry, foreshore slope etc.) were estimated by analyzing time-averaged images of different tide levels between high and low tide bands as the same manner of Takewaka (2005) for both the typhoon periods. Sequence of individual echo images were averaged to yield a time-averaged image or so called time-exposure image. Figure 7.6 shows images averaged over 17 min (512 images) during a high tide and a low tide observed under stormy conditions on 2005/Jul/26 (typhoon BANYAN period) and also on 2005/Sept/25 (typhoon SAOLA period). Individual waves vanish in the time-averaged images, and a horizontal edge corresponding to the shoreline appears as shown in Fig. 7.6.

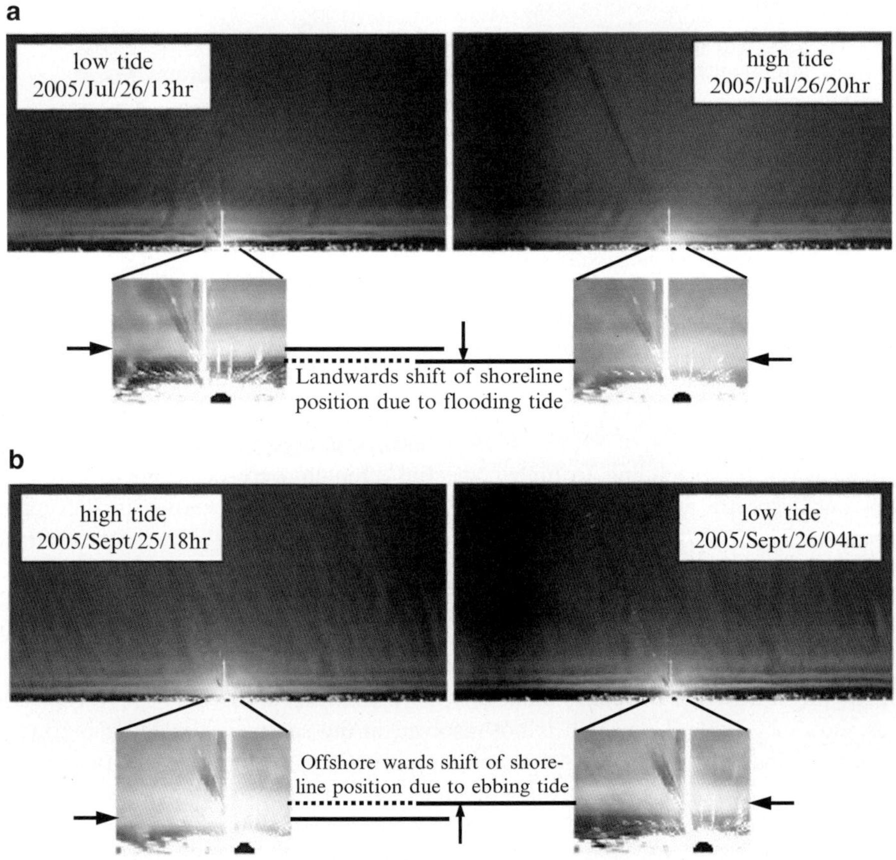

Fig. 7.6 Averaged images at low tide and high tide condition for the period of (**a**) typhoon BANYAN (T7) and (**b**) typhoon SAOLA (T17) in 2005. Series of averaged images within this low tide and high tide band was used to estimate the foreshore profile

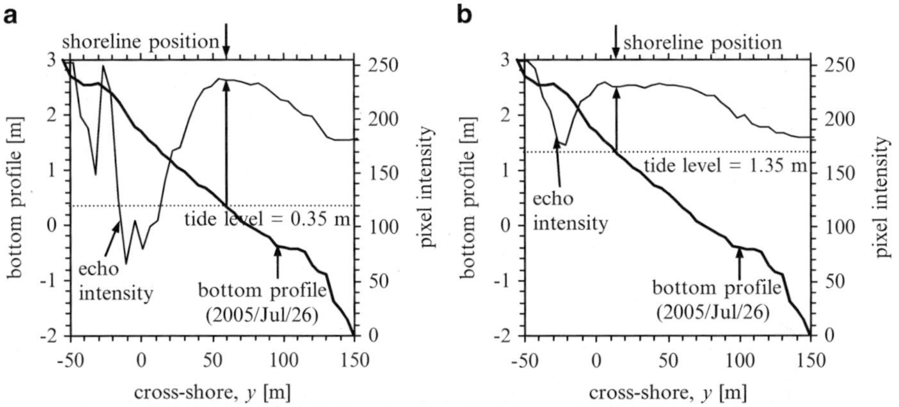

Fig. 7.7 Cross-shore pixel intensities extracted from time-averaged images at $x = -47$ m displayed with the bottom profile along the pier ($x = 0$) during (**a**) low tide (2005/Jul/26/13 h) and (**b**) high tide (2005/Jul/26/20 h) condition for the period of typhoon BANYAN

A close-up view around the pier is included in this figure in which shifting of shorelines due to changing tide levels are clearly visible.

Pixel intensities extracted along a cross-shore line from time-averaged images are shown in Fig. 7.7 along with the tidal elevation and bottom profiles. Location of the broad peak in the echo intensity variation coincides with the position of the intersection of the tidal elevation and bottom profiles, which is the shoreline position (i.e., Takewaka 2005). Thus the shorelines can be determined by locating peaks in the cross-shore pixel intensity distribution.

For the estimation of foreshore profile, horizontal position of the shoreline was determined from radar measurements, and vertical position was taken from corresponding tide records. A period of large tidal changes was chosen for the inter-tidal bathymetry preparation to obtain a wide coverage for the particular analyzing periods of run-up observations. During typhoon BANYAN, 2005/Jul/26/13 h as low tide (tide level: 0.348 m) and 2005/Jul/26/20 h as high tide (tide level: 1.348 m) were considered as the inter-tidal band. Similarly, inter-tidal range between 2005/Sept/25/18 h (as high tide, tide level: 1.578 m) and 2005/Sept/26/04 h (as low tide, tide level: 0.598 m) was chosen for the estimation during the other typhoon period. Tide levels recorded by the JMA at the Choshi Fishery Port are shown in Fig. 7.8 with indication of inter-tidal range used for the profile estimation. Shoreline positions were digitized manually from hourly time-averaged images within the inter-tidal band. Then parabolic curves were fitted for the shoreline positions with the help of corresponding tidal elevations. The estimated bathymetries from radar measurements close to the pier ($y = -47$ m) are depicted in Fig. 7.9, which agrees reasonably well with the results of the bathymetric survey along the pier ($y = 0$ m). Note that $x = -47$ m is the nearest available position from the radar measurements which is the reason for the horizontal shift of the comparison. Foreshore profile has been changed between the two typhoon events which can be observed from the comparison.

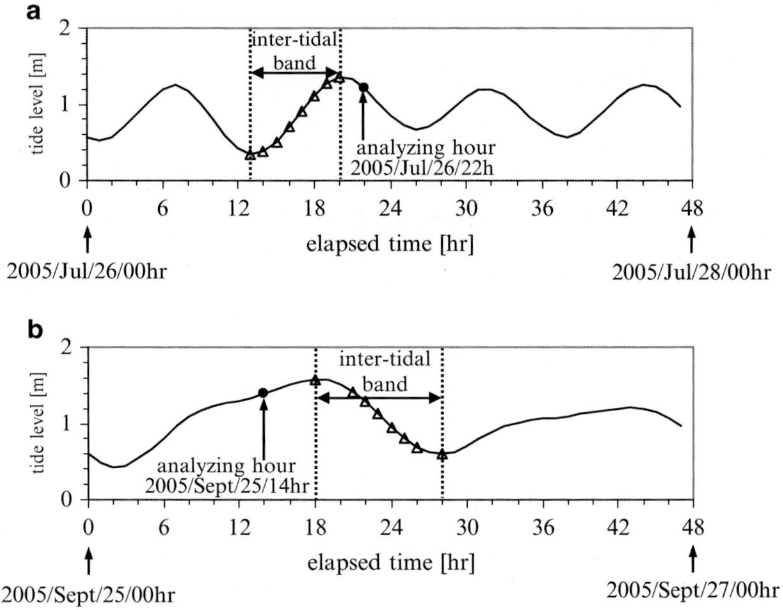

Fig. 7.8 Tide records at Choshi Fishery Port measured by the Japan Meteorological Agency (*JMA*) during (**a**) typhoon BANYAN (T7) and (**b**) typhoon SAOLA (T17). *Dotted lines* indicate the inter-tidal range used for estimating foreshore profiles and also open symbols indicate the periods considered for digitization of the time-averaged images

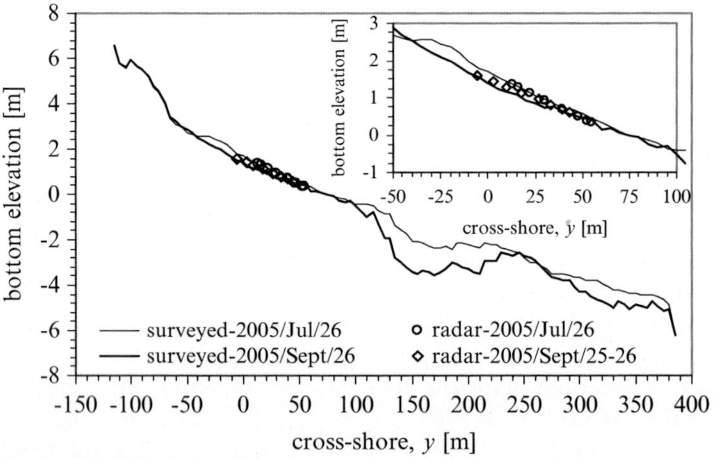

Fig. 7.9 Comparison of bottom profiles between radar measurements ($x = -47$ m) and surveyed results along the pier ($x = 0$ m) for both the typhoon periods (e.g., typhoon BANYAN: 2005/Jul/26 and typhoon SAOLA: 2005/Sept/25–26)

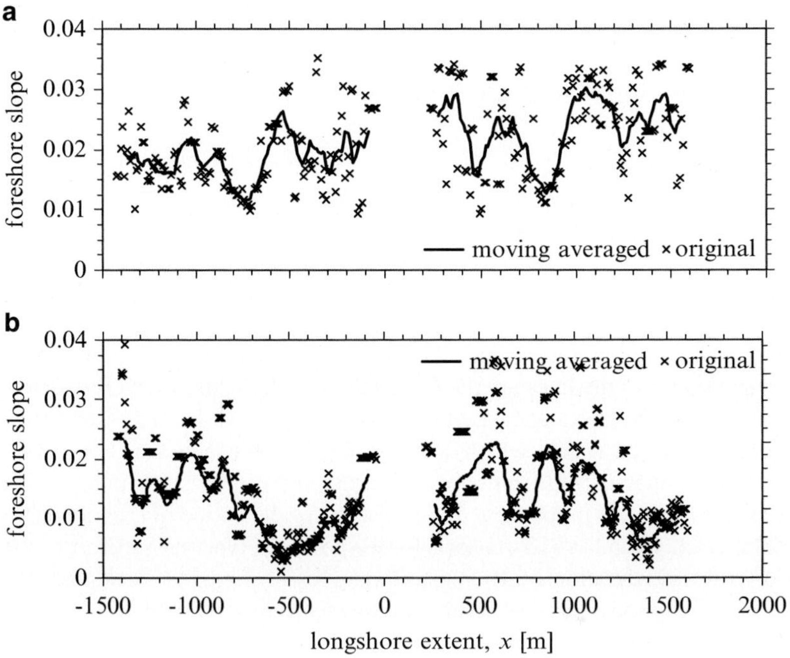

Fig. 7.10 Longshore variation of foreshore slope for the period of (**a**) typhoon BANYAN (2005/Jul/26/22 h) and (**b**) typhoon SAOLA (2005/Sept/25/14 h)

7.3.2 Foreshore Slope

Foreshore slopes were estimated from the fitted foreshore profiles as the same manner of Hasan and Takewaka (2009). Mean foreshore slope is defined at every longshore position from average of foreshore slopes at three cross-shore locations – swash-up, swash-down and mean swash position (locations are explained in the next section). The foreshore slopes at those locations were estimated from the results of the parabolic curve fitting applied to the digitized inter-tidal foreshore profile. Longshore distributions of the mean foreshore slopes are shown in Fig. 7.10, where symbols indicate the original variation and continuous lines indicates moving average of them with an averaging width of 86.8 m. A wavy structure with length of order of 500 m along the shoreline configuration is observed. Longshore-averaged foreshore slope during typhoon BANYAN (2005/Jul/26/22 h) is 0.022 (~1/45), while the foreshore slopes varied from 0.011 to 0.03 over the area. The estimated foreshore slopes also varied from 0.004 to 0.033 for the other typhoon period (2005/Sept/25/14 h) with milder longshore-averaged slope 0.0125 (~1/80).

7.4 Swash Front Analyses

7.4.1 Digitization of Swash Motion

Pixel intensities along a cross-shore line are extracted from a sequence of echo images and stacked in time to produce a cross-shore time stack image. Time stack is a composite image whose one axis represents time and coastal extent is the other axis. Local gradients of the oblique white streaks in the time stack image correspond to wave propagation speeds in the cross-shore direction. Changes in local inclinations of these lines in the vicinity of the shore reflect retardation of wave celerity during shoaling, and merging of waves occurs at intersections of the lines in the swash zone. Lower end points of the lines are the maximum swash points, which fluctuate in time suggesting that low frequency motion exists in the swash zone. Landward most identifiable edge of water, which is the boundary between land and water, was determined by manual digitization of the cross-shore time stack images. Series of time stack images was prepared at 256 transects for the period of typhoon BANYAN (2005/Jul/26/22 h) and at 512 transects for the period of typhoon SAOLA (2005/Sept/25/14) with time duration of 1024.0 s. Examples of digitized stacks are shown in Fig. 7.11 which reveal large scale up and down variation in time.

The digitization was done by tracking the up-rush of individual waves from offshore points up to the swash peaks and subsequent down-rush. Unfortunately, the down-rush is less distinct and not able to detect clearly from the time stack

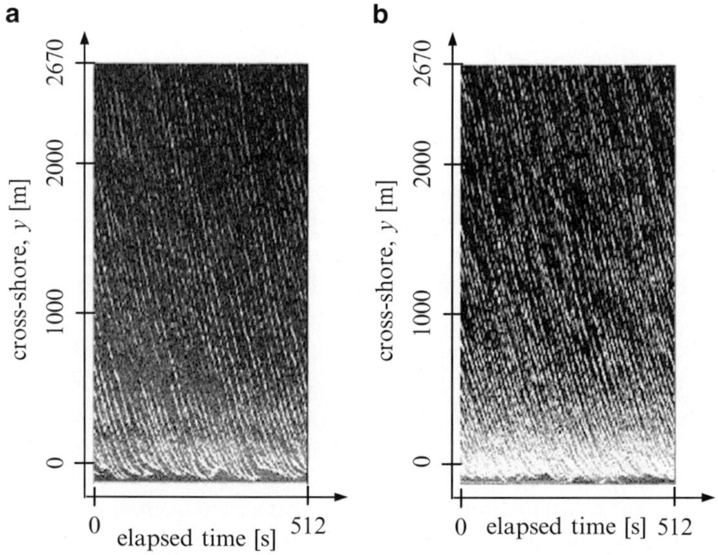

Fig. 7.11 Cross-shore time stack images for the period of (**a**) typhoon BANYAN (2005/Jul/26/22 h) and (**b**) typhoon SAOLA (2005/sept/25/14 h) at $x = -166$ m

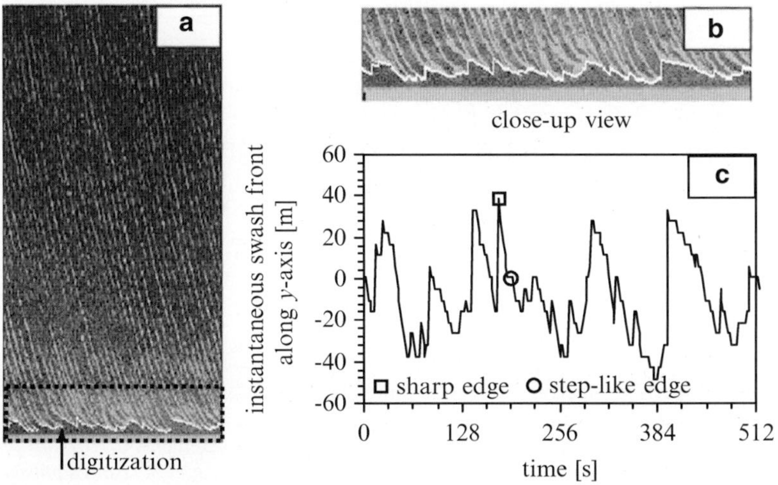

Fig. 7.12 (**a**) Manual digitization of the swash front, (**b**) close-up view of the digitization process, and (**c**) time variation of instantaneous swash front S at a cross-shore transect ($x = -166$ m) during high tide condition (2005/Jul/26/22 h). Open symbols on (**c**) are indicating a sharp and a step-like edge

images, which is a limitation of radar measurements and explained in Hasan and Takewaka (2009) in details. An example of a digitized stack is depicted in Fig. 7.12a with a close-up view in Fig. 7.12b and a variation of digitized swash front S (x, t) in Fig. 7.12c. The swash front shows large scale up and down variation with an amplitude of approximately 50 m. Sharp edges probably appeared during manual digitization from not following the actual down-rush pattern. Step-like edges also become visible in the variation, most likely due to the non-updated parts of the image sequence. Figure 7.12c depicts a sharp and a step-like edge with open symbols as an example. Troughs in the variation curve are defined as individual swash-ups correspond to run-up positions, whereas peaks are termed as individual swash-downs.

Extreme values (troughs in S) located further land-wards than the temporal mean of the swash variation at a cross-shore transect were identified as swash-ups. Swash-downs were also estimated in a similar manner from extreme values (peaks in S) located further offshore-wards than the temporal mean. The positions of swash-up, swash-down and mean swash S_m at each cross-shore transect were used to estimate foreshore slope as explained in the previous section.

7.4.2 Validation of Swash Front

The digitization of the swash motion is dependent on the subjective judgments of the individual operators. Accuracy of the digitization was checked by comparison with in situ measurements. A number of wave gauges were placed along the pier to observe

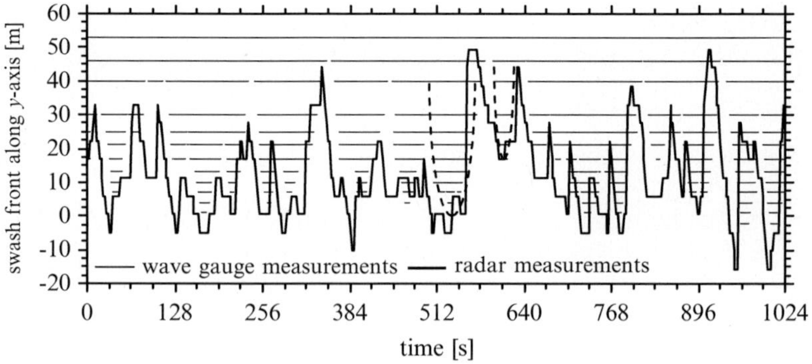

Fig. 7.13 Comparison between swash front *S* digitized from *X*-band radar images at $x = -47$ m and water traces detected with ultra-sonic wave gauge system at $x = 0$ m (along the pier) during a low tide (2005/Jul/27/02 h, tide level: 0.66 m). Wave gauge measurements (*horizontal lines*) are displayed at the location of gauges, whose positions are shown in Fig. 7.3c. *Dotted parabolas* are schematically drawn the up-rush and down-rush pattern of the wave gauge records

run-up fluctuations during the typhoon BANYAN as explained in Sect. 7.2.1. The location of the gauges are shown in Fig. 7.4c with the most landward one placed at $y = -23$ m. The digitized water-front variations *S* at $x = -47$ m were compared with wave gauge records at $x = 0$ m as shown in Fig. 7.13. $x = -47$ m is the nearest available position from the digitization which is the reason for the horizontal shift of the comparison. The horizontal lines in the figure represent the times and places where water was detected by the wave gauge system. The overall comparison of the water fronts detected by both systems shows an acceptable agreement, particularly for up-rush patterns. Smooth down-rush patterns can be identified from the wave gauge records, whereas the variation is somewhat abrupt for radar estimates, resulting in discrepancies in the swash-down phases.

It is widely accepted that the pattern of up-rush and down-rush of the waterfront can be approximated with a parabola (e.g., Shen and Meyer 1963), and this is observed from the wave gauge record in Fig. 7.13; however the digitized swash front is a bit skewed due to its less distinctive swash-down pattern. Since the down-rush pattern of the radar measurements does not follow an actual parabolic path and the digitization was down by jumping to the next wave at an offshore location from the swash peak, a small offshore-ward shift of the digitized swash-down position from its actual position occurs.

7.4.3 Shoreline Position

The instantaneous swash variation *S* at $x = -47$ m is displayed along with a surveyed bottom profile along the pier ($x = 0$ m) in Fig. 7.14 to cross check the accuracy of the digitized swash. Temporal mean of the swash variation S_m should close to the shoreline position, which can also be determined from the intersection

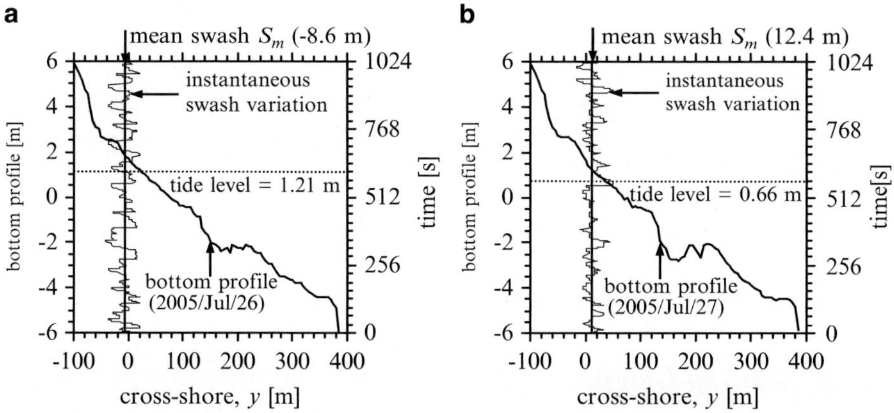

Fig. 7.14 Bottom profile along the pier ($x = 0$) displayed with time variation of instantaneous swash front S close to the pier ($x = -47$ m) during (**a**) high tide (2005/Jul/26/22 h) and (**b**) low tide (2005/Jul/27/02 h) condition

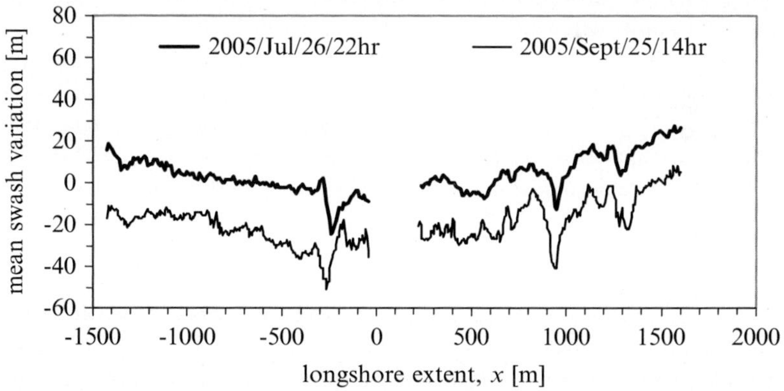

Fig. 7.15 Comparison between longshore distribution of mean swash S_m for the period of typhoon BANYAN (2005/Jul/26/22 h, tide level = 1.21 m) and typhoon SAOLA (2005/Sept/25/14 h, tide level = 1.39 m)

of the surveyed bottom profile and tide level (as explained through Fig. 7.7). Figure 7.14 also plotted simultaneously mean swash S_m, tide level and bottom profile. Small discrepancy between the mean swash and the shoreline position defined from the bottom profile is observed, which probably arises from two sources: (i) the mean swash determined from the radar data includes wave setup at the shoreline, whereas the measured tidal elevation does not, and (ii) a horizontal shift of 47 m between the points of comparison.

Longshore distribution of mean swash variation S_m (x) is shown in Fig. 7.15. The figure compares S_m between the period of typhoon BANYAN (2005/Jul/26/22 h) and typhoon SAOLA (2005/Sept/25/14 h). The mean position is gradually shifting landward towards the pier in negative longshore locations, whereas wavy distribution is observed in positive longshore locations and the pattern is almost similar in both the periods. Cross-shore transition of the shoreline position observed from the figure is basically due to difference of the tide levels, which is 1.39 m during 2005/Sept/25/14 h and 1.21 m during 2005/Jul/26/22 h.

7.5 Longshore Structure of Wave Run-up

7.5.1 Run-up Height

Pixel intensities along a cross-shore transect are extracted from a sequence of echo image and stacked in time to produce a cross-shore time stack image. Cross-shore time stack, a composite image whose one axis represents time and cross-shore extent is the other, is useful to understand maximum swash positions and its fluctuation in time. Landward most identifiable edge of water, i.e., the boundary between land and water, was determined by manual digitization of the time stack images. Series of cross-shore time stack images in alongshore direction were digitized to examine the longshore structure of the low frequency run-up motion.

Time stack images were digitized at 256 longshore locations with an interval of 10.85 m (2 pixel spacings) for the period of typhoon BANYAN (2005/Jul/26/22 h), whose first half falls in the negative longshore region (-1425 m $< x <-47$ m) and rest in the positive region (225 m $< x <$1,603 m). Longshore coverage by all 256 transects is approximately 3.0 km. For the period of typhoon SAOLA (2005/Sept/25/14 h), time stack images were digitized with a similar manner at 512 longshore locations with an interval of 5.42 m (1 pixel spacing). Longshore coverage by all 512 transects are $-1{,}425$ m $< x <-41$ m and 219 m $< x <$1,603 m. The gap of the digitization (-36 m $< x <$214 m) in front of radar is due to avoid signal saturation, which was difficult to digitize and this study neglects this region for subsequent analyses. Digitized swash fronts $S(x, t)$ during the analyzing periods are displayed in Fig. 7.16 along the cross-shore transect $x = -166$ m.

Swash fluctuation digitized from cross-shore time stack images at each transect and were converted to vertical elevation with the help of foreshore profile. Time history of the estimated water front elevation R_f at a cross-shore transect $x = -166$ m is shown in the upper panel of Fig. 7.17. Still water level (SWL) is considered as the position of tidal elevation and its level is also shown in the same figure. The vertical elevation basically consists with two components: a super elevation known as wave setup which is the mean of the estimated elevation, and fluctuation about that mean. Statistical analyses of the time history of estimated elevation and their comparison between the two analyzing periods were done and discussed in the following sections.

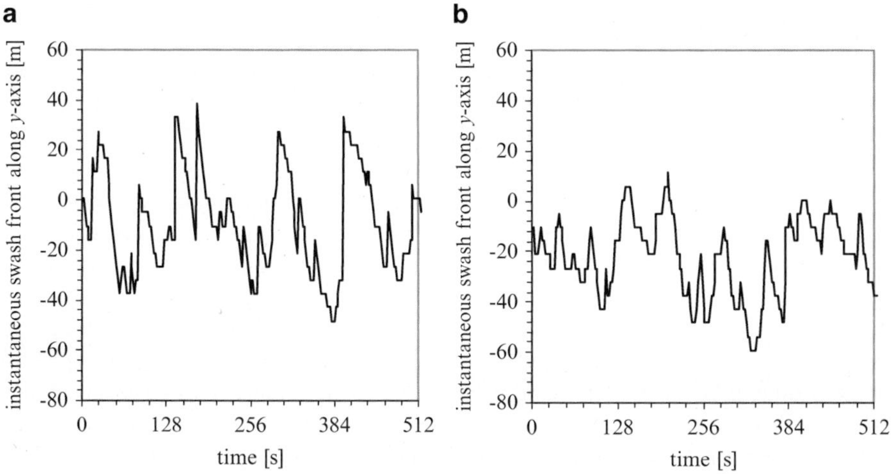

Fig. 7.16 Time variation of digitized swash front S for the period of (**a**) typhoon BANYAN (2005/Jul/26/22 h) and (**b**) typhoon SAOLA (2005/Sept/25/14 h) at a cross-shore transect $x = -166$ m

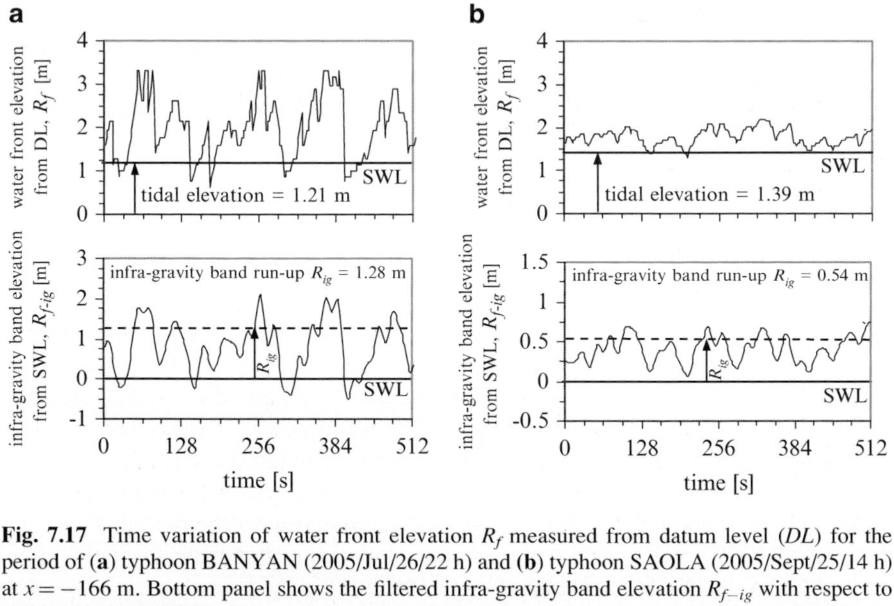

Fig. 7.17 Time variation of water front elevation R_f measured from datum level (DL) for the period of (**a**) typhoon BANYAN (2005/Jul/26/22 h) and (**b**) typhoon SAOLA (2005/Sept/25/14 h) at $x = -166$ m. Bottom panel shows the filtered infra-gravity band elevation R_{f-ig} with respect to still water level (SWL). R_{f-ig} variation was used to estimate the infra-gravity band run-up R_{ig}

7.5.2 *Infra-gravity Run-Up Distribution*

Energy in the incident and infra-gravity frequency bands generally contributes varying amounts of total run-up (e.g., Guza and Thornton 1982) and therefore separation into two bands is required. In order to check the relative role of the infra-gravity band run-up variation, the water front elevations R_f were band pass filtered with frequency range of $0.004 < f < 0.05$ Hz. The separation frequency used for filtering the infra-gravity variation is somewhat subjective, which was also chosen by several researchers for run-up analyses (e.g., Holland and Holman 1999; Ruessink et al. 1998; Ruggiero et al. 2004; Hasan and Takewaka 2009 etc.). The reference level of the water front elevations was then shifted to still water level for run-up analyses. Time histories of the filtered infra-gravity band run-up variations R_{f-ig} along $x = -166$ m are shown in lower panel of Fig. 7.17. Hasan and Takewaka (2009) analyzed the relative contribution of both the infra-gravity band and incident band run-up during typhoon BANYAN. They observed that energy of infra-gravity band run-up was dominant and energy of incident band was negligible.

Run-up heights of the infra-gravity band R_{ig} at each transect was estimated from the time history of R_{f-ig} variation as the same manner of Hasan and Takewaka (2009). Mean of the discrete local maxima above still water level in the R_{f-ig} variation is defined as $R_{ig}(x)$. Longshore distribution of the R_{ig} is shown in Fig. 7.18, where the crosses indicate the original estimates and the solid line indicates moving average of the estimates with an 86.8 m averaging width (nine grid spacing during 2005/Jul/26/22 h and 17 grid spacing during 2005/Sept/25/14 h). The figure simultaneously shows the distribution of the foreshore slopes for both the periods. R_{ig} has a tendency to be relatively high in regions where the foreshore slopes are steep and vice versa. Longshore distributions of R_{ig} were ranging from 0.33 to 1.67 m and from 0.37 to 1.78 m, for the period of typhoon BANYAN (2005/Jul/26/22 h) and typhoon SAOLA (2005/Sept/25/14 h), respectively.

7.5.3 *Spectra of Water Front Elevation*

Spectra of the water front elevation R_f was estimated at each cross-shore transect using the Fast Fourier Transform (FFT). Data length and interval for the spectra estimation were 1024.0 s and 2.0 s. The estimated spectra were alongshore averaged and depicted in Fig. 7.19. Single peak frequency of 0.0078 Hz (128.0 s) corresponding to highest spectral density is observed in the spectrum for the period of typhoon BANYAN (2005/Jul/26/22 h), whereas two dominant peaks of 0.0078 Hz (128.0 s) and 0.0107 Hz (93.1 s) are observed for the period of typhoon SAOLA (2005/Sept/25/14 h). Spectral characteristics of the water front elevations resemble the spectra of the incident wave field for both periods.

Frequency spectra of the offshore wave field for the period of typhoon BANYAN (2005/Jul/26/22 h) show dominant waves with 15.1 s period (0.066 Hz) were approaching towards shore, whereas two dominant waves with 16.0 s (0.0625 Hz) and 10.5 s (0.095 Hz) were observed from the frequency spectra for the period of typhoon SAOLA (2005/Sept/25/14 h). Energies of these incident waves are not found in the R_f spectra for both the periods, suggesting a saturated condition of the incident waves.

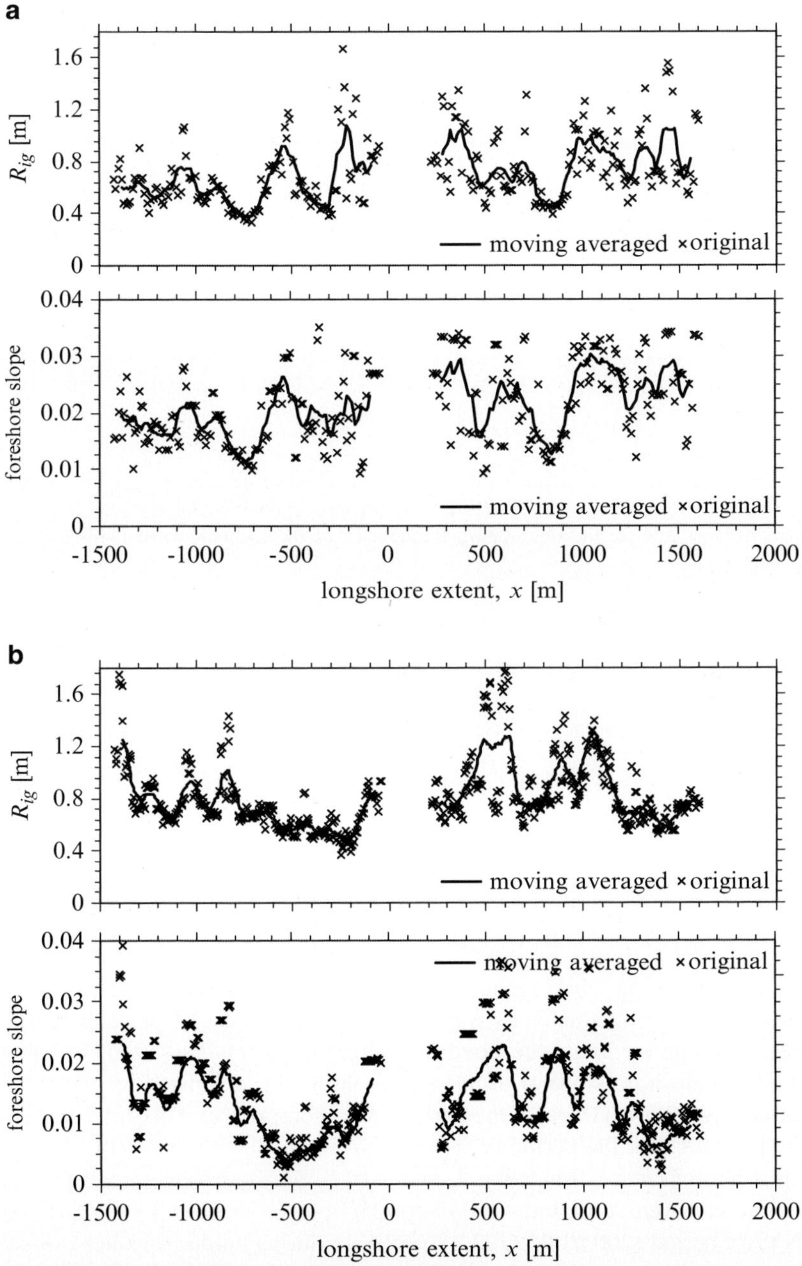

Fig. 7.18 Longshore distribution of infra-gravity band run-up R_{ig} with corresponding foreshore slope for the period of (**a**) typhoon BANYAN (2005/Jul/26/22 h: $H_s = 3.69$ m, $T_s = 12.1$ s, tide level = 1.21 m) and (**b**) typhoon SAOLA (2005/Sept/25/14 h: $H_s = 4.87$ m, $T_s = 9.6$ s, tide level = 1.39 m)

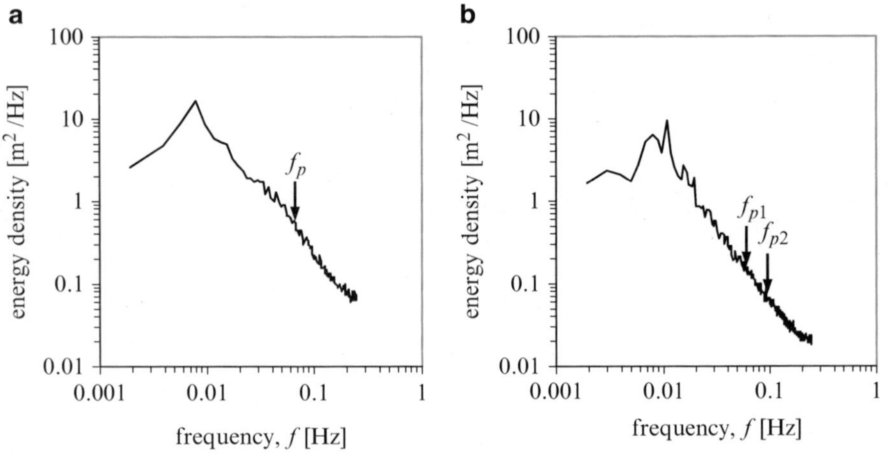

Fig. 7.19 Longshore-averaged energy density spectra of the water front elevation R_f for the period of (**a**) typhoon BANYAN (2005/Jul/26/22 h) and (**b**) typhoon SAOLA (2005/Sept/25/14 h). *Vertical arrows* indicate the dominating frequencies of the incident wave fields (discussed in Sect. 7.2.4)

7.5.4 Dependence of Run-Up Height on Slope

Surf similarity parameter ξ_0, a non-dimensional parameter, which is frequently used for examining beach condition and is defined as

$$\xi_0 = \tan\beta/(H_s/L_0)^{1/2} \tag{7.1}$$

where, β is the beach slope, H_s is the wave height, L_0 is the deep water wave length define as $L_0 = gT_s^2/2\pi$ by linear theory, g is the acceleration due to gravity and T_s is the wave period. Low surf similarity parameters ($\xi_0 \leq 0.3$) typically indicate dissipative condition while higher values suggest more reflective condition.

Surf similarity parameter ξ_0 was estimated using Eq. 7.1 at each cross-shore transect considering sea state conditions for the particular periods as shown in Table 7.2. Foreshore slope at each transect tan β was also used for the estimates. ξ_0 varied from 0.07 to 0.28 for the period of typhoon BANYAN (2005/Jul/26/22 h) and 0.02 to 0.18 for the period of typhoon SAOLA (2005/Sept/25/14 h) indicating beach conditions were dissipative. Lower surf similarity parameters were observed for typhoon SAOLA period (2005/Sept/25/14 h) compared to that of typhoon BANYAN period (2005/Jul/26/22 h), probably due to milder foreshore slope.

Correlation between the infra-gravity run-up R_{ig} normalized by significant wave height H_s and surf similarity parameter $\xi_0, R_{ig}/H_s = a\xi_0 + b$ where a and b are regression coefficients, was assessed and is depicted in Fig. 7.20 for both the analyzing periods. Obtained slopes of the best fit lines (i.e., 1.068 during 2005/ Jul/25/22 h and 1.265 during 2005/Sept/25/14 h) are in the same order. Several

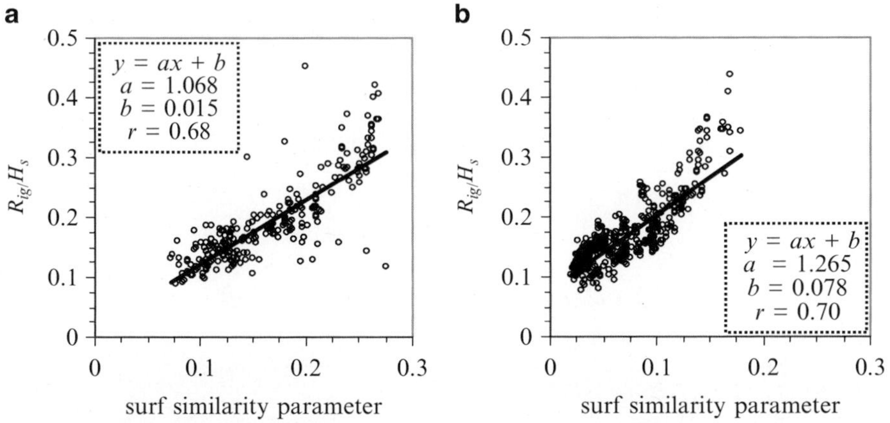

Fig. 7.20 Infra-gravity band run-up R_{ig} normalized by the significant wave height H_s versus surf similarity parameter ξ_0 for the period of (**a**) typhoon BANYAN (2005/Jul/26/22 h) and (**b**) typhoon SAOLA (2005/Sept/25/14 h). Best fit lines are $R_{ig}/H_s = 1.068\ \xi_0 + 0.015$ and $R_{ig}/H_s = 1.265\ \xi_0 + 0.078$ (*solid lines*)

studies (e.g., Holman and Sallenger 1985; Raubenheimer and Guza 1996; Ruessink et al. 1998; Ruggiero et al. 2001; Hasan and Takewaka 2009 etc.) have confirmed this linear dependence between R_{ig}/H_s and ξ_0, however, the constant of proportionality is dependent on specific site and sea condition. The linear relationship obtained at Hasaki beach during the energetic sea state is reasonable compared to results obtained at other sites and under different sea conditions.

7.5.5 Propagation of Low Frequency Motion

Propagations of large-scale swash-up and -down in the longshore direction were observed visually in the sequence of radar images. A two-dimensional diagram, with the vertical axis for time, and the lateral for longshore extent, was processed to visualize this over the area as shown in Fig. 7.21. The digitized swash front S of all transects were first converted to elevation R_f and then separated into infra-gravity band elevation $R_{f\text{-}ig}$ as described above. The time varying $R_{f\text{-}ig}$ was stacked line by line in the diagram to visualize this low frequency motion. Brighter intensities in this diagram correspond to higher run-up values. Inclined streaks are indicating the propagation direction of the low frequency run-up motion. The streaks are clear and inclined for the period of typhoon BANYAN (2005/Jul/26/22 h) as shown in Fig. 7.21a, indicating the run-up variation was propagating in the negative longshore direction. On the contrary, pattern of the streaks are complex for the period of typhoon SAOLA (2005/Sept/25/14 h) as shown in Fig. 7.21b. The figure is suggesting propagations into both negative longshore as well as

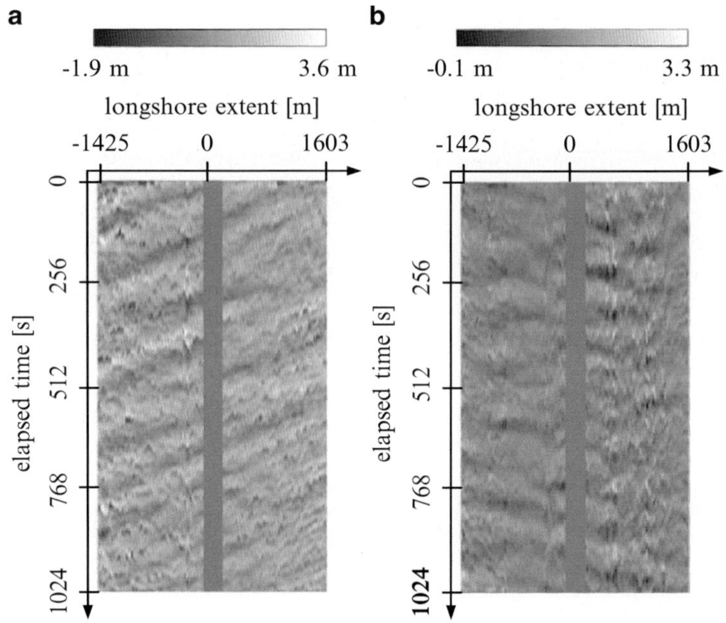

Fig. 7.21 Longshore propagation of infra-gravity band run-up height variation R_{f-ig} displayed as image view for the period of (**a**) typhoon BANYAN (2005/Jul/26/22 h) and (**b**) typhoon SAOLA (2005/Sept/25/14 h). *Vertical gray stripe* in the middle of the images indicates saturated part in front of the pier which was unable to digitize

positive longshore directions. From these results, we may conclude that the behavior of the low frequency motion is dependent on the incident wave field.

Wavenumber-frequency spectra in the time and longshore domain were estimated for the water front elevations R_f with the two-dimensional fast Fourier transform (2D-FFT) to investigate the spectral characteristics of the run-up motion over the area. The 2D-FFT was applied to data over a spatio-temporal extent of 2778.0 m × 1024.0 s (256 transects × 512 images, referred to as a sub-area) for the period of typhoon BANYAN (2005/Jul/26/22 h). The estimation was done by shifting the longshore position of the sub-area by 24 times with an interval of 10.85 m (two grid spacing), and then the spectra were spatially averaged in the longshore direction. The size of the sub-area for the period of typhoon SAOLA (2005/Sept/25/14 h) was 2778.0 m × 1024.0 s (512 transects × 512 images) with shifting interval was 5.42 m (one grid spacing). The spatially averaged spectra during both the analyzing periods are shown in Fig. 7.22. Each sub-area includes deficient data due to signal saturation close to the pier, which is shown as the vertical gray stripe in Fig. 7.21. This may interfere with the spectral estimation; however, the result shown in Fig. 7.22 is smooth and has a dominant energy concentration, and it seems that the stripe does not make any severe mischief.

The majority of the energy was concentrated at low wave numbers. The spectrum for the period of typhoon BANYAN (2005/Jul/26/22 h) is clear with distinct

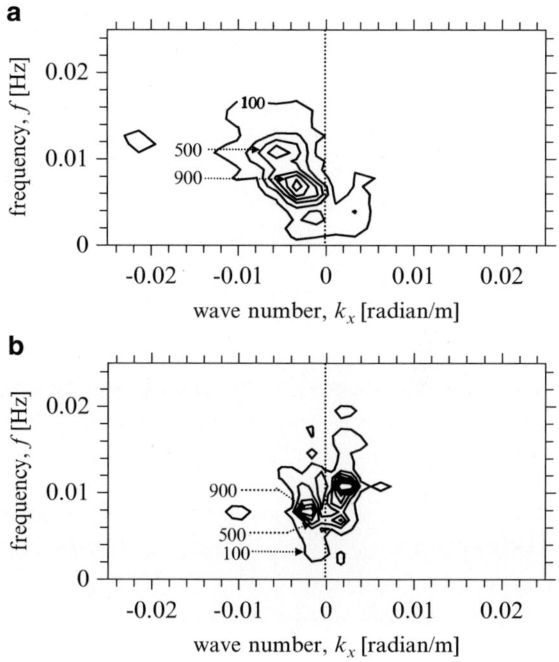

Fig. 7.22 Wavenumber-frequency spectra of run-up variation R_f for the period of (**a**) typhoon BANYAN (2005/Jul/26/22 h) and (**b**) typhoon SAOLA (2005/Sept/25/14 h). Unit of the spectral intensities is in m^2/Hz/rad/m

spectral peak in the negative wave number, whereas energy concentrations in both negative and positive wave numbers are observed in the spectrum for the other period (2005/Sept/25/14 h). From the spectrum, estimated peak period and peak wave length was 128.0 s and 1389.0 m during 2005/Jul/26/22 h, whereas the values were 93.1 s and 1389.0 m during 2005/Sept/25/14 h. The order of propagation speed of the low frequency motion in the longshore direction was $\boldsymbol{O}$ (15 m/s).

Spectral peak obtained at negative wave number for the period of typhoon BANYAN (2005/Jul/26/22 h) corresponds to the propagation of the low frequency run-up motion in the negative longshore direction. Whereas bidirectional propagation (both in the negative and position longshore direction) was observed for the period of typhoon SAOLA (2005/Sept/25/14 h).

7.6 Concluding Remarks

X-band nautical radar measurements were conducted during several storm events in 2005 at the research pier HORS located in Hasaki, Japan. Analyses on echo images were done to estimate the longshore distribution of the mean shoreline positions and inter-tidal foreshore profiles, temporal and spatial variations of the wave run-up and their bulk statistics. Swash fronts were digitized manually from the sequence of radar images and converted to elevations with the help of the foreshore profile.

Low frequency run-up motions during two different stormy wave fields were analyzed and compared in this chapter.

The run-up data were filtered to determine the role of the infra-gravity band component since this band contained most of the total run-up variances, which was observed from their spectral characteristics. The infra-gravity band run-up along an almost 3.0 km length dissipative stretch of coast demonstrates strong dependence on the foreshore slope for both the typhoon periods. Low frequency run-up motion in the longshore direction was observed from the run-up variation. Analyses of the incident wave field revealed single dominant wave incidences for the period of typhoon BANYAN (2005/Jul/26/22 h) and multiple wave incidences for the period of typhoon SAOLA (2005/Sept/25/14 h). The structure of the low frequency motion was clear with organized propagation in the negative longshore direction for the period of typhoon BANYAN (2005/Jul/26/22 h), whereas, complex structure with bi-directional propagation for the period of typhoon SAOLA (2005/Sept/25/14 h).

References

Bell PS (1999) Shallow water bathymetry derived from an analysis of *X*-band marine radar images of waves. Coast Eng 37:513–527

Borge JCN, Rodriguez GR, Hessner K, Gonzalez PI (2004) Inversion of marine radar images for surface wave analysis. J Atmos Ocean Technol 21:1291–1300

Butt T, Russell P (2000) Hydrodynamics and cross-shore sediment transport in the swash zone of natural beaches: a review. J Coast Res 16:255–268

Elfrink B, Baldock T (2002) Hydrodynamics and sediment transport in the swash zone: a review and perspectives. Coast Eng 45:149–167

Gurgel K-W, Antonischki G, Essen H–H, Schlick T (1999) Wellen Radar (WERA): a new ground-wave HF radar for ocean remote sensing. Coast Eng 37:219–234

Guza RT, Thornton EB (1982) Swash oscillations on a natural beach. J Geophys Res 87(C1):483–491

Guza RT, Thornton EB (1985) Observations of surf beat. J Geophys Res 90(C2):3161–3172

Hasan GMJ, Takewaka S (2007a) Observation of a stormy wave field with *X*-band radar and its linear aspects. Coast Eng J 49(2):149–171

Hasan GMJ, Takewaka S (2007b) Depth-current inversion from a stormy wave field with *X*-band radar. In: Proceedings of the Asian and Pacific Coasts on CD-Rom (APAC 2007), China, pp 1808–1821

Hasan GMJ, Takewaka S (2009) Wave run-up analyses under dissipative condition using *X*-band radar. Coast Eng J 51(2):177–204

Herbers THC, Elgar S, Guza RT (1995) Infragravity-frequency (0.005–0.05 Hz) motions on the shelf. Part II: free waves. J Phys Oceanogr 25:1063–1079

Holland KT, Holman RA (1999) Wavenumber-frequency structure of infra-gravity swash motions. J Geophys Res 104(C6):13479–13488

Holland KT, Raubenheimer B, Guza RT, Holman RA (1995) Runup kinematics on a natural beach. J Geophys Res 100(C3):4985–4993

Holman RA (1981) Infra-gravity energy in the surf zone. J Geophys Res 86(C7):6442–6450

Holman RA (1986) Extreme value statistics for wave run-up on a natural beach. Coast Eng 9:527–544

Holman RA, Boyen AJ (1984) Longshore structure of infra-gravity wave motions. J Geophys Res 89(C4):6446–6452

Holman RA, Sallenger AH (1985) Setup and swash on a natural beach. J Geophys Res 90 (C1):945–953
Hunt IA (1959) Design of seawalls and breakwaters. J Waterw Harb Coast Eng Div ASCE 85:123–152
Inman DL, Guza RT (1982) The origin of swash cusps on beaches. Mar Geol 49:133–148
Miche R (1951) Le Pouvoir reflechissant des ouvrages maritimes exposes a l'action de la houle. Ann Ponts Chaussees 121:285–319
Mizuguchi M, Seki K (2007) Field observation of waves and topographical change near the shoreline. In: Proceedings of Asian and Pacific Coasts on CD-Rom (APAC 2007), China, pp 30–40
Oltman-Shay J, Howd PA, Birkemeier WA (1989) Shear instabilities of the mean longshore current 2: field observations. J Geophys Res 94:18031–18042
Osborne PD, Rooker GA (1999) Sand re-suspension events in a high energy infragravity swash zone. J Coast Res 15:74–86
Plant NG, Holman RA (1997) Inter-tidal beach profile estimation using video images. Mar Geol 140:1–24
Raemer HR (1996) Radar systems principles. CRC Press, Boca Raton, Florida, 544 pp
Raubenheimer B, Guza RT (1996) Observations and predictions of run-up. J Geophys Res 101 (C11):25575–25587
Raubenheimer B, Guza RT, Elgar S, Kobayashi N (1995) Swash in a gently sloping beach. J Geophys Res 100(C5):8751–8760
Ruessink BG, Kleinhans MG, van den Beukel PGL (1998) Observations of swash under highly dissipative conditions. J Geophys Res 103(C2):3111–3118
Ruessink BG, Miles JR, Feddersen F, Guza RT, Elgar S (2001) Modeling the alongshore current on barred beaches. J Geophys Res 106:22451–22464
Ruggiero P, Komar PD, Marra JJ, McDougal WG, Beach RA (2001) Wave run-up, extreme water levels and the erosion of properties backing beaches. J Coast Res 17:407–419
Ruggiero P, Holman RA, Beach RA (2004) Wave run-up on a high-energy dissipative beach. J Geophys Res 109:C06025
Sallenger AH (2000) Storm impact scale for barrier islands. J Coast Res 16:890–895
Shen MC, Meyer RE (1963) Climb of a bore on a beach, Part 3: run-up. J Fluid Mech 16:113–125
Stockdon HF, Holman RA (2000) Estimation of wave phase speed and nearshore bathymetry from video imagery. J Geophys Res 105(C9):22015–22033
Takewaka S (2005) Measurements of shoreline positions and intertidal foreshore slopes with X-band marine radar system. Coast Eng J 47(2 & 3):91–107
Takewaka S, Nishimura H (2005) Wave run-up analyses during a storm event with nautical X-band radar. Proceedings of Asian and Pacific Coasts on CD-Rom (APAC 2005), Korea, pp 780–788
Thornton EB, Kim CS (1993) Longshore current and wave height modulation at tidal frequency inside the surf zone. J Geophys Res 98:16509–16519
Wilson GW, Ozkan-Haller HT, Holman RA (2010) Data assimilation and bathymetric inversion in a two-dimensional horizontal surf zone model. J Geophys Res 115:C12057

Part II
Advances in the Study and Interpretation of Coastal Oceans, Estuaries, Sea-Level Variation, and Water Quality

Chapter 8
Digital Ocean Technological Advances

Xin Zhang, Xiaoyi Jiang, Suixiang Shi, and Tianhe Chi

Abstract Digital Ocean (DO) is a new research domain of Digital Earth. Because of the spatio-temporal, three-dimensional and intrinsically dynamic nature of ocean data, it is more difficult to make a breakthrough in this domain. In this chapter, the DO technological advances were introduced in detail. Firstly, the technological advances for DO from six aspects. (1) Introduction to DO data sources, which comprise Digital Elevation Model data of the seafloor and coast, in situ observational data, remote sensing data, computational model results and data stored in databases or other medium. (2) The three-dimensional ocean data integration platform. (3) The dynamic tide data visualization. (4) Remote sensing information products integration and sharing. (5) Computational ocean model data integration and service. (6) Spatio-temporal model of marine disasters. Secondly, the DO system initial architecture was introduced, which include the data acquire layer, standard data layer, data service layer, function layer, and application layer, respectively. Thirdly, a case study of DO was introduced, which include the China Digital Ocean Prototype System and the DE in support of an online oceanic educational public service and popularization system named iOcean. At last, the future work relating to DO should be carried out in the seven research areas were introduced as follows. (1) Aiding integrative oceanic scientific research. (2) Constructing regional DO systems. (3) Studying multiple spatio-temporal scales of ocean factors.

X. Zhang (✉)
State Key Laboratory of Remote Sensing Science, Institute of Remote Sensing and Digital Earth, Chinese Academy of Sciences, Beijing 100101, China

Hebei University of Engineering, Handan, Hebei Province 056038, China
e-mail: zhangx@irsa.ac.cn

X. Jiang • S. Shi
National Marine Data & Information Service, State Oceanic Administration, Tianjin 300171, China

T. Chi
State Key Laboratory of Remote Sensing Science, Institute of Remote Sensing and Digital Earth, Chinese Academy of Sciences, Beijing 100101, China

C.W. Finkl and C. Makowski (eds.), *Remote Sensing and Modeling: Advances in Coastal and Marine Resources*, Coastal Research Library 9,
DOI 10.1007/978-3-319-06326-3_8,

(4) Studying the relationship between ocean elements at different times. (5) Studying the spatial relationship between different ocean elements or components. (6) Studying the architecture of the DO. (7) Studying the coastal river basin non-point pollution landscape source and assemblage pattern remote sensing parsing.

8.1 Digital Earth and Digital Ocean

Technological advances over the past few years have demonstrated that many aspects of the Digital Earth (DE) that Al Gore envisioned in 1992 and discussed in his 1998 speech are now technically feasible (Goodchild 2008). Chen and Genderen (2008) believed that DE should be an integrated approach to building the next level of the scientific infrastructure that supports global change research. They presented a number of relevant research areas that would benefit from such a system, including forest and grassland fires, desertification and sandstorms, deforestation, forest carbon sequestration, wetlands conservation, monitoring migratory birds for the spread of avian influenza (bird flu), the Tibet Plateau uplift, sea level rise and underground coal fires. In terms of DE's potential applications, Guo et al. (2009, 2010) studied the DE prototype system DEPS/CAS and, through this research, defined DE systems as either scientific systems (such as those of the World Wind of USA, the DE Prototype system/Chinese Academy of Sciences (DEPS/CAS) of China, the Blue Link and the Glass Earth of Australia, and the Earth Simulator (ES) of Japan, among others) or as commercial systems (such as Skyline and Google Earth). They proposed that DE should eventually be a comprehensive platform for the integration of future information resources. Wright et al. (Wright et al. 2009) presented a methodology for reconstructing plumes using virtual globes, such as Google Earth, which allows the user to animate the evolution of a gas plume and easily display and share the results on a common platform. Yasuko et al. (2010) developed a visualization system (KML generators) for multidisciplinary geoscience data that visualizes seismic tomographic models, geochemical rock datasets, and geomagnetic field models by utilizing Google Earth technologies. Chen et al. (2009) put forward a solution for rendering the vertical profiles of atmospheric data from A-Train satellite formation in Google Earth, using data from the NASA CloudSat satellite as a proof-of-concept. However, these authors have not yet faced the issues involved with the exploration and visualization of three-dimensional ocean data on visual globes.

As far as we know, the Earth is the only blue planet in the universe. Approximately 71 % of the earth's surface is covered by ocean, which operates as a fundamental component of the global life support system and helps balance the impacts of human resource extraction on the environment. Oceans and, indeed, the entire planet are changing at faster rates and over broader scales than before, and they are changing in fundamentally new ways. In a very short period of time, the bounty of the oceans has been depleted, and ocean ecosystems have become seriously disrupted (Wright et al. 2007). The bounty and circulation of oceans makes studying sea level rise, warming seawater, increasing storm intensities and other issues important from a global perspective. Since 2005, the Global Earth

Observation System of Systems (GEOSS) has been implemented to achieve comprehensive, coordinated and sustained observations of the Earth to improve knowledge of the Earth's condition, increase the understanding of the Earth's processes and enhance the predictive power regarding the behavior of the Earth system (GEO 2005). As the oceanographic component of GEOSS, one objective of the Global Ocean Observing System (GOOS) is to foster the development of data management systems that allow users to exploit multiple data sets from many different sources through "one stop shopping" (Thomas 2003). The purpose of the Integrated Ocean Observing System (IOOS) is to make more effective use of the existing resources, new knowledge and advances in technology to provide the data and information required for scientific global or regional study (Ocean US 2002). As the National Science Foundation's contribution to the U.S. IOOS, the Ocean Observatories Initiative (OOI) will construct a networked infrastructure of science-driven sensor systems to measure the physical, chemical, geological and biological variables of the ocean and seafloor. Greater knowledge of these variables is vital for the improved detection and forecasting of environmental changes and their effects on biodiversity, coastal ecosystems and climate (COL 2009). In the future, ocean information will become an increasingly valuable commodity worldwide because of the role of maritime commerce and new ocean-related investments, the vulnerability to ocean-related natural disasters, the need to provide security for coastal populations, and the challenges of providing food and water for increasing populations (Interagency Ocean Observation Committee 2013).

The term Digital Ocean (DO) emerged after Al Gore proposed the DE program (1998). "DO" is the embodiment and re-innovation of "DE" theory and technology for the oceans of the world (Hou 1999). Patrikalakis et al. (2000) studied a knowledge network of distributed heterogeneous data and software resources for multidisciplinary ocean research that brought together advanced modeling, observation tools and field estimation methods. However, these authors did not discuss how to integrate these technologies for a global visualization of the earth. Su et al. (2006a, b, c) studied marine GIS technologies and proposed the benchmarks and key technologies for a Chinese DO Prototype System. As a public application platform, Google Ocean (www.googleearth.com) can provide a great deal of educational information, such as videos of marine life, and can allow the public to watch unseen footage of historic ocean expeditions, but the site has many shortcomings for scientific research and for governmental applications. Many visualization systems have been developed for regional ocean data using OpenSceneGraph, including the Regional Ocean Modeling System (ROMS) and others (Shen et al. 2007), but many other technologies should be researched from the DO perspective.

8.2 Technological Advances for Digital Ocean

8.2.1 Introduction to Digital Ocean Data Sources

Digital data and information are the foundation of the DO application system. Due to the nature of available data types, ocean datasets generally comprise Digital

Elevation Model (DEM) data of the seafloor and coast, in situ observational data, remote sensing data, computational model results and data stored in databases or other medium (Zhang et al. 2011).

The DEM data are acquired from single- or multi-beam ship-borne echo sounders, which are the classical systems used to map seafloor topography with highly accurate results. Additionally, data are produced using airborne AIRSAR/POLSAR synthetic aperture data or other methods (Maged et al. 2009). Subsurface ocean DEM data are used to construct a three-dimensional seafloor model and a model of the sea surface. The DEM data from the coast are used in land surface three-dimensional modeling, using remote sensing data at different spatial resolutions to represent surface texture.

The in situ observational data are collected from different types of sensors on a range of time scales, such as the data produced by the Argo buoy. The data-collecting devices include oceanic optical buoys, seafloor moored instrumentation (for dynamic factors), self-locating sub-water tide monitors, sea sound detection buoys, sub-water tide comprehensive admeasuring apparatuses, and cruises. Real-time and delayed-mode observed data collected from the above equipment are transmitted to a local data center. A major characteristic of in situ observational data is that it is directly linked to a fixed x, y, z location in space and has attributes that are attached or related to that location. Most of the data have the required temporal attributes.

The remote sensing data are used in two ways. As they pertain to islands or coastal areas, the remote sensing data are overlaid on the DEM data to represent the geographical environment. In relation to seawater, the remote sensing data are used as information products to demonstrate physical or chemical characteristics, such as sea surface temperature (SST) after technical processing. Satellite remote sensing data are collected from the MODIS (Moderate Resolution Imaging Spectroradiometer), QuikSCAT (NASA's Quick Scatterometer), and other sensors. For example, the products derived from NASA's MODIS include 250 m RGB, chlorophyll and chlorophyll fluorescence.

The model output data are derived from ocean numerical modeling, which is used to predict parameters such as currents, sea water temperatures, salinity, water levels, and sea state in real or near-real time. The model data are usually stored as a multilayer stack of column and row data. The structure can be defined as regularly or irregularly spaced grid units with discrete node locations defined in the x, y and z dimensions. The grid nodes or cell values vary over time, and each layer represents a unique slice in space with a given depth.

The ocean data and information described above are archived or stored in more than one department because they are usually collected or produced on a project-specific basis at different times. However, interdisciplinary research requires different types of data at different times. To effectively and efficiently integrate a substantial amount of ocean data, a single global visualization system should be used to facilitate the sharing, accessing, exploration and visualization of these datasets for scientists, users and decision-makers.

To construct a DO system, many different technologies will have to be synthesized to efficiently acquire, store, manage, visualize, analyze, share, update, and serve data to different types of users.

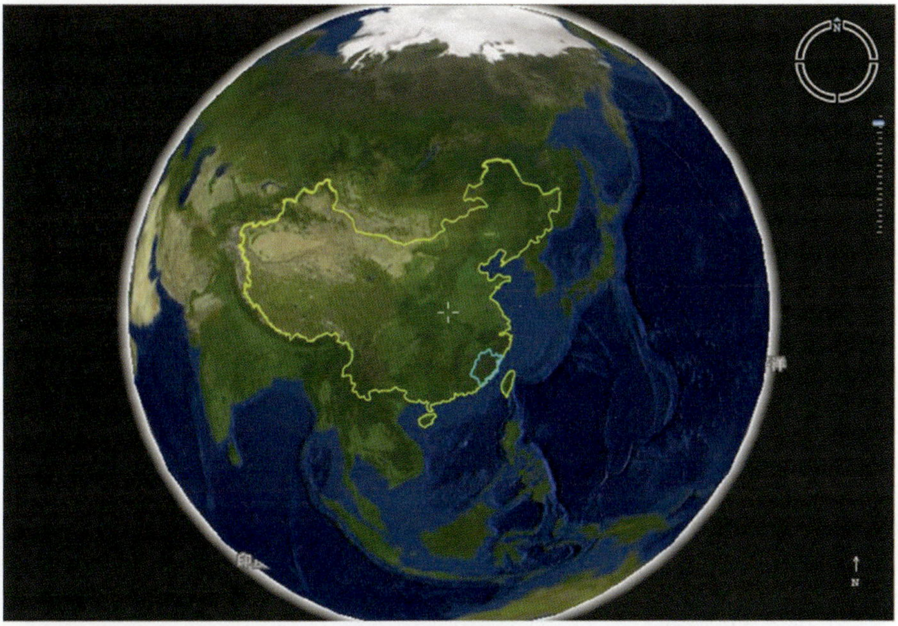

Fig. 8.1 The view result from the global scale

8.2.2 The Three-Dimensional Ocean Data Integration Platform

The three-dimensional ocean data integration platform is the basis for the construction of DO systems, and it is composed of three parts: the three-dimensional coast, the ocean surface and the three-dimensional ocean floor (Zhang et al. 2011). The three-dimensional coast is constructed from DEM data and multi-resolution terrestrial remote sensing data. The ocean surface is constructed from pictures with lumpy textures taken from commercial platforms such as Skyline and Virtual Earth. The three-dimensional ocean floor is constructed from DEM data of the seabed with gradual color rendering. In the three-dimensional ocean environment, a user can zoom in from space all the way down to street level on the coast (with images that are often sharp enough to show individual houses or people) and through the ocean surface to the ocean floor. Some three-dimensional biological or man-made establishment models can also be added between the ocean surface and the three-dimensional ocean floor. Figure 8.1 shows a view from the global scale. Figure 8.2 shows a view over the coast from space. Figure 8.3 shows a view over the ocean surface. Figure 8.4 shows a view over the ocean floor.

Because commercial modeling platforms are mainly used for land environments, they have some shortcomings in DO applications, especially in the construction of the three-dimensional ocean environment. First, because the surface is constructed from pictures with a lumpy texture, the region representing the borders of different

Fig. 8.2 The view result over coast

Fig. 8.3 The view result over the ocean surface

Fig. 8.4 The view result over the ocean floor

pictures contains discontinuities. Second, the visual resolution can change when users zoom in from space to sea surface level. Third, the system efficiency decreases with the accumulation of pictures. Fourth, a DO application cannot simulate a scene to an extent where features such as the ocean current direction are presented. The reason for the shortcomings listed above is that these commercial systems do not provide programming interfaces for the ocean application. An independent scientific DE system can make up for these shortcomings through special arithmetic research and software development. Figure 8.5 shows a scientific DE system that includes the global ocean current presentation function developed by Beijing University of China.

Because the component object model (COM) has nothing to do with programming languages and can be issued by a binary specification, the scientific DE system with the global ocean current presentation function is designed to be developed using COM technology. Consequently, this system can be integrated with a commercial system such as Skyline. The Beijing University development team has provided a COM that can visualize global ocean current along with software integration that has already been developed into a system framework.

The construction of a real three-dimensional ocean environment should be realized with these virtual reality technologies. The technologies stated above provide a three-dimensional ocean scene that can integrate ocean data representing marine physical phenomena such as whirlpools, tides and the like.

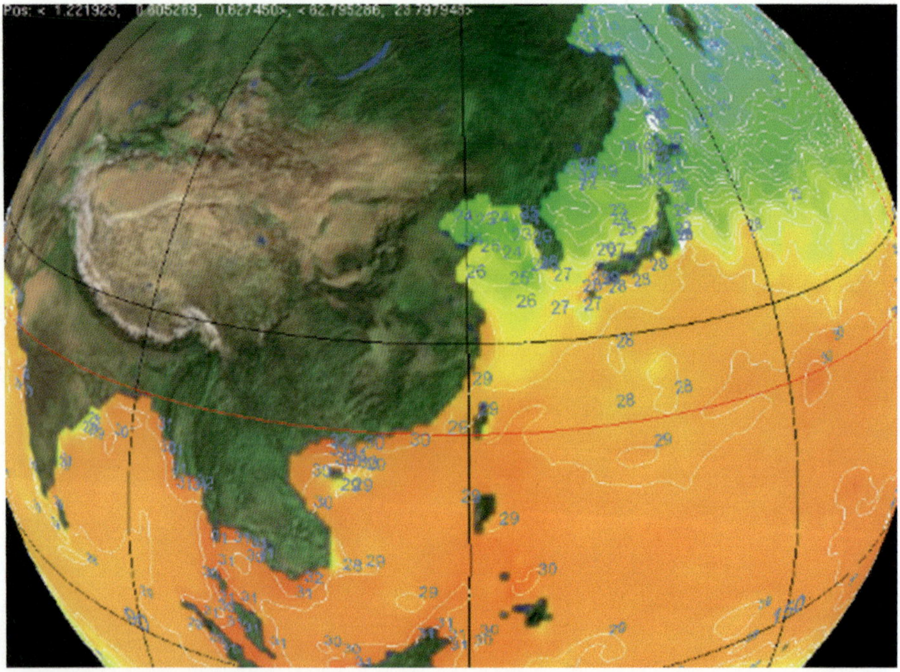

Fig. 8.5 The view result over the ocean floor

8.2.3 The Dynamic Tide Data Visualization

Tide forecast data includes high tide frequency and time information, as well as associated geographical locations. In the past, the data were usually displayed as numbers in a chart or in the form of a two-dimensional graph. In the DO system, the geographical location of the tide station can be presented as a three-dimensional symbol in the visualized scene, which is constructed by DEM and high resolution remote sensing data. The dynamic nature of a tidal event can then be presented with a video that simulates the character of the coast and animates the change from spring tide to ebb-tide. Specific tide data at different times can be shown in a two-dimensional graph. If a user is interested in tidal information relating to a particular time, he can click the three-dimensional symbol on the virtual globe, and a two-dimensional graph containing the desired data will be displayed. A view of the dynamic tide result is shown in Fig. 8.6.

8.2.4 Remote Sensing Information Products Integration and Sharing

For remote sensing data, a number of current real-time and archived products for the different regions can be overlaid on the global visualization platform. For

Fig. 8.6 The view result of the dynamic tide

example, the products collected from NASA's MODIS include 250 m RGB images, sea surface temperature imagery, chlorophyll imagery and chlorophyll fluorescence. For information sharing and interoperation, remote sensing information products should conform to the standards set by the Open Geospatial Consortium (OGC). Two standard web service interfaces, including Web Map Services (WMS) and Web Feature Services (WFS), have been used to provide widely distributed access to various ocean data. Conversely, different types of remote sensing information products are integrated and shared with other methods. Either way, the most important point to note is the magnitude of the data product. Figure 8.7 displays the sea surface temperature (SST) products from MODIS.

8.2.5 Computational Ocean Model Data Integration and Service

Model output data are divided into two types: global data and regional data. The various visualizing technologies were discussed in a paper published in the International Journal of DE in 2011 (Zhang et al. 2011). The grid data can also be visualized on the globe in the form of grid lines. The number of each grid cell or node can be queried by clicking on its geographic location on the globe. Furthermore, the different types of model result data can be viewed at varying sea depths. The model result data are updated according to the ocean model. Figure 8.8 shows a view of the model result data.

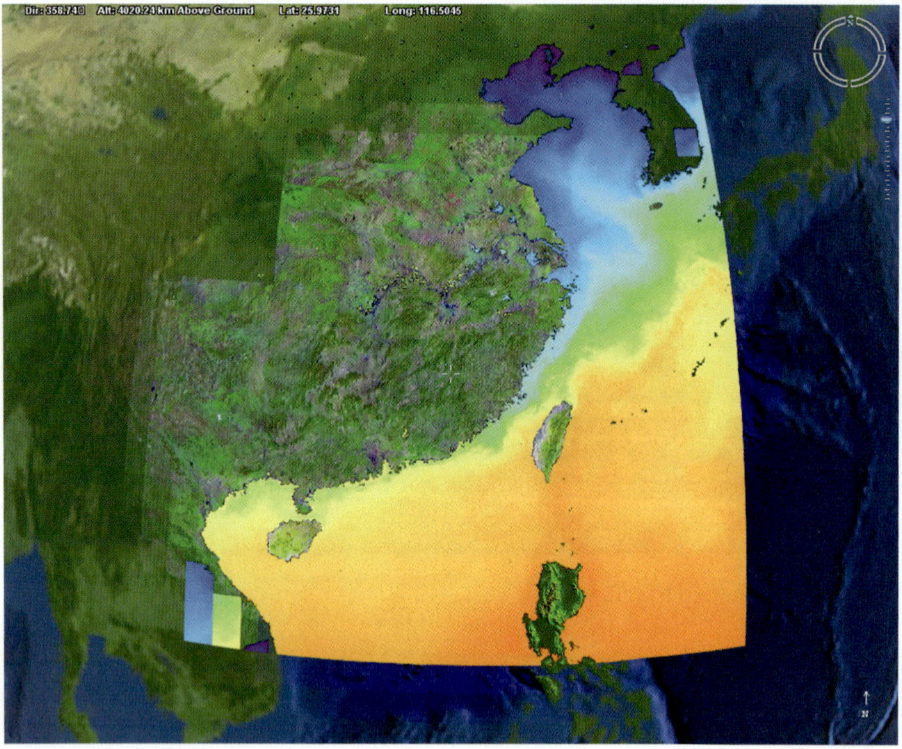

Fig. 8.7 The sea surface temperature (SST) products of MODIS

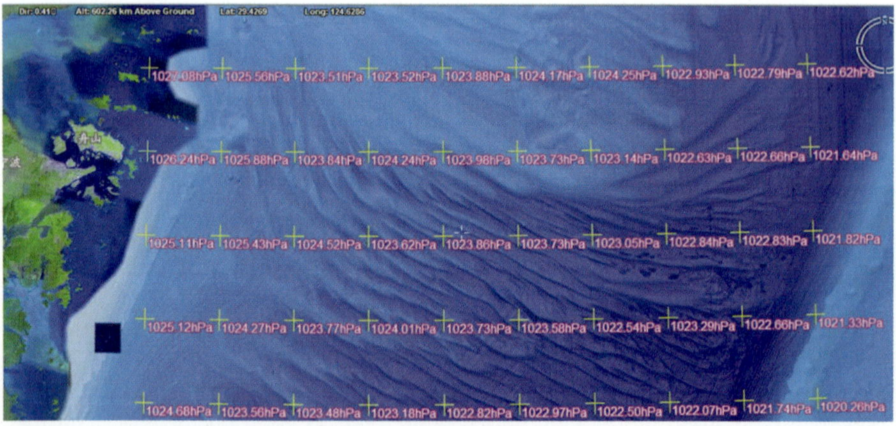

Fig. 8.8 The view result of the model computing result data

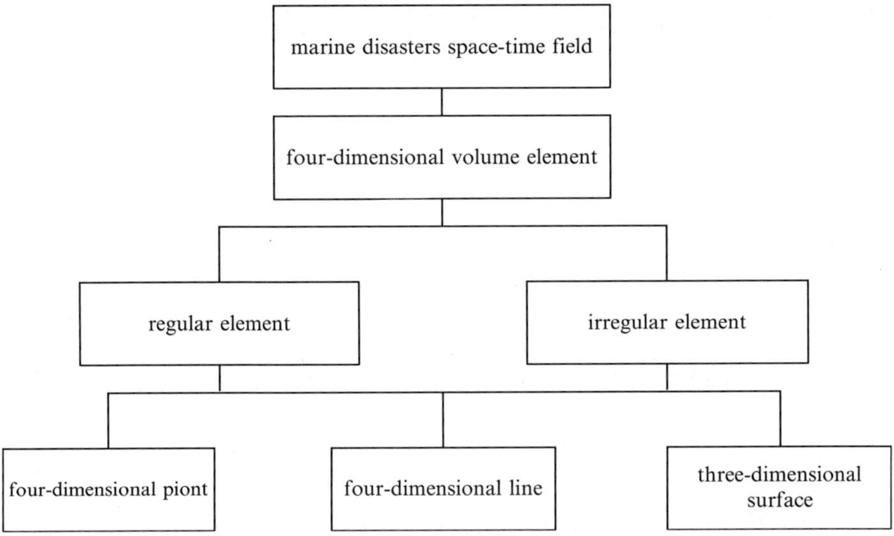

Fig. 8.9 Marine four-dimensional space-time field structure expression

8.2.6 Spatio-temporal Model of Marine Disasters

The formation, development, and consequences of marine disasters are typically affected by changes in the three-dimensional tide field and velocity field during storm tides, typhoons, and rising sea level disasters; spatial diffusion and three-dimensional changes in concentration; sea-surface winds, waves, tides, and coastal landforms; and the distribution of coastal residents and coastal land use. In this paper, we propose that the spatio-temporal pattern of marine disasters is a function of the marine four-dimensional space-time field. We have therefore developed a marine disaster data model based on the four-dimensional space-time field that can be described as follows. The four-dimensional space-time field includes a tide field, a temperature field, a salinity field, a density field, a velocity field, a sea-surface wind field, a wave field, and other fields:

$$(\mathrm{st1}, \mathrm{st2}, \mathrm{pg}, \mathrm{dp}, \{\mathrm{a1}, \mathrm{a2}, \ldots\}, \{\mathrm{r1}, \mathrm{r2}, \ldots\}, \{\mathrm{ir1}, \mathrm{ir2}, \ldots\})$$

where st1 represents spatio-temporal state 1, st2 represents spatio-temporal state 2, pg represents temporal granularity, dp represents the marine dynamics forecast principle, the set {a1, a2, ...} represents the attributes of the marine disaster, the set {r1, r2, ...} represents the influence rules of the marine disaster, and the set {ir1, ir2, ...} represents the influence result of the marine disaster.

The base of the marine disaster space-time state description is the structure of the marine four-dimensional space-time field. Figure 8.9 shows the structure and relationships within the marine four-dimensional space-time field. This structure is conducive to the integration and expression of marine disaster vector factor and scalar data. As shown in Fig. 8.10, a node can be used to represent (a) a vector or

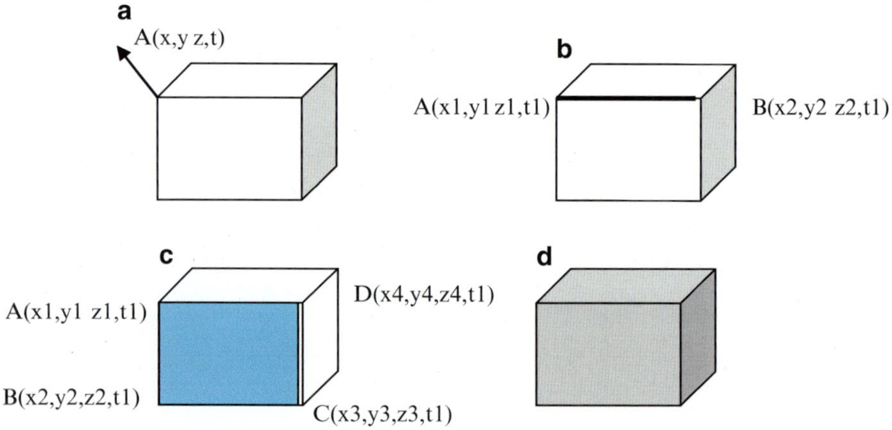

Fig. 8.10 Expression of marine four-dimensional space-time field structure elements

scalar (e.g., flow velocity, flow direction, or temperature), (b) a line (such as an ocean wave object trajectory), (c) a face (such as the temperature profile), or (d) a volume element (such as the concentration of pollutants).

8.3 Digital Ocean System Initial Architecture

The DO system's initial architecture consists of five distinct layers, as illustrated in Fig. 8.11. From the top down, these include the data acquire layer, standard data layer, data service layer, function layer, and application layer, respectively.

8.3.1 Data Acquire Layer

The data acquire layer consists of many types of ocean observation data resources. The integrated ocean observation devices include an oceanic optical buoy, a large-scale environment observing buoy, a seabed based comprehensive dynamic factors automatic observing system, a self-locating sub-water tide monitor, a sea sound detection buoy, a sub-water tide comprehensive admeasuring apparatus, a ship based ocean environment observing system, and a land based ocean environment observing system. In general, the integrated ocean observation data are derived from many sources, which include the in-situ fixed platforms' observation data, the in-situ mobile platforms' observation data, and the remote sensing platforms' observation data. The real-time and delayed-mode observation data collected from the equipment listed above are transmitted to data distribution centers through by way of a communication network. Observation data may be transmitted hourly or even more frequently. The observation data are stored in XML, text files or other formats. Additionally, the data from remote sensing satellites and an extensive array of high frequency ground wave radar systems are already diverse and continuing to expand.

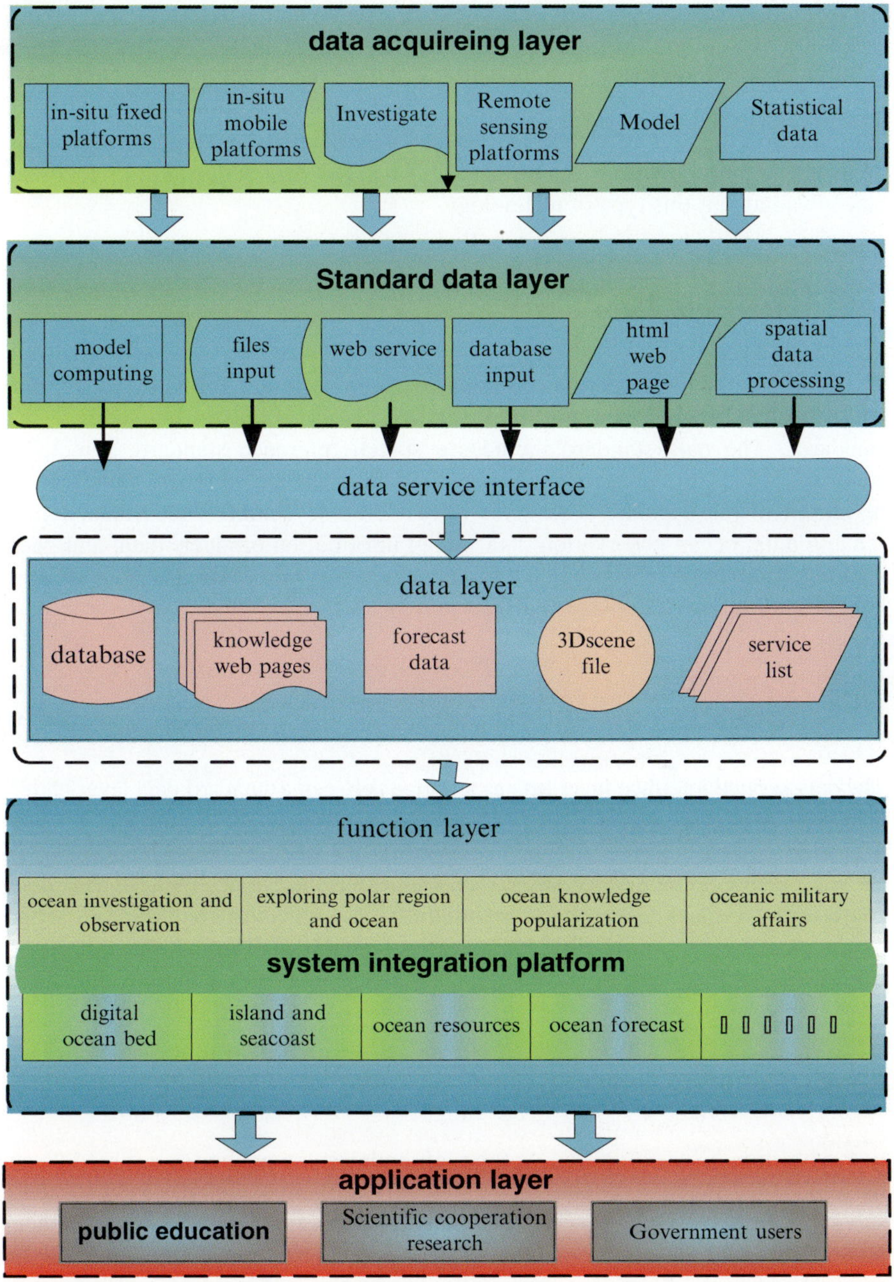

Fig. 8.11 Marine four-dimensional space-time field structure expression. The Digital Ocean system initial architecture

For example, satellite remote sensing data can be collected from the MODIS (Moderate Resolution Imaging Spectroradiometer) and QuikSCAT (NASA's Quick Scatterometer) platforms. The products derived from NASA's MODIS include 250 m RGB images, sea surface temperature, chlorophyll and chlorophyll fluorescence. With the development of technologies such as the Internet of Things, data can also be acquired through web-based wireless clients. The data acquisition protocol will take advantage of this type of technology.

8.3.2 Standard Data Layer

The data acquire layer consists of data in many different formats. To facilitate data storing, sharing and application, the data should conform to a series of data standards. The data standard includes a spatial reference standard, a metadata standard, a data formatting standard, and a data exchange standard. Data that can be stored in the standard data layer format include georeferenced data, remote sensing data, ocean observation data, ocean information products data, and ocean business data. For data stored in file formats, metadata information is encoded using Extensible Markup Language (XML).

8.3.3 Data Service Interface Layer

The data service interface layer acts as a bridge between a standard data layer and the data users. It is based on a series of data accessing interfaces. For example, web-based data sharing should be executed in a standard format, such as Web Mapping Service (WMS), Web Feature Service (WFS), or Web Coverage Service (WCS), and should conform to the Open Geospatial Consortium (OGC) specifications. The data service industry can be developed using Web Service technology. Through the data service interface layer, users can view data through standard interfaces and protocols.

8.3.4 Function Layer

In the function layer, there are two types of modules: the public service module and the application function module. The public service module includes a 3D-scene file service, a data processing service, an ocean data service and a business data integration interface module. The public service module uses pre-processed data and prepares it for visualization or exploration or for the application function module. According to the application requirements, the application function module can be divided into different modules. For example, this module may consist of data regarding basic geographical information, island information, maritime space management, ocean economy, ocean disasters, ocean environment, ocean supervising, ocean rights and interests, or polar regions.

8.3.5 Application Layer

In the application layer, different types of users are provided with different usage methods. Some users may simply want to receive particular information from function providers, such as disaster prediction information or drinking water quality announcements. These users may not understand technical terms or need to know how the system works; they just need to obtain accurate information in a timely manner. As it pertains to the web application, these users will simply need to visit the home page of the website, propose their requests and then receive the results. For users who have their own data centers and need to perform special analyses for their own uses, the system provides some basic functions from the national center in the form of application programming interfaces (APIs). These users can use the APIs to build their own application system, which can then use all of the data in the national center and all of the data in their own data center. Using this method, local departments can build local professional systems easily. They can develop new functions based on APIs while investing less money and time. The interface between the application layer and the function layer include the COM interface and the HTTP protocol. If scientists want to acquire the data, the application can be developed into an ocean data clearinghouse. Then, scientists in different domains can obtain data in different space-time formats.

8.4 A Case Study of Digital Ocean

8.4.1 A Digital Ocean Prototype System

There are more than 20 oceanic administration units along the coast of China. In the process of investigating oceanic data integration, many types of data from different departments were stored, integrated, shared and visualized. The China Digital Ocean Prototype System (CDOPS) was developed and was available on the China Digital Ocean Special Network in 2009, a special network within the departments of the State Bureau of Oceanic Administration of China.

The most obvious virtue of the CDOPS is its ability to cater to different types of users. Users with security access to the 22 nodes can browse, upload, and analyze a variety of ocean data through the web. They can even construct their own professional systems with no investment.

The national center is located in the National Marine Data & Information Service Center of China. The data types integrated in CDOPS include geographical vector data with multiple scales from 1:100,000,000 to 1:50,000, remote sensing data with multiple spatial resolutions from 250 to 0.61 m, ocean physical data, ocean chemical data, ocean biological data, ocean geological data and other types of data. The formatting includes Digital Elevation Model (DEM) data of the seafloor and coast, the in situ observational data, remote sensing data and model output data.

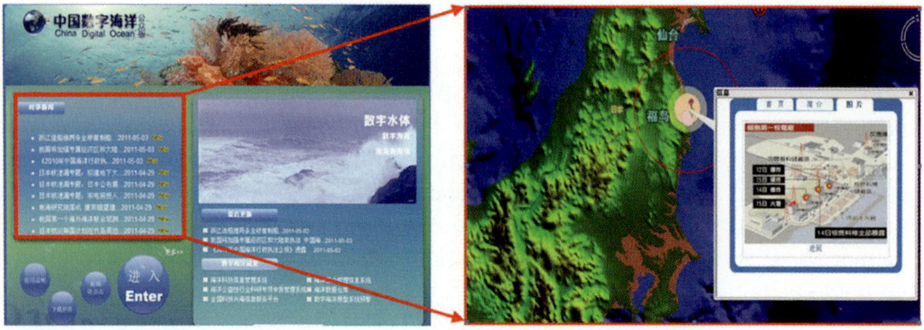

Fig. 8.12 Expression of marine four-dimensional space-time field structure elements. The function interface of the iOcean and the news report of Fukushima Daiichi nuclear power plant Explosion

8.4.2 DE in Support of an Online Oceanic Educational Public Service and Popularization System

Based on the initial vision of DE, the public should constitute a significant proportion of DE users. The iOcean system was made available on the web at "www.iocean.net.cn" in the main center located in the National Marine Data & Information Service, State Oceanic Administration of China on June 12, 2009. The data types integrated in iOcean include geographical vector data with multiple scales from 1:100,000,000 to 1:2,500,000, remote sensing data with multiple spatial resolutions from 250 to 0.61 m, ocean physical data, ocean chemical data, ocean biological data, and ocean geological data. To date, more than half a million members of the public have visited the website. The system has been updated continuously. Figure 8.12 shows the function interface of the iOcean system during the news report of the Fukushima Daiichi nuclear power plant explosion (Zhang et al. 2013).

8.4.3 The Evaluation of Initial DO Application

The CDOPS was an instructive attempt at progress in DO research. The CDOPS provides us with a new data and information integration and application model. With the CDOPS, users at various locations can view, query, download and analyze different types of oceanic data in a virtual globe environment. The CDOPS has proved helpful in accelerating cooperation between scientists and government managers, and it has demonstrated the possibility and potential uses of constructing a DO system.

The application results of iOcean indicate that the use of a DE-supported online oceanic educational public service and popularization system may support a more thorough use of ocean information services and provides a new method for increasing public awareness of the ocean.

In the future, many issues must be researched and resolved to make DO a reality.

8.5 Future Work with Digital Ocean

Future work relating to DO should be carried out in the following seven research areas.

8.5.1 Aiding Integrative Oceanic Scientific Research

Oceans play an important role in other natural Earth processes, such as substance and energy exchange. For example, the study of global warming and global sea level rise requires integrative study by many scientists from different research domains. As an integrative research platform, the DO provides many advantages to scientists wishing to share and exchange data, ideas, and other achievements. Additionally, the DO can provide a virtual environment that simulates past and future events. Furthermore, the DO can provide a global, web-based research platform for scientists from different countries and different research domains to collaborate.

8.5.2 Constructing Regional DO Systems

To advance the application potential of DO, many regional application systems should be constructed. The demands on these systems will be different, so each oceanic region should construct their own system. Here, we take the Taiwan Strait region as an example to illustrate this problem.

The Taiwan Strait, a channel connecting the East China Sea and the South China Sea, is located between Taiwan and southeastern Mainland China. It is approximately 200 km wide and 400 km long, with an average depth of 60 m (Lin et al. 2005). The Taiwan Strait and its adjacent region is a flourishing economic zone, but it is troubled by serious natural disasters, such as typhoons and storm surges. This region's DO system should therefore concentrate more on ocean disaster prevention and forecasting.

8.5.3 Studying Multiple Spatio-temporal Scales of Ocean Factors

Oceans have many different spatio-temporal scales. The relationship between ocean data at these varying spatio-temporal scales should be studied in the DO. Studying ocean data at different spatio-temporal scales can reveal new information and trends across multiple oceanic disciplines, so changes in oceanic factors should be studied at many spatio-temporal scales. The DO can integrate data types at different spatio-temporal scales, so any study of ocean elements at multiple spatio-temporal scales should be studied with the support of the DO.

8.5.4 Studying the Relationship Between Ocean Elements at Different Times

Oceans element change continuously with the passing of time. Studying the relationship between ocean elements at different times can help researchers analyze how and why these elements change. Furthermore, we can predict future development trends of ocean elements. The DO can integrate vast time-series data, so any study of the relationship between ocean elements at different times should be carried out with the support of the DO.

8.5.5 Studying the Spatial Relationship Between Different Ocean Elements or Components

Different ocean elements or components are influenced by each other; therefore, the spatial relationship between different ocean elements or components should be studied. DO can integrate multiple resolution data, so any study of the spatial relationship between different ocean elements or components should be carried out with the support of the DO.

8.5.6 Studying the Architecture of the DO

The specific architecture of a DO system determines its potential applications, so the system architecture should be studied in detail to understand the most appropriate uses of a DO. The study of a system's architecture should include its hardware, software, user interface, and data storage.

8.5.7 Studying the Coastal River Basin Non-point Pollution Landscape Source and Assemblage Pattern Remote Sensing Parsing

The coastal river basin non-point pollution landscape source and assemblage pattern and its spatio-temporal process analysis is the scientific theory basis for studying the coastal river water and marine water pollution's producing, transporting and prevention and cure policy making. Aiming at mining the spatio-temporal character and advantages of the remote sensing data, the definition schedule of the landscape of river basin non-point pollution source and assemblage should be studied firstly. Then the river basin non-point pollution landscape source and assemblage pattern remote sensing parsing method should be put forward from pixel, sub-pixel, and time renewing levels. In this field, taking Jiulong river basin as example, the impervious surface area (ISA) information extraction as the non-point pollution source pattern in 2010 has been achieved, and the renewing and spatial analysis are also concluded in 2005 and 2000. The result shows the ISA increased 33.38 % from 2000 to 2010. This technology is also a reference for the environment change study for the Zhang river basin in China.

Acknowledgments The study is funded by the free exploring program of the State Key Laboratory of Remote Sensing Science, Institute of Remote Sensing and Digital Earth, Chinese Academy of Sciences, the Special Research Project for the Commonwealth of the Ministry of Water Resources of the People's Republic of China (grant no. 201201092), the 908 Project of the State Oceanic Administration, China (No. 908-03-03-02), the National Natural Science Foundation of China (grant no. 61074132, 41076001, and 61375002), and the International Science & Technology Cooperation Program of China (grant no. 2010DFA92720). All members of the Research Team are gratefully acknowledged for their contributions to the work carried out for this paper in past years. In particular, special thanks are given to Director Suixiang Shi, Director Jiye Jin, Director Feng Zhang, Director Sihai Li, Dr. Jian Liu, Dr. Wen Dong, Miss Bing Jiang, Dr. Wei Wang, Mr. Jin Liu, Dr. Xiansan Liu, Mr. Haitao Gao, and Mr. Rongqiang Hu.

References

Al Gore (1998) The DE: understanding our planet in the 21st century [online]. Available from: http://portal.opengeospatial.org/files/?artifact_id=6210. Accessed 14 Jul 2010

Chen SP, Genderen J (2008) Digital Earth in support of global change research. Int J Digit Earth 1(1):43–65

Chen AJ, Leptoukh G, Kempler S, Lynnes C, Savtchenko A, Nadeau S, Farley J (2009) Visualization of A-Train vertical profiles using GoogleEarth. Comput Geosci 35(2):419–427

Consortium for Ocean Leadership (COL) (2009) Project Execution Plan [online]. Available from: http://www.oceanleadership.org/wp-content/uploads/2009/04/. Accessed 14 July 2010

GEO (2005) The Global Earth Observation System of Systems (GEOSS) 10-year implementation plan [online]. Available from: http://www.earthobservations.org/documents/. Accessed 14 July 2010

Goodchild MF (2008) The use cases of digital earth. Int J Digit Earth 1(1):31–42

Guo HD, Fan XT, Wang CL (2009) A digital earth prototype system: DEPS/CAS. Int J Digit Earth 2(1):3–15
Guo HD, Liu Z, Zhu LW (2010) Digital Earth: decadal experiences and some thoughts. Int J Digit Earth 3(1):31–46
Hou WF (1999) Tentative idea for development of "digital ocean". Aviso of Ocean 18(6):1–10
Interagency Ocean Observation Committee (2013) U.S. IOOS summit report: a new decade for the integrated ocean observing system [online]. Available from: http://www.iooc.us/wp-content/uploads/2013/01/U.S.-IOOS-Summit-Report.pdf. Accessed 14 Nov 2013
Lin SF, Tang TY, Jan S, Chen CJ (2005) Taiwan strait current in winter. Cont Shelf Res 25:1023–1042
Maged M, Mazlan H, Cracknell AP (2009) 3-D reconstruction of coastal bathymetry from AIRSAR/POLSAR data. Chin J Oceanol Lirrmol 27:117–123
Ocean US (2002) Building consensus: toward an integrated and sustained ocean observing system [online]. Available from: http://www.ocean.us/documents/docs/Core_lores.pdf Accessed 14 July 2010
Patrikalakis NM, Bellingham JG, Mihanetzis KP (2000) The digital ocean. Computer Graphics International 2000 (CGI'00). (1):45–48
Shen YZ, Austin JA, Crouch JR, Dinniman MS (2007) Interactive visualization of regional ocean modeling system. In: Proceedings of the IASTED international conference on graphics and visualization in engineering, pp 74–82
Su FZ, Du YY, Pei XB (2006a) Constructing digital sea of China with the datum of coastal line. Geo Inf Sci 8(1):12–15
Su FZ, Yang XM, Xu J (2006b) Basic theory and key technologies for marine geographic information system. Acta Oceanol Sin 25(2):80–86
Su FZ, Zhou CH, Zhang TY (2006c) Constructing a raster-based spatial-temporal hierarchical data model for marine fisheries application. Acta Oceanol Sin 25(1):57–63
Thomas CM (2003) The coastal module of the Global Ocean Observing System (GOOS): an assessment of current capabilities to detect change. Ocean Pol 27(3):295–302
Wright DJ, Blongewicz MJ, Halpin PN, Breman (2007) Arc marine: GIS for a blue planet. ESRI Press, Redlands
Wright TE, Burton M, Pyle DM, Caltabiano T (2009) Visualising volcanic gas plumes with virtual globes. Comput Geosci 35(3):1837–1842
Yasuko Y, Yanaka H, Suzuki K, Tsuboi S, Isse T, Obayashi M, Tamura H, Nagao H (2010) Visualization of geoscience data on GoogleEarth: development of a data converter system for seismictomographic models. Comput Geosci 36(2):373–382
Zhang X, Dong W, Li S, Luo J, Chi T (2011) China digital ocean prototype system. Int J Digit Earth 4(3):211–222
Zhang X, Dong W, Xiaoyi J et al (2013) "Digital earth" in support of an online oceanic educational public service and popularization. Acta Oceanol Sin 32(5):82–86, http://www.iocean.net.cn

Chapter 9
A New Statistical-Empirical Hybrid Based Model to Estimate Seasonal Sea-Level Variation in the Gulf of Paria from River Discharge

Carol Subrath-Ali

Abstract The study presents new insight into the quantitative role of the world's third largest discharging river, the Orinoco of South America as modulating coastal water levels in the vicinity of its outflow. The case study is in a semi-enclosed sea, the Gulf of Paria, located in the southern extreme of the Caribbean Sea. The discharge – coastal water level relationship has been investigated and the water levels exhibit a high correlation ($R^2 = 0.92$) to the trends in actual discharge. The relationship is non-linear and there is a lower threshold value across the months of the year below which the water levels are characterised by large variability around a mean linear trend showing independence of the Orinoco's discharge. There is also an upper threshold value where the maximum amplitude of variation is 21.4 cm. The study utilises a vertically integrated 2D numerical modelling suite to execute a series of experiments to ascertain the variation of the coastal water levels from the variation in the river discharge. The other drivers are wind, salinity, oceanic currents and tidal forcing. The results are finally utilised to develop a third order model function to estimate the average monthly river-driven water level in the Gulf of Paria dependent only on the parameter of river discharge.

List of Symbols

C_D	Drag coefficient
d	Still water depth (m)
D	Eddy diffusion coefficient
f	Coriolis parameter (s^{-1})
F_T, F_s, F_c	Horizontal diffusion terms
g	Acceleration due to gravity (m/s^2)

C. Subrath-Ali (✉)
Department of Civil and Environmental Engineering, Faculty of Engineering, The University of the West Indies, St. Augustine Campus, St. Augustine, Trindad and Tobago
e-mail: subrath_ali@yahoo.com; carol.subrath.ali@gmail.com

C.W. Finkl and C. Makowski (eds.), *Remote Sensing and Modeling: Advances in Coastal and Marine Resources*, Coastal Research Library 9,
DOI 10.1007/978-3-319-06326-3_9,

h	Total water depth (m)
H	Height of Planetary Boundary Layer (m)
$\widehat{H}$	Source term due to heat exchange with atmosphere
p	Pressure (pa)
R	River discharge
s	Salinity (psu)
S	Magnitude of source discharge (m^3/s)
t	Time (s)
T	Temperature (°C)
x,y,z	Cartesian coordinates
u,v,w	Flow velocity components
η	Surface elevation (m)
ρ	Density of water (kg/m^3)
U	Wind speed
WL	Water level

Subscripts

a	Atmospheric component
h	Horizontal component
o	Initial condition (reference value)
s	Source parameter
v	Vertical component

9.1 Introduction

The purpose of this chapter is to present a newly derived formulation to estimate the contribution to coastal water levels in a semi-enclosed sea, from driving effects of seasonal, river driven, freshwater flux into the coastal environment.

It is well known that higher coastal water levels (WLs) from coastal setup can exacerbate shoreline erosion, reduce bottom friction with the inland movement of the water line and in so doing, amplify the tidal range and serve to alter the coastal morpho-dynamics. Coastal setup, generally defined as the temporary elevation of coastal WLs may be observed as the slope on the water surface or the rise in the mean sea level towards the shore from the combined effects of wave, wind and barometric pressure (Allen 1997; Loveless et al. 1998). These short term sea level variations of coastal WLs have been shown to have several orders of magnitude greater than the trend (Kolker and Hameed 2007).

Previous researchers have examined the effects of these varying coastal processes mainly along the individual veins of wave set-up [temporary sea level elevations from breaking waves] (Dean and Walton 2010; Stockdon et al. 2006; Gourlay 1992), wind set-up [temporary sea level elevations from transfer of momentum by wind stress effects] and a pressure effect from a decrease in atmospheric loading [the inverse barometer effect].

While the major impacts from these effects may be more noticeable during an extreme event, the climatic characteristics of the individual variables however usually introduce an element of seasonality. One such other seasonal but infrequently considered driver is the freshwater flux into the coastal environment from river discharge. Although the term "river setup" is not defined in the literature, the concept has been examined although quantitative studies are not quite common. It is however acknowledged that river runoff can induce large-amplitude seasonal variation of the sea level in regions located near river outflows (Tsimplis and Woodworth 1994; Meade and Emery 1971). These elevated WLs with varying residence times can impact salinity stratification, effluent dispersion rates and engineering activities.

It is acknowledged that each coastal site will host varied coastal processes due to the native environment (for example: bathymetry, topography and shoreline configuration); as such, riverine discharge influences are not expected to be globally homogeneous. This paper then, examines a case study site where the coastline is affected by the world's third largest discharging river – the Orinoco, which discharges from north-eastern Venezuela of the South-American continent, directly in part into a semi-enclosed bay – the Gulf of Paria (GoP). The formulation is based on numerical, experimental methods utilising a vertically integrated 2D hydrodynamic model – the MIKE 21 Flow Model HD FM model, as well as incorporating statistical methods by the use of k-means cluster and regression analyses.

9.1.1 Global Freshwater Influence

Freshwater flow has been assessed for its contribution to coastal WLs in various parts of the world. Some illustrative examples of river discharge contribution to seasonal sea level change include the Chanjiang Estuary in China which has an annual sea level variation from 40 to 70 cm. The most significant factor in causing this variation is the river runoff (Baocan et al. 1992). In the Bay of Bengal, river runoff produces a WL variation of 100 cm within a year (Pattullo 1966) and along the eastern United States, variations in annual runoff inflow account for 7–21 % of the total sea level variation from 1 year to the next (Meade and Emery 1971). Varying levels of the Black Sea bounded by Europe, Anatolia and the Caucasus, are closely related to river discharge from the Danube. In Varna Bay in the Black Sea, the maximum amplitude of variation is 16.1 cm (Trifonova and Demireva 2003). River runoff has also been described as a considerable factor causing sea level changes both in an estuary as well as the adjacent coast. The largest reported variation at 100 cm/year occurs in the Bay of Bengal (Pattullo 1966).

Since steric effects are usually considered as the combination of salinity and temperature effects, the changes from these may manifest themselves from either internal oceanic changes such as upwelling, or the entrainment of freshwater into the more saline sea water. Some of the difficulty of estimating the forcing from river discharge lies in the separation of these processes. Nonetheless, as shown above, there have been studies which have examined the effect of riverine, freshwater flux on coastal WLs. The effect however, has been particularly dominant in estuaries with large salinity slopes and the variation has been described as being dominant on a suggested time scale of a few days (Bjork et al. 2000).

While volumetric freshwater flow has been discussed as one of the factors affecting sea level, the variation has also been ascribed to being heavily dependent on the density of sea water (Nomitsu and Okamoto 1927). This density variation has generally been defined in terms of the seasonal fluctuations in the specific volume of sea water (Lisitzin 1974) and can be expected to exhibit significant changes in regions where there is an influx of freshwater such as in the vicinity of river outflow.

With regard to the relationship between seasonal river discharge and WL variations, there are comparatively few studies that have informed us of quantitative relationships to enable the use of the volumetric discharge for these non-tidal elevation estimates. While earlier researchers (Nomitsu and Okamoto 1927; Galerkin 1960; Lisitzin and Pattullo 1961; Pattullo 1966; Meade and Emery 1971; Svensson and Jones 2004) have noted the river contributions, quantification of the discharge – sea level variation relationship has been less forthcoming; perhaps because of the vast number of signals that contribute to the overall variation of the coastal WL making the task far from straightforward in most cases.

In addressing the relationship between river discharge variations and coastal WL elevations, it is useful to quantify the relationship so that an assessment of the elevation can be made in a predictive mode. A relationship of this kind is not expected to be universally applicable to estimate sea level variation from freshwater flux due to the variations in site-specific characteristics. The development of the relationship can however, provide a useful tool for a given region and will further provide an economic means to estimate the seasonal, freshwater contribution to the coastal WL in the relevant coastal environment. Applicable areas are coastal engineering, port and harbour development and the provision of baseline information to coastal flood defence strategies.

The chapter next presents an overview of the study area and the data and methodology employed to execute the study. The results and discussion then follow and the chapter then concludes on the results of the findings.

9.1.2 Study Area

The site of interest is located in the extreme south-east of the Caribbean Sea, off the South American continent (Fig. 9.1), situated between extreme north-eastern Venezuela and the southernmost Caribbean island – Trinidad. The semi-enclosed water-body, the GoP, covers an area of roughly 7,600 km^2 and at its widest, measures approximately 150 km, at its narrowest about 14 km. It is generally shallow with average depths of about 20 m (van Andel and Postma 1954; Gade 1961).

The largest river which discharges directly into the GoP from Trinidad is the Caroni with an average flow rate of 12 m^3/s (Environmental Management Authority 1998). The Orinoco River originates in Venezuela and discharges waters amassed

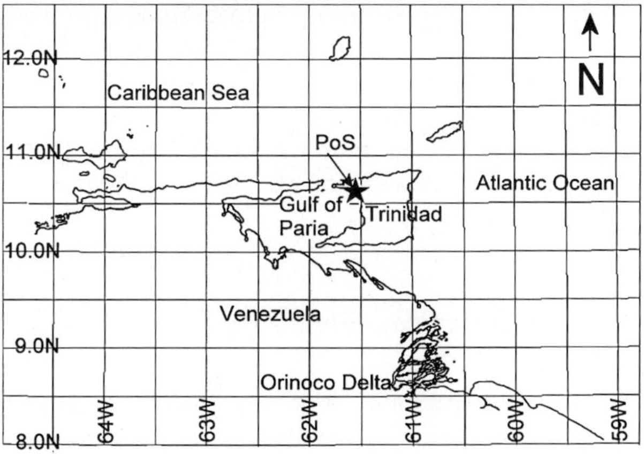

Fig. 9.1 The location of the Gulf of Paria in the southern Caribbean

from an approximate 31 major and 2,000 minor tributaries into the western tropical Atlantic Ocean largely through the Orinoco Delta (Odriozola et al. 2007). It is the third largest river volume discharge in the world with an average of about 3.6×10^4 $m^3\ s^{-1}$ (Meade et al. 1983; Müller-Karger et al. 1989) to the Atlantic Ocean. Since the Orinoco directly empties in part into the GoP, there are a variety of coastal and dynamic influences from its discharge. The effects have been examined with respect to rainfall over the Orinoco watershed region (Schot et al. 2001), seasonal control of salinity in the eastern Caribbean (Froelich et al. 1978), river water plume dispersal by examining ocean colour (Hu et al. 2004) and sediment input into the eastern Caribbean (Bowles and Fleischer 1985). There have been no comprehensive studies however, on the quantitative effect of the freshwater flux from the Orinoco and the forcing of the WL in this coastal environment. What makes this study of particular interest is the fact that the largest distributary which accounts for a near 85 % of the discharge, does not flow directly into the GoP, but discharges at some 114 km away.

The climate of this tropical area is strongly reflected as being bi-seasonal. The two seasons differ primarily by precipitation with the migratory Inter-Tropical Convergence Zone (ITCZ) playing a key role in the precipitation patterns. The dry season (January to May) is characterised by relatively lower precipitation and the wet season (June to December) is characterised by heavier precipitation. The latter hosts the Atlantic hurricane season (June to November). The dominant wind regime is primarily governed by the north easterly trade winds.

While it is well noted that common sea level forcing mechanisms include drivers such as tides, winds, offshore waves and river discharge as well as their combination (Brown and Kraus 2007); there are however, other additional parameters that may or more aptly, should, be considered such as the bathymetry and topographic configuration, But as Svensson and Jones (2004) point out, to address the site specific nature of river flow characteristics a dense network of gauges would be required.

To adequately address the site specific effects of a given river outflow using such a dense network is however, not easily accomplished. For example, in the study of the oceans, a conservative, economic estimate for tidal observations at a one degree resolution to study the oceans at a needed 45,360 observation sites would cost nearly half a billion dollars per year, with an enormous time budget and at least a 45 year timeline for the project (Parker 1991). To overcome such problems, the use of numerical models has, within recent time, become increasingly popular to investigate the behavior and effects of coastal processes.

9.1.3 Numerical Modelling

The 2D hydrodynamic, propriety "MIKE 21 Flow Model HD Flexible Mesh" model of the Danish Hydraulic Institute is employed to execute a series of numerical experiments to assess the coastal WL response in the GoP to the Orinoco's discharge. This model is designed to simulate flows and WLs not only in bays and estuaries but in other coastal areas. The model is capable of constant and varying complex flow conditions (in terms of depth, average velocity and flow direction) which may be laterally unsteady across the width of the flow and resolves flows with variable directions (unlike 1-D models). Additionally, the model's suitability for the study area is based on the physical characteristics of the site; such as its shallow nature (20 m average depths). The 2D model also reduces the computational expenditure that would otherwise be incurred with the use of a 3D model.

The basic, governing equations are all based on the 3D Navier-Stokes equation and the main equations which are integrated over depth are:

$$\frac{\partial u}{\partial x} + \frac{\partial v}{\partial y} + \frac{\partial w}{\partial z} = S \tag{9.1}$$

where (9.1) is an expression of the local continuity equation.

The x component of the horizontal momentum equation is given by:

$$\begin{aligned}\frac{\partial u}{\partial t} + \frac{\partial u^2}{\partial x} + \frac{\partial vu}{\partial y} + \frac{\partial wu}{\partial z} &= fv - g\frac{\partial \eta}{\partial x} - \frac{1}{\rho_0}\frac{\partial p_a}{\partial x} - \frac{g}{\rho_0}\int_z^\eta \frac{\partial \rho}{\partial x}dz + F_u \\ &+ \frac{\partial}{\partial z}\left(v_t\frac{\partial u}{\partial z}\right) + u_s S\end{aligned} \tag{9.2}$$

the y component of the horizontal momentum equation:

$$\begin{aligned}\frac{\partial v}{\partial t} + \frac{\partial v^2}{\partial y} + \frac{\partial vu}{\partial x} + \frac{\partial wv}{\partial z} &= -fu - g\frac{\partial \eta}{\partial y} - \frac{1}{\rho_0}\frac{\partial p_a}{\partial y} - \frac{g}{\rho_0}\int_z^\eta \frac{\partial \rho}{\partial y}dz + F_v \\ &+ \frac{\partial}{\partial z}\left(v_t\frac{\partial v}{\partial z}\right) + v_s S\end{aligned} \tag{9.3}$$

the general transport-diffusion equation for temperature and salinity:

$$\frac{\partial T}{\partial t} + \frac{\partial uT}{\partial x} + \frac{\partial vT}{\partial y} + \frac{\partial wT}{\partial z} + \frac{\partial wT}{\partial z} = F_t + \frac{\partial}{\partial z}\left(D_v \frac{\partial T}{\partial z}\right) + \widehat{H} + T_s S \tag{9.4}$$

$$\frac{\partial s}{\partial t} + \frac{\partial us}{\partial x} + \frac{\partial vs}{\partial y} + \frac{\partial ws}{\partial z} = F_s + \frac{\partial}{\partial z}\left(D_v \frac{\partial s}{\partial z}\right) + s_s S \tag{9.5}$$

The definition of the horizontal diffusion terms is as follows:

$$(F_T, F_s) = \left[\frac{\partial}{\partial x}\left(D_h \frac{\partial}{dx}\right) + \frac{\partial}{\partial y}\left(D_h \frac{\partial}{dy}\right)\right](T, s) \tag{9.6}$$

The reader can refer to DHI's scientific documentation (Danish Hydraulic Institute 2011) for the full description of the governing equations.

9.2 Data and Method

This section is discussed in two parts. The first part presents the data (as referenced to the mesh in the second part). The second part of the method looks at the generation of the mesh, data treatments and execution procedures of numerical model experiments. It uses the results to develop the statistical model to estimate riverine driven WLs from a given discharge.

9.2.1 Data

Since no observed tidal data were available along the open boundaries (south, east and north) (Fig. 9.2) on the model domain, tidal boundaries were created using the tidal prediction capacity of the inbuilt MIKE 21 Tidal Toolbox by using the defined boundaries of the model. The tidal values along all open boundaries were created in this manner and the forcing was verified using measured data inside of the domain at Port of Spain (PoS, Fig. 9.1). Measured, hourly tidal values at PoS for 1987 are available.

The Guiana current data were input as a monthly 1×1 degree spatial u and v velocity component grid derived by an Objective Analysis (OA) scalar subroutine (Mariano and Brown 1992) represented by the Mariano Global Surface Velocity Analysis (MGSVA) (Mariano et al. 1995). This resolution is justified on the basis on the strength and persistence of the Guiana current (Boisvert 1967) with the majority (80–85 %) of scalar currents found in the range (0.41–1.2) m/s (Febres-Ortega and Herrera 1976) and another estimate in the (0.1–2.2) m/s range found by Boisvert (1967).

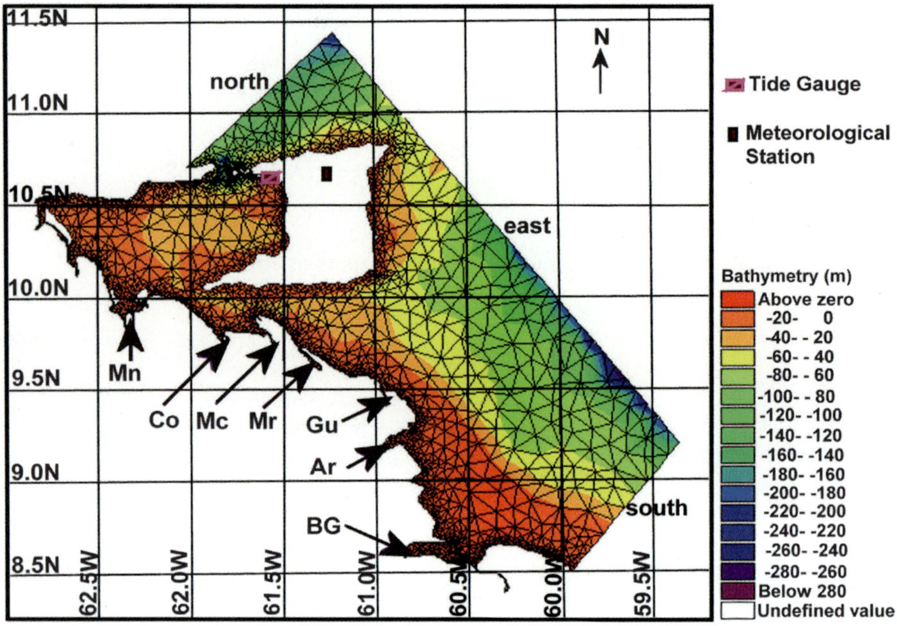

Fig. 9.2 Mesh created for the model domain. The seven major distributaries are designated: BG-Boca Grande; Ar-Araguao; Gu-Guiniquina; Mr-Mariusa; Mc-Macareo; Co-Cocuina; Mn-Manamo

Depth averaged oceanic salinity data were extracted from the World Ocean Atlas (WOA09) (Antonov et al. 2010) database for the monthly variations utilizing data along the boundary and where not available, the nearest gridded value from an objectively analyzed (1 × 1 degree grid) climatological salinity field.

Hourly observed wind data were sourced from the island of Trinidad (TTMS 2010) and were taken to be representative of the wind field and allowed to vary in time but taken as constant in the domain but with a varying Coriolis force to allow veering of the vectoral wind. This domain representation was based on the persistent nature of the north-easterly trade winds (meteorological convention). The wind drag coefficient of 0.001255 was applied (over water wind speeds of average 7 m/s).

The bathymetry used, is a combination of data extracted from MIKE C-Map for the wider oceanic region of the model domain and locally surveyed depths in the GoP by a local consultant company. MIKE C-Map is a DHI software tool for extracting depth data and predicted tidal elevation from the world-wide Electronic Chart Database CM-93 Edition 3.0 from C-Map Norway (Danish Hydraulic Institute 2011).

The river discharge data were sourced from a hydrological database: *The Global Runoff Data Centre, 56068, Koblentz, Germany*, as monthly data from gauging stations at Musinacio (7.69 N 64.76 W) for the period (1970–1992) and daily data from Puente Angostura (8.15 N, 63.6 W) for (1973–1989). Data for each of the distributaries are not available and an appropriate procedure, as discussed in the data treatment method was used to distribute the waters according to the seven (as guided by literature) major outflows.

9.2.2 Mesh Generation and Boundary Conditions

An unstructured, triangular mesh (Fig. 9.2) was generated using digital bathymetric, shoreline and water data with the vertical datum as mean sea level and the map projection system UTM20. The defined area was selected such that the major outflow channel of the Orinoco Delta (the Boca Grande) towards the southern end is accommodated within the model domain. The outflow from the GoP into the Caribbean Sea at the northern end was similarly accommodated.

The mesh boundary conditions were defined so that there are three open boundaries (south, east and north) and one land boundary (the continental coastline, bounding Venezuela). The varying triangular elements are such that on the open boundaries the resolution is at 10 km on the flexible mesh, to a smaller grid spacing of 250 m inside of the GoP. This spacing was chosen in consideration of the large scale forcing on the boundaries as well as in keeping the number of elements to a minimum as possible to ensure reasonable run-time on the available computing system (a finer mesh of a 10 m grid size increased the computational time from 5 to 32 h per run). The entire mesh domain area is greater than, but estimated near 48,000 km^2.

The south boundary defines the major inflow boundary into the model domain with regard to forcing from the oceanic currents. This boundary is aligned at a quasi-perpendicular orientation to the closed boundary and the inflowing major dominant current – the large-scale Guiana current regime which exhibits a persistent flow from south-east to north-west throughout the year. The seaward extent of the south boundary is near the edge of the continental shelf at the 200 m isobath depth where it connects to the east boundary.

The east boundary is aligned along the major dominant flow of the Guiana currents (towards the northwest) and is approximately along the shelf break at the 200 m isobath.

The northern boundary is oriented in a similar fashion to the south boundary but is essentially an outflow boundary with regard to the main current flows and as such is quasi-perpendicular to the north-westward, heading Guiana current.

The seven major sources of freshwater discharge from the Orinoco Basin defined within the model domain, account for the distribution of the cumulative discharge from the many distributaries which enable the river water to reach the Atlantic. This distribution procedure is further discussed in the second part of the data treatment method.

9.2.3 Data Treatment Method

Ideally, the wind data along the boundaries are needed for wind forcing. As such, wind parameters over the sea were calculated according to an expression developed by (Hsu 1981) which uses wind speed (U), height of the planetary boundary layer (H) and wind drag coefficient according to Eq. 9.7:

$$\frac{U_{sea}}{U_{land}} = \left[\frac{H_{sea} \times C_{D\ land}}{H_{land} \times C_{D\ sea}}\right]^{1/2} \tag{9.7}$$

The subscripts “sea” and “land” denote the offshore and onshore states respectively. Values for the $\frac{U_{sea}}{U_{land}}$ ratio have been suggested as 1.7 (SethuRaman and Raynor 1980) and 1.6 ± standard deviation (Hsu 1981), it is also noted that the ratio varies with distance offshore with an average value given as 1.6 at a distance of 20 km offshore at a 10 m height. The correction by such a ratio did not however, address the condition of calm. The regression model developed by (Hsu 1985) was used and the corresponding wind directions were linearly interpolated.

$$U_{sea} = 1.62 + 1.17 U_{land} \tag{9.8}$$

The use of such a constructed dataset produced erratic variations on the model results are the constructed winds were clearly not representative of the actual vectoral wind. The use of the meteorological data was reverted to, and is taken as the driving wind field throughout the domain but varying in time.

Since the distributary flow rates are not available except for a fractional estimate at the Boca Grande, an approximation of the hourly flows at Barrancas (Fig. 9.3), at the top of the delta, was utilised. A river flow, disaggregation method (Acharya and Ryu 2014) was employed to increase the resolution on the Musinacio time series (longer running dataset) using 3 month daily historical window on data at Puente Angostura. A correction of 11 % (Lewis and Saunders 1989) was added for the significant contribution from the Caroni River (Venezuela) and the other runoff and direct precipitation were considered negligible.

The streamflow estimates at Barrancas were distributed among the seven largest caños (Boca Grande, Araguao, Guiniquina, Mariusa, Macareo, Cocuina, Manamo) (Fig. 9.2) each according to their carrying capacity as a fraction of the total Barrancas flow rates.

Since the Boca Grande (the primary distributary) is known to account for about 85 % of the river flow, the major remaining secondary distributaries account for fractions of the remaining 14 % (−1 % for unaccounted for processes).

The discharge was distributed according to the cross sectional carrying capacity of each of the remaining six major channels (using the channel cross-sectional width near outflow and average bathymetric depth). Each channel’s (Boca Guanipa, Cocuina, Macareao, Mariusa, Guanipa, Araguao) outflow capacity was estimated and assigned a fraction of the total of the six combined so that each carries a fraction

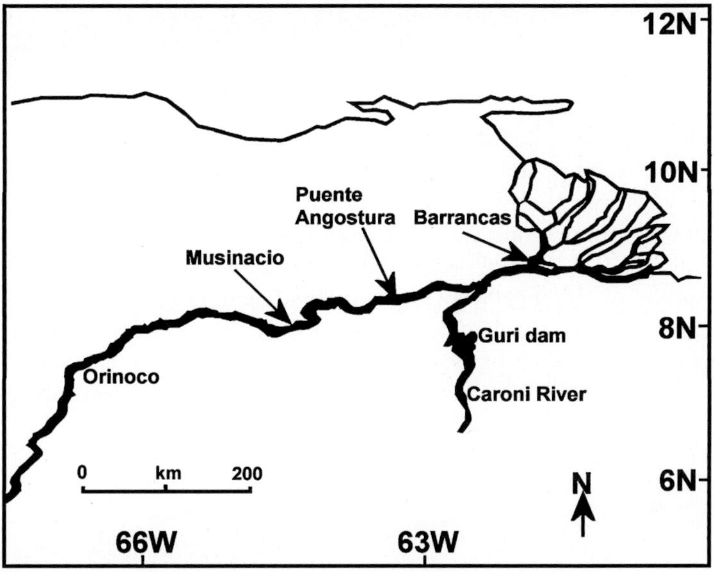

Fig. 9.3 The approximate locations of the gauging stations at Musinacio, Puente Angostura and Barrancas

of the remainder of the waters, not discharged through the Boca Grande, flowing out to the ocean. In this way, the bulk volume of water that is estimated at Barrancas is distributed such that it collectively reaches the Atlantic.

9.2.4 Model Execution Procedures

The hydrodynamic model is driven with data for the entire year of 1987. This particular year has been selected because of the availability and gap-free attributes of the all the required datasets of tides, winds, salinities and oceanic currents. The statistical model is developed with monthly river discharge data for a 23 year period (1970–1992). The open east and south boundaries are forced with changes in the parameters of tides, oceanic currents and salinity, except for the north boundary where currents are not specified. This allows the movement of water through the domain to flow out guided only by the inflow attributes. The closed, land boundary is set to a zero-normal flux velocity.

MIKE 21 offers two choices for time integration and space discretization in the solution of the shallow water equations. These are a lower order, fast algorithm and a higher order solution. At the expense of increased computational time, the higher order was chosen since the flow is convection-type and also to enables a higher level of accuracy. Stability in the transport equations was maintained by holding the critical CFL number to 0.8 (the recommended value is less than 1). To facilitate turbulence in the water body from mixing, due to large velocity gradients as produced by the high flow rates from the Orinoco, the condition of turbulent velocity is accounted for by the Smagorinsky eddy viscosity formulation of a constant value of 0.28. A low bed resistance was applied with a Manning's coefficient of 40 at the higher range of recommended values (20–40 $m^{1/3}/s$) as the output of river-borne sediments consists of fine material, about 85–95 % suspended silt and clay (Warne et al. 2002).

The model's simulations are scenario analyses, executed for a full year (1987). Each month's WL in the GoP was simulated by gradations of 10 % (−50 to 50 %). Monthly initial water levels were also simulated by turning off all sources (that is a zero discharge). To establish a given month's flow, the model was given a soft start and allowed to initialize or enter a convergence phase by execution from the previous month. In this way any 1 month's simulation is the latter month's result from a 2 month run at a computational time of approximately 5 h for each execution.

9.2.5 Calibration and Validation

The river discharge data which were disaggregated using Acharya and Ryu's (2014) method allow compatibility on hourly estimates and enable hourly flow rates to facilitate WL estimates in the GoP that are as close as possible to the measured tide gauge values.

Model runs are done in parallel at the calibration stage using the hourly disaggregated flows as well as the monthly averaged values interpolated with a cubic, shape – preserving function. When the model outputs are compared to the observed tide gauge values, the use of the monthly river discharge data results a lower RMSE value than the use of the hourly disaggregated flow consistently across the majority of the months (Fig. 9.4) with the same RMSE for February and marginally higher values for April, September and December. Since the goal of parameter estimation for numerical modelling is to ultimately reduce the error between modelled and observed values, the monthly, interpolated values are utilised for the simulation routine.

The calibration exercise uses the spring/neap cycles for the months of low (March) and high (September) discharge. Measured tide gauge values are utilised inside of the model domain at the tide gauge location in the north GoP at Port of Spain.

The model's results are adjusted for phase lag (GMT/LST) and a WL adjustment at a 0.71 m (the vertical datum is 0.73 m below MSL). The model is validated for all the months of 1987.

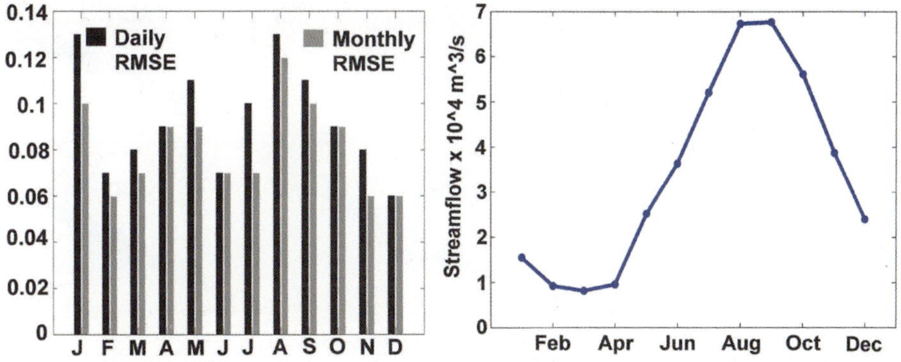

Fig. 9.4 Error variation in the use of the 1987 daily vs. monthly streamflows; the 1987 hydrograph is shown in the *right panel*

9.2.6 Development of the Statistical Model

Based on the output of the model runs, the WLs which reflect the monthly influence of the driving stream-flow are analysed by regression procedures to obtain a function for the discharge/WL relationship for that month for 1987.

The functions are then used to estimate the WLs across each month driven by discharge for the period (1970–1992) and the resulting WLs were phase lagged and used to develop a statistical relationship between both parameters. The result is a statistical model which can adequately estimate the riverine-driven variation of WL based on the discharge value only.

9.3 Results and Analysis

9.3.1 Validation of Modelled Water Levels

The validation data to assess the model's ability to adequately represent the real world hydrodynamic environment across all the months for 1987 are presented as scatterplots in Fig. 9.5 and the RMSE (root mean square errors) of the observed vs. the modelled WLs are shown.

The monthly WL averages, associated Pearson's correlation coefficient (R) coefficient of determination (R^2) and RMSE values between the modelled and the measured WLs for the composite scatterplots in Fig. 9.5 are shown in Table 9.1.

The largest measure of scatter as indicated by an R^2 value of 0.81 occurred during August with May and September also exhibiting high scatter with the R^2 values of 0.88 and 0.82 respectively. The largest measure of error occurred during August assessed by a RMSE of 0.12, with September and January also having

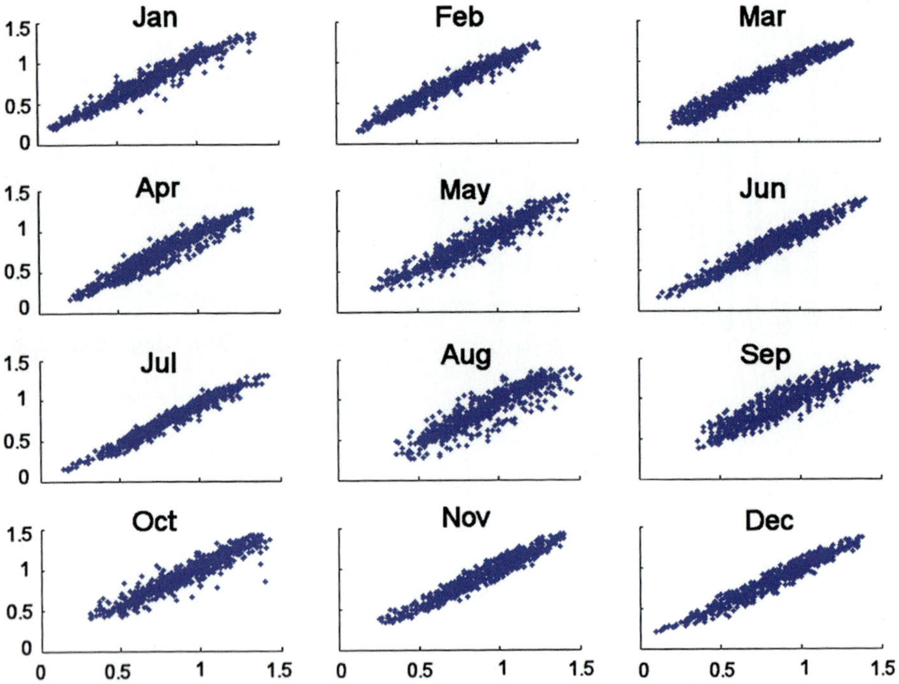

Fig. 9.5 Scatterplots of the observed vs. the modeled WLs for 1987

Table 9.1 Measured and modelled monthly mean water level statistics for scatterplot panel in Fig. 9.4

Month	Measured	Modelled	R	R^2	RMSE
Jan	0.72 ± 0.26	0.80 ± 0.26	0.97	0.94	0.10
Feb	0.71 ± 0.26	0.73 ± 0.25	0.98	0.96	0.06
Mar	0.73 ± 0.27	0.72 ± 0.26	0.97	0.94	0.07
Apr	0.79 ± 0.26	0.75 ± 0.26	0.96	0.92	0.09
May	0.87 ± 0.26	0.85 ± 0.25	0.94	0.88	0.09
Jun	0.81 ± 0.25	0.80 ± 0.25	0.97	0.94	0.07
Jul	0.84 ± 0.25	0.78 ± 0.25	0.98	0.96	0.07
Aug	0.92 ± 0.24	0.87 ± 0.25	0.90	0.81	0.12
Sep	0.91 ± 0.25	0.94 ± 0.24	0.91	0.82	0.11
Oct	0.90 ± 0.25	0.95 ± 0.24	0.95	0.90	0.09
Nov	0.89 ± 0.25	0.91 ± 0.25	0.97	0.94	0.06
Dec	0.83 ± 0.26	0.84 ± 0.25	0.97	0.94	0.06

RMSEs of 0.11 and 0.10 respectively. It is noted that August and September are months where maximal flows occur, with August being a near stationary point on the hydrograph (Fig. 9.4, panel B).

Based on the spread of the RMSE across the months, the model is assessed as adequately representing the hydrodynamic conditions within the GoP despite the lower R^2 and higher RMSEs across some months.

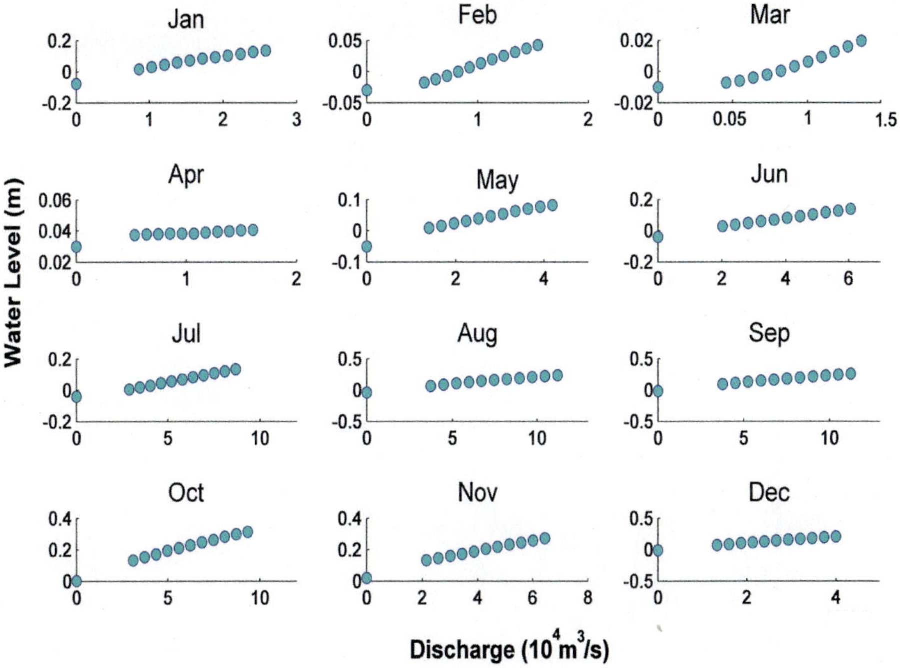

Fig. 9.6 Results of water level response to variation in river discharge

9.3.2 Statistical Water Level Estimates

The model setup, when used to execute the series of numerical experiments where the river discharge was varied at increments of 10 %, provides results (Fig. 9.6) of the numerical simulations thus enabling the monthly formulation of the river-driven, WL/discharge relationship. The results represent the changes in the WL for corresponding changes in the discharge.

Regression analysis by a method of least squares was utilised to formulate the relationship and the lowest order polynomial was selected according to the highest coefficient of determination that is, for any R^2 over 0.99 a quadratic fit was made. April was the only month that did not enable a fit over 0.99 for a second order polynomial ($R^2 = 0.94$) and a third order polynomial was employed. The May's scatter values produced a relationship that presented averaged WL values which were anomalously high. This result was deemed an outlier and the values were re-adjusted to conform to the trend of the monthly plots. The composite scatter for the WL/discharge relationship from the model simulations are shown in Fig. 9.6.

The resulting equations from each month, when applied to estimate the monthly WL variation over the 23 year period (1970–1992) for which the Orinoco's flow rates are available, produced a heterogeneous plot. This is demonstrated by the scatterplot, top left panel in Fig. 9.7 which shows the discharge vs. WL for each

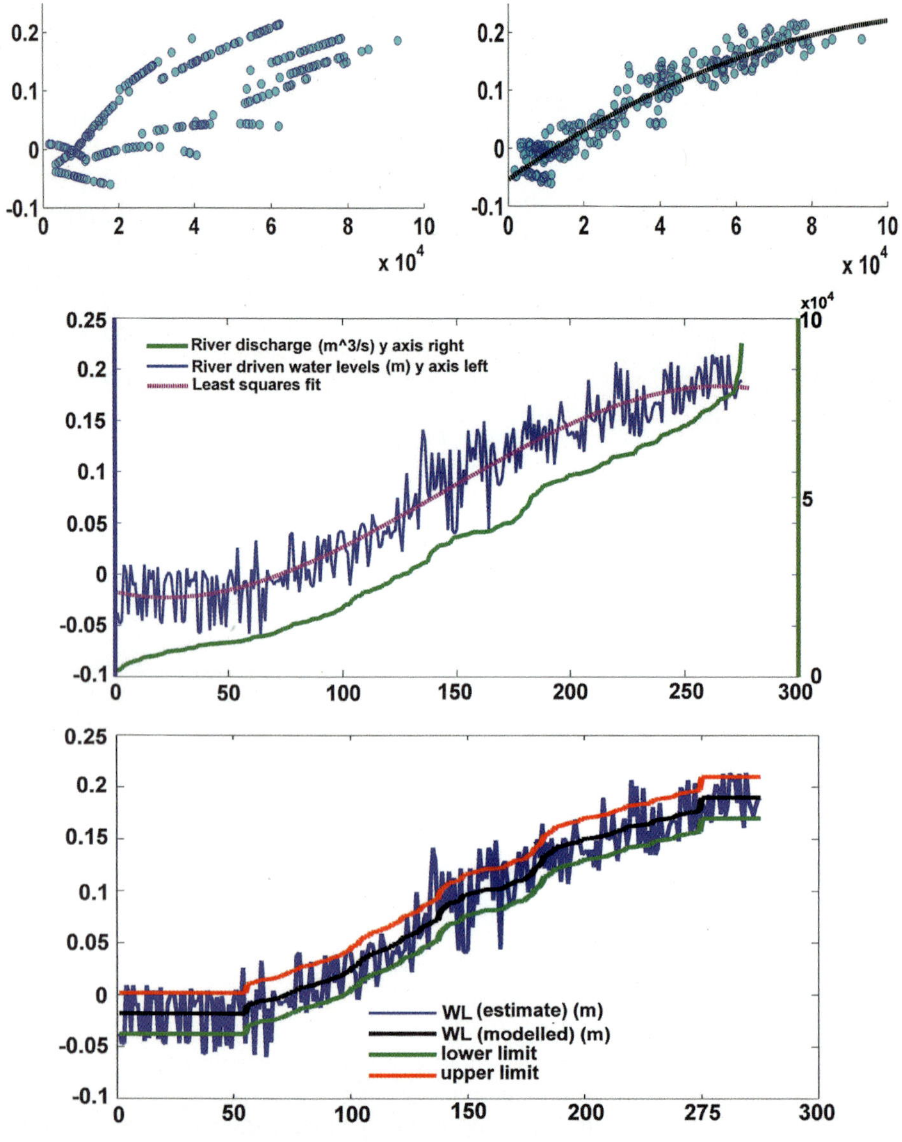

Fig. 9.7 Discharge/water level relationship (1970–1992). *Top panels* shows the relation between the river discharge ((m^3/s), x axis) and the WLs ((m), y axis) with cross correlated signals *top right*. *Middle panel* shows the river discharge and the corresponding phased WLs for ascending values of river discharge. *Bottom panel* shows upper and lower limits on WLs for the lower threshold, rising limb and upper threshold of the modelled function

month maintaining it's integrity with some overlap. When however, the WLs are phased by 1 month's lag as shown in the top right panel there is a clear relationship between the WLs and the previous month's discharge with a high regression coefficient (0.92).

All of the calculated WLs, when sorted according to the increases in the river discharge produced a third order polynomial fit ($R^2 = 0.90$) across the 23 years. The sorted WLs/river discharge data as shown in the bottom panel in Fig. 9.7 indicate that there is a lower threshold value below which the variations in the coastal WLs are independent of the river flow rates and there is also some upper threshold upper limit where the curve levels.

K-means clustering for three clusters suggested by the last regression fit provides the three separate cluster means of −0.01 cm, 0.05 cm and 0.16 cm although the silhouette plot indicates that the first two clusters are not well separated. By accounting for 98 % of the WLs below 0.01 m the threshold value for discharge is 10,000 ($+2.3 \times 10^3$) m^3/s, below which there is no significant trend. At the upper limit, the curve flattens out with the upper threshold at 70,000 ($+5.2 \times 10^3$) m^3/s.

Thus, 90 % of the WL values above the lower threshold can be then estimated by third order polynomial Eq. 9.9, with a standard deviation of ±0.08 m.

$$y = \left[\left(-4E(-16)R^3\right)\right] + \left[3E(-11)R^2\right] + \left[3E - 06R\right] - \left[0.0422\right] \qquad (9.9)$$

There is an upper threshold value of 0.19 ± 0.024 for a discharge rate of 70,000 m^3/s (+5,000 m^3/s). From the use of the equation, typical WL values from the lower threshold to the upper threshold, based on the flow-rate's data range from 1 to 21.4 cm.

9.4 Discussion

9.4.1 Freshwater Effects

The fact that the hourly streamflow data produced generally higher RMSEs than the monthly dataset indicates that the GoP's water levels are more influenced by bulk flow from the Orinoco and less by the near instantaneous flows. This may be best explained as the GoP is not directly in the river mouth, but is at some 114 km away from the main Boca (the Boca Grande). It is noted however, that there is less substantial discharge directly into the GoP. Thus, as the volumetric discharge outflows into the ocean, it loses its instantaneous characteristics due to mixing and dispersal.

Of importance to the study is the presence of other freshwater streams which may contribute to the volume of freshwater in the area, but have not been included in the modelling scenarios. The basins that provide the unaccounted for streamflows are primarily the Guarapiche Basin on the North-eastern South American continent and the Caroni River basin of Trinidad. These basins directly empty into the GoP.

The Guarapiche basin is described as the smallest, exorreica (waters reach the ocean) basin in Venezuela with an average size of about 21,000 km^2 (Machado-Allison 2012). In comparison to the Orinoco which is a significantly larger basin of about 880,000 km^2 (Sánchez-Villagra et al. 2010), the Guarapiche basin contributes a mere 2 % of freshwater by basin discharge to the ocean and the Caroni/Oropuche combined contribution from cumulative average basin sizes of about 1,616 km^2 (Environmental Management Authority 1998) – about 0.18 % compared to the Orinoco. Even at a combined contribution of near 3 % at the upper threshold of 70,000 m^3/s the rivers input an average of 2,100 m^3/s, well below the lower threshold of 10,000 m^3/s as indicated by the modelled results.

The largest external freshwater streamflow into the model domain is the Amazonian plume. The liquid volumetric discharge of the Amazon River has not been considered as the river is a considerable distance away from the GoP and while river runoff can induce large-amplitude seasonal variation of sea level in regions located near river outflows (Tsimplis and Woodworth 1994), the geographical distance of over 1,500 km, precludes the river as near. Further, the main oceanic current, the North Brazil Current (NBC) which drives the Amazonian plume northwards, retroflects before entering the model domain, this retroflection is part of the dispersal mechanism of the plume in the Atlantic. The dispersal seems more extended during the second half of the year when a large fraction of the Amazon's water flow out to deep sea as the plume flows around the NBC and is driven into the eastern Atlantic out towards Africa by the North Equatorial Counter Current (Muller-Karger et al. 1988). To account for the influence of the plume's presence as it nears the inflow boundary of the numerical model, the freshwater lenses' characteristics have however been catered to by the inclusion of the salinity changes along the boundaries.

9.4.2 Sources of Error

From the varying measures of scatter in Fig. 9.5, it is suggested that the reducing stream-flow from a stationary point (at maximum) to the decreasing limb on the hydrograph may introduce variability in phase shifts. The same is suggested for the larger variability of May as the flows increase from near minimal flows during Jan to April, although May is climatologically the windiest month of the year. These rationales may not necessarily hold for January, suggesting another driver for the degree of scatter. The largest RMSE in August also suggests the influence of another driver. It is noted that as the wind field slackens in August in the eastern Caribbean, there is a trend towards a relative maximum in sea level in the southern Caribbean due to changes in the upwelling processes. It is noted however, that these processes may not have the same effect on the GoP as there is a lack of coastal landmass between the island of Trinidad and the northeastern-most coastline of Venezuela – the Paria peninsula.

It is possible that incorporated errors may have found an avenue from not utilizing the actual discharge values from each of the water-ways. This is perhaps overcome to some extent by the utilization of the collective distribution of the discharge data to ensure that the main discharge reached the ocean. It would also be ideal to have measured data at the boundaries but the lack of availability has been addressed by using acquired, measured datasets as close as possible (spatially) to the ideal required data points.

9.4.3 Significance of Results

The results indicate that the WLs exhibit distinct seasonality mirrored by the volumetric river discharge and correlates directly to the trends on the hydrograph. Although each month's discharge/WL relationship is represented by different functions reflected in each of the scatterplots in Fig. 9.6, the lagged, bundled relation for increasing river discharge presents a definitive correlation (Fig. 9.7) indicative of the significant role of the Orinoco's discharge. The significance of the relation indicates that other individual parameters which vary across the months may not necessarily be weighted on par with the volumetric streamflow and thus may not be as influential as the stream flow.

The quasi-linear, zero sloping nature of the mean WL at the lower threshold limit indicates that the coastal WLs are largely independent of the volumetric river discharge below the threshold. The large variability on the WL therefore is more likely governed by other factors. The similar WL distribution, evident beyond the upper threshold indicates that additional discharge has little influence on the WL in the GoP. It is postulated that the reason for this is a change in the flow regime for exceedingly large volumes (approximately 70,000 m^3/s and greater). Beyond this threshold, as larger volumes of water enter the Atlantic, the momentum may take the water a greater distance across the continental shelf before the Guiana currents drives it northward, such that a smaller volume may be transported through the GoP.

The salient feature of the rising limb of the third order regression fit on the WL is the coherence of trend with the discharge rates. An increasing monthly discharge produces a corresponding increase in the WL which is signalled into the following month (but not necessarily in 1 month's time). This is not surprising as the travel time along the coast is expected to be slower than the speeds out in deeper waters. The current speeds are fastest along the edge of the continental shelf with the lower end of the range being 10 cm/s (Boisvert 1967), for a speed of 5 cm/s closer to the shore the travel time from the Boca Grande is approximately 23 days. This will of course vary with the strength of the driving currents which vary throughout the year. The effect may very well be earlier or later and this variability is reflected in the measure of scatter of the discharge/WL plot. The fairly high regression coefficients indicates that the river discharge closely drives the WL variation and although there is variation on the other parameters of salinity, wind and oceanic currents, the range of variability is fairly small enough across the months so that the discharge effect is dominant.

9.5 Conclusion

The Orinoco River significantly modulates the water levels in the Gulf of Paria with a sigmoid-type relationship. These water levels are largely driven by bulk flow from the Orinoco with lesser dependence on the near instantaneous or daily values. The water level variation is signalled for approximately 1 month later for a given month's discharge from the river and there is a lower threshold discharge rate of $(10{,}000 \pm (2.3 \times 10^3))$ m^3/s where the water levels do not contribute more than 0.01 m of water to the variation. Above the lower threshold, the contribution may be estimated according to the derived, third order function up to an upper threshold discharge value of $(70{,}000 \pm (5 \times 10^3))$ m^3/s where the maximum amplitude of variation is estimated at 21.4 cm. This function signifies that between both lower and upper thresholds, the water levels increase for values of increasing discharge.

The study also concludes that the smaller, neighbouring rivers which have an average discharge of less than 10,000 m^3/s do not modulate the GoP's coastal water levels.

Acknowledgements The author thanks the University of the West Indies, Faculty of Engineering for the resources made available to execute the study. Acknowledgements also to the following organisations and persons for their contributions and guidance: The Global Runoff Data Centre (GRDC), The Trinidad and Tobago Meteorological Service, Dr. Julio Zyserman and the Danish Hydraulic Institute (DHI), Dr. Deborah Villarroel-Lamb, Mr. Nazeer Gopaul, Professor Arthur Mariano for his assistance with direct provision of data, Professors Jae Ryu and Lewis Williams for their guidance on watershed issue resolution.

References

Acharya A, Ryu J (2014) Simple method for streamflow disaggregation. J Hydrologic Eng 19(3):509–519

Allen PA (1997) Earth surface processes. Blackwell Science Ltd, Oxford

Antonov JI, Seidov D, Boyer TP, Locarnini RA, Mishonov AV, Garcia HE, Baranova OK, Zweng MM, Johnson DR (2010) World Ocean Atlas 2009, volume 2: salinity. NOAA Atlas NESDIS, Washington, DC, p 69

Baocan W, Jian S, Shenliang C (1992) Analysis of seasonal variation along Changjiang Estuary China. In: Woodworth PL, Pugh DT, Deronde JG, Warrick RG, Hannah J (eds) Sea level changes: determination and effects. American Geophysical Union, Washington, DC

Bjork G, Olof L, Lars R (2000) Net circulation and salinity variations in an open-ended Swedish Fijord system. Estuaries 26:367–380

Boisvert WE (1967) Major currents in the North and South Atlantic Oceans between 64°N and 60°S. U.S. Naval oceanographic technical report no. 193. United States Navy, Washington, DC

Bowles FA, Fleischer P (1985) Orinoco and Amazon River sediment input to the eastern Caribbean Basin. Mar Geol 68:53–72
Brown EB, Kraus NC (2007) Tips for developing bathymetry grids for coastal modelling systems applications, coastal and hydraulics laboratory engineering technical note ERDC/CHL CHETN-IV-69. United States Army Corps of Engineers, U (ed). U.S. Army Engineer Research and Development Center, Vicksburg
Danish Hydraulic Institute, D (2011) Mike 21 and Mike 3 flow model FM hydrodynamic and transport module scientific documentation. Danish Hydraulic Institute, Denmark
Dean RG, Walton TL (2010) Wave setup. In: Young KC (ed) Handbook of coastal and ocean engineering. World Scientific, California
Environmental Management Authority, E (1998) Trinidad and Tobago: state of the environment 1998 report. Environmental Management Authority, St. Clair
Febres-Ortega G, Herrera LE (1976) Caribbean Sea circulation and water mass transports near the Lesser Antilles. Bol Inst Oceanogr Univ Oriente 15:14
Froelich PN Jr, Atwood DK, Giese GS (1978) Influence of the Amazon River discharge on surface salinity and dissolved silicate concentration in the Caribbean Sea. Deep-Sea Res II:735–744
Gade HG (1961) On some oceanographic observations in the southeastern Caribbean Sea and the adjacent Atlantic Ocean with special reference to the influence of the Orinoco River. Bol Inst Oceanogr Univ Oriente 1:287–342
Galerkin LI (1960) On the physical basis of the forecast of the seasonal variations of sea level in the sea of Japan. Tr Inst Okeanol SSSR 37:73–91 (in Russian) in Lisitzin, E. 1974. Sea level changes. Elsevier Scientific Publishing Co, Amsterdam
Gourlay MR (1992) Wave set-up, wave run-up and beach water table: interaction between surf zone hydraulics and groundwater hydraulics. Coast Eng 17:93–144
Hsu SA (1981) Models for estimating offshore winds from onshore meteorological measurements. Bound Layer Meteorol 20:341–351
Hsu SA (1985) Correction of land-based wind data for offshore applications: a further evaluation. J Phys Oceanogr 16:390–394
Hu C, Montgomery ET, Schmitt RW, Muller-Karger FE (2004) The dispersal of the Amazon and Orinoco River water in the tropical Atlantic and Caribbean Sea: observation from space and S-PALACE floats. Deep Sea Res Part II Top Stud Oceanogr 51:1151–1171
Kolker AS, Hameed S (2007) Meteorologically driven trends in sea level rise. Geophys Res Lett 34:L23616
Lewis WM (1988) Primary production in the Orinoco River. Ecology 69:679–692
Lewis WMJ, Saunders JFI (1989) Concentration and transport of dissolved and suspended substances in the Orinoco River. Biogeochemistry 7:203–240
Lisitzin E (1974) Sea level changes, vol 8, Elsevier oceanography series. Elsevier Scientific Publishing Company, Amsterdam
Lisitzin E, Pattullo JG (1961) The principal factors influencing the seasonal oscillation of sea level. J Geophys Res 66:845–852
Loveless J, Debski D, Macleod A (1998) Sea level set-up behind detached breakwaters. In: International conference on coastal engineering, Copenhagen
Machado-Allison A (2012) Environmental impact of oil spill in Monagas unprecedented in the history of the country. Universidad Central de Venezuela, Venezuela
Mariano AJ, Brown OB (1992) Efficient objective analysis of dynamically heterogeneous and nonstationary fields via the parameter matrix. Deep-Sea Res 39:1255–1271
Mariano AJ, Ryan EH, Perkins BD, Smithers S (1995) The Mariano Global Surface Velocity Analysis 1.0 (MGSVA) U.S. Coast Guard Technical Report. CG-D-34–95
Meade RH, Emery KO (1971) Sea level as affected by river runoff, Eastern United States. Science 173:425–428
Meade RH, Nordin CF Jr, Hernandez DP, Meija A, Godoy JMP (1983) Sediment and water discharge in Rio Orinoco, Venezuela and Colombia. In: Proceedings of second international

symposium on river sedimentation. Water Resources and Electric Power Press, Beijing, pp 1134–1144
Müller-Karger FE, Mcclain CR, Richardson PL (1988) The dispersal of the Amazon water. Nature 56–59
Müller-Karger FE, Mcclain CR, Fisher TR, Esaias WE, Varela R (1989) Pigment distribution in the Caribbean Sea: observations from space. Prog Oceanogr 23:23–64
Nomitsu T, Okamoto M (1927) The causes of the annual variation of the mean sea level along the Japanese coast. Mem Coll Sci Kyoto Univ 10:125–161
Odriozola AL, Varela R, Hu C, Astor Y, Lorenzoni L, Muller-Karger FE (2007) On the absorption of light in the Orinoco River plume. Cont Shelf Res 27:1447–1464
Parker BB (ed) (1991) Tidal hydrodynamics. Wiley, New York
Pattullo J (1966) Seasonal changes in sea level. In: Hill MN (ed) The sea. Inter Science, New York
Sánchez-Villagra MR, Aguilera OA, Carlini AA (eds) (2010) Urumaco and Venezuelan paleontology: the fossil record of the Northern Neotropics. Indiana University Press, Bloomington
Schot PP, Poot A, Vonk GA, Peeters WHM (2001) A surface water model for the Orinoco River basin. A technical report. Utrecht University
Sethuraman S, Raynor GS (1980) Comparison of mean wind speeds and turbulence at a coastal site and offshore location. J Appl Meteorol 19:15–21
Stockdon HF, Holman RA, Howd PA, Sallenger AH Jr (2006) Empirical parameterization of setup, swash, and runup. Coast Eng 53:573–588
Svensson C, Jones DA (2004) Dependence between sea surge, river flow and precipitation in south and west Britain. Hydrol Earth Syst Sci 8:973–992
Trifonova E, Demireva D (2003) An investigation of sea level fluctuations in Varna and Bourgas. Proc Inst Oceanol 4:3–9
Tsimplis MN, Woodworth PL (1994) The global distribution of the seasonal sea level cycle calculated from coastal tide gauge data. J Geophys Res 99:16031–16039
TTMS TATMS (2010) MSL and wind speeds. Service TATM (ed). Ministry of Public Utilities, Government of the Republic of Trinidad and Tobago, Piarco
Van Andel TH, Postma H (1954) Recent sediments in the gulf of Paria. Reports of the Orinoco shelf expedition. Verh, der Konink-lijke Nederlandse Akad van Wetensch Afd Natuurkunde, Eerste Reeks 20:244
Warne AG, Meade RH, White WA, Guevara EH, Gibeaut J, Smyth RC, Aslan A, Tremblay T (2002) Regional controls on geomorphology, hydrology, and ecosystem integrity in the Orinoco Delta, Venezuela. Geomorphology 44:273–307

Chapter 10
Advances in Modeling of Water Quality in Estuaries

Isabella Ascione Kenov, Francisco Campuzano, Guilherme Franz, Rodrigo Fernandes, Claudia Viegas, João Sobrinho, Hilda de Pablo, Andreia Amaral, Ligia Pinto, Marcos Mateus, and Ramiro Neves

Abstract Water quality models are in great demand to complement studies about the status of estuarine waters. However, local models do not perform well when boundary conditions are not properly defined and when biogeochemical processes are not described with adequate detail. This chapter presents advanced modeling applications to perform water quality studies in Portuguese estuaries. Boundary conditions for hydrodynamics and biogeochemistry are provided by the Portuguese Coast Operational Model, downscaled by using nested domains with increasing resolution from the regional to the local scale. The nested models of the estuaries are described, and case studies are presented for specific estuaries to compute sediment transport (Tagus estuary), to calculate residence time of water (Mondego estuary), to forecast quality of bathing waters (Estoril Coast), and to quantify nutrient fluxes between estuaries and the open ocean (Ria de Aveiro). The level of detail used to represent biological processes in water quality models is also addressed, including the description of a case study about modeling of species vulnerable to water quality, such as *Zostera noltii* in Ria de Aveiro. The need for high level of detail to represent microbial loop and carbon cycle in estuaries is discussed with the application of a complex biological model to the Tagus estuary.

10.1 Introduction

Estuaries are among the most productive environments on earth; they are considered coastal marine areas because they share more features with marine systems than with freshwater environments. Estuaries are the land-ocean link, transition

I. Ascione Kenov (✉) • F. Campuzano • G. Franz • R. Fernandes • C. Viegas • J. Sobrinho • H. de Pablo • A. Amaral • L. Pinto • M. Mateus • R. Neves
Marine Environment Technology Center (MARETEC), Instituto Superior Técnico, Universidade de Lisboa, Av. Rovisco Pais 1, 1049-001 Lisbon, Portugal
e-mail: isabella.ascione@tecnico.ulisboa.pt

C.W. Finkl and C. Makowski (eds.), *Remote Sensing and Modeling: Advances in Coastal and Marine Resources*, Coastal Research Library 9,
DOI 10.1007/978-3-319-06326-3_10, © Springer International Publishing Switzerland 2014

areas between striking different environments such as land and sea, but also between fresh and seawater. Estuaries are highly dynamic in nature, with significant spatial and temporal variation on their physical and biogeochemical processes, and strong gradients in water properties. Estuaries have productive habitats with high ecological and economic importance.

Many of the world's largest cities are in coastal zones and more than 75 % of people are expected to live within 100 km of a coast by 2025 (Bulleri and Chapman 2010). Most estuaries are characterized by proximity with large cities, and receive anthropogenic inputs coming from agriculture and urban wastes with consequences on water quality. Estuarine waters are used for aquaculture farming, fishing, recreation, and navigation. The European Union (EU) lays down rules for monitoring, assessment, and management of the quality of waters, and commits EU member states to achieve good qualitative and quantitative status of all water bodies by 2015 (2000/60/EC). Huge investments have been done by EU member states to monitor, assess, and improve status of water quality of the coastal zone, including estuaries.

The assessment of the water quality is done by using both monitoring and modeling tools. In the last decades there has been a growing demand for modeling tools to support water quality management in estuarine areas. Water quality models are effective tools to simulate and predict transport of organic and inorganic substances, contaminants, and pollutants. Model results could contribute to monitoring cost savings in areas which are inaccessible to conduct sampling campaigns and experiments. Furthermore, water quality models could be used to simulate and then to evaluate different pollution and contamination scenarios, which are used for environmental impact assessment. Modeling tools are available for the modeling of rivers (Viero et al. 2013; Chibole 2013; Lai et al. 2013), lakes (Ali et al. 2013), reservoirs (Deus et al. 2013), coastal areas (Viegas et al. 2012), and estuaries (Trancoso et al. 2005; Sohma et al. 2008).

MOHID (acronym of MOdelo HIDrodinâmico) is a water modeling system developed at the Marine Environment Technology Center (MARETEC) at Instituto Superior Técnico (IST), which belongs to the Universidade de Lisboa (ULISBOA), Portugal. MOHID incorporates modules for modeling of hydrodynamics, sediment transport, and water quality, with state-of-the-art process kinetics. In this chapter, an overview of recent advances in estuarine water quality modeling with MOHID is provided. The main equations and concepts of the MOHID system are described in Sect. 10.1.1. In Sect. 10.2, advanced applications with MOHID are presented, including case studies about Portuguese estuaries. Modeling of these estuaries requires the definition of boundary conditions at the interface between the estuary and the ocean. The definition of these boundary conditions is challenging because it requires the integration between the regional scale (ocean) and the local scale (estuary). The integration of different spatial scales is achieved by the Portuguese Coast Operational Model System (PCOMS), described in Sect. 10.2.1. PCOMS is a 3-D hydrodynamic-biological operational regional model of the Portuguese coast. It includes several nested models which are used to simulate hydrodynamics and water quality of estuaries, including Tagus, Mondego, and Ria de Aveiro, among others. As such, PCOMS can be considered as a global platform feeding boundary conditions to local scale models. These local scale models are used to support water quality studies, to answer common management question, and to improve the overall understanding of physical and ecological processes. The model of the Tagus estuary, which is nested in PCOMS, and presented in Sect. 10.2.2, is used to support water quality studies related to concerns about water clarity, pollutant distribution and sediment transport. The quality of estuarine and coastal waters used for recreational purposes is addressed in Sect. 10.2.3, with the presentation of the case study of the Estoril coast, adjacent to the Tagus mouth. The model of the Mondego estuary, described in Sect. 10.2.4, is another example for a local scale model integrated in PCOMS. It was used to solve common management questions, such as residence time of water. MOHID is a flexible tool, which can be integrated with watershed models, used to calculate freshwater inflows and nutrient inputs to estuarine systems. This multi-model integration enables the construct of an integrated and comprehensive understanding of nutrient exchanges between land, estuaries and ocean. This understanding is of importance to define the trophic status of estuarine waters in the context of the European Water framework Directives (EU WFD). The case of Ria de Aveiro, described in Sect. 10.2.5, is representative of this multi-model integration. It was applied to quantify the nutrient budget between the estuary and the ocean. Furthermore, Ria de Aveiro represents one of the most important seagrass ecosystems of Portugal. Seagrasses are very sensitive to water quality changes. For this reason, they have been recognized as important indicator species, which reflect the overall health of coastal ecosystems. Recently, a new seagrass model was developed in MOHID, to represent seagrasses in estuarine systems. The seagrass model, presented in Sect. 10.2.6, was applied to Ria de Aveiro to calculate the spatial and temporal distribution of *Zostera noltii*. Finally, the description of estuarine system often requires the use of complex models to enable a comprehensive representation of biological processes. The role of complex models for the study of estuarine processes is addressed in Sect. 10.2.7.

10.1.1 MOHID Water Modeling System

The current MOHID structure was designed in the 1990s based on the experience gained by MARETEC in the 1980s, by using the new developments of the FORTRAN 90 language, and the establishment of new standards of hierarchical data formats (HDF). To date, the developments and applications of the MOHID system include hydrodynamics (Miranda et al. 2000), wave modeling (Neves and Silva 1991), ecological modeling (Trancoso et al. 2005; Mateus et al. 2012c), Eulerian and Lagrangian transport (Pando et al. 2013; Deus et al. 2013). Presently, MOHID can handle both 2-D and 3-D simulations, Eulerian, Cartesian, and Lagrangian vertical coordinates, and a number of biogeochemical models to simulate ecological processes in water and sediment. In this section, a short description of equations and main processes simulated in MOHID is provided. Further information can be found at www.mohid.com.

10.1.1.1 Model Equations

MOHID is formulated by using the integral approach (Eq. 10.1) according to which the rate of accumulation inside a control volume equals to the sum of input and output fluxes, plus sources minus sinks:

$$\frac{\partial}{\partial t}\left(\iiint_{CV}\beta dV\right)=-\iint_{surface}\left[\left(\beta\vec{u}\cdot\vec{n}\right)+\left(-\vartheta\vec{\nabla}c\cdot\vec{n}\right)\right]dA+\iiint_{CV}(S_O-S_I)dV \tag{10.1}$$

where CV is the Control Volume, u is the fluid velocity relative to the surface A, ϑ is the diffusivity, β is a generic water property (such as phytoplankton, and nutrients), and S_O-S_I is the difference between sources and sinks of the property β per unit volume. By applying the Gauss's theorem to Eq. 10.1, the differential conservation equation obtained is:

$$\frac{\partial\beta}{\partial t}=-\frac{\partial}{\partial x_j}\left(u_j\beta-\vartheta\frac{\partial\beta}{\partial x_j}\right)+(S_O-S_I) \tag{10.2}$$

Following Eq. 10.2, the rate of change of a property in a point equals the divergence of the advection, plus diffusion at this point, plus the difference between production and consumption rates per unit volume. In a Lagrangian reference, the Eq. 10.2 can be rewritten as:

$$\frac{d\beta}{dt}=-\frac{\partial}{\partial x_j}\left(\vartheta\frac{\partial\beta}{\partial x_j}\right)+(S_O-S_I) \tag{10.3}$$

Following Eq. 10.3, the rate of change in a volume element equals the diffusion across its surfaces plus the difference between sources and sinks. The form of Eq. 10.3 is obtained when the surface of the control volume used in Eq. 10.1 is moving at the fluid velocity. In this case, the total volume inside the surface is constant, even if the surface is deformed. Due to random velocity associated to Brownian movement or turbulent movement, the volume will expand as a result of diffusion.

10.1.1.2 Water Processes

The conceptual diagram of MOHID water environment is described in Fig. 10.1. The water quality model used in MOHID calculates sources and sinks terms specific for each water property, in each volume element and in each time instant. The sources and sinks of a property depend on the chemical and biological processes which are involved and are associated with the biogeochemical cycles of carbon, nitrogen and phosphorus.

Air-water fluxes are computed by considering air temperature, moisture, wind speed and direction, cloud cover, precipitation, and solar radiation. At each time

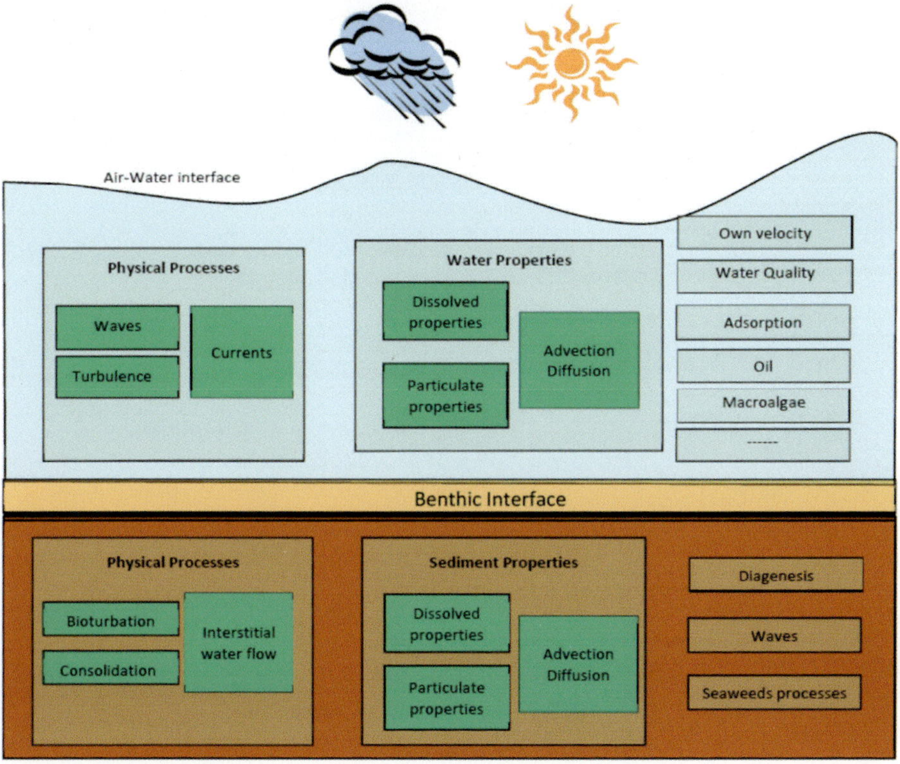

Fig. 10.1 Conceptual scheme of the MOHID aquatic environment. The water column is separated from the atmosphere by the air-water interface, and from the sediment by the benthic interface (Neves 2013)

instant the hydrodynamic model simulates the free surface elevations and velocities of the fluid necessary for calculating the terms of advection-diffusion.

The combination of transport, chemical and biological reactions in the water column, produces the spatial distribution of the water properties. MOHID offers different degrees of complexity in the description of ecological processes. According to the required level of complexity, several species of primary producers, secondary producers and decomposers in the water can be simulated. Processes at the sediment-water interface are handled in the benthic layer, where particulate matter can accumulate in a "fluff layer", and fluxes of dissolved properties are computed. Benthic organisms such as microphytobenthos, filter feeders, and deposit feeders, are assumed to be living fixed to a substrate, and interact with the environment at the water-sediment interface. Seagrasses uptake nutrients from water and sediment, while macroalgae uptake nutrients from water only, and can be transported by currents as drifting macroalgae. In the sediment, the interstitial water is transported by advection, groundwater exfiltration, and it is affected by sediment consolidation. Molecular diffusion induced by bioturbation is simulated in the upper 10 cm of sediment.

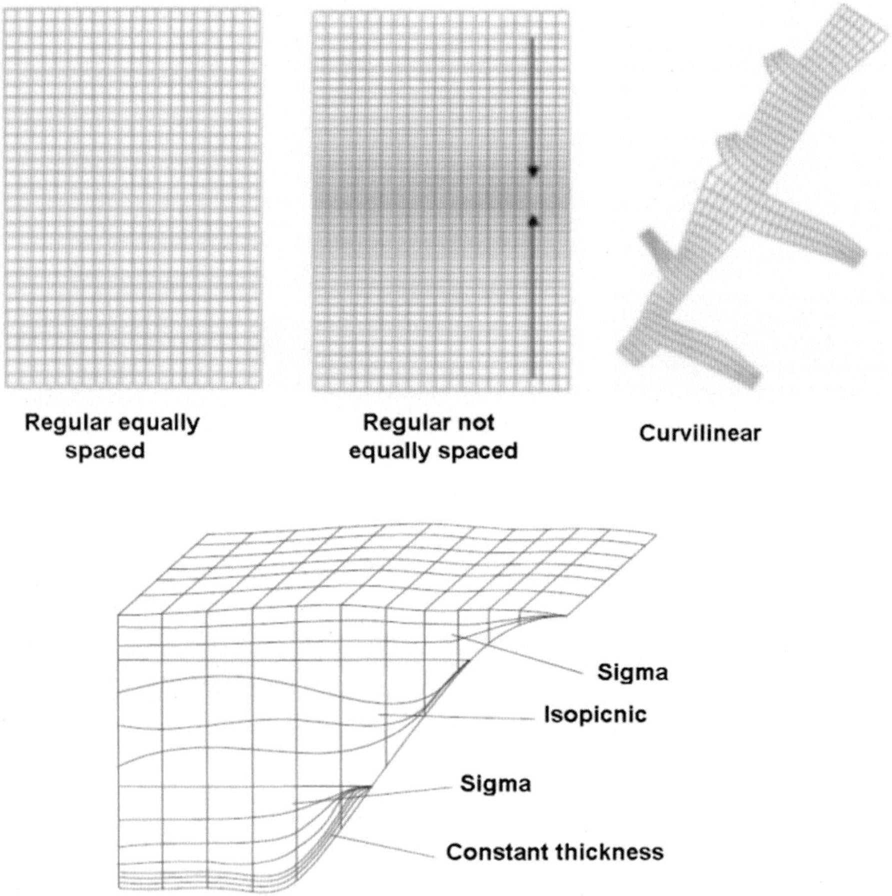

Fig. 10.2 Horizontal and vertical geometries handled by MOHID

Organic and inorganic particulate material can be transported vertically at a defined settling velocity, while oil moves vertically in the water column until it reaches a stability depth (usually the surface). Diurnal vertical migrations of fish larvae and dinoflagellates can be simulated as well.

10.1.1.3 Horizontal Geometry

Mathematical models describe nature by using discrete values computed in a grid. MOHID can handle constant spacing grids, irregular grids and curvilinear grids (Fig. 10.2). Regular equally spaced grids have squared cells, and are used to give the same importance to all the areas of the represented geographic domain. These grids can be rotated to adapt the simulated domain to the inclination of the coast. Regular not-equally spaced grids are used when some areas have to be described at different resolution. As an example, a regular not-equally spacing grid can be used with low resolution at the boundary with the ocean, and with higher resolution in proximity of the coast, to represent horizontal geomorphological gradients in more detail. Finally, to handle complex geometries, such as estuaries and lagoons with intricate channels, curvilinear grids can be used.

10.1.1.4 Vertical Coordinates

Presently, MOHID can handle different vertical coordinates: Sigma, Cartesian, Lagrangian (based on Sigma or based on Cartesian), "Fixed spacing", and Harmonic (Fig. 10.2). These coordinate systems can be used individually or in a combination. Cartesian coordinates are adequate when the flow is horizontal. Harmonic coordinate are similar to the Cartesian coordinate, with the main difference being that the horizontal faces close to the surface expand and collapse depending on the variation of the surface elevation. The sigma coordinates are expressed as a percentage of the total depth, and are convenient when the pressure gradient is barotropic. The Lagrangian coordinates can act as sigma and Cartesian, because it moves the upper and the lower faces of the volume element with the vertical flow velocity.

10.1.1.5 Eulerian and Lagrangian Approaches

MOHID can handle both Eulerian and Lagrangian approaches. The Eulerian approach is implemented by defining a point in a flow field (x, y, z), and by observing how the properties (e.g., velocity, phytoplankton, temperature) change as the fluid passes through this specific location. This approach is the most used in fluid mechanics, because it enables to produce high-resolution pictures of water properties in a given region. The use of the Eulerian approach can prove difficult to label water masses and to monitor their position from the origin over space and time, whereas in the Lagrangian approach, particles can be labelled with information about their releasing point or origin. MOHID's Lagrangian module uses the concept of lagrangian tracers. The most important property of a tracer is its position (x,y,z). A tracer can be a water body, a sediment particle, a group of organisms, or a molecule. The movement of the tracers can be influenced by the velocity field, by the wind, and by random component of velocity. Fecal coliforms can be considered as tracers, and their movement can be tracked down to identify and assess fecal contamination of coastal waters (Sect. 10.2.3). Water masses can be labelled at their origins, and their movement can be tracked down over time. This feature enables to calculate residence time of water in estuaries and bays (Sect. 10.2.4). Oil particles coming from accidental spills can also be labelled and tracked, enabling to assess their displacement over time.

10.2 Advanced Applications with MOHID

10.2.1 Downscaling: From the Large Scale to the Local Scale

In this section, the concept of downscaling is introduced, as a technique to provide boundary conditions from regional to local scale models. Advanced applications of this concept are presented for the Portuguese Coast Operational Model. Operational models are traditionally used in the field of regional oceanography for ocean forecasts and do not include the necessary spatial detail to answer water quality management questions on the local scale. For this reason, downscaling techniques are used to transfer information from regional to local models, enabling a comprehensive description of hydrodynamics and water quality processes. Furthermore, these techniques enable to extend the operational forecasts to the local scale as well.

The idea of downscaling consists of simulating hydrodynamics and water quality on a local scale, based on information provided by larger-scale models. The geographic area covered by a model is named as domain. A domain included inside the domain of another model is named as nested domain, or subdomain. A domain can have one or more subdomains. Models with larger domains are called father models, and provide boundary conditions to their nested models, named as son models. Son models are used to represent estuaries, harbors, and lagoons at a spatial resolution higher than this of the father model.

In the last years, downscaling techniques have been successfully applied in the field of coastal oceanography to simulate hydrodynamics and water quality, and to solve common management questions on the local scale (Leitão et al. 2005; Ascione Kenov et al. 2012; Mateus et al. 2012b). However, the application of these techniques is challenging because it requires reliable and consistent atmospheric and oceanic boundary conditions. Measured data are often difficult to retrieve for the description of boundary conditions. For this reason, meteorological models and large scale ocean models are used to provide boundary conditions to regional models. Additionally, operational modeling requires that model forecasts are timely available to be analyzed by coastal managers and decision makers. This aspect poses additional challenges such as the need to reduce simulation time and the need to simulate several nested domains.

When several nested domains are used, the simulation time is defined by the innermost domain, usually the one with higher spatial and temporal resolution, corresponding to the slower model.

PCOMS is a 3-D hydro-biogeochemical model of the Iberian Western Atlantic region (Fig. 10.3). This model was implemented by MARETEC at IST, Lisbon, Portugal. Ocean boundary conditions are provided by the Mercator-Ocean PSY2V4 North Atlantic and by tidal levels computed by a 2-D version of MOHID, forced by FES2004, and running on a wider region (Mateus et al. 2012b). PCOMS has a horizontal resolution of 6.6 km and a vertical discretization of 50 layers with increasing resolution from the sea bottom upward, reaching 1 m at the surface. Atmospheric forcing is provided by Penn State–NCAR fifth-generation Mesoscale

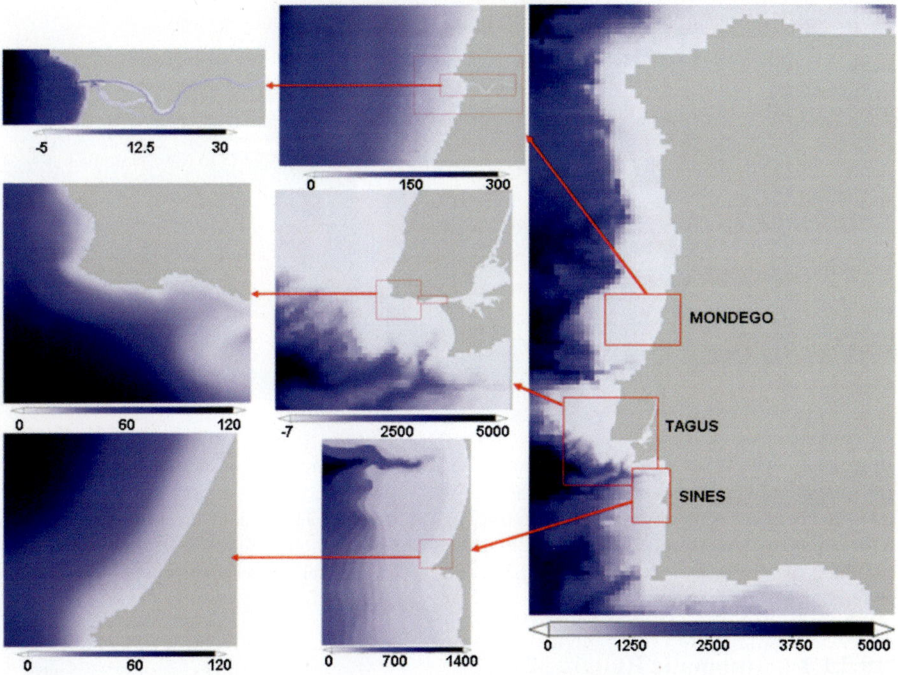

Fig. 10.3 Examples of local models nested in PCOMS. Local applications are systems of nested models, obtained by downscaling the solution of the regional model from 6.6 km up to a 100 m resolution

Model (MM5) (Grell et al. 1994) with a resolution of 9 km (http://meteo.ist.utl.pt). PCOMS integrates different spatial scales and different sources of forcing and boundary conditions, thus improving local-scale model results.

Presently, PCOMS includes several nested models to simulate hydrodynamics and water quality of Portuguese estuaries and coastal areas, including Tagus, Mondego, Ria de Aveiro, and Douro estuaries, among others. Examples of models nested in PCOMS are indicated in Fig. 10.3.

To reduce the simulation time, a delayed mode (offline) technique was designed for PCOMS: the Window Downscaling Technique (WDT). This technique consists of saving model results as pre-defined windows containing results extracted from the regional model of the Portuguese coast. These windows have high temporal resolution, 900 s, enough to represent the main processes coming from the open ocean including the tide signal. The spatial domain of the window must be larger than this of the nested domain. Once the father model simulation is concluded, a window is loaded and applied to a nested model. The described technique enables to run the local model independently, saving time and minimizing redundancy, while improving results. This technique also does not affect the running time of the father domain and enables simultaneous nested model simulations. The WDT enables to nest any coastal or estuary model located in the Portuguese continental zone. An example of nested model is the Tagus estuary, presented in Fig. 10.4 and described in Sect. 10.2.2.

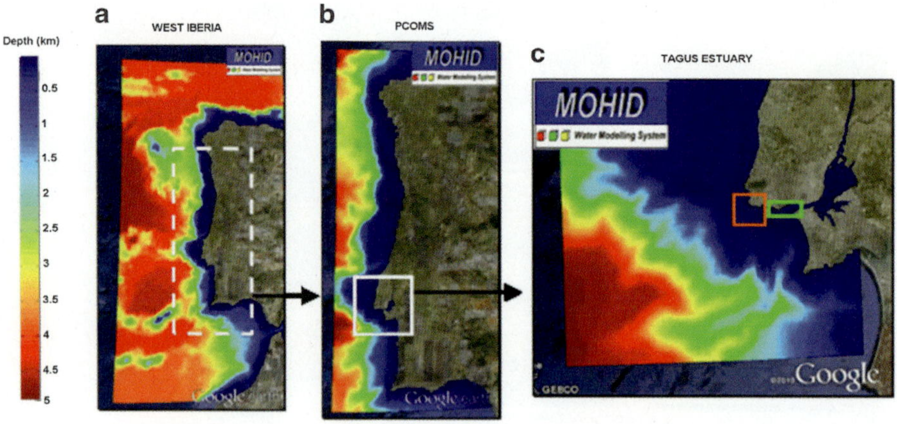

Fig. 10.4 Nested domains used to implement the Tagus model. The domain on the left (**a**) provides tidal boundary conditions to the PCOMS model (**b**), which supplies hydrodynamic and bio-geochemical boundary conditions to the Tagus model (**c**). This model also provides boundary conditions to high resolution local models such as the Tagus mouth (*yellow box*), the Estoril coast model (*green box*), described in Sect. 10.2.3, and the Guia outfall model (*orange box*) (color figure online)

10.2.1.1 Automatic Running Tool

An Automatic Running Tool (ART) was developed to handle preprocessing, running, and post-processing of MOHID simulations, including storing, plotting, and distribution of model results via OPeNDAP, smartphone, and Web pages (Fig. 10.5). The results of the PCOMS and nested models can be found as published maps at http://forecast.maretec.org/.

During the pre-processing phase, ART adapts different data sources to the model domain, including atmospheric model data sources, global circulation model results (i.e. Mercator-Ocean, http://www.mercator-ocean.fr), and measured data from monitoring stations. Atmospheric data sources include MM5 and WRF (Weather Research and Forecasting Model, http://www.wrf-model.org) models.

ART enables running models in a cascade scheme, where son models are waiting for a signal indicating that the father model simulation is concluded. This signal triggers the next model simulation, reducing the time requested for preprocessing. Furthermore, different models can run in different computers at the same time.

10.2.2 The Tagus Operational Model

This section describes a modeling application used to study water quality of the Tagus estuary, Portugal. The modeling application is nested in PCOMS by using the downscaling concept described in Sect. 10.2.1. The geographic domain of the

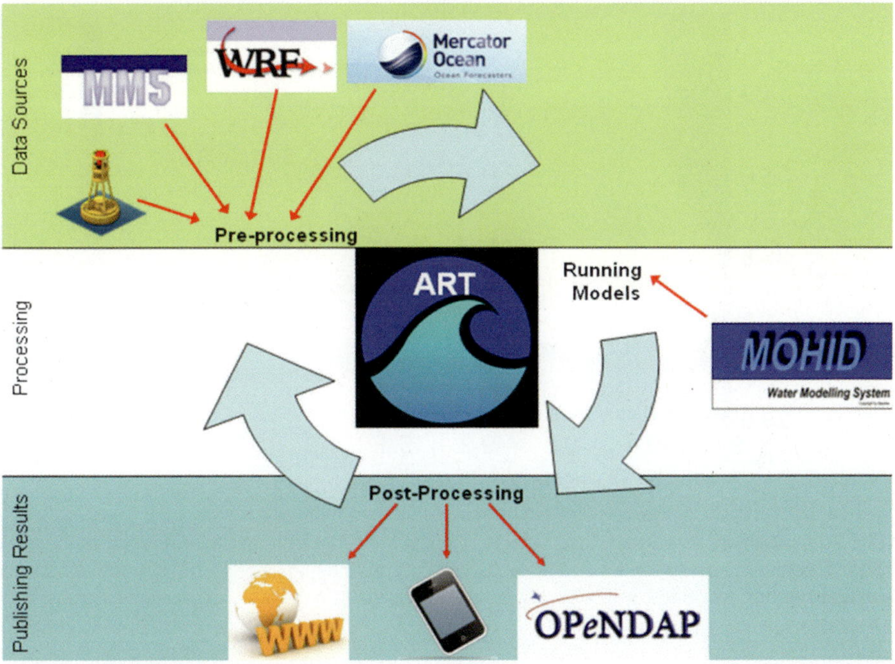

Fig. 10.5 General scheme of the Automatic Running Tool (*ART*)

Tagus model is indicated in Fig. 10.4 (white box). The Tagus estuary is the largest in Portugal and one of the largest estuarine systems in Europe. More than 2.5 million people (25 % of the total Portuguese population) live on the margins of this estuary, with the highest density in Lisbon, the Portuguese capital.

Following the delimitation of the Water Framework Directive (2000/60/EC), the Tagus estuary covers an area of 368 km^2, with an average volume of about 2,700 hm^3. The estuary is mesotidal, dominated by semidiurnal tide. Freshwater inflows are coming mainly from the rivers Tagus, Sorraia, and Trancão (Fig. 10.6). The Tagus River provides the main freshwater input to the estuary. Freshwater inflows data were retrieved by the Portuguese National Information System of Water resources (SNIRH – www.snirh.pt). According to data from the Almourol hydrometric station (39.28°N, −8.22°E) for the period 1973–2012, the annual mean freshwater inflow is 300 m^3s^{-1}. The average and the maximum water fluxes in summer are 100 m^3s^{-1} and 400 m^3s^{-1}, respectively. In winter, the average and maximum fluxes are 700 m^3s^{-1} and 3,500 m^3s^{-1}, respectively. Sorraia and Trancão, the other two most important tributaries, have an annual mean flow of 39 m^3s^{-1} and 6 m^3s^{-1}, respectively. The bathymetry of the Tagus estuary, shown in Fig. 10.6, was generated by merging data collected during several bathymetric surveys, and provided by the Portuguese Hydrographical Institute (www.hidrografico.pt/).

MARETEC has a 3-D hydro-biogeochemical operational model for the Tagus estuary since 2011. The horizontal resolution ranges from 2 km in the ocean to

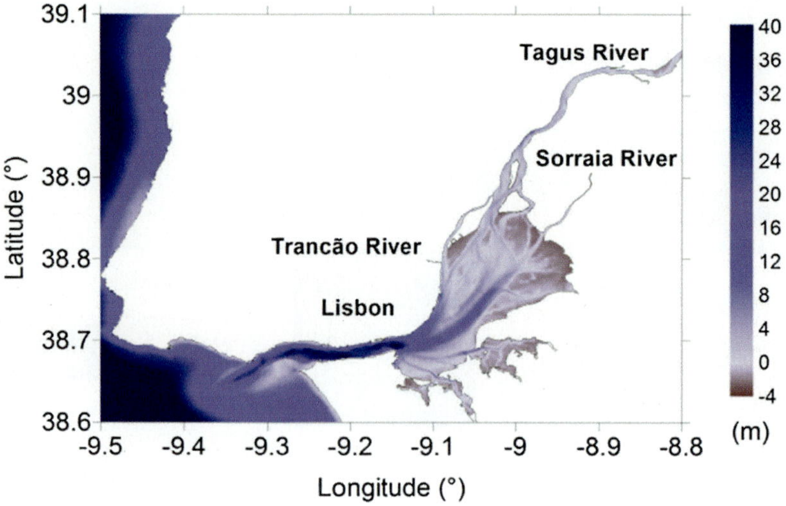

Fig. 10.6 Bathymetry of the Tagus estuary. Depths are referred to the hydrographic zero, set to 2.08 m below the mean sea level. The areas with negative depths are intertidal zones, which are emerged during low tides

300 m around the estuary mouth. The vertical discretisation consists of 43 Cartesian layers overlapped by seven Sigma layers. The vertical resolution is about 1 m near the water surface. Boundary conditions are provided by PCOMS (Sect. 10.2.1). The atmospheric forcing is provided by WRF model results produced at IST with a 3 km resolution (http://meteo.ist.utl.pt/).

In 2013, a new subdomain was developed inside the Tagus estuary operational model. This subdomain has a horizontal resolution of 200 m and covers the mouth and the inner part of the estuary. The vertical discretisation is the same as the one of the father domain. The two nested domains of the Tagus estuary operational model are shown in Fig. 10.7.

The Tagus operational model currently produces daily results and 48 h forecasts. Every day, the forecast of the previous day are simulated again, by using the best ocean boundary conditions and atmospheric forcing available, and hourly freshwater inflow measurements from the Almourol hydrometric station. The discharges of biogeochemical properties are defined upstream of the Tagus River by using climatological data. For Sorraia and Trancão rivers, both freshwater river inflow and biogeochemical properties discharges, are defined by using climatological data. Average discharges from urban waste water treatment plants are included in the model as well. Apart from the hydrodynamic results (currents and water level), the Tagus estuary operational model generates forecasts for the properties described in Table 10.1.

The Tagus model is used to assess the complex dynamic of the water properties in the estuarine system, including the cohesive sediment transport. Simulated velocity modulus and cohesive sediment concentrations during one tidal cycle are

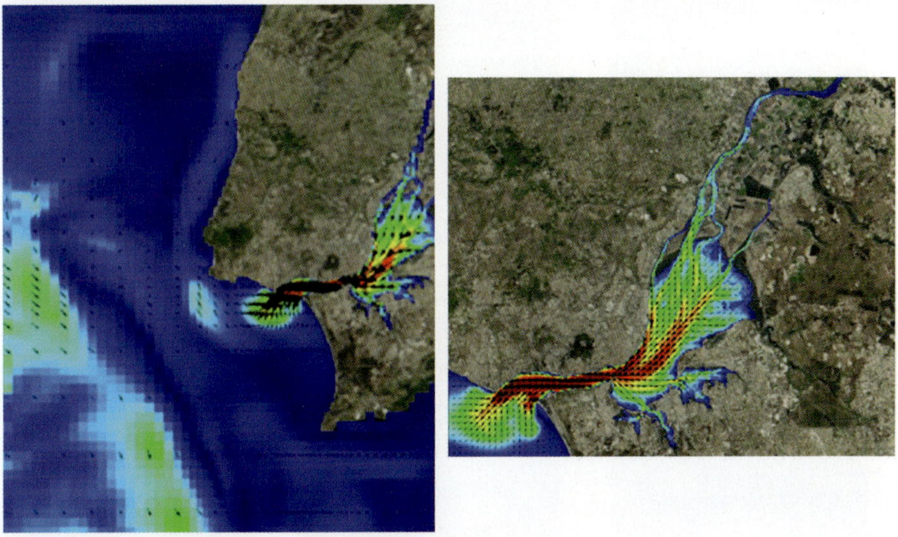

Fig. 10.7 The two domains of the Tagus estuary operational model. The horizontal resolution ranges from 2 km in the ocean to 300 m around the estuary mouth in the first domain (*left*). The second domain has a constant horizontal resolution of 200 m (*right*)

Table 10.1 Properties simulated by the Tagus estuary operational model

Water properties
Salinity
Temperature
Dissolved oxygen
Cohesive sediments
Phytoplankton
Zooplankton
Ammoniacal nitrogen
Nitrate
Nitrite
Refractory dissolved organic nitrogen
Non-refractory dissolved organic nitrogen
Particulate organic nitrogen
Inorganic phosphorous
Refractory dissolved organic phosphorous
Non-refractory dissolved organic phosphorous
Particulate organic phosphorous

shown in Figs. 10.8 and 10.9, respectively. The results show cohesive suspended sediments are eroded from the bottom when the velocities are high. The currents carry upstream the cohesive suspended sediments during flood tide, and downstream during ebb tide. The cohesive sediment concentrations in water decrease in the slack tide period, when the deposition rate is the highest, due to low velocity intensities.

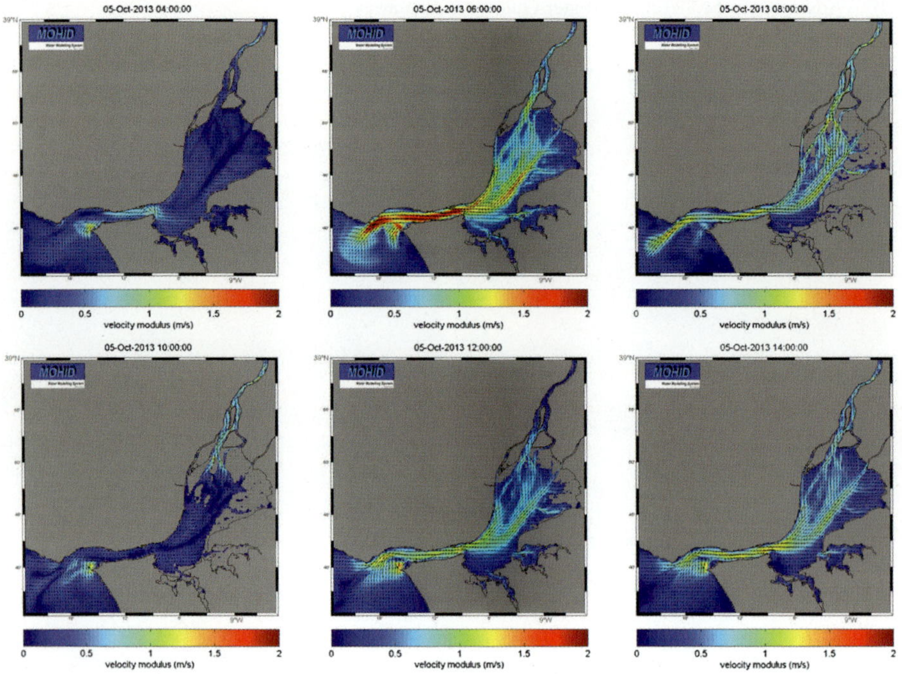

Fig. 10.8 Velocity modulus in the Tagus estuary during one tidal cycle

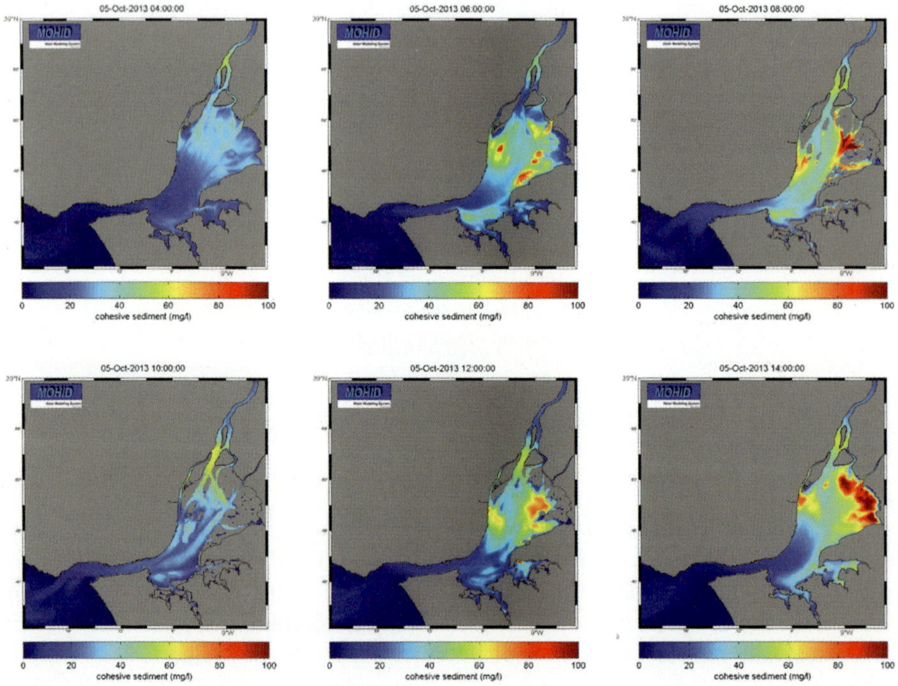

Fig. 10.9 Cohesive suspended sediments in the Tagus estuary during one tidal cycle

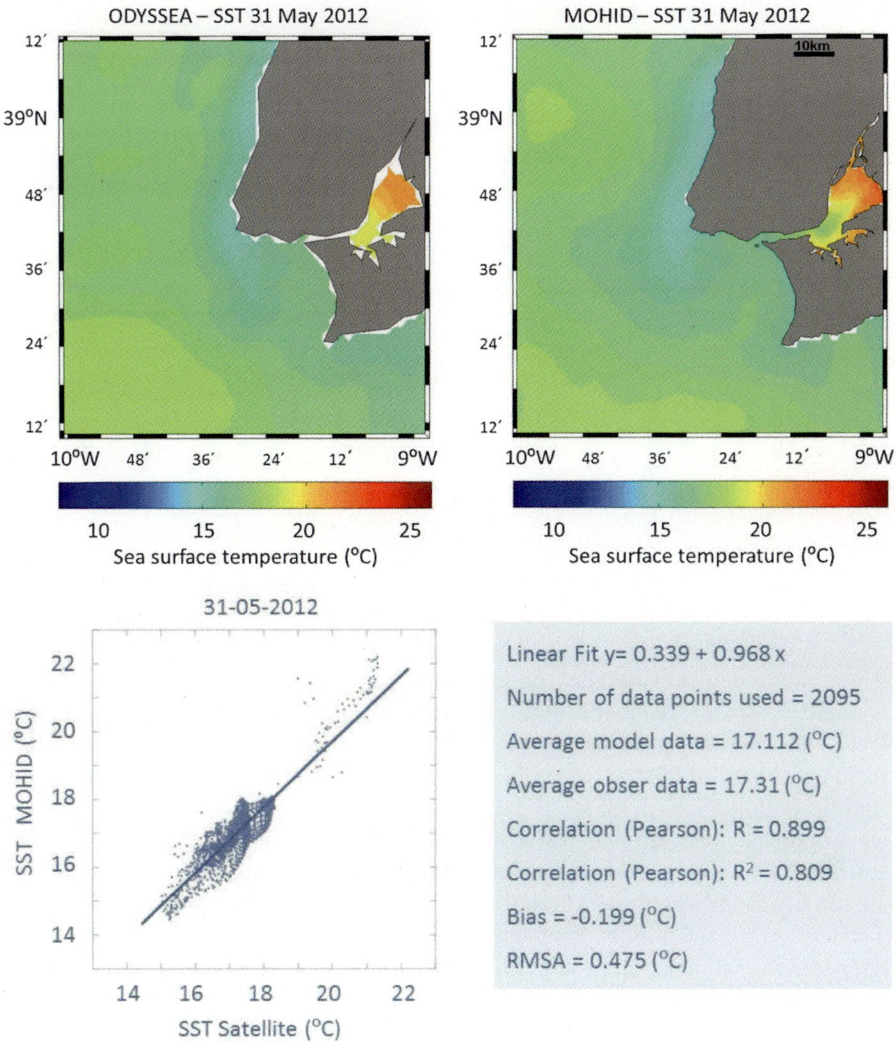

Fig. 10.10 Remote sensing data for sea surface temperature (*left panel*) and model predictions (*right panel*) for 31 May 2012

10.2.2.1 Model Validation

The validation of the Tagus estuary model relies on both satellite images and in-situ data from moored stations in the inner estuary and monitoring programs in the coastal adjacent area. An example of modeled surface temperature validation with remote sensing data is presented in Fig. 10.10 for May 2012. A simple comparison of both figures shows a remarkable resemblance, denoting that the model is able to reproduce some of the major physical patterns in the estuarine and coastal areas,

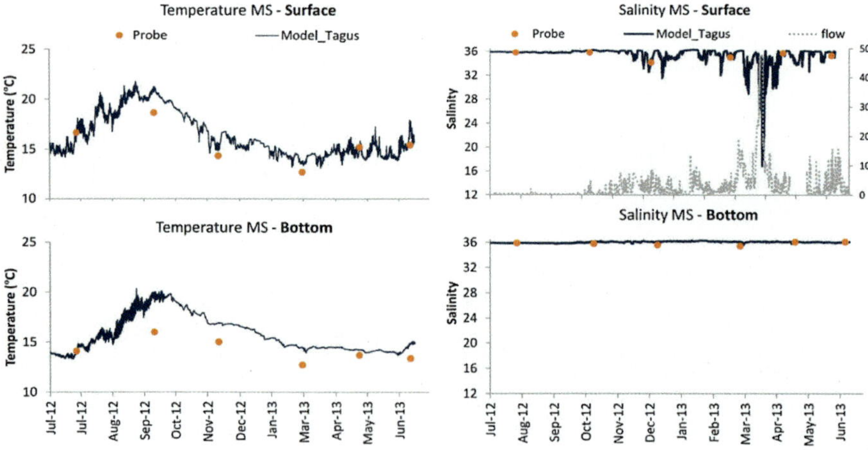

Fig. 10.11 Model results (*line*) and field data (*markers*) for temperature and salinity at different depths for a monitoring station located in the adjacent area of the Tagus estuary (Station MS in Fig. 10.10), from July 2012 to July 2013

such as, the cold water mass near the coast as a result of upwelling, and the higher temperature waters inside the estuary.

Field data retrieved with multi-parameter probes or discrete samples (3 depth levels) during monitoring programs provide a description of the vertical structure in the coastal area around a submarine outfall. Monitoring programs, however, have significant limitations in the temporal characterization of the processes, although they are important nonetheless because they can provide a view of the tri-dimensional structure of the estuarine adjacent area, where strong horizontal and vertical gradients are usually found. Figures 10.11 and 10.12 illustrate the model validation with in-situ data for surface and bottom depth levels. The comparison of simulated and measured temperature, salinity, nitrate and chlorophyll concentrations shows good fit, with model results following the seasonal trend seen in field data.

These examples of estuarine model validation denote the advantages of comparing field data with model results, but also expose some of its limitations. Modelers trying to validate their models of such dynamic systems must bear in mind the existence of discrepancy between the spatial and temporal resolution of both data sets; while model results cover fully the studied domain and can have small temporal resolutions, in-situ data sets are usually scarce because of methodological and equipment constraints, and an intrinsic economic restrictions in the monitoring programs. If temporal and spatial coverage of field data is too sparse for the scales of the processes, the comparison of such data sets must be made on a qualitative, rather than quantitative, basis. The ability of the model to reproduce general features such as seasonal trends and major vertical patterns, as in the case presented in Figs. 10.10 and 10.11, stand as a good example of validation that can be attained in such cases.

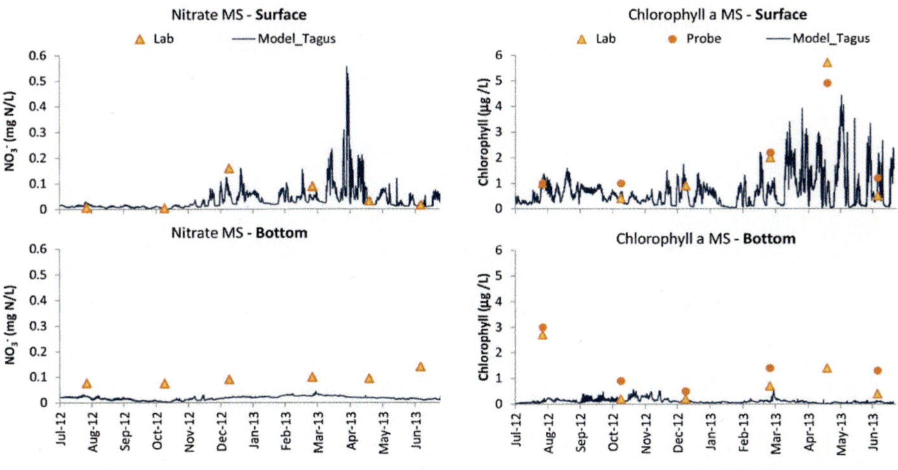

Fig. 10.12 Model results (*line*) and field data (*markers*) for nitrate and chlorophyll-a at different depths for a monitoring station located in the adjacent area of the Tagus estuary (Station MS in Fig. 10.10), from July 2012 to July 2013

10.2.3 Bathing Water Quality

This section describes a modeling application used to study bathing water quality (BWQ) of the Estoril coast, located in proximity to the Tagus estuary, Portugal. An operational model was implemented to provide daily BWQ forecasts. The model was used to simulate and study contaminant plume dispersion along the coastline. Using the Lagrangian methodology and automatic hydrometric data from four streams discharging in the area, the model simulated streams discharges (that can have fecal contamination), and provided BWQ forecast. The model configuration was based on an implementation originally developed by Fernandes (2005) and used in several studies of the Tagus Estuary (Fernandes et al. 2006; Campuzano et al. 2010, 2012; Pinto et al. 2012).

The model included three nested domains. The first domain is PCOMS (Sect. 10.2.1). The second domain (Fig. 10.13b) is the Tagus estuary model described in Sect. 10.2.2, which provided hydrodynamic fields, and density gradients (temperature and salinity) to the next level. The third domain (Fig. 10.13c), included the area between Carcavelos and Belém, on the Estoril coast (from −9.356°W to −9.2285°W and from 38.669°N to 38.705°N). The model was implemented by using a 3-D configuration, with a constant horizontal spatial step of 30 m (460 × 120 cells), and a time step of 6 s. The vertical geometry included seven sigma layers between the surface and 8.68 m, and five Cartesian layers between 8.68 and 35 m (maximum depth).

Atmospheric forcing was provided by MM5 and WRF models (3 km resolution) (www.meteo.ist.utl.pt), including wind (modulus and direction), air temperature, relative humidity, solar radiation and wind stress.

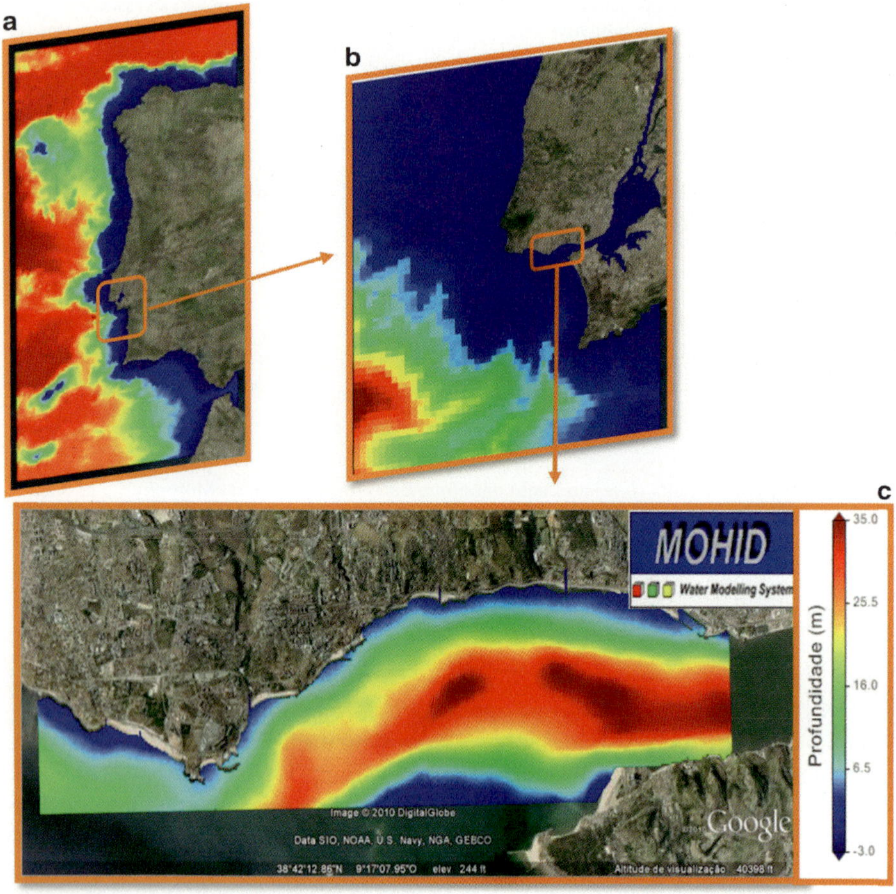

Fig. 10.13 Nested domains. (**a**) PCOMS, (**b**) Tagus, and (**c**) Estoril coast

Freshwater discharges were obtained from historical data and INAG[1] automatic stations (in Jamor), SANEST[2] and IST stations (in Sassoeiros, Marianas, Lage and Barcarena streams). The automatic stations were used to collect daily hydrometric levels which were converted into stream discharges, using level/discharge gauges. Due to the lack of up-to-date information, discharges for Algés, Junça, and Porto Salvo streams were estimated from historical data.

The lagrangian model was coupled to the hydrodynamic model, to simulate stream discharges and their effect on the beaches. After the hydrodynamic simulation was carried out, model results were stored at a high temporal resolution. Using

[1] INAG - Instituto da Água, http://www.apambiente.pt

[2] SANEST - Saneamento Costa do Estoril, http://www.sanest.pt

an off-line method, Lagrangian transport was efficiently and rapidly computed by using previously generated results of the hydrodynamic model.

10.2.3.1 Fecal Decay Model

The model assumed that fecal bacteria do not reproduce in the aquatic environment and that their decay is linear. The bacteria concentration c is described by a first order equation:

$$\frac{\partial c}{\partial t} = -kc \tag{10.4}$$

The mortality rate k was computed as a function of solar radiation intensity, salinity and temperature based on Canteras et al. (1995) and described in Viegas et al. (2012).

10.2.3.2 Bathing Water Quality Assessment

The new European Bathing Water Directive (NBWD) classifies BWQ by using maximum thresholds and exceeding probability. If the maximum threshold is respected for longer than 90 % of the time water is "acceptable" and if it is respected for longer than 95 % of the time, water is classified as "excellent". These percentages can be read as "the probability of a bather to be in bathing water complying with the maximum threshold". This probability is measured as the probability of weekly sampling compliant water in a fixed point assumed as representative of the whole bathing water.

To assess BWQ it was necessary to define the area of bathing water, to identify polluted water masses inside the bathing water (and their origin), and to calculate bulk concentration of the bathing water. The spatial distribution of bathing waters was defined by dividing the study area into sub-areas considered as monitoring boxes, following the methodology described in Braunschweig et al. (2003) for tracking water masses and for the computation of water residence time. A bathing water was defined as the monitoring box delimited by the beach line and with its cross dimension in the order of the swimming length (up to 200 m).

To identify polluted masses inside bathing water, the tracers were labelled by their origin, in such a way that their position at any time instant could be linked to their release points (origins). At each time step, the model calculates the volume of each monitoring box and the volume of the tracers inside each monitoring box. The contamination probability was computed as the ratio between the volume of contaminated water and the volume of the monitoring box. The fecal contamination was quantified trough the geometric mean of all contaminated tracers in each monitoring box. To simulate the plume of each pollution source, twenty Lagrangian

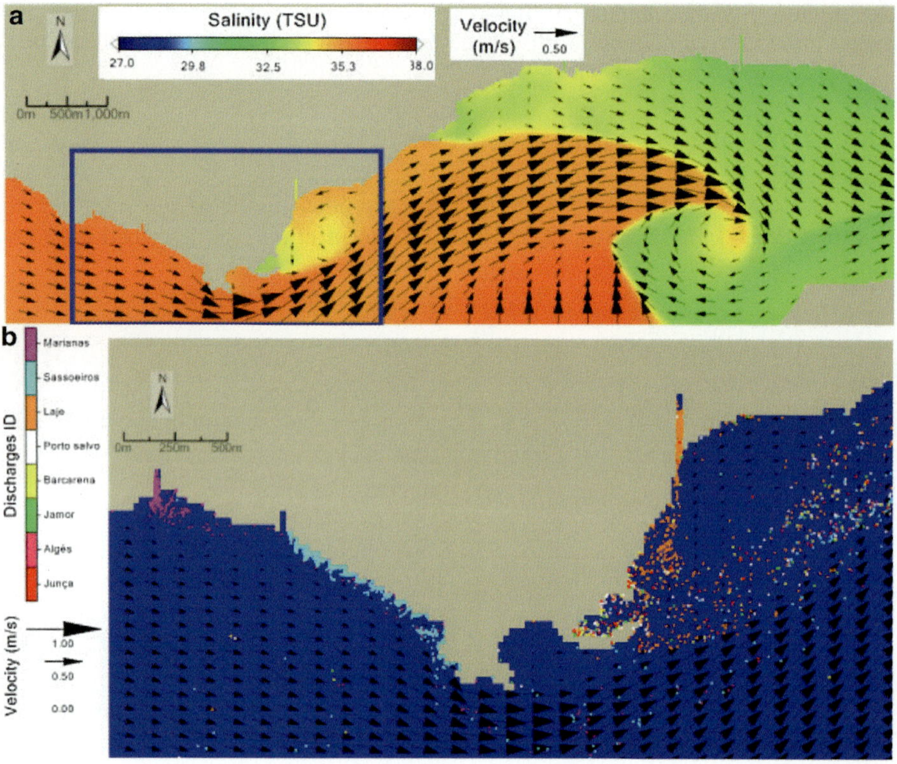

Fig. 10.14 Salinity and velocity fields for Estoril Coast domain during flood tide (**a**), and a detailed image for bathing waters for the same time with lagrangian tracers representing stream discharges ID with different colours (**b**)

tracers were released per minute, at each discharge point (Fig. 10.14). The load associated to tracers released at the mouth of a stream under normal conditions was obtained by using the geometric average of historical data.

The average concentration in a generic monitoring box “x” was calculated as:

$$C_x = 10^e \tag{10.5}$$

$$e = \frac{\sum_{NPollutedTracers} \log C_i}{NPollutedTracers} \tag{10.6}$$

C_i is the concentration of a tracer computed as the ratio between the number of bacteria transported by the tracer and its current volume Vol_i. Tracers were considered polluted if their fecal contamination reached the NBWD limit of 102UFC/100 ml. *NPollutedTracers* is the number of polluted tracers.

The probability P of a bather to be in contact with polluted water was computed as the ratio between the total volume of polluted tracers and the volume of the monitoring box “x”:

$$P = \frac{\sum_{NPollutedTracers} Vol_i}{Vol_x} \tag{10.7}$$

Finally, contamination risk was evaluated by combining (1) fecal contamination, and; (2) the probability of a bather to be in contact of contamination water. The bathing water was considered contaminated if:

- the concentration was above the maximum threshold, and;
- the probability of a bather to be in contact with that water was above 1 %.

To estimate contamination risk, the probability results were analyzed at intervals of 3 h, considering the maximum probability value of each interval. Following this methodology, it was possible to predict the contamination probability in bathing water (Viegas et al. 2012).

The proposed methodology enables the answering of common management questions related to BWQ such as:

- Is there contaminated water in bathing water? What is the concentration?
- What is the probability of a bather being exposed to such that water?
- What is the origin of the contamination?

The methodology was applied, tested, calibrated and validated, during the 2011 and 2012 bathing seasons. Results showed that the Lagrangian model approach using the box methodology was able to forecast the contamination risk in the bathing water, and to predict fecal contamination. The proposed methodology also enabled the identification and quantification of contaminant plume, the identification of major pollution sources, and the establishment of minimization or mitigation actions.

Model simulations evidence the temporal and spatial variability of fecal contamination, revealing the importance of using modeling tools to explain and predict bathing water quality. For the performed tests, the modeling system was crucial to estimate the probability of the bathers being in contact with contaminated water. Results evidence the role of the local hydrodynamic fields on the stream plumes distribution along bathing waters, showing that contamination risk and values can change during the day on the same beach.

The implemented risk prediction methodology allows responding to NBWD statements, giving timely predictions about short-term pollution events, when they occur after raining events, or water level rising in streams. With the implementation of this methodology beach managers can predict the risk of contamination events, prevent bathers from being exposed to pollution, and in accordance with bathing water profile definition (IST and SANEST 2004; Viegas et al. 2009; Neves et al. 2010; Leitão et al. 2012) to minimize by up to 15 % the use of monitoring samples. This methodology was applied to specific bathing waters, but it can be implemented to other coastal areas around the world.

10.2.4 Residence Time of Water

This section describes a modeling application used to calculate the residence time of water in the Mondego estuary, Portugal. To implement the model of the Mondego estuary, four domains were nested inside PCOMS by using downscaling concept described in Sect. 10.2.1. The geographic area of the four domains is indicated in Fig. 10.3.

Residence time of water is the average time that a water parcel spends inside a system such as an estuary, a harbor, or a lagoon (Braunschweig et al. 2003). The residence time of water provides useful information in estuarine water quality studies because it can be used to determine for how long discharged pollutants will remain in the system before to be flushed away. Numerical models often contain advanced tools to calculate residence time of water (Braunschweig et al. 2003; Choi and Lee 2004; Cucco and Umgiesser 2006; Ascione Kenov et al. 2012). MOHID includes tools to calculate residence time of water by using Lagrangian models coupled with 2-D and 3-D hydrodynamic models. The methodology used in MOHID to calculate residence time of water in estuaries is divided into four steps (Braunschweig et al. 2003; Ascione Kenov et al. 2012):

- The estuary is divided into sub-areas, named as boxes, characterized by similar geomorphology and hydrodynamics;
- The boxes are filled with Lagrangian tracers in such a way that the volume of the tracers matches the total volume of the estuary;
- The Lagrangian tracers are transported out of the estuary by currents;
- The residence time of water is computed as the time required for 80 % of the initial tracers to leave the estuary.

The methodology enables to calculate water residence time of the whole estuary and of the single boxes as well. Each tracer is identified by its own origin, so that for a given water mass at a point x and at a time t it is possible to know from where the particle was released. Using the Lagrangian approach, the water fraction f_{ij} inside a box i at a time instant t with origin from box j is calculated as (Eq.10.8):

$$f_{i,j}(t) = \frac{V_{i,j}(t)}{V_{i,i}(0)} \tag{10.8}$$

$V_{ij}(t)$ is the volume of tracers released in box j, present inside box i at time t, and $V_{i,i}(t=0)$ is the water volume in box i at the beginning of the simulation. By integrating the water fraction and normalizing it by time, the integrated water fraction is obtained:

$$F_{i,j}(T) = \frac{1}{T}\int_0^T \frac{V_{i,j}(t)}{V_{i,j}(0)} dt \tag{10.9}$$

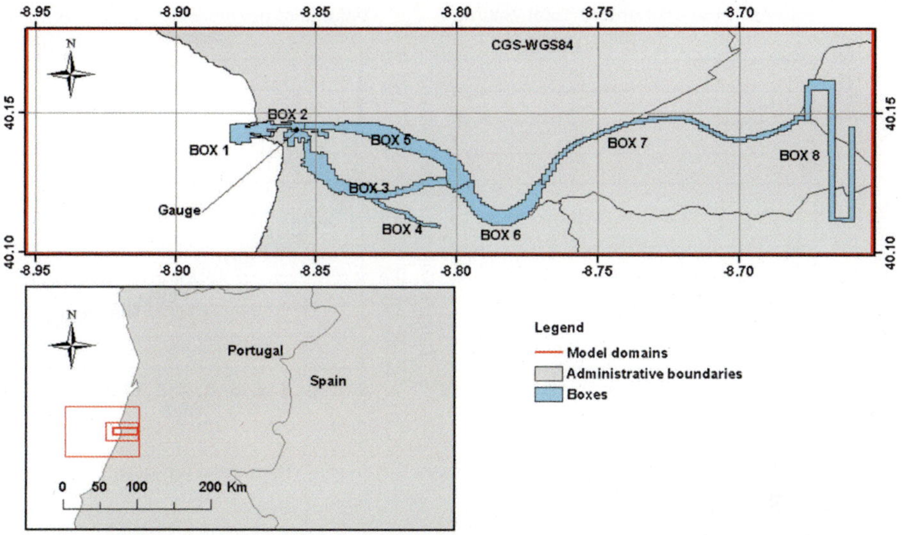

Fig. 10.15 Geographic location of the Mondego estuary. Boxes, model domains, and velocity gauges are indicated as well

The parameter $F_{i,j}$ in Eq. 10.9 measures the integrated influence of box j over box i during the time T. For the special case $i = j$ this parameter is related to the residence time of water. At the beginning of the simulation, F is equal to 1. As water is renewed, the contribution of the initial water to the total volume of the box decreases and F tends to 0. The time at which $F = 0$ is the residence time of water. However, in estuaries characterized by tidal influence, some of the water particles remain trapped inside intertidal areas and never leave the box. For this reason, following Braunschweig et al. (2003), the residence time of water is considered as the time at which 80 % of the tracers have already left the box. The same concept used for a box is applied to calculate the residence time of water for an estuary.

10.2.4.1 Residence Time of Water in the Mondego Estuary

The Mondego Estuary (40.15°N, −8.85°E) is located on the Portuguese Atlantic coast, and it is divided by the Murraceira Island into two arms (Fig. 10.15). The northern arm has depths varying with tides (between 5 and 10 m during flood tide, with respect to mean sea level), and receives water inflows from the Mondego River. The southern arm is characterized by lower depths compared to the northern arm (between 2 and 4 m during flood tide, with respect to mean sea level), and receives water inflows from the Pranto river. The tidal range varies between 0.35 and 3.3 m with respect to the mean sea level. Freshwater inflows range between 800 m^3/s in the winter and 10 m^3/s in the summer. To calculate water residence time, the estuary was divided into boxes. These boxes were filled with Lagrangian

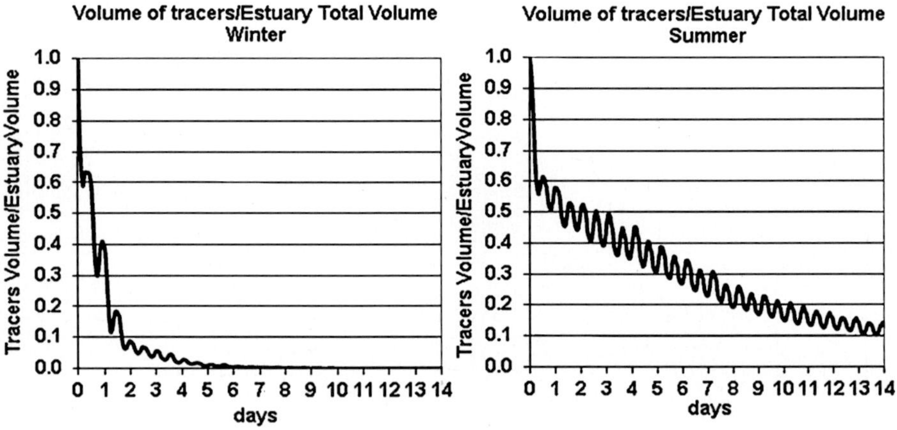

Fig. 10.16 Temporal evolution of the ratio between the tracers' volume and the estuary volume in winter and in summer

tracers, in a way that the total volume associated to the tracers matched the total volume of the estuary. The boxes were used both to release tracers, and to monitor the path of the particles passing through them. Each tracer was identified by its own origin, so that for a given water mass at a point *x* and at a time *t* it was possible to know from where it was released.

The evolution of the tracers' fraction in the estuary was calculated by considering river inflows in winter and in summer. The model was forced by using atmospheric data provided by the MM5 model (http://meteo.ist.utl.pt) at a resolution of 3 km. River upstream freshwater inputs were defined by using daily freshwater inflows from the database of Portuguese National Hydrological Service (SNIRH-www.snirh.pt).

For modeling the hydrodynamics in the estuary, the downscaling technique described in Sect. 10.2.1 was used. A model configuration was used with four domains nested in PCOMS (Fig. 10.3), to transfer boundary conditions from the regional to the local scale.

PCOMS provided fields of velocity, temperature, salinity and water levels with a fixed resolution of 6.6 km. The results of the hydrodynamic model were validated by comparing simulated and measured velocities collected by the buoy of the SIMPATICO monitoring system, located in the estuary at 40.14°N −8.85°E. Model implementation is described in Ascione Kenov et al. (2012).

The ratio between the volume of the tracers and the volume of the river is described in Fig. 10.16. This ratio decreases over time, indicating that the water is gradually renewed. The distribution of lagrangian tracers is described in Fig. 10.17. The results showed that the residence time of water was about 1 day in winter, and about 9 days in summer. This result is explained by the seasonal variation of the freshwater inflows, which values are usually higher in winter than in summer. The tide did not show a relevant effect on water renewal, concluding that the freshwater inflow was the main factor affecting water residence time in the study area.

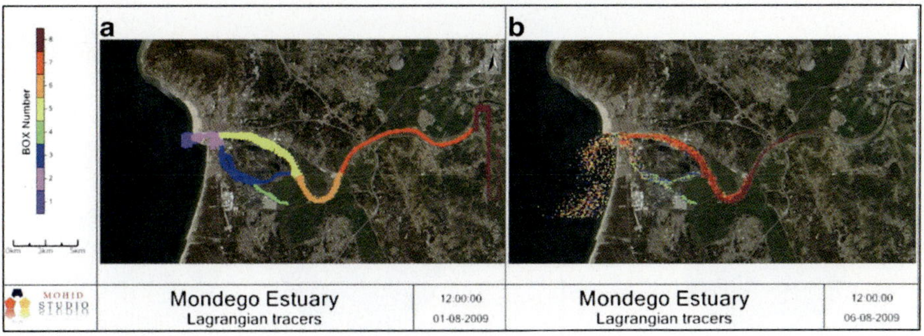

Fig. 10.17 Spatial distribution of Lagrangian tracers at the beginning of summer simulation (**a**) and after 6 days (**b**)

10.2.5 Nutrient Exchanges Between Estuaries and the Ocean

A 3-D hydrodynamic-ecological model of the Aveiro coastal area, Portugal, was nested in PCOMS, following the concept of downscaling described in Sect. 10.2.1. The model was used to quantify nutrient exchanges between the estuarine system and the ocean, considering nutrient inputs coming from land and from a submarine outfall.

The nutrient exchanges between estuaries and the ocean influence the trophic status of water bodies because they may affect primary production and oxygen renewal rates. The assessment of nutrient exchanges is challenging because it requires the identification and quantification of diffuse and point sources. Marine eutrophication is often heightened by anthropogenic inputs and the nutrients released by submarine outfalls could trigger algal blooms in coastal areas. The Vouga estuary (hereafter Ria de Aveiro), located on the Portuguese coast (40°N, −9°E), is a mesotidal estuary with a complex morphology and a wide intertidal area. The estuary is considered a high productive system, classified as a "sensitive area" in terms of eutrophication (91/271/EC). Little is known about nutrient fluxes between Ria de Aveiro and the ocean. Main freshwater inputs come from five tributaries (Caster, Antuã, Vouga, Boco and Mira channel). Additionally, nutrients are discharged into the coastal water by the São Jacinto submarine outfall located at a distance of about 3 km from the Aveiro coast. The outfall carries treated wastewater from three wastewater treatment plants and a paper mill, and it has a 300 m long diffuser located at a depth of 17 m. The average depth of the estuary is about 1 m, but in some channels, depths are kept between 4 and 7 m through constant dredging. The estuary's mouth is 350 m wide and about 20 m deep, with a large tidal prism (Dias et al. 2000). The study area is located in the northern part of the northern-hemisphere subtropical high pressure belt, and its climate is affected by the Azores anti-cyclone (Lopes et al. 2009). Due to this high pressure system, winter is dominated by weak southerly and westerly winds. In summer, the atmospheric current is characterized by northerly and north-westerly winds of about 5–6 m/s (Silva et al. 2001).

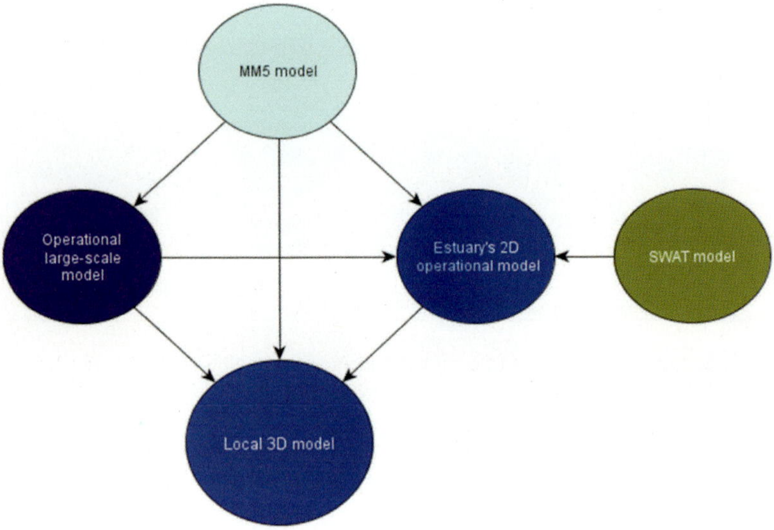

Fig. 10.18 Conceptual scheme of the models used to compute nutrient fluxes between the Ria de Aveiro and the ocean

Three different sources of nutrients were considered: the estuary, the ocean, and the submarine outfall. To account for exchanges between the estuary and ocean, the 3-D model was coupled to a 2-D hydrodynamic-ecological model of the estuary. The 3-D model has a horizontal grid of 1.5×1.5 km resolution (56×48 cells), covering an area of about 2,500 km^2. The 2-D model was used to represent the inner estuary, with a variable resolution between 1 km offshore and 200 m inside the estuary. The 2-D model of the estuary was coupled to the Soil and Water Assessment Tool (SWAT) applied to the Vouga catchment, to account for the exchanges between the land and the estuary. Figure 10.18 shows the conceptual scheme used to integrate the different models and the information flow between them. The simulation covered the period between February and November 2011. The model was validated by comparing model results with satellite images of chlorophyll, and with observed chlorophyll and nitrate concentrations. The spatial variation of nitrate and phytoplankton was characterized by subdividing the study area into boxes (Fig. 10.19). Fluxes of water, nitrate, and chlorophyll between the boxes and the surrounding waters were computed.

Atmospheric forcing was imposed by using the MM5 model results with 9 km resolution (http://meteo.ist.utl.pt). All bathymetries were built from the EMODnet Gridded Bathymetry (EGB) database (http://www.emodnet-hydrography.eu/). The submarine outfall's discharge was implemented as an increment of volume and included flow, nutrient concentrations, temperature and salinity.

The 2-D model simulated hydrodynamics and water quality inside the estuary, providing fields of velocity, temperature, salinity and biogeochemical properties as output over time. Hourly fluxes of biogeochemical properties provided by the 2-D

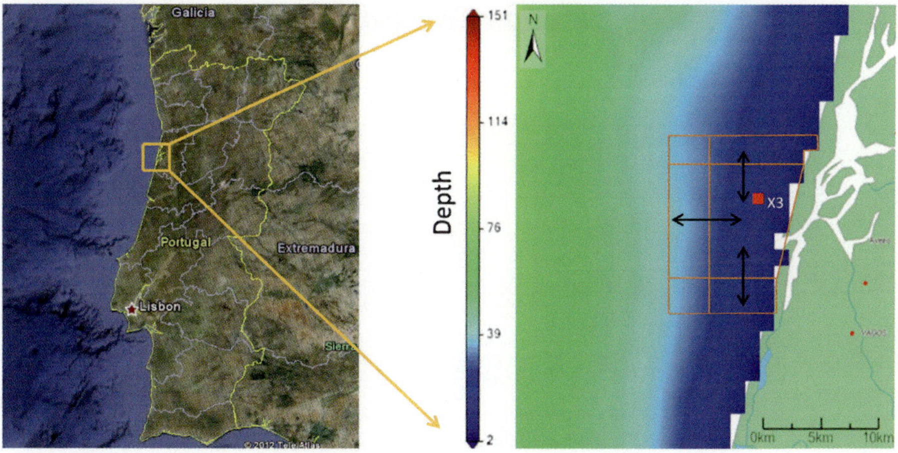

Fig. 10.19 Location of boxes used to compute nitrate and phytoplankton fluxes. The *red square* indicates the location of both the submarine outfall and the monitoring station. The Ria de Aveiro tributaries are indicated as well (color figure online)

model were used to determine nutrient concentrations exiting the estuary. Additionally, the 2-D model provided velocity and direction of the flow exiting the estuary, which was used to calculate the momentum discharge, important for the simulation of the Coriolis effect on coastal hydrodynamics.

To evaluate the importance of the three different nutrient sources (the ocean, the estuary, and the submarine outfall), three different scenarios were built: the first one considered the ocean as the only nutrient source; the second one included the ocean and the estuary, and; the third one included all the three sources.

Results in Fig. 10.20 show the comparison between simulated and measured concentrations of nitrate and chlorophyll at station X3, located close to the submarine outfall and indicated in Fig. 10.19. In this station, the maximum measured chlorophyll *a* concentration was 7 μg/l, with an average value of 2.9 μg/l at the surface and 2.2 μg/l in the bottom waters. Following the classification of the 91/676/CEE directive, these waters can be considered as mesotrophic (<10 μg/l) and yet very close to oligotrophic (<2.5 μg/l). Simulated nitrate concentrations were about 0.019 mg N/l (average values) in surface and bottom waters. Following the same directive, the waters can be classified in the first eutrophication level (0–2 mg N/l).

In overall, measured and simulated nitrate and chlorophyll *a* concentrations showed qualitative agreement. The concentrations of chlorophyll *a* decreased in August and stabilized at the end of the summer. Simulated nitrate concentrations had low agreement with data, with small variations over the simulated period, but almost always below 0.04 mg N/l.

Fluxes of water, nitrate, and phytoplankton between the estuary and the ocean are described in Table 10.2, for the month of July. The directions are taken by considering the box in the mouth of the estuary as a reference point. As an example, Northward in Table 10.2 indicates the direction of the flow going North with respect

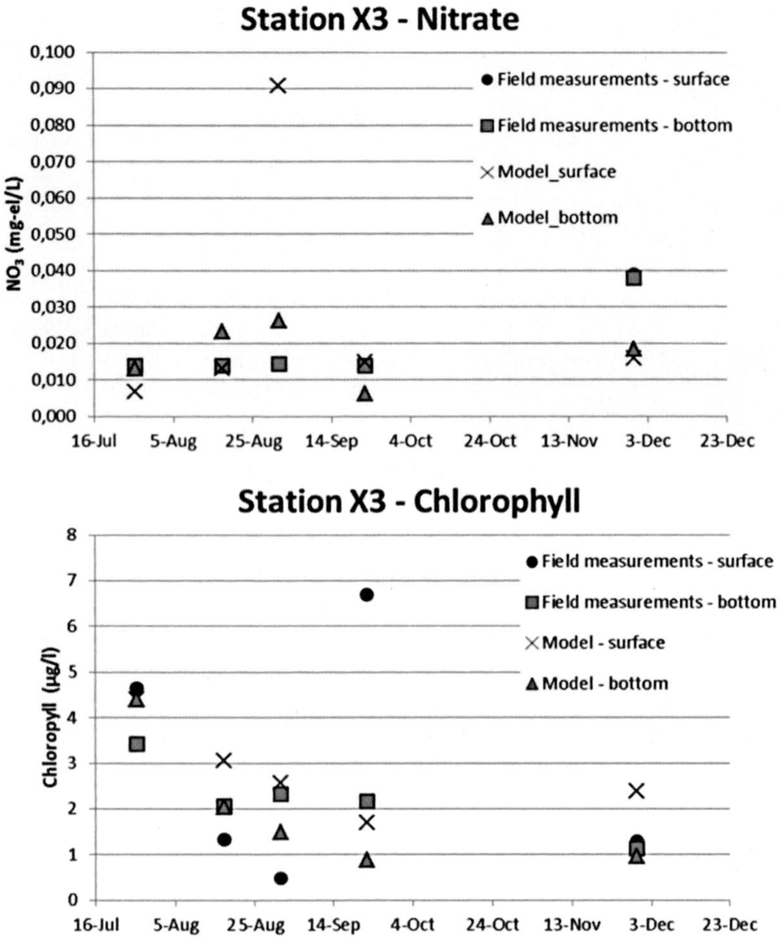

Fig. 10.20 Comparison between simulated and measured concentrations of nitrate and chlorophyll *a* at station X3, located close to the submarine outfall (Fig. 10.19)

to the box in the mouth of the estuary. In the three scenarios the dominant fluxes were from North to South and to West, as expected during upwelling conditions (July).

Results in Table 10.2 show that the estuarine discharge affects the fluxes considerably, with high monthly volume of water carried southward (from $32{,}100 \times 10^6$ m^3 to $35{,}833 \times 10^6$ m^3) and westward (from $15{,}500 \times 10^6$ m^3 to $20{,}220 \times 10^6$ m^3). The results show an increase of phytoplankton fluxes from 4,770 t of carbon in the scenario without estuarine discharge to 5,542 t of carbon in the scenario with estuarine discharge. The nitrate fluxes increase from 81 in the scenario without estuarine discharge to 236 t of nitrogen in the scenario with estuarine discharge. The reason for this increase of flux is directly related to the

Table 10.2 Simulated water, nitrate and phytoplankton fluxes in July 2011

Flux direction	Ocean			Ocean + estuary			Ocean + estuary + outfall		
	Water	Nitrate	Phy	Water	Nitrate	Phy	Water	Nitrate	Phy
	10^6 m^3	ton N	ton C	10^6 m^3	ton N	ton C	10^6 m^3	ton N	ton C
Northward	−32,100	−81	−4,770	−35,833	−236	−5,542	−35,843	−233	−5,535
Southward	16,500	37	2,622	15,599	98	2,418	15,630	103	2,471
Westward	15,500	−18	2,533	20,220	78.4	3,328	20,202	81	3,361

Phy indicates phytoplankton

near-horizontal angle of the discharge, which made the water from the estuary to head West before to be moved northward by the Coriolis force. Results in Table 10.2 show that high fluxes of phytoplankton were associated to low nitrate fluxes in the same direction, because phytoplankton consumes the nitrate. This is particularly evident in the westward direction (main flow direction of the estuary's discharge): high fluxes of phytoplankton are transported westward, and the nitrate along this direction is quickly consumed, resulting in low nitrate fluxes.

The most important result shown in Table 10.2 is the very low contribution of the submarine outfall on phytoplankton and nitrate fluxes. The fraction of the fluxes represented by this source was below 6 % with a maximum of 5.1 % in the southward flux from the estuary's mouth. On the contrary, the Ria de Aveiro was responsible for more than a 31 % increase of nitrate and phytoplankton fluxes. Thus, the estuary can be regarded as the most important source of nutrients in the coastal area. The nitrate concentration contributions from the ocean, the estuary, and the submarine outfall near the estuary's mouth were below 38 %, over 65 % and below 6 %, respectively, which means that near the mouth of the Ria de Aveiro, nitrate fluxes were strongly associated with freshwater discharge. It can be concluded that the submarine outfall has little influence on the fluxes of nitrate and phytoplankton in the Aveiro coast. These modeling results are consistent with previous studies (Silva et al. 2002) according to which the eutrophication conditions in Ria de Aveiro is unlikely to change as a result of the operation of the submarine outfall. From the management perspective, this methodology can support decision making in the implementation of treatment processes of the water discharged by the S. Jacinto outfall.

10.2.6 Seaweed Modeling

Seagrasses have a significant ecological importance in estuaries because they provide habitat for species which use them as a site for breeding, feeding and sheltering. Degradation of seagrass habitats negatively affects populations of species relevant for fishing.

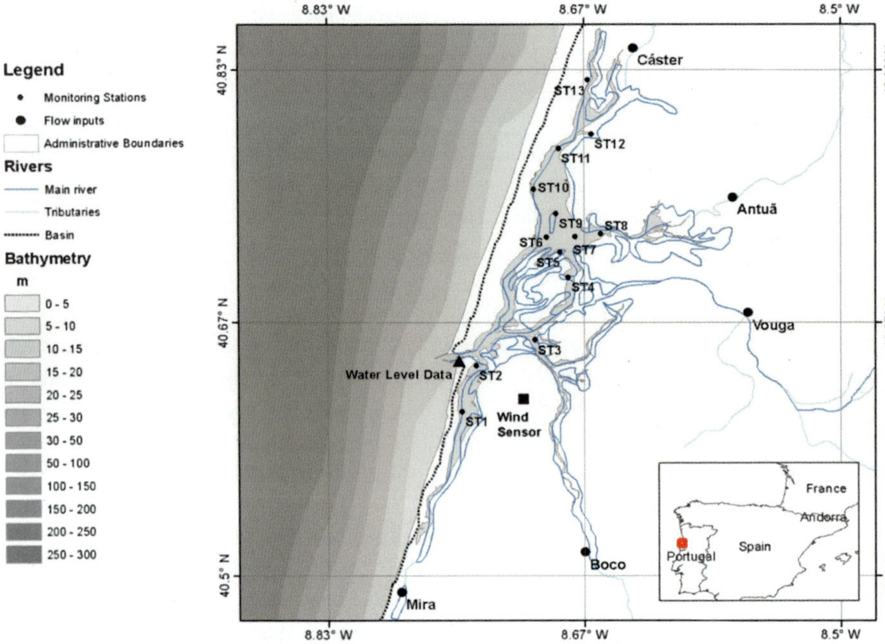

Fig. 10.21 Geographic location of Ria de Aveiro. Bathymetry of the study area, main rivers, flow inputs, wind sensor, and monitoring stations are indicated as well

In Portugal, seagrass habitat experienced a decline in the last 20 years (Cunha et al. 2013), causing biodiversity loss and degradation of coastal fisheries and water quality. *Zostera noltii* coverage in Ria Aveiro, Portugal, was about 8 km^2 in 1984, and it decreased down to 3 km^2 in 2004 (Silva et al. 2009; Cunha et al. 2013). The location of the Ria de Aveiro is described in Fig. 10.21. This decline in Ria Aveiro was attributed to a combination of factors such as dredging, deepening of channels, loss of fine sediments, siltation, nutrient washing, increasing tidal wave penetration, and increasing water currents. In Ria de Aveiro, the reduction of areas covered by seagrasses was followed by an increase of the areas of uncovered sediment, supporting the growth of sparse macroalgae populations only (Silva et al. 2009). Nutrient inputs from anthropogenic sources along Ria de Aveiro may trigger blooms of opportunistic macroalgae which could compete with other primary producers such as *Zostera noltii*.

A seagrass model was coupled to MOHID to assess the distribution of *Zostera noltii* in Ria de Aveiro. In the model, plant's growth depends on light, temperature, internal nutrient content, external nutrient availability, photoperiodicity, and space availability. Leaves and roots dynamics were expressed as:

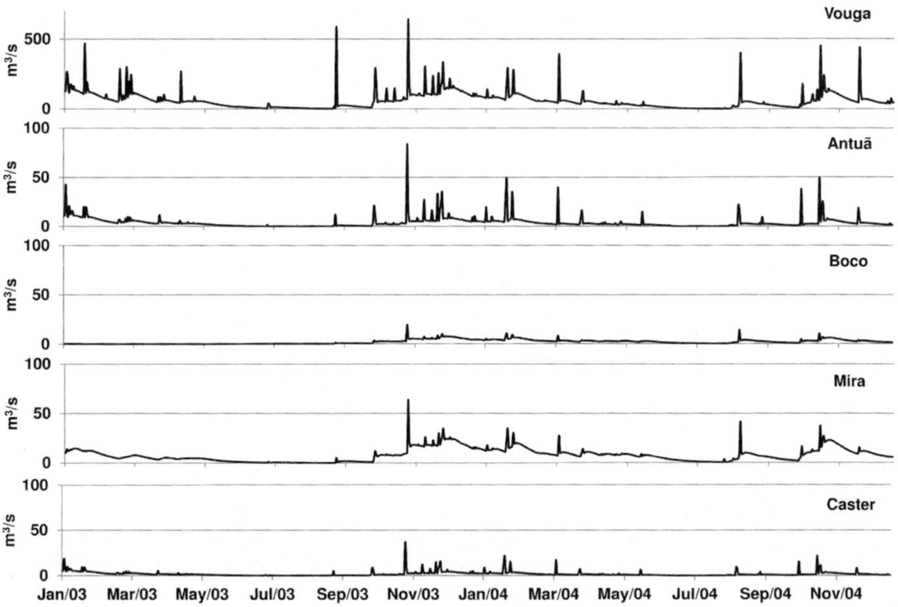

Fig. 10.22 Freshwater inflows at the main river canals, calculated by SWAT model applied to the Vouga catchment

$$\frac{dL}{dt} = (1 - tr)G \cdot L - m_l L \tag{10.10}$$

$$\frac{dR}{dt} = tr \cdot G \cdot L - m_r \cdot R \tag{10.11}$$

where L is the biomass of leaves (kg DW/m^2), R is the roots biomass (kg DW m^{-2}), m_r is the roots' decay rate (day^{-1}), and m_l is the leaves decay rate (day^{-1}). G is the growth rate (day^{-1}), and is a function of internal nutrient content. The full description of the seagrass model equations can be found in Ascione Kenov et al. (2013).

A 2-D hydrodynamic-biogeochemical model for Ria de Aveiro was used. The grid of the 2-D model had 87×81 cells and a variable resolution between 0.2 and 1 km (Fig. 10.21). The model was forced with average daily water inflows (Fig. 10.22) and nutrients coming from the main canals of Ria de Aveiro. These discharges were calculated by the Soil and Water Assessment Tool (SWAT), applied to the Vouga catchment.

At the open boundary, constant values were assumed for physical and biogeochemical properties. The model simulated hydrodynamics and water biogeochemistry, and provided fields of velocity, temperature, salinity, biogeochemical properties, and seagrass biomass, as output over time. The simulation time span

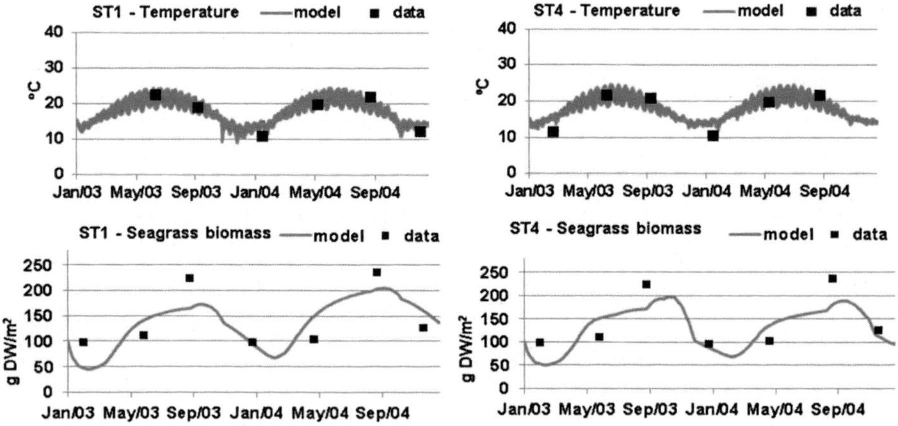

Fig. 10.23 Comparison between measured and simulated temperature and seagrass biomass (*Zostera noltii*)

was 2 years (between 01 January 2003 and 01 January 2005), with a time step varying between 10 and 15 s. Wind data used to force the hydrodynamic model were retrieved from the SNIRH database. Wind velocity and direction were measured by a Thies Clima/Young installed at a height of 2 m with respect to the ground level, at the Gafanha da Nazaré station (40.616°N, −8.706°E). The geographic location of the sensor is displayed in Fig. 10.21.

Temperature and *Zostera noltii* biomass data used to verify the model were collected by the University of Aveiro (Silva et al. 2009) in ten sampling points located in the intertidal areas of Ria de Aveiro with *Zostera noltii* beds, during the period between October 2002 and December 2004. A complete description of sampling methods and laboratory analysis used to retrieve *Zostera noltii* biomass can be found in Silva et al. (2009).

Results in Fig. 10.23 show comparison between data and model results for stations ST1 and ST4. The seagrass model outputs can be easily analyzed in the form of time series and maps, enabling easy calibration (Trancoso et al. 2005). The inclusion of the seagrass model in MOHID opened new possibilities to study interactions between primary producers (phytoplankton, macroalgae, and seagrasses) in response to natural and anthropogenic factors (nutrient inputs variations, temperature variations due to climate change, storms occurrence, among others). The flexibility of the model and its open source format enables to include new characteristics. For example, in the future the model may become more complex by adding a feedback effect by seagrasses over suspended sediment (seagrasses are capable to retain sediment), or over the bottom drag coefficient (seagrasses may alter the bed rugosity).

10.2.7 The Role of Complex Biogeochemical Algorithms in Estuarine Modeling

Estuaries are complex environments with their diverse biotic and abiotic conditions. Modeling approaches to simulate estuarine processes in water frequently rely on algorithms based on basic assumptions about the processes and constituents of marine systems. The advances which marine models underwent over the last decade (Vichi et al. 2003; e.g., Vichi et al. 2007a, b; Mateus et al. 2012a; Mateus 2012) are usually absent in estuarine model applications and leave out important components such as the microbial loop, phytoplankton composition, multiple nutrient limitation and the explicit modeling of the carbon cycle.

Despite the increase in the amount of model output in complex models, they provide additional knowledge and insights on the functioning of the system which are impossible to tackle with simpler algorithms. This section presents examples of the information provided by complex models. These examples are taken from published results of complex ecological models applied to the Tagus estuary, Portugal. The model is a biomass-based pelagic biogeochemical model (Mateus et al. 2012c) based on the ERSEM biochemical modeling philosophy (Baretta-Bekker et al. 1997; Baretta et al. 1995); it has 12 major components: producers, consumers, decomposers, organic matter (particulate, dissolved labile and semi-labile), nutrients (nitrate, ammonium, phosphate, silicate), biogenic silica and oxygen. The model has a decoupled nutrient and carbon dynamics with explicit parameterization of carbon, nitrogen, phosphorus, silica and oxygen cycles, and all living groups have variable stoichiometry and chlorophyll synthesis is accounted for in producers, allowing variable C:Chla ratios.

Complex ecological models can capture some of the complexity of estuarine system and provide reasonable estimates of the trends of the studied system. However, complex model generate a significant amount of results, and the calibration-validation effort can become a difficult task because just a few state-variables can be compared with field data. The results, even when not evaluated against data, and can still provide relevant information on the processes they address, helping to test hypotheses or raise additional relevant questions.

10.2.7.1 Nutrient Availability

The dynamics of nutrients depend on a number of physical, biological and chemical processes. Physical processes include mixing, flushing and sedimentation. Biological processes include fixation of dissolved and particulate nutrients, primarily by bacteria and phytoplankton, and release of inorganic nutrients through mineralization, mostly by bacterioplankton (decomposers).

Nutrients can be present in two major forms: inorganic (or mineral) and organic (both living and detrital). Nitrogen and phosphorus are the most significant nutrients, and their main species include dissolved (nitrate, nitrite, ammonium,

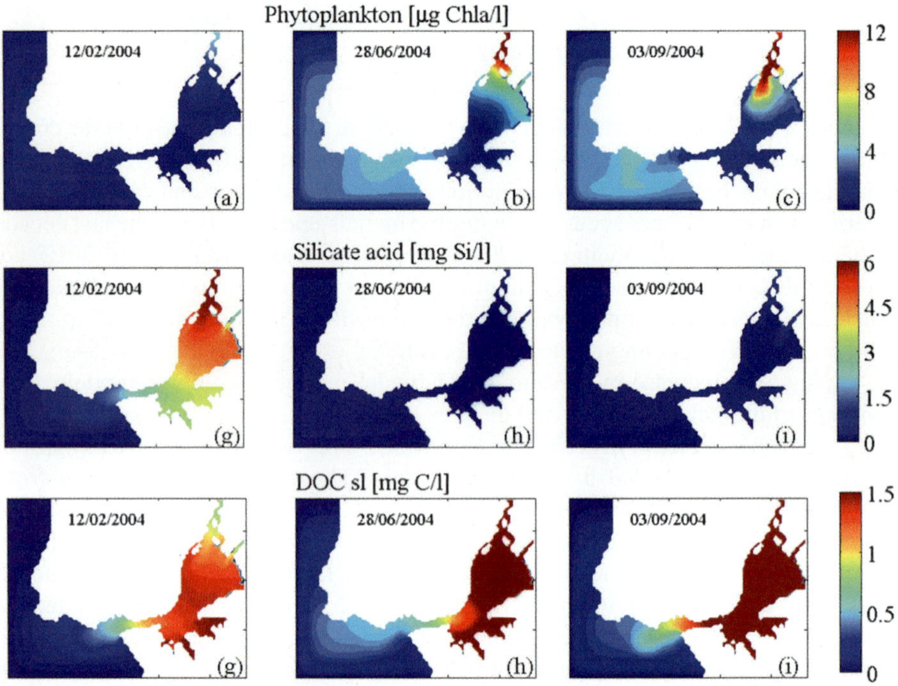

Fig. 10.24 Spatial maps for chlorophyll, silicate acid and semi-labile organic matter in a mesotidal estuary

organic N, phosphate, organic P) and particulate (organic N, organic P) components. Silica is also a fundamental nutrient in coastal systems because diatoms, a predominant phytoplankton group in estuaries and coastal systems, require silica in their structural composition. Complex model frequently address all of these properties, accounting for the explicit parameterization of several element cycles, thus tracing each nutrient and carbon source in the system. Figure 10.24 illustrates some of these results, with an example of spatial maps for chlorophyll, silicate acid and semi-labile organic matter in a mesotidal estuary.

There are no general rules to predict which nutrients limit estuarine primary producers, and when. Frequently nitrogen is the limiting nutrient, but even so, the limitation imposed by this nutrient may have significant temporal and spatial variations. Complex models are usually based on the assumption of variable internal nutrient composition, allowing to trace the evolution of C: Nut ratios in time for phytoplankton, and providing relevant clues on the influence of nutrient deficiency.

10.2.7.2 Light Conditions

Many estuaries are relatively shallow. In some systems this can provide optimal underwater light climate for primary production, but the high concentrations of suspended sediment greatly reduces the penetration of light through the water column. If the sediment is systematically suspended, then the consequent high turbidity and low light levels will reduce bloom conditions, regardless of nutrient levels. The result is that in many estuaries light is a key limiting factor for primary production (Cloern 2001), and there is a natural tolerance to eutrophication, i.e., a lack of a significant response to the nutrient loading in their inflowing rivers.

Phytoplankton responds to the variation in the light regime and nutrient availability across the temporal and spatial scales in estuaries. One of the most important physiological responses is the adjustment of chlorophyll content to maximize the use of available radiation. Phytoplankton will react to low light availability by increasing chlorophyll synthesis and this, in turn, will affect the C:Chla ratio of the cells. Models with this adjustment mechanisms (e.g., Mateus et al. 2012a) have proven useful for the understanding of how different ambient conditions influence the fluctuation of the C:Cha ratio. This is exemplified by model results in Fig. 10.25a, b, where the temporal and spatial C:Chla ratio of diatoms and autotrophic flagellates is shown to vary, as a response of ambient conditions (nutrient and light availability).

10.2.7.3 Relevant Ecological Groups

Most marine models rely on the assumption of a simple NPZD (nutrient-phytoplankton-zooplankton-detritus) chain, frequently addressing one group of phytoplankton and ruling out explicit parameterization of bacterioplankton. Complex models, on the other hand, usually have a functional approach for organisms, with more than one phytoplankton group and with bacterioplakton as a state-variable.

Regarding phytoplankton composition, estuaries show a significant diversity of species, but this variety can be addressed by considering two functional groups: diatoms (silica dependent) and autotrophic flagellates (silica independent). A simple approach with two phytoplankton groups provides significant insights on the nutrient limitation patterns inside estuaries, mostly related to the decrease in the concentration of some nutrients, such as phosphorus and silica, from the inner estuary to the estuary mouth. The fact that diatoms depend also on silica can explain in some cases the change in the phytoplankton composition, like the shift in diatom dominance from the inner estuarine areas to autotrophic flagellate dominance near the coast. Complex models can attempt to reproduce this shift, as in the example provide in Fig. 10.25c where diatom dominance decreases towards the estuary mouth, and silica-independent producers become dominant. These results may point to systematic shortage of dissolved silica in the lower estuarine areas, especially when flushing from the river is reduced.

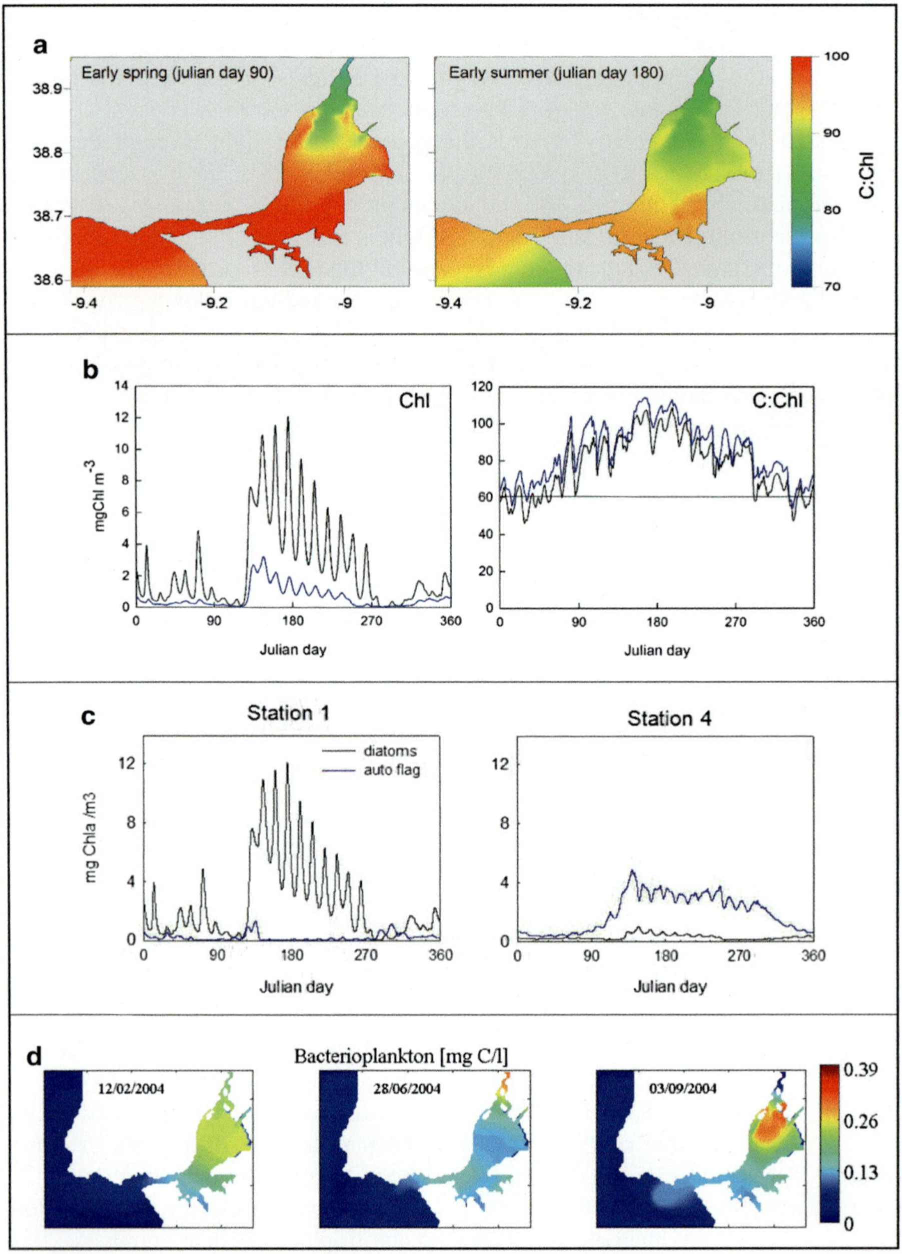

Fig. 10.25 Spatial patterns at different times of year obtained with a complex model that includes algorithms to address the bacterial loop

Bacterioplankton has a crucial role on the nutrient and carbon cycles in estuaries. Estuarine zones are usually dominated by allochtonous inputs of organic matter which promote the growth of heterotrophic organisms like bacteria, and for that reason estuarine waters are 1–3 orders of magnitude richer in bacterioplankton than the open ocean (Borsheim 2000). Since the availability of organic substrates for bacterioplankton growth varies significantly in estuaries, bacteria abundance will also vary, thus having a different impact on oxygen concentrations and nutrient release by mineralization of organic matter in different areas of the estuary. For their important role, bacterioplankton should be included in ecological models for estuaries, with all the complexity of the microbial loop and carbon-source substrates. Figure 10.25d shows significantly distinct spatial patterns at different times of year obtained with a complex model which includes algorithms to address the bacterial loop.

10.3 Conclusions

This chapter presented advanced modeling applications to address relevant water quality management questions related to estuarine water quality. Several case studies related to water quality assessment and management in Portuguese estuaries were presented. The described downscaling techniques enabled the integration of regional operational models with local scale models, improving open boundary conditions for estuarine models. The integration of regional and local scale models enabled the extension of operational forecasts to the local scale as well, and to the answering to common management questions, such as residence time of water. The presented modeling tools were used to complement the assessment of bathing water quality, and to answer NBWD statements. A new risk prediction methodology was proposed and applied to the Estoril coast to give timely predictions of fecal contamination of coastal waters. With the implementation of this methodology beach managers can predict the risk of contamination events and to prevent bathers from being exposed to pollution. The presented case studies show an effort to integrate different scales and to transfer information between hydrodynamic-ecological models, meteorological models, and watershed models, providing advancements to previous research carried out in this area (Dias et al. 2000; Trancoso et al. 2005; Lopes et al. 2009; Bierman et al. 2011; Mateus et al. 2012b). This approach can strongly improve the knowledge of the interconnections between land, ocean and atmosphere, moving forward the development of integrated models.

Acknowledgments Professor J.F. Da Silva of the Aveiro University for kindly providing monitoring data collected in Ria de Aveiro. Dr. Pedro Chambel, for kindly providing data of nutrients and freshwater inflows, calculated by using the SWAT model applied to the Vouga catchment. SANEST Project, S.A. (Saneamento Básico da Costa do Estoril) – Guia Submarine Outfall Monitoring Program. RECONNECT research project funded by Fundação para a Ciencia e

Tecnologia (FCT), under contract n. PTDC/MAR/64627/2006. LENVIS Project (FP7-ICT-2007/2/223925) financed by 7 FP7 of the European Commission. EASYCO Project, financed by the Atlantic Area Transnational Programme of the European Commission (EC), priority 2, through the European Regional Development Fund (ERDF), contract n. 2008-1/002. EMoSEM Project, contract n. SD/ER/11, financed under the network program ERANET SEAS-ERA "Towards Integrated Marine Research Strategy and Programmes".

References

Ali A, Lemckert CJ, Zhang H, Dunn RJK (2013) Sediment dynamics of a very shallow subtropical estuarine lake. J Coast Res. doi:10.2112/JCOASTRES-D-12-00255.1

Ascione Kenov I, Garcia AC, Neves R (2012) Residence time of water in the Mondego estuary (Portugal). Estuar Coast Shelf Sci 106:13–22. doi:10.1016/j.ecss.2012.04.008

Ascione Kenov I, Deus R, Alves CN, Neves R (2013) Modelling seagrass biomass and relative nutrient content. J Coast Res. doi:10.2112/JCOASTRES-D-13-00047.1

Baretta JW, Ebenhoh W, Ruardij P (1995) The European-regional-seas-ecosystem-model, a complex marine ecosystem model. Neth J Sea Res 33(3–4):233–246

Baretta-Bekker JG, Baretta JW, Ebenhoh W (1997) Microbial dynamics in the marine ecosystem model ERSEM II with decoupled carbon assimilation and nutrient uptake. J Sea Res 38(3–4):195–211

Bierman P, Lewis M, Ostendorf B, Tanner J (2011) A review of methods for analysing spatial and temporal patterns in coastal water quality. Ecol Indic 11(1):103–114. doi:10.1016/j.ecolind.2009.11.001

Borsheim KY (2000) Bacterial production rates and concentrations of organic carbon at the end of the growing season in the Greenland Sea. Aquat Microb Ecol 21(2):115–123

Braunschweig F, Martins F, Chambel P, Neves R (2003) A methodology to estimate renewal time scales in estuaries: the Tagus Estuary case. Ocean Dyn 53(3):137–145. doi:10.1007/s10236-003-0040-0

Bulleri F, Chapman MG (2010) The introduction of coastal infrastructure as a driver of change in marine environments. J Appl Ecol 47:26–35

Campuzano F, Nunes S, Malhadas MS, Neves R (2010) Modelling hydrodynamics and water quality of Madeira Island. GLOBEC Int Newsl 16(1):40–42

Campuzano F, Fernandes R, Leitâo P, Viegas C, De Pablo H, Neves R (2012) Implementing local operational models based on an offline downscaling technique: the Tagus estuary case. In: 2.as Jornadas de Engenharia Hidrográfica, Lisbon, 20–22 June 2012, pp 105–108

Canteras JC, Juanes JA, Pérez L, Koev KN (1995) Modelling the coliforms inactivation rates in the Cantabrian Sea (Bay of Biscay) from in situ and laboratory determinations of t90. Water Sci Technol 32(2):37–44. doi:10.1016/0273-1223(95)00567-7

Chibole OK (2013) Modeling River Sosiani's water quality to assess human impact on water resources at the catchment scale. Ecohydrol Hydrobiol 13(4):241–245. doi:10.1016/j.ecohyd.2013.10.003

Choi KW, Lee JHW (2004) Numerical determination of flushing time for stratified water bodies. J Mar Syst 50(3–4):263–281. doi:10.1016/j.jmarsys.2004.04.005

Cloern JE (2001) Our evolving conceptual model of the coastal eutrophication problem. Mar Ecol Prog Ser 210:223–253

Cucco A, Umgiesser G (2006) Modeling the Venice Lagoon residence time. Ecol Model 193(1–2):34–51. doi:10.1016/j.ecolmodel.2005.07.043

Cunha AH, Assis JF, Serrão EA (2013) Seagrasses in Portugal: a most endangered marine habitat. Aquat Bot 104:193–203

Deus R, Brito D, Ascione KI, Lima M, Costa V, Medeiros A, Neves R, Alves CN (2013) Three-dimensional model for analysis of spatial and temporal patterns of phytoplankton in Tucuruí reservoir, Pará, Brazil. Ecol Model 253:28–43

Dias JM, Lopes JF, Dekeyser I (2000) Tidal propagation in Ria de Aveiro Lagoon, Portugal. Phys Chem Earth B 25(4):369–374

Fernandes R (2005) Modelação operacional no estuário do Tejo. Instituto Superior Técnico, Technical university of Lisbon, Lisbon

Fernandes L, Saraiva S, Leitão PC, Pina P, Santos A, Braunschweig F, Neves R (2006) Mabene deliverable D4.3d-code and description of the benthic biogeochemistry module-managing benthic ecosystems in relation to physical forcing and environmental constraints. MaBenE-Managing benthic ecosystems in relation to physical forcing and environmental constraints

Grell G, Dudhia J, Stauffer D (1994) A description of the fifth-generation Penn State/NCAR Mesoscale Model (MM5)

IST, SANEST (2004) Bacteria mortality tests considering different environmental parameters

Lai YC, Tu YT, Yang CP, Surampalli RY, Kao CM (2013) Development of a water quality modeling system for river pollution index and suspended solid loading evaluation. J Hydrol 478(0):89–101. doi:10.1016/j.jhydrol.2012.11.050

Leitão P, Coelho H, Santos A, Neves R (2005) Modelling the main features of the Algarve coastal circulation during July 2004: a downscaling approach. J Atmos Ocean Sci 10(4):421–462. doi:10.1080/17417530601127704

Leitão P, Galvão P, Aires E, Almeida L, Viegas C (2012) Faecal contamination modelling in coastal waters using a web service approach. Environ Eng Manag J 11(5):899–906

Lopes JF, Cardoso AC, Moita MT, Rocha AC, Ferreira JA (2009) Modelling the temperature and the phytoplankton distributions at the Aveiro near coastal zone, Portugal. Ecol Model 220 (7):940–961. doi:10.1016/j.ecolmodel.2008.11.024

Mateus M (2012) A process-oriented model of pelagic biogeochemistry for marine systems. Part I: model description. J Mar Syst 94:S78–S89. doi:10.1016/j.jmarsys.2011.11.008

Mateus M, Leitão PC, de Pablo H, Neves R (2012a) Is it relevant to explicitly parameterize chlorophyll synthesis in marine ecological models? J Mar Syst 94:S23–S33. doi:10.1016/j.jmarsys.2011.11.007

Mateus M, Riflet G, Chambel P, Fernandes L, Fernandes R, Juliano M, Campuzano F, de Pablo H, Neves R (2012b) An operational model for the West Iberian coast: products and services. Ocean Sci 8(4):713–732. doi:10.5194/os-8-713-2012

Mateus M, Vaz N, Neves R (2012c) A process-oriented model of pelagic biogeochemistry for marine systems. Part II: application to a mesotidal estuary. J Mar Syst 94(Supplement):S90–S101. doi:10.1016/j.jmarsys.2011.11.009

Miranda R, Braunschweig F, Leitâo P, Martins F, Santos A (2000) MOHID2000 – a coastal integrated object oriented model. In: Hydraulic engineering software VIII. WIT Press

Neves R (2013) The MOHID concept. In: Mateus M, Neves R (eds) Ocean modelling in coastal management. IST Press, Lisbon, pp 1–11

Neves R, Silva A (1991) An extension of the Boussinesq equations to deep water. A case study. Paper presented at the computer modelling in ocean engineering, Barcelona

Neves R, Viegas C, Fernandes R, Nunes S, Pina P, Carvalho C, Granger C (2010) Ferramentas matemáticas de suporte à definição de perfis de água balnear: 2 casos de estudo, vol 30, 90th edn, Recursos Hídricos. Associação Portuguesa dos Recursos Hídricos, Lisbon

Pando S, Juliano MF, García R, de Jesus Mendes PA, Thomsen L (2013) Application of a lagrangian transport model to organo-mineral aggregates within the Nazaré canyon. Biogeosciences 10(6):4103–4115. doi:10.5194/bg-10-4103-2013

Pinto L, Campuzano F, Fernandes R, Fernandes L, Neves R (2012) An operational model for the Portuguese coast. In: 2.as Jornadas de Engenharia Hidrográfica, Lisbon, 20–22 June 2012, pp 85–88

Silva JF, Duck RW, Anderson JM, McManus J, Monk JGC (2001) Airborne observations of frontal systems in the inlet channel of the Ria de Aveiro, Portugal. Phys Chem Earth B Hydrol Oceans Atmos 26(9):713–719. doi:10.1016/S1464-1909(01)00075-2

Silva J, Duck RW, Hopkins TS, Rodrigues M (2002) Evaluation of the nutrient inputs to a coastal lagoon: the case of the Ria de Aveiro, Portugal. Hydrobiologia 475–476(1):379–385. doi:10.1023/A:1020347610968

Silva JF, Duck RW, Catarino JB (2009) Nutrient retention in the sediments and the submerged aquatic vegetation of the coastal lagoon of the Ria de Aveiro, Portugal. J Sea Res 62:276–285

Sohma A, Sekiguchi Y, Kuwae T, Nakamura Y (2008) A benthic-pelagic coupled ecosystem model to estimate the hypoxic estuary including tidal flat-model description and validation of seasonal/daily dynamics. Ecol Model 215:10–39

Trancoso AR, Saraiva S, Fernandes L, Pina P, Leitão PC, Neves R (2005) Modelling macroalgae using a 3D hydrodynamic-ecological model in a shallow, temperate estuary. Ecol Model 187:232–246

Vichi M, Oddo P, Zavatarelli M, Coluccelli A, Coppini G, Celio M, Umani SF, Pinardi N (2003) Calibration and validation of a one-dimensional complex marine biogeochemical flux model in different areas of the northern Adriatic shelf. Ann Geophys Germany 21(1):413–436

Vichi M, Masina S, Navarra A (2007a) A generalized model of pelagic biogeochemistry for the global ocean ecosystem. Part II: numerical simulations. J Mar Syst 64(1–4):110–134

Vichi M, Pinardi N, Masina S (2007b) A generalized model of pelagic biogeochemistry for the global ocean ecosystem. Part I: theory. J Mar Syst 64(1–4):89–109

Viegas C, Nunes S, Fernandes R, Neves R (2009) Streams contribution on bathing water quality after rainfall events in Costa do Estoril – a tool to implement an alert system for bathing water quality. J Coast Res Proc 10th Int Coast Symp ICS 2009 II(Special Issue No. 56):1691–1695

Viegas C, Neves R, Fernandes R, Mateus M (2012) Modelling tools to support an early alert system for bathing water quality. Environ Eng Manag J 11(5)

Viero DP, D'Alpaos A, Carniello L, Defina A (2013) Mathematical modeling of flooding due to river bank failure. Adv Water Resour 59(0):82–94. doi:10.1016/j.advwatres.2013.05.011

Chapter 11
Advances in Video Monitoring of the Beach and Nearshore: The Long-Term Perspective

Ana Nobre Silva and Rui Taborda

Abstract It is widely recognized that video monitoring systems are excellent tools in coastal morphodynamic studies as they can capture, simultaneously, beach morphological changes and the forcing mechanisms. Over the last years this remote sensing technique has experience huge developments related not only to technology advances but also to the success of numerous applications reported by the scientific community. Since the first steps of video monitoring of the coastal zone, made by the Coastal Imaging Laboratory of Oregon University in the 1980s, several years have passed by enabling the existence of long-term imagery records. The existence of extended high-resolution time-series make possible to broaden typical video application studies from short-term studies to annual (or longer) time scales as morphological inter-annual variability of the coastal systems is still largely unknown. This work summarizes recent developments on the use of video systems in the understanding of yearly to decadal beach morphological changes and describes the application of a video system deployed at Nazaré, Portugal. The system, operational since December 2008, allowed a detailed description of the coastal evolution at a pluriannual timescale. Results, which are in agreement with previous works, indicate that system variability depends not only on the forcing characteristics that occur at different time-scales (storm, seasonal and inter-annual) but also on antecedent morphology. The use of video systems arises as an optimal data acquisition method to capture this variability and thus support the fully understanding of beach morphodynamics at these wide range of spatio-temporal scales.

A.N. Silva (✉) • R. Taborda
Instituto Dom Luiz, Geology Department, Science Faculty, University of Lisbon, Campo Grande, 1749-016 Lisbon, Portugal
e-mail: amasilva@fc.ul.pt

C.W. Finkl and C. Makowski (eds.), *Remote Sensing and Modeling: Advances in Coastal and Marine Resources*, Coastal Research Library 9,
DOI 10.1007/978-3-319-06326-3_11,

11.1 Introduction

Beaches are, by nature, extremely dynamic systems. The processes acting on these environments range from short spatio-temporal scales, as waves or storm events, to longer ones, related to beach seasonality, interannual variability or even long-term coastal evolution. In order to understand this dynamics (including those related with climate change effects), and to support adequate management policies, it has been long recognized the importance of developing long-term monitoring programs.

Until recently, long-term data sets typically consisted of low frequency and spatially limited topographic surveys that aimed to characterize beach seasonality and evolution, but often fail in the characterization of high-frequency events such as storms effects or post-storm recovery because of its low temporal resolution. This could have strong consequences in the understanding and interpretation of system dynamics as the continuous interaction between the forcing mechanisms and the morphology is added to system complexity. In fact "snapshot" monitoring approach can be of little added value in the understanding of beach dynamics, as system behavior depends on the antecedent morphology (Wright and Short 1984; Alegria-Arzaburu and Masselink 2010; Quartel et al. 2008) and on the non-linear system dynamics that can lead to chaotic behavior (Vriend and Capobianco 1993; Southgate et al. 2003; van Enckevort and Ruessink 2003b). Thus, beach dynamics cannot be fully understood using only low-frequency data.

Currently the importance of developing long-term high-resolution monitoring programs is rather consensual (Larson et al. 2003; Larson and Kraus 1994; Smit et al. 2007; Southgate et al. 2003; Harley et al. 2011). A review of multi-decadal and high resolution coastal monitoring programs is presently in (Harley et al. 2011).

Long-term monitoring can be supported by a wide range of techniques that, over the last years, have experienced large developments. From the plethora of available methods, video monitoring has emerged as one of the most effective techniques. Video monitoring is supported on the fact that most of the nearshore phenomena that can be visually discerned can be quantified thought video image processing techniques (Holman et al. 1993; Holland et al. 1997; Aarninkhof 2003; Holman and Stanley 2007). Considering these features and the ease and safety of operation (only minor logistic and financial commitments is needed compared to in situ measurements in a hostile surf zone environment). These characteristics makes video systems adequate to serve long-term monitoring programs with high-resolution in time (Smit et al. 2007).

The main objective of this work is to summarize recent developments on the use of video systems in the understanding of long-term (yearly to decadal) beach morphological changes and to demonstrate the use of such a video-monitoring system deployed at Nazaré (Portugal).

11.2 Pluriannual Video Monitoring Programs

Since the first steps of video monitoring of the coastal zone, made by the Coastal Imaging Laboratory (CIL) of Oregon University in the 1980s, several years have passed by enabling the existence of long-term imagery data that relates the oceanographic processes with the beach response.

The existence of extended high-resolution time-series makes possible to broaden typical video application studies from short-term studies to annual (or longer) time scales, which is a major improvement for coastal region planning and management (e.g. Liu et al. 2012; Davidson et al. 2007), for the correct formulation of conceptual behavior models (e.g. Kroon et al. 2008) and even for better understand the processes and improve the modelling strategies (e.g. Siegle et al. 2007; Southgate et al. 2003; Larson et al. 2003).

Despite the importance of pluriannual studies, the number of published works that have been supported in more than 3 years video data still is relatively small (Table 11.1). Most of the references found in literature refers to ARGUS monitoring system, the pioneer system from CIL, even though nowadays several similar coastal monitoring systems exists (see Taborda and Silva 2012).

Table 11.1 Pluriannual monitoring programs based on more than 3 years of video data

	Video monitoring		
Aim	Time span (years)	System	References
Beach cusps	8.9	Argus	Holland (1998)
Transverse sandbars	10	Argus	Konicki and Holman (2000)
Sandbar morphology	3.4	Argus	van Enckevort and Ruessink (2003a)
Nearshore sandbars	3.4	Argus	van Enckevort and Ruessink (2003b)
Sandbar location; shoreline location	–	Argus	Alexander and Holman (2004)
Outer sand bar migration	9	Argus	Haxel and Holman (2004)
Nearshore morphodynamics	~4	Argus	Ranasinghe et al. (2004)
Sandbar morphodynamics	3.2	Argus	Siegle et al. (2007)
Shoreline evolution	4.5 (with video)	Argus	Smit et al. (2007)
Rips	3	Argus	Turner et al. (2007)
Beach cusps	3	–	Almar et al. (2008)
Beach users	4	Argus	Guillén et al. (2008)
Nearshore bars	4 (with video)	Argus	Kroon et al. (2008)
Storm response beach morphology – shoreline	3	Argus	Alegria-Arzaburu and Masselink (2010)
Outer bar	4.5	Argus	Armaroli and Ciavola (2011)
Sand bars	4.3	Argus	Ojeda et al. (2011)
Nearshore morphology	4.3	–	Liu et al. (2012)
Shoreline variability and beach rotation	3	Camera	Foster (2012)
Shoreline variability	3.4	Cosmos	Silva et al. (2013)

The first long-term work (Holland 1998) refers to the study of the timing of cusps formation and its relation with the environmental forcing from nearly 9 years of daily video images.

Soon after, new long-term studies started to develop mainly focusing the study of sand bars and/or nearshore morphology where 10 and 9 years period of daily sandbars monitoring can be found (Konicki and Holman 2000) amongst several other authors that investigate bar dynamics in a less continuous approach (e.g. Siegle et al. 2007; Armaroli and Ciavola 2011) quantifying particular episodes occurring during the time span of the video monitoring.

For instance, Siegle et al. (2007) use the video images to extract the intertidal morphology which supports the morphological modelling at a frequency that is dependent on the model required resolution, whereas Armaroli and Ciavola (2011) propose a morphological classification of outer bars based on its plan shape visible on TIMEX images spanning from February 2004 to May 2007.

Almar et al. (2008) performed a long-term study analyzing beach cusps through discrete episodes by selecting 24 beach cusps episodes (from 1999 to 2001) and analyzing its evolution.

The shoreline pluriannual evolution and variability is also a theme starting to rise in video monitoring searches by means of the shoreline response to storm events (Alegria-Arzaburu and Masselink 2010) or to address the interannual coastline tendencies and behaviors (Smit et al. 2007; Foster 2012; Silva et al. 2013).

11.3 Case Study

In order to demonstrate the applicability of a video system in the understanding of long-term of beach morphodynamics, a case study of a pluriannual monitoring program, at Nazaré (Portugal), is presented. It is clear that the coastline position and the beach volume are ultimately related to a fragile balance between relative sea-level positioning, sediment supply and wave forcing, acting at different time scales. Masselink and Pattiaratchi (2001) classify the beach change into two categories the "seasonal" that comprehend the winter/summer characteristic beach behavior and the "cyclic" where the main changes are related to the storm and post-storm changes on the beach, nevertheless in both cases the morphological change is primarily due to changes in incident wave energy. A "chaotic" category can be added into this list, where the beach behavior is also deeply dependent on the antecedent morphology of the beach (cf. Wright and Short 1984; Wright et al. 1985; Quartel et al. 2008; Alegria-Arzaburu and Masselink 2010).

The join contribution of these three beach behaviors can led to an extremely high dynamic system that is often difficult to fully understand.

The presented case study aims to describe and interpret video-derived beach volume variations along nearly three and half years of monitoring and investigate its relation with incident waves, in order to contribute to the understanding of beach behavior at different temporal and spatial scales.

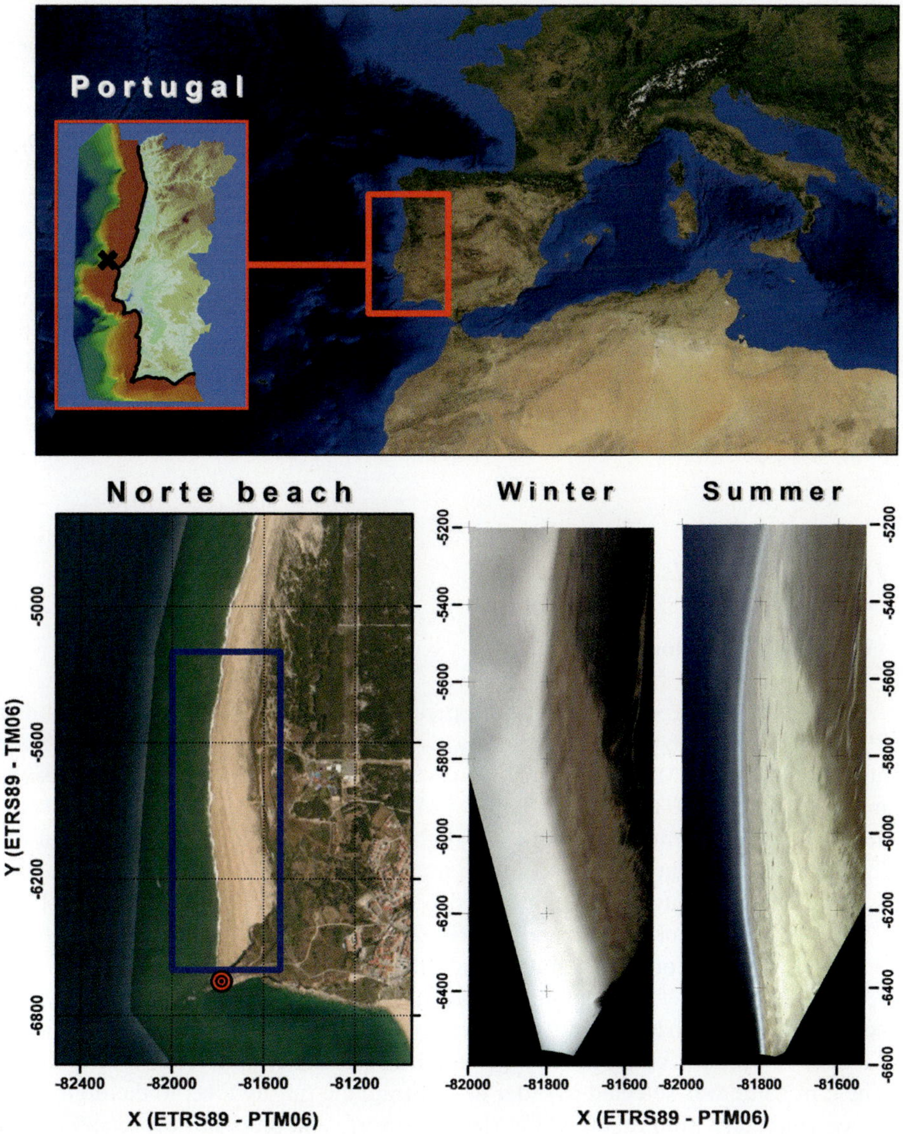

Fig. 11.1 Study site with reference to the position off the offshore wave buoy (*black cross*), the video monitoring extent (*blue rectangle*) and video camera position (*red circle*) (Imagery Source: Esri, DigitalGlobe, GeoEye, i-cubed, USDA, USGS, AEX, Getmapping, Aerogrid, IGN, IGP, swisstopo, and the GIS User Community); The *bottom right* images represent examples of video derived rectified images of Norte beach in winter and summer condition (color figure online)

11.3.1 Study Site

Norte beach (Nazaré) is located at the west Portuguese coast (Fig. 11.1), is roughly N-S oriented and subject to a high energetic wave regime, generated at the Northeast Atlantic. Wave propagation over the Nazaré canyon head is responsible for unusual high waves (that can exceed 20 m) near the beach. Norte beach is a reflective beach characterized by a wide berm (or multiple wide berms at summer) and a steep beach face. It is subject to a mesotidal regime, with tidal amplitudes ranging from 1.5 to 3.5 m, at neap and spring tides, respectively. Average offshore wave regime, obtained by 56 years hindcast data for the northwest Portuguese coast (Dodet et al. 2010) can be described by 2 m significant wave heights, 11 s of average peak period and mean wave direction approaching from ENE-NE (308°). Sediments are characterized by quartzic sand with dimensions variable from $-2.6\ \phi$ to $1.79\ \phi$ and mean value $0.18\ \phi$ (Cascalho et al. 2012).

In what concerns morphodynamics, Norte beach can be individualized into two main sectors, one in the vicinity of the Nazaré headland where the seasonal variability, expressed through the position of the shoreline, is extremely high (seasonal changes exceed 160 m); and one sector less variable and less seasonal imprinted where the amplitude of coastline changes does not exceeds 75 m (Silva et al. 2013).

11.3.2 Methods

11.3.2.1 Video Monitoring

A video monitoring system has been operational in Norte beach since December 2008. The COSMOS coastal monitoring system (Taborda and Silva 2012) installed at Forte S. Miguel Arcanjo, Nazaré, consists of a 3.1 megapixel resolution camera mounted at about 50 m high relatively to MSL which is acquiring snapshot images at 1 Hz, during 20 min interval each daylight hour. The geometric correction of the images, including the correction of the image distortions and image rectification, followed respectively the Heikkila and Silvén (1997) distortion model and the collinearity condition (Wolf and Dewitt 2000), a detailed description of these procedures can be found in Taborda and Silva (2012) and Silva et al. (2009).

TIMEX images were computed using 10 min block of snapshot imagery. To access shoreline changes between December 2008 and May 2012, a monthly TIMEX image was carefully chosen based on the same total water level (TWL). The 2 m value was selected in order to constrain the run-up level to the high slope beach (therefore minimizing the horizontal errors) and to assure comparable conditions. Despite some gaps, due to technical failures, a total of 31 images where analyzed during the 3.5 years of monitoring presented in this study.

The shoreline herein defined as the 2 m height contour line was detected through image processing and analysis on the selected TIMEX images. Sand and water where automatically classified using the maximum likelihood method. Due to different image luminosity conditions, related to atmospheric conditions and the high variable extent of the breaking zone, four different spectral signatures were used. Shoreline detection results were visually validated and about 6 % of the initially selected images were rejected and replaced by similar ones.

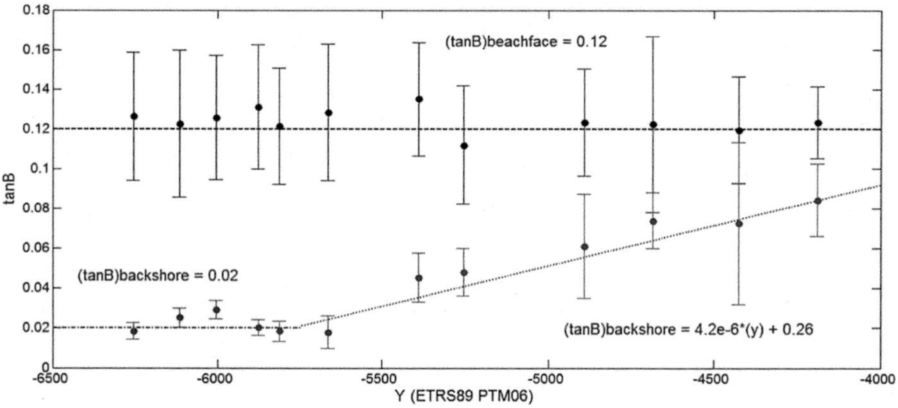

Fig. 11.2 Alongshore beach face slope (*black*) and backshore slope (*blue*) variability (the equation represent the model fitted to the alongshore backshore slope (y coordinate of the ETRS – PTM06 coordinate system) (color figure online)

The estimation of the shoreline elevation (or total water level), comprehensively described in (Silva et al. 2013), was done adopting an empirical site specific formulae which estimates the run-up height in function of the offshore significative wave height (Hs), the offshore wave direction (Dir) and the astronomical tide (AT).

11.3.2.2 Beach Volume

To estimate the beach volume from the shoreline position it was necessary to assume a representative beach profile configuration along the study area. This profile model was based on the fact that the configuration of the surveyed transverse beach profiles was found to be highly uniform in what concerns the beach face and backshore slope (beach profile terms follows Shore Protection Manual, CERC 1984).

The mean beach face slope (tanβ) was found to be spatially uniform, ranging from 0.10 to 0.16, with an average value of 0.12 along the entire surveyed stretch. The backshore slope exhibits longshore variations that enables to individualize the beach in two sectors: a southern sector where the backshore slope is rather mild and uniform (0.02); a northern sector where this slope increases northward from a value of 0.02 up to 0.08 at about 2 km north of Nazaré headland (Fig. 11.2).

The profile model, represented in Fig. 11.3, was developed considering two fixed points: one that represents the maximum seaward extent of the profile (the shoreline position, i.e. 2 m MSL contour line) and another that is the landward limit of the beach (the dune base represented by the 8 m MSL contour line). Below the 2 m contour and up to the closure depth the profile was extended using the equilibrium Dean profile (Dean 1991, with an A parameter that best fitted the submarine profile data acquired from LiDAR data).

From the knowledge of shoreline position (xS(y,t) acquired by video), dune foot coordinates (xD(y), zD acquired by a GPS survey and considered steady), representative foreshore slope (tanF) and backshore slope (tanB(y)), the berm position

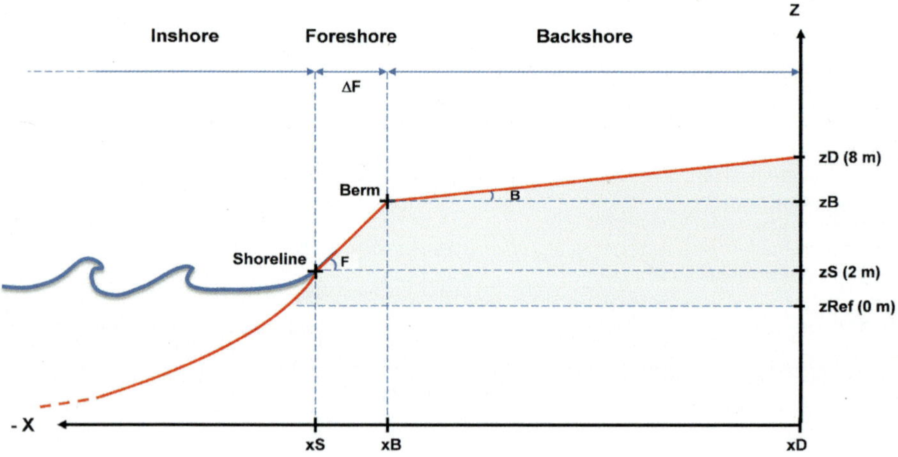

Fig. 11.3 Schematics of the Norte beach profile model

(xB(y,t); zB(y,t)) along Norte beach at given time (t) can be estimated from simple geometric considerations using Eqs. 11.1 and 11.2.

$$\mathrm{xB}(\mathrm{y},\mathrm{t}) = \mathrm{xS}(\mathrm{y},\mathrm{t}) + \Delta\mathrm{F}(\mathrm{y},\mathrm{t}) \tag{11.1}$$

$$\mathrm{zB}(\mathrm{y},\mathrm{t}) = \mathrm{zS} + \mathrm{tanF} \times \Delta\mathrm{F}(\mathrm{y},\mathrm{t}) \tag{11.2}$$

where

$$\Delta\mathrm{F}(\mathrm{y},\mathrm{t}) = ((\mathrm{zD} - \mathrm{zS}) - \mathrm{tanB}(\mathrm{y}) \times (\mathrm{xD}(\mathrm{y}) - \mathrm{xS}(\mathrm{y},\mathrm{t})))/(\mathrm{tanF} - \mathrm{tanB}(\mathrm{y})) \tag{11.3}$$

and ΔF(y,t) is the foreshore width and zS the shoreline elevation (2 m) (Fig. 11.2).

The comparison of the above profile model and the surveyed profiles is exemplified in Fig. 11.4, using winter and summer profiles at two Norte beach sites. Both profiles extend between the dune foot and the 0 m contour. The profile model seems to be a good estimator of the real profile area even when the beach presents a multiple berms system (Fig. 11.4).

Using the profile model and the video-derived shoreline, beach volume was estimated, at a monthly basis, from December 2008 to May 2012. Total volume estimation, for the investigated area (between −6350 and −5210 coordinates, Y ETRS89-PTM06), was based on trapezoidal numerical integration with a longshore discretization of 5 m.

Normalized beach volume (*VD* (*y*, *t*)) was computed at each transect (y) by:

$$VD(y,t) = V(y,t) - \overline{V(y)} \tag{11.4}$$

where *V*(*y*, *t*) is the volume estimated for each alongshore position (*y*) at a given time (*t*) and $\overline{V(y)}$ the average volume for the y alongshore position.

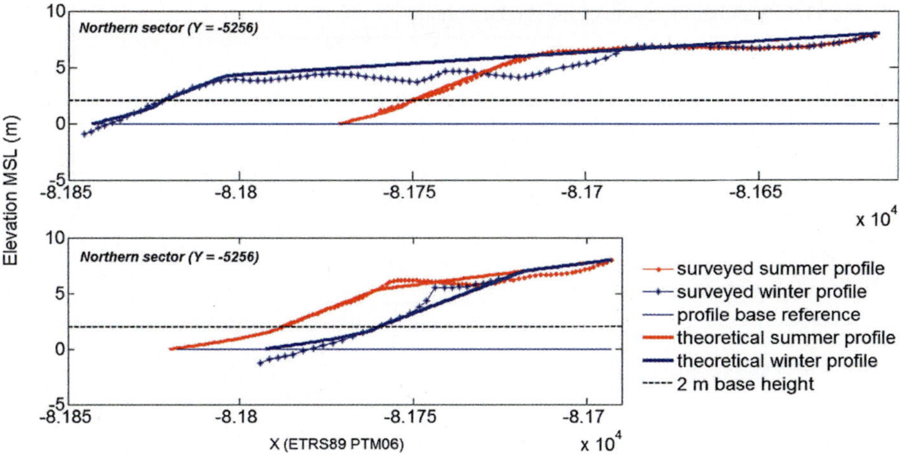

Fig. 11.4 Comparison between theoretical and surveyed profiles at two different locations at Norte beach (*top* and *bottom*)

Table 11.2 Digital terrain models (DTMs) used in volume validation

Date	DTM description	Scale/ resolution	Credits
2001	Aerial photogrammetric-derived	1:2000	INAG – Instituto da Água
2008	Aerial photogrammetric-derived	2 m	IGP – Instituto da Água
2011 (May)	RTK-DGPS topographic survey-derived	5 m	Beach to Canyon Project
2011 (Jun–Dec)	Topo-bathymetric LiDAR-derived	2 m	IGP – Instituto Geográfico Português

Model Validation

The beach volume model validation was performed by comparing the results with the ones obtained from methods that allows complete beach coverage: airborne LiDAR (Light Detection And Ranging) data, four-wheel all-terrain vehicle RTK-DGPS topographic surveys and aerial photogrammetry-derived digital terrain models (DTMs) (Table 11.2).

Due to some inconsistencies found in LiDAR data, particularly in the area where the bathymetric and terrestrial LiDAR data was merged, beach volumes were only estimated up to the 2 m height reference. The same height reference was adopted for the May 2011 DTM because the survey did not reached the 0 m (MSL). Beach volume was estimated, using ArcMap© surface volume tool, for the four DTMs, above 2 m height and above the 0 m height plane for the DTMs of 2001 and 2008 (Table 11.3).

The comparison between model and measured volumes show a maximum overestimation lower than 7 % and an underestimation that does not exceed 4 %. These results, which also depend on the quality of the measured data, validate the volume model used in the present work.

Table 11.3 Validation results of the model volume

MDT	Reference height (MLS)	Volume (m^3) Modeled	Measured	Difference (%)	Length of beach compared (m)
2001	0 m	11.99×10^5	1.12×10^5	6.67	1,140
	2 m	7.11×10^5	7.36×10^5	−3.32	
2008	0 m	11.56×10^5	10.88×10^5	6.29	
	2 m	7.02×10^5	7.21×10^5	−2.67	
2011 (May)	2 m	6.01×10^5	5.66×10^5	6.10	1,010
2011 (Jun–Dec)	2 m	6.33×10^5	6.36×10^5	−0.47	1,140

11.3.2.3 Wave Forcing

The characterization of the wave regime during the monitoring period was performed based on offshore wave data acquired by MONICAN01, an oceanographic buoy deployed by the Portuguese Hydrographic Institute (MONICAN Project – http://monican.hidrografico.pt/) at c.a. 2,000 m depth (11.1). The gaps in the wave data record (about 8 % of the records) were filled with NOAA (National Oceanic and Atmospheric Administration) hindcast for the nearest node, and wave time series was resample to meet the NOAA resolution (3 h interval between samples). In result, a continuous offshore wave time series of the main wave statistics (Hs – Significant wave height; Tp- Peak period and Dir – wave direction), at a 3 h interval was obtained, for the monitoring period. An improved description of the comparison between hindcast and buoy data can be seen in Silva et al. (2013), whereas the root mean square error is 0.39 m, 1.82 s and 23° for Hs, Tp and Dir, respectively.

The characterization of high energetic events during the monitoring period, was performed through the description of storms events (defined by a significant wave height threshold of 5 m – if two consecutive storm occurrences were spaced less than 24 h they were merged into a single storm). Total storm wave energy (E) was estimated using Dolan and Davis (1992) approximation:

$$E = \int_{t1}^{t2} Hs^2 dt, Hs > H_{threshold} \tag{11.5}$$

where t_1 and t_2 are the start and end times of the storm event and $H_{threshold}$ the significant wave height threshold (5 m).

11.3.3 Results

11.3.3.1 Wave Forcing

The reconstructed wave record extends from 1st December 2008 to 1st May 2012, and can be characterize by mean significant wave height of 2.36 m, mean peak

Table 11.4 Offshore wave parameters during the monitoring period (overall), and discretization of winter and summer conditions

	Hs (m)		Tp (s)		Dir (°)	
	Mean	Maximum	Mean	Maximum	Mean	Mean power
Annual	2.36	10.25	11.05	21.90	319	316
Winter (DJF)	3.00	9.10	12.59	21.30	311	310
Summer (JJA)	1.77	4.51	8.76	14.40	328	334

period of 11.05 s and mean wave direction of 319°, for the overall wave series (Table 11.4 and Fig. 11.5a). The seasonality of wave regime was also analyzed by the individualization of winter months (December, January and February; DJF) and summer months (June, July and August; JJA). Winter is characterized by 3.00 m height waves, mean peak period of near 13 s and waves approaching from 310°, although a wider direction spreading is visible at those months (see Fig. 11.5a). Summer months have mean significant wave height of 1.77 m, mean peak period around 9 s and mean wave direction of 334°, with a narrower direction spreading (Table 11.4 and Fig. 11.5a).

The offshore wave conditions synoptically related to each beach volume estimation were also investigated through the directional distribution of Hs (Fig. 11.5b); each plot includes the waves that occurred in the time interval between two consecutive dates of volume estimation (the represented date refers to the end of the interval).

These data depicts the seasonal wave energy modulation: summer is dominated by low energetic north-northwest waves with small directional spreading, while in winter the higher energetic waves have a much wide directional spreading. Amongst these directional wave plots some events stand out due to its high waves and wide directional spreading: 12-Fev-09, 19-Fev-10, 18-Mar-10, 08-Jan-11 and 15-Fev-11.

A total of 44 storm events were recognized over the 3.5 year of wave data, the average parameters of the storm conditions is expressed in Table 11.5. The average duration of the storms is slightly longer than a day (1.37 days) with a mean wave direction of 310°.

11.3.3.2 Beach Morphodynamics

During the monitoring period, total beach volume (above 0 m MSL) ranged from 0.88×10^6 to 1.22×10^6 m^3; the amplitude of volume variation was circa 3.4 10^5 m^3 which represents about 27 % of the total volume above mean sea level.

Minima values were found to occur in late winter months, Feb/09, Mar/10 and Feb/11, while annual maxima occurred in summer or early autumn (Aug/10 and Sep/11) (Fig. 11.6). Superimpose to this behavior a strong interannual signal is also present: for example in the Nov/09 beach volume is relatively low while in subsequent years the volume at the same month reaches a value closed to the annual maxima.

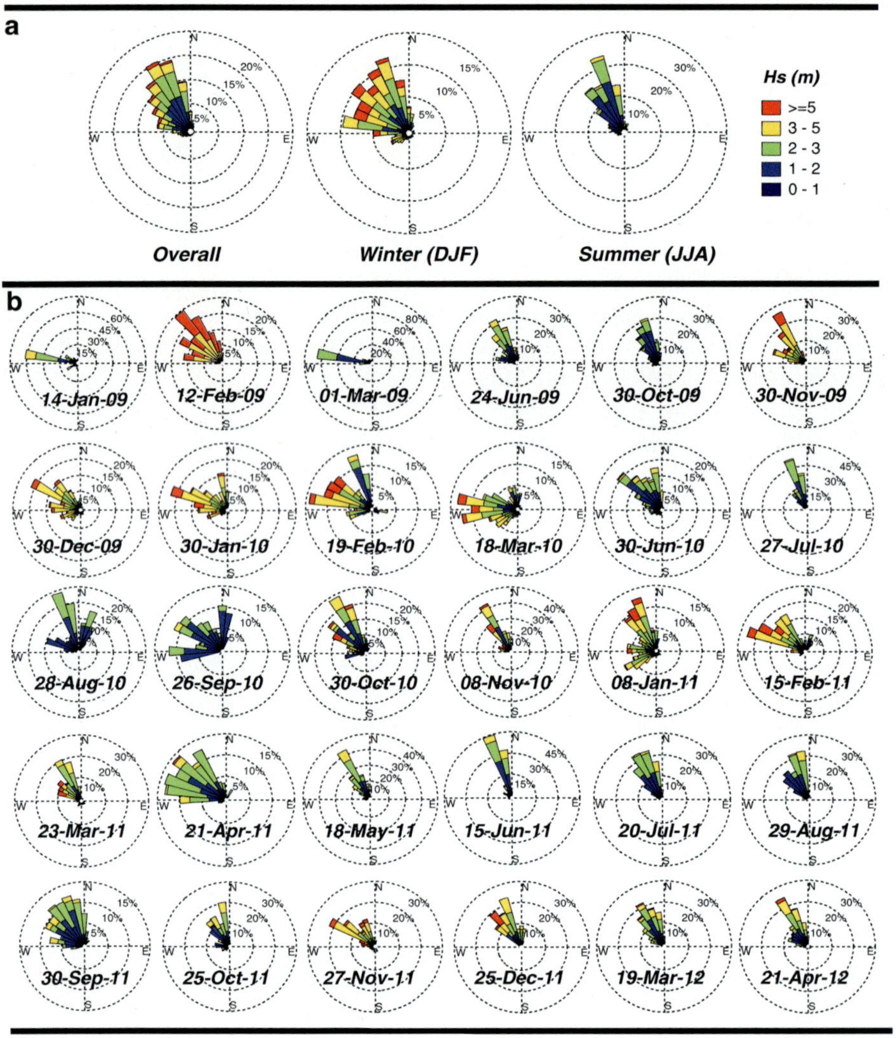

Fig. 11.5 Offshore wave conditions during the overall monitoring period, winter and summer months (**a**) and offshore wave conditions between each beach volume estimation (**b**) (The dates refer to the date of the shoreline from which the volume was estimated)

Table 11.5 Average characteristics of the 44 identified storm events

Hs (m)		Tp (s)		Dir (°)	Duration (days)	Energy (m^2h)	
Mean	Maximum	Mean	Maximum	Mean	Mean	Mean	Maximum
5.67	6.54	14.90	19.32	310	1.37	1,204	6,201

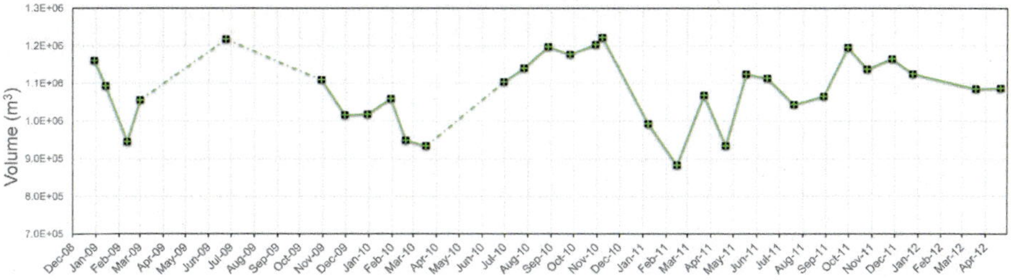

Fig. 11.6 Total beach volume time series (*dashed lines* represent monitoring interruptions longer than 2 months)

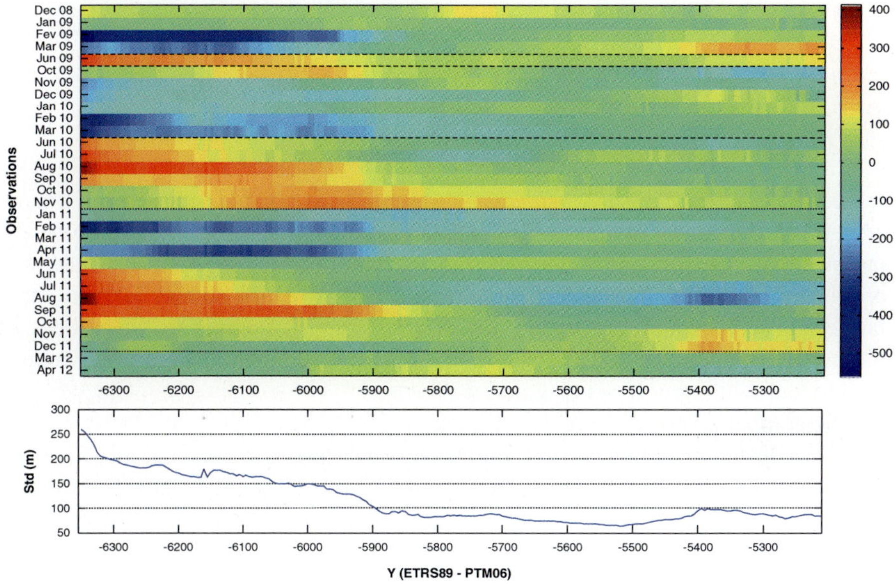

Fig. 11.7 Time stack of alongshore volume deviation (m^3) (*top*) and standard deviation (Std) of alongshore beach volume (*bottom*). *Long dashed lines* represent interruptions longer than 2 months while *short dashed lines* represent minor (less than 2 months) interruptions

Norte beach volume variations also show strong spatial gradients with a complex spatio-temporal pattern which are represented in Fig. 11.7 as time stack of volume deviations. The analysis of the alongshore standard deviation of the beach volume considering a cut off value of 100 m (Fig. 11.7 – Bottom) allows to differentiate the beach into a, more variable, southern sector and a northern sector that is more stable.

The southern sector, adjacent to Nazaré headland ($\approx -6{,}300$ m coordinate), displays a clear seasonal signal similar to the one found for the entire beach: the larger positive deviations are found in August while the minima occurred in February and March. Volume variations at the southern sector are clearly affected

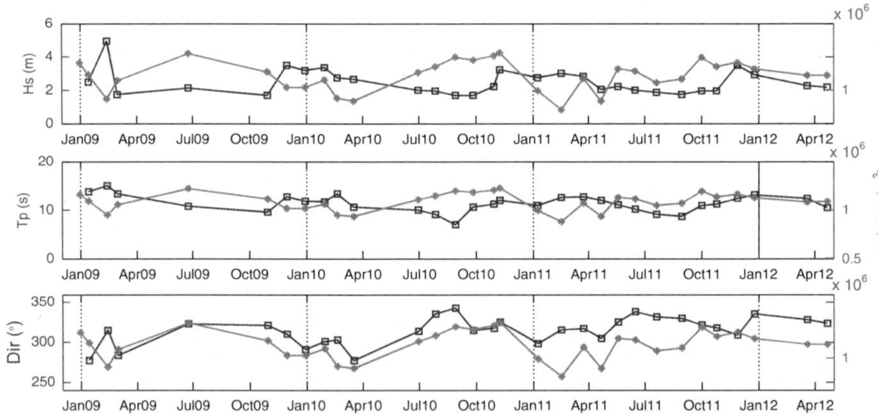

Fig. 11.8 Time series of total beach volume (*green*) overlaid with offshore significant wave height (Hs), peak period (Tp) and wave direction (Dir) (color figure online)

by the presence of the Nazaré headland which acts as a gated sedimentary boundary. At this sector, the morphological responses tend to have a larger magnitude and lag ahead northern changes. In fact, the seasonal signal progressively fades out northward. Additionally, the late summer maxima reveals a spatial delayed phase lag so that at −5,900 m coordinate occurs in autumn (September to November). At the northern sector the beach volume variability is dominated by the interannual component while the seasonal signal weakens.

The comparison of the total beach volume time series with the offshore incident wave conditions is represented in Fig. 11.8; weak negative correlation was found for Hs and TP ($r = -0.38$ in both cases) and a positive, slightly higher, correlation for the wave Dir ($r = 0.47$). When considering the volume estimated at each sector (Figs. 11.9 and 11.10), the correlation improved at the southern sector ($r = -0.60$, $r = -0.69$ and $r = 0.57$ for Hs, Tp and Dir, respectively) and worseness at the northern one ($r = 0.17$, $r = 0.32$ and $r = 0.06$ for Hs, Tp and Dir, respectively). When average wave conditions between volume estimations are considered is possible to get further insights on the North beach morphodynamics response. For example, minimum beach volumes, observed at Feb/2009, Feb/10, Mar/10 and Feb/11, coincide with the more energetic incident wave conditions (Fig. 11.5). In contrast, the beach is wider in later summer/early autumn in response to northwest low amplitude waves (for example the period immediately before 28-Aug-10 or 30-Sep-11).

These results support the existence of a relation between forcing and morphologic response, more evident, in the southern sector of the beach. In this sector, beach volume increases in relation with the low energetic, short period, northwest summer waves as they promote southward longshore drift and sedimentary accumulation updrift of the Nazaré headland. On the other hand, the northern sector

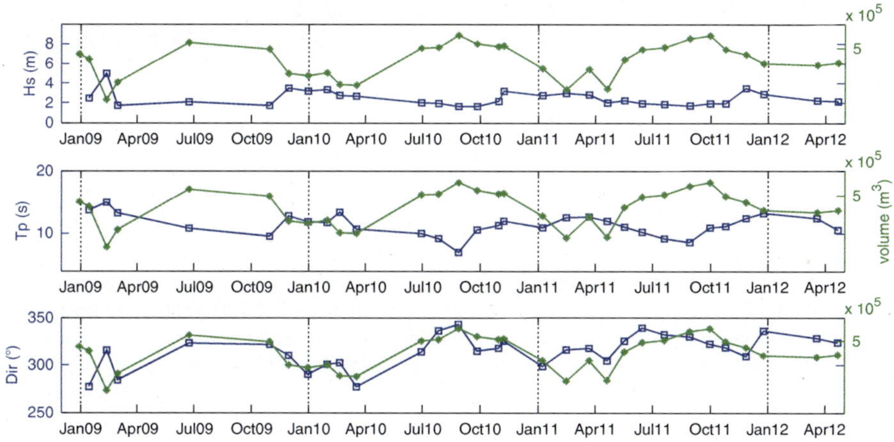

Fig. 11.9 Time series of beach volume (*green*), estimated for the Norte beach southern sector, overlaid with offshore significant wave height (Hs), peak period (Tp) and wave direction (Dir) (color figure online)

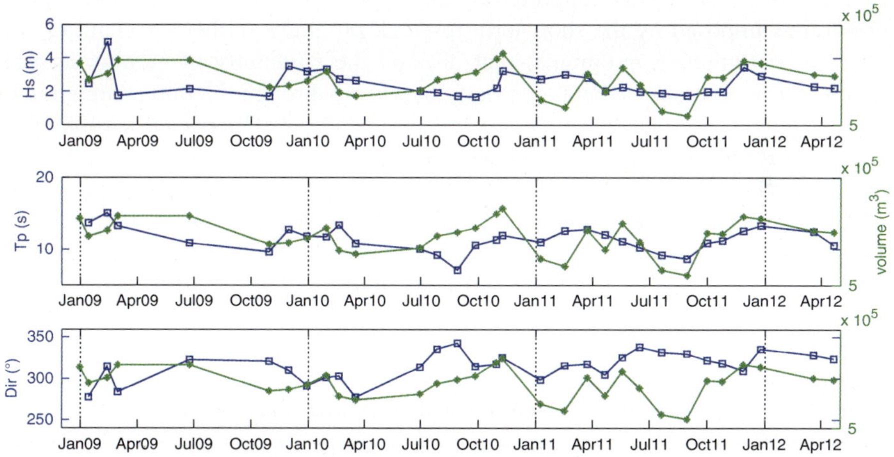

Fig. 11.10 Time series of beach volume (*green*), estimated for the Norte beach northern sector, overlaid with offshore significant wave height (Hs), peak period (Tp) and wave direction (Dir) (color figure online)

exhibits a more stable behavior that is not linearly related to the wave regime. Low frequency phenomena associated to interannual variability and short-term high energetic storm events may play a higher relative influence in this sector of the beach (Fig. 11.11).

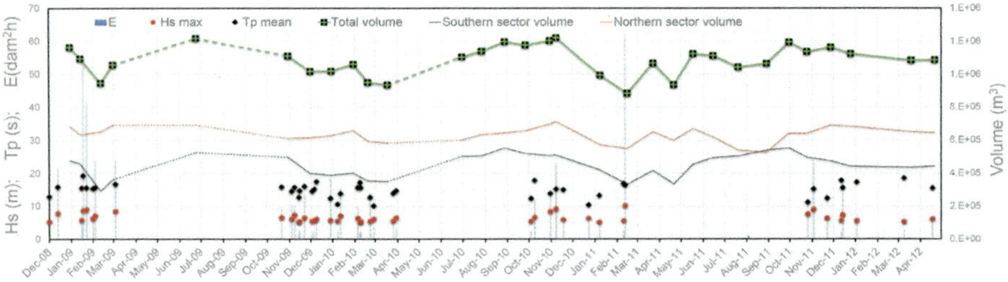

Fig. 11.11 Norte beach volume time series (total, southern sector and northern sector symbolized by *green*, *black* and *red lines* respectively) and characteristics of storm events (E – Energy, Hs maximum and mean Tp represented by *blue lines*, *red circles* and *black diamonds*, respectively) (*dashed lines* represent monitoring interruptions longer than 2 months) (color figure online)

11.4 Conclusions

This work summarizes recent developments on the use of video systems in the understanding of yearly to decadal beach morphological changes. While it is unanimously recognized that these systems have a huge potential, represented for example by the large number of existing systems, the number of long-term studies is still scarce (and mostly supported by the Argus system). This may be related to constraints imposed by the short-term research programs (either in scientific projects or post-graduate programs) that difficult the continuity of monitoring programs. These limitations can be partially overcome through the implementation of light weighted systems (in what concerns set-up and maintenance costs) combined with simple data exploitation techniques, as exemplified in the case study presented in this work.

In this study the morphodynamics of the Norte beach, Nazaré, Portugal was investigated coupling the wave data acquired by an offshore buoy with beach morphological data derived from video monitoring. The results, encompassing 3 and a 1/2 years of data, enabled to acknowledge the spatio-temporal dependence of the beach evolution in response to incident waves at a large range of temporal scales (storm to interannual). This study corroborate previous studies findings which highlight the importance of antecedent morphology in storm response and the connection between longshore and cross-shore processes not only at a seasonal scale but also at an interannual scale.

Acknowledgments The authors acknowledge the funding of the PhD grant by FCT (Portuguese Foundation for Science and Technology, ref. SFRH/BD/41762/2007) and the Portuguese Navy for the support in allowing the installation of COSMOS video monitoring system in the Nazaré lighthouse. This work was done in the framework of the PTDC/MAR/114674/2009 program (project B2C – Beach to Canyon Head Sedimentary Processes), financed by FCT.

References

Aarninkhof SGJ (2003) Nearshore bathymetry derived from video imagery. Delft University
Alexander P, Holman RA (2004) Quantification of nearshore morphology based on video imaging. Mar Geol 208(1):101–111. doi:10.1016/j.margeo.2004.04.017
Almar R, Coco G, Bryan KR, Huntley DA, Short AD, Senechal N (2008) Video observations of beach cusp morphodynamics. Mar Geol 254:216–223. doi:10.1016/j.margeo.2008.05.008
Armaroli C, Ciavola P (2011) Dynamics of a nearshore bar system in the northern Adriatic: a video-based morphological classification. Geomorphology 126(1–2):201–216. doi:10.1016/j.geomorph.2010.11.004
Cascalho J, Bosnic I, Taborda R, Ribeiro M, Lira C, Carapuço M (2012) Beach sediment grain size variability based on image analysis. In: Proceedings 2as Jornadas de Engenharia Hidrográfica, Lisboa, pp 287–290
de Alegria-Arzaburu AR, Masselink G (2010) Storm response and beach rotation on a gravel beach, Slapton Sands, U.K. Mar Geol 278(1–4):77–99. doi:10.1016/j.margeo.2010.09.004
De Vriend H, Capobianco M (1993) Approaches to long-term modelling of coastal morphology: a review. Coastal. Retrieved from http://www.sciencedirect.com/science/article/pii/037838399390051 9
Dean RG (1991) Equilibrium beach profiles: characteristics and applications. J Coast Res 7(1):53–84
Dodet G, Bertin X, Taborda R (2010) Wave climate variability in the North-East Atlantic Ocean over the last six decades. Ocean Model 31(3–4):120–131. doi:10.1016/j.ocemod.2009.10.010
Foster RA (2012) Shoreline variation and beach rotation of Pauanui beach. University of Waikato
Guillén J, García-Olivares A, Ojeda E, Osorio A, Chic O, González R (2008) Long-term quantification of beach users using video monitoring. J Coast Res 246:1612–1619. doi:10.2112/07-0886.1
Harley MD, Turner IL, Short AD, Ranasinghe R (2011) Assessment and integration of conventional, RTK-GPS and image-derived beach survey methods for daily to decadal coastal monitoring. Coast Eng 58(2):194–205. doi:10.1016/j.coastaleng.2010.09.006
Haxel J, Holman RA (2004) The sediment response of a dissipative beach to variations in wave climate. Mar Geol 206(1–4):73–99. doi:10.1016/j.margeo.2004.02.005
Heikkila J, Silvén O (1997) A four-step camera calibration procedure with implicit image correction. In: IEEE computer society conference on Computer Vision and Pattern Recognition (CVPR'97), San Ruan, Puerto Rico, pp 1106–1112
Holland KT (1998) Beach cusp formation and spacings at Duck, USA. Cont Shelf Res 18 (10):1081–1098. doi:10.1016/S0278-4343(98)00024-7
Holland KT, Holman RA, Lippmann TC (1997) Practical use of video imagery in nearshore oceanographic field studies. IEEE J Ocean Eng 22(1):81–92
Holman RA, Stanley J (2007) The history and technical capabilities of Argus. Coast Eng 54 (6):477–491
Holman RA, Sallenger J, Lippmann TC, Haines JW (1993) The application of video image processing to the study of nearshore processes. Oceanography 6(3):78–95
Konicki K, Holman RA (2000) The statistics and kinematics of transverse sand bars on an open coast. Mar Geol 169(1–2):69–101. doi:10.1016/S0025-3227(00)00057-8
Kroon A, Larson M, Möller I, Yokoki H, Rozynski G, Cox J, Larroude P (2008) Statistical analysis of coastal morphological data sets over seasonal to decadal time scales. Coast Eng 55(7–8):581–600. doi:10.1016/j.coastaleng.2007.11.006
Larson M, Kraus N (1994) Temporal and spatial scales of beach profile change, Duck, North Carolina. Mar Geol 117:75–94, Retrieved from http://www.sciencedirect.com/science/article/pii/0025322794900078
Larson M, Capobianco M, Jansen H (2003) Analysis and modeling of field data on coastal morphological evolution over yearly and decadal time scales. Part 1: Background and linear techniques. J Coast Res 19:760–775. Retrieved from http://www.jstor.org/stable/10.2307/ 374

Liu H, Arii M, Sato S, Tajima Y (2012) Long-term nearshore bathymetry evolution from video imagery: a case study in the Miyazaki coast. Coast Eng Proc. Retrieved from http://journals.tdl.org/icce/index.php/icce/article/view/6496

Masselink G, Pattiaratchi CB (2001) Seasonal changes in beach morphology along the sheltered coastline of Perth, Western Australia. Mar Geol 172(3–4):243–263. doi:10.1016/S0025-3227(00)00128-6

Ojeda E, Guillén J, Ribas F (2011) Dynamics of single-barred embayed beaches. Mar Geol 280(1–4):76–90. doi:10.1016/j.margeo.2010.12.002

Quartel S, Kroon A, Ruessink BG (2008) Seasonal accretion and erosion patterns of a microtidal sandy beach. Mar Geol 250(1–2):19–33. doi:10.1016/j.margeo.2007.11.003

Ranasinghe R, Symonds G, Black K, Holman RA (2004) Morphodynamics of intermediate beaches: a video imaging and numerical modelling study. Coast Eng 51 (7):629–655. doi:10.1016/j.coastaleng.2004.07.018

Siegle E, Huntley DA, Davidson MA (2007) Coupling video imaging and numerical modelling for the study of inlet morphodynamics. Mar Geol 236(3–4):143–163. doi:10.1016/j.margeo.2006.10.022

Silva AN, Taborda R, Antunes C, Catalão J, Duarte J (2013) Understanding the coastal variability at Norte beach, Portugal. J Coast Res SI 65:2173–2178. doi:10.2112/SI65-367.1

Smit MW, Aarninkhof SGJ, Wijnberg KM, González M, Kingston KS, Southgate HN, Medina R (2007) The role of video imagery in predicting daily to monthly coastal evolution. Coast Eng 54(6–7):539–553. doi:10.1016/j.coastaleng.2007.01.009

Southgate H, Wijnberg K, Larson M, Capobianco M, Jansen H (2003) Analysis of field data of coastal morphological evolution over yearly and decadal timescales. Part 2: Non-linear techniques. J Coast Res 19:776–789. Retrieved from http://www.jstor.org/stable/10.2307/4299220

Taborda R, Silva A (2012) COSMOS: a lightweight coastal video monitoring system. Comput Geosci 49:248–255. doi:10.1016/j.cageo.2012.07.013

Turner IL, Whyte D, Ruessink BG, Ranasinghe R (2007) Observations of rip spacing, persistence and mobility at a long, straight coastline. Mar Geol 236:209–221. doi:10.1029/2005/JC002965

Van Enckevort IMJ, Ruessink B (2003a) Video observations of nearshore bar behaviour. Part 1: alongshore uniform variability. Cont Shelf Res 23(5):501–512. doi:10.1016/S0278-4343(02)00234-0

Van Enckevort IMJ, Ruessink BG (2003b) Video observations of nearshore bar behaviour. Part 2: alongshore non-uniform variability. Cont Shelf Res 23(5):513–532. doi:10.1016/S0278-4343(02)00235-2

Wolf P, Dewitt B (2000) Elements of photogrammetry with applications in GIS, 3rd edn. McGraw-Hill Book Company, Boston, 607 pp

Wright L, Short A (1984) Morphodynamic variability of surf zones and beaches: a synthesis. Mar Geol 56:93–118, Retrieved from http://www.sciencedirect.com/science/article/pii/0025322784900082

Wright L, Short A, Green M (1985) Short-term changes in the morphodynamic states of beaches and surf zones: an empirical predictive model. Mar Geol 62:339–364, Retrieved from http://www.sciencedirect.com/science/article/pii/0025322785901239

Chapter 12
Application of Advanced Remote Sensing Techniques to Improve Modeling Estuary Water Quality

A.K.M. Azad Hossain, Yafei Jia, Xiaobo Chao, and Mustafa Altinakar

Abstract Estuaries, the interface between terrestrial and coastal waters are an important component of complex and dynamic coastal watersheds. They are usually characterized by abrupt chemical gradients and complex dynamics, which can result in major transformations in the amount, chemical nature and timing of the flux of material along these river–sea transition zones. The ecological functioning of these areas is considered to be of major concern, as estuaries offer the last opportunity to manage water quality problems before they become uncontrollable in the coastal waters.

Numerical models can provide hydrodynamically computed water quality data to study estuary water quality, but they have problems with initializations, boundary conditions, calibration, and validation. Another way is to use remote sensing technology, but they provide only surface observations and there are challenges related to cloud coverage, ground truthing, and variable spatial and temporal resolution.

Although both methods have weaknesses when used together, they can become a powerful tool to study water quality in estuary. This has been demonstrated through recent application of this capability to study water quality problem in Lake Pontchartrain. This study evaluated the use of Landsat 5 TM multispectral imagery to generate spatially distributed water quality data for use in CCHE2D Water Quality model to improve its performance in this estuary to simulate sediment transport, predict algal bloom and monitor salinity after the Bonnet Carré Spillway opening event in 1997. The outcome of this research clearly indicates that the application of remote sensing techniques for estuarine water quality study can be advanced by integrating them with numerical water quality models.

A.K.M.A. Hossain (✉) • Y. Jia • X. Chao • M. Altinakar
National Center for Computational Hydroscience and Engineering (NCCHE), The University of Mississippi, Oxford, MS, USA
e-mail: ahossain@ncche.olemiss.edu

C.W. Finkl and C. Makowski (eds.), *Remote Sensing and Modeling: Advances in Coastal and Marine Resources*, Coastal Research Library 9,
DOI 10.1007/978-3-319-06326-3_12,

12.1 Background and Introduction

The impacts and implications of water quality perturbations related to consistent human population growth and industrialization, pressures continue to increase in coastal and inland areas (Caraco 1995; National Research Council 2000; World Resources Institute 2003). As a result the need for effective water quality monitoring has become very critical for water resource management programs. Lacking accurate, intensive and long-term data acquisition capability, the state of the world's water resources cannot be adequately assessed, effective preservation and remediation programs cannot be run, and program success cannot be properly evaluated (Glasgow et al. 2004).

Typically, estuaries are semi-enclosed areas along the coastal shoreline where fresh water enters the oceans, at the end of its journey over the land surface via rivers, lakes and/or wetlands. Estuaries are biologically very productive and are also vital to commerce, transportation, and recreation activities. Many estuaries, which are located close to growing urban areas, industrial zones, and/or commercial agricultural activities, are the downstream recipients of sediments, nutrients, and other discharges generated from inland sources (Hu et al. 2004). Some of the largest estuaries in the U.S. have experienced significant ecosystem changes over the past 40 years (Hu et al. 2004). It is very important to understand the impacts and implications of these stresses on the ecological functions of estuaries, and how various natural and anthropogenic factors affect the health of these areas. However, it has been very challenging to develop methods and strategies that facilitate monitoring and assessment of estuaries, particularly over synoptic scales (Hu et al. 2004).

Availability of reliable measured data are invaluable for helping decision makers to identify the scope of the environmental problems. Observations are the only way to know the real characteristics of the ecosystem. For decades, field measurements for water quality evaluation have depended upon costly, time consuming and labor-intensive on-site sampling and data collection, and transportation to land-based or shipboard laboratories for evaluation. These research and monitoring efforts are episodically intensive, however, they generally have been too limited on temporal and spatial scales to adequately address factors influencing development of events such as harmful algal blooms, oxygen depletion (Hypoxia), fish kills, and contamination of shellfish beds by enteric bacteria. In addition, data quality may be compromised due to inadequate quality-control and quality-assurance protocols such as extended holding times before conducting the analysis and use of non-standardized methodologies. On the other hand, data are frequently susceptible to recording and geo-referencing errors during data acquisition and transcription (Pettinger 1971; Teillet et al. 2002).

Over the past few decades, there have been significant advances in the development of space-borne systems/tools that offer rapid, repeated, and concurrent synoptic assessment of environmental parameters in both aquatic and terrestrial environments (Hu et al. 2004). For more than 25 years, satellite remote sensing technology has represented an opportunity for synoptic and multitemporal viewing of water quality. Estimation of water quality parameters (WQPs) from different satellite imagery can be achieved by three different approaches (Cracknell et al. 2001; Dekker et al. 1995). First, the empirical approach, which is based on the development of bivariate or multivariate regressions between remote sensing data and measured WQPs. Digital numbers (DN) or radiance values at the sensor, as well as their band combinations, are correlated with in situ measurements of different WQPs, usually acquired at near real time of the sensor overpass. A summary of empirical approaches for lakes can be found in Lindell et al. (1999). Second, the semi-empirical approach, which may be used when spectral characteristics of the parameters of interest are known. This knowledge is included in the statistical analysis by focusing on well-chosen spectral areas and appropriate spectral bands used as correlates. An example of a semi-empirical approach with different sensors is reported by Harma et al. (2001) over Finnish lakes. Third, the analytical approach, where WQPs are related to the bulk Inherent Optical Properties (IOPs) via the Specific Inherent Optical Properties (SIOPs). The IOPs of the water column are then related to the Apparent Optical Properties (AOPs) and hence to the Top of Atmosphere (TOA) radiance such as described by the radiative transfer theory (Mobley 1994; Vermote et al. 1997). The analytical method involves inverting all above relations (WQPs→IOPs→AOPs→TOA radiances) to determine the WQPs from remote sensing data. An example of such approach, using Landsat data over lakes, can be found in Dekker et al. (2001) for the total suspended matter retrieval.

However, due to frequent cloud coverage issue, variable spatial and temporal resolution, and dependency of field measurements it is not always possible to ensure generation of necessary water quality data over the entire water bodies at desired spatial and temporal resolution using only satellite observations. In addition, the measurements are always limited to surface waters and no prediction in water quality is also not possible with current remote sensing applications.

Numerical water quality modelling is the most reasonable option in terms of helping decision makers providing computed (and predicted) water quality data at desired spatial and temporal resolutions (in both horizontal and vertical directions) if appropriate field derived boundary conditions and calibration data (initial condition) are available. Observation is the only way to know the real characteristics of the ecosystem and to provide the basis for theoretical analysis and numerical modeling. Only after certain observations are made can theoretical analysis and numerical modeling help the understanding of hydrodynamic and water quality processes and provide reliable results for supporting decision making. These processes, in many cases, cannot be described well in mathematical models before they are measured in real water bodies.

Since obtaining large amount of data for validating flood, pollutant transport and sediment transport models are very difficult, time consuming and expensive, traditionally numerical models' simulation results are validated using limited in situ measurements. For example, Chao et al. (2008), used measured data from three stations in a 8 ha size oxbow lake (Deep Hollow Lake in MS). In case of flood event, model validation data from routine observation stations is not even possible to obtain for inundated areas outside channel. Initial condition data is also very critical for the realistic simulation of water quality by 2D/3D WQ models. It has been noticed that using only few available in-situ measurements it is not possible to generate acceptable initial condition data.

As satellite observations can provide water quality estimation for large extend it is expected that above described issues of numerical models can be resolved by remote sensing technology and the applications of numerical models in real cases significantly enhanced. Therefore, it is strongly believed that remote sensing applications for estuarine water quality modeling can be advanced by integrating them with numerical models. The integration can be done through generating different estuarine water quality data from satellite observations for initialization, calibration, and validation of numerical water quality models.

In recent years, several studies were conducted to evaluate the potential of Landsat 5 TM imagery to estimate spatio-temporal water quality data in the Lake Pontchartrain and use them to initialize, calibrate, and validate CCHE2D WQ model developed at NCCHE to study the impact of the Bonnet Carré Spillway (BCS) release events on the lake water quality. This chapter summarizes the case studies on mapping and monitoring suspended sediments, salinity and chlorophyll *a* concentration in the lake water after the BCS release event in 1997.

12.2 Case Study Site: Lake Pontchartrain

Lake Pontchartrain (Fig. 12.1) is a brackish estuary located in southeastern Louisiana. It is not a true lake but an estuary connected to the Gulf of Mexico via Rigolets strait, to Lake Borgne via Chef Menteur Pass, and to Lake Maurepas via

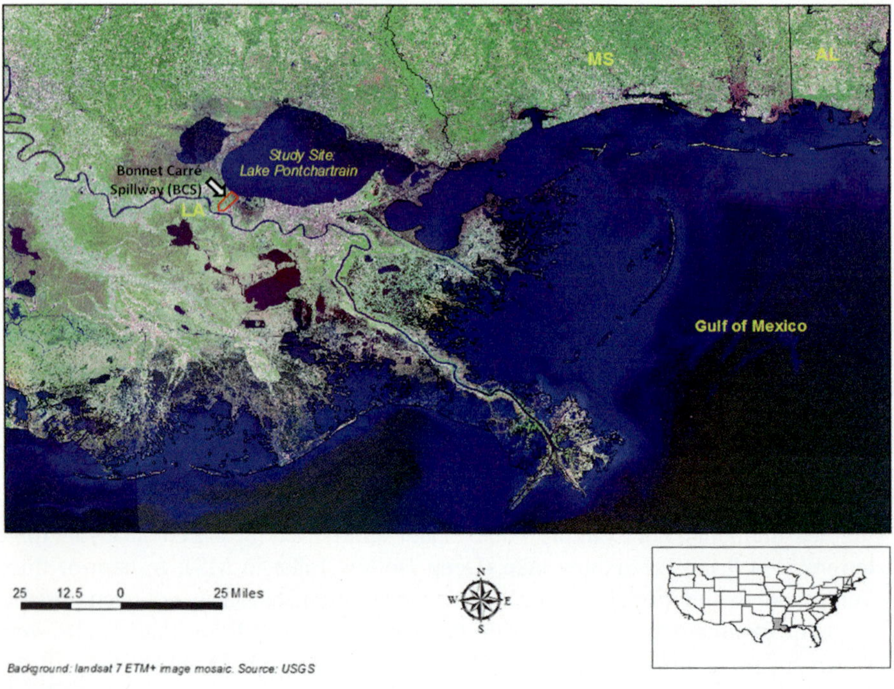

Fig. 12.1 Location and regional setup of the study site: Lake Pontchartrain

Pass Manchac. These lakes form one of the largest estuaries in the Gulf Coast region. It is the second-largest saltwater lake in the U.S. The lake covers an area of 1,630 km^2 with a mean depth of 4.0 m. It is an oval-shaped quasi-enclosed water body with the main east-west axis spanning 66 km, while the shorter north–south axis is about 40 km. It receives fresh water from a few rivers located on the north and northwest of the lake. The estuary drains the Pontchartrain Basin, an area of over 12,000 km^2 situated on the eastern side of the Mississippi River delta plain.

Lake Pontchartrain has served the surrounding communities for more than two centuries. The coastal zone of the lake and its basin has offered opportunities for fishing, swimming, boating, crabbing and other recreational activities. This lake basin is Louisiana's premier urban estuary and nearly one-third of the state population lives within this area. In response to the high flood stage of the Mississippi River and to protect the city of New Orleans, the Bonnet Carré Spillway (BCS) (Fig. 12.1) was built to divert Mississippi River flood waters to the Gulf of Mexico via Lake Pontchartrain.

The construction of the spillway was completed in 1931. It is located in St. Charles Parish, Louisiana, about 19 km west of New Orleans. The spillway consists of two basic components: a 2.4 km long control structure along the east bank of the Mississippi River and a 9.7 km floodway that transfers the diverted flood waters to the lake. The design capacity of the spillway is 7,080 m^3/s and opens when the Mississippi river water levels in New Orleans approached the flood stage of 5.2 m. It was first operated in 1937 and nine times thereafter (1945, 1950, 1973, 1975, 1979, 1983, 1997, 2008 and 2011) (USACE 2011; GEC 1998).

During the BCS opening, a large amount of fresh water and sediment discharged from the Mississippi River into Lake Pontchartrain and then into the Gulf of Mexico. The flow discharge over the spillway produces significant effects on the lake hydrodynamics. It also changes the distributions of salinity, nutrients and suspended sediment concentration (SSC) in the lake dramatically. During a flood releasing event, the fresh water dominates the whole lake and the lake salinity reduces significantly. A lot of sediment deposits into the lake or is transported into the Gulf of Mexico. The contaminated sediment from Mississippi River brings a lot of pollutants to the lake, and cause environmental problems (Dortch et al. 1998; Penland et al. 2002).

To evaluate the environmental impacts of a flood release event from BCS on lake ecosystems, it is important to understand the hydrodynamics as well as sediment transport and salinity distribution in the lake. Lake Pontchartrain is a large shallow lake and the water column is well mixed. In general, the water movements within the lake are primarily wind and tide-induced. However, during a flood releasing event, the flow discharge over the spillway produces significant effects on the lake hydrodynamics. In this study, a numerical model, CCHE2D integrated with remote sensing technology, was applied for simulating the flow circulations, salinity, sediment distributions, and algal biomass in Lake Pontchartrain during the 1997 BCS opening for flood release.

12.3 Description of Numerical Model Used (CCHE2D)

CCHE2D, developed at NCCHE, is a 2D depth-averaged model that can be used to simulate unsteady turbulent flows with irregular boundaries and free surfaces. It is a finite element model utilizing a special method based on the collocation approach called the "efficient element method". In order to improve the local mass conservation, mass fluxes through cells are computed using finite volume method. This model is based on the 2D Reynolds-averaged Navier-Stokes equations. By applying the Boussinesq approximation, turbulent stress can be simulated by the turbulent viscosity and time-averaged velocity. There are several turbulence closure schemes available within CCHE2D, including the parabolic eddy viscosity, mixing length, k–ε and nonlinear k–ε models. Detail technical description of this model can be found in Jia and Wang (1999), and Jia et al. (2002).

In general, circulation in Lake Pontchartrain is an extremely complicated system. It is affected by tide, wind, fresh water input, etc. The lake has a diurnal tide with a mean range of 11 cm. Higher salinity waters from the Gulf of Mexico can enter the lake through three narrow tidal passes: the Rigolets, Chef Menteur, and a man-made Inner Harbor Navigation Canal (IHNC). Freshwater can discharge into the lake through the Tchefuncte and Tangipahoa Rivers, the adjacent Lake Maurepas, and from other watersheds surrounding the lake. Therefore, the salinity distribution of the lake is governed by the saltwater fluxes from the three tidal passes, and the freshwater inputs. Based on the bathymetric data, the computational domain was descritized into an irregular structured mesh with 224×141 nodes using the NCCHE Mesh Generator (Zhang and Jia 2009), and shown in Fig. 12.2a.

As shown in Fig. 12.2a, there are two inlet boundaries located at the northwest of the lake, and three tidal boundaries located at the south and east of the lake. The flow discharges at Tchefuncte and Tangipahoa Rivers were set as two inlet boundary conditions, and the daily discharge data can be obtained from USGS. For the tidal boundaries, the hourly surface elevation data at the Rigolets Pass can be obtained from USGS. Due to the lack of measured elevation data at Chef Menteur Pass, the Rigolets data was used (McCorquodale et al. 2005). The flow discharges at Tchefuncte and Tangipahoa Rivers, and the water surface elevations at Rigolet, Chef Menteur and IHNC were set as boundary conditions. The wind speeds and directions at the New Orleans International Airport obtained from the National Climatic Data Center, National Oceanographic and Atmospheric Administration (NOAA), were used for model simulation. The period from March 1–31, 1998, was selected for model calibration. Figure 12.2b–d show the model calibration for water surface elevation and flow velocity for this period. In this period, the water movements within the lake are primarily wind and tide-induced.

The model was integrated with remote sensing technology to improve its prediction capability and make the simulation results more reliable. Satellite imagery and digital image processing techniques were used to estimate suspended sediments, chlorophyll-*a*, and salinity over the entire lake at 30 m grid spacing for use in model initialization and validation.

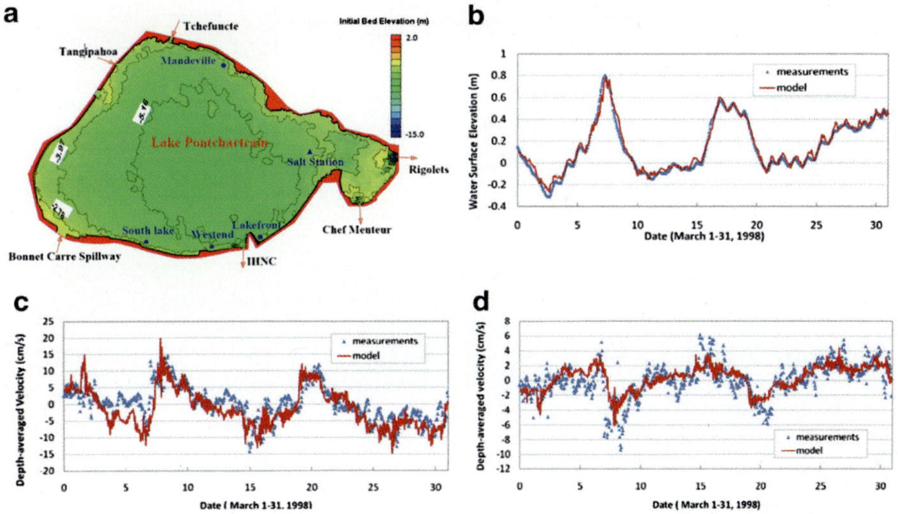

Fig. 12.2 (**a**) The bathymetry of Lake Pontchartrain, (**b**) Simulated and measured water surface elevations at the Mandeville Station, (**c**) Simulated and measured velocities in west-east direction at the South Lake, (**d**) Simulated and measured velocities in south-north direction at the South Lake

12.4 Satellite Data Used

The use of remote sensing techniques has been increasing exponentially in surface water quality studies. Earth observation satellites, such as Landsat, which repeatedly covers target estuaries at regular interval, provide information on solar radiation at various wavelengths reflected by surface water, can be correlated to water quality parameters. Thus remote sensing technology constitutes an alternative means of estimating water quality, and also offers significant advantages over ground sampling (Hellweger et al. 2004). The near-continuous spatial coverage of satellite imagery allows for synoptic estimates over large areas; the global coverage of satellites allows for the estimation of water quality in remote and inaccessible areas; and the relatively long record of archived imagery allows estimation of historical water quality, when no ground measurements can possibly be performed.

A time series of Landsat 5 Thematic Mapper (TM) visible and near infra red (VNIR) and shortwave infra red (SWIR) data acquired over Lake Pontchartrain were used in this study to estimate suspended sediment concentration (SSC), chlorophyll-*a* concentration and salinity in the lake surface water. Figure 12.3 shows obtained image time series in 4,3,2 false color composite. All data were obtained from the United States Geological Survey (USGS) Global Visualization Viewer (GLOVIS) (http://glovis.usgs.gov/) as Level 1 product. The imagery were radiometrically corrected and resampled for geometric correction and registration to geographic map projections. The georectified and atmospherically corrected DN imagery were processed to obtain the true surface reflectance values. The DN values were converted to radiance values from which path radiance values were

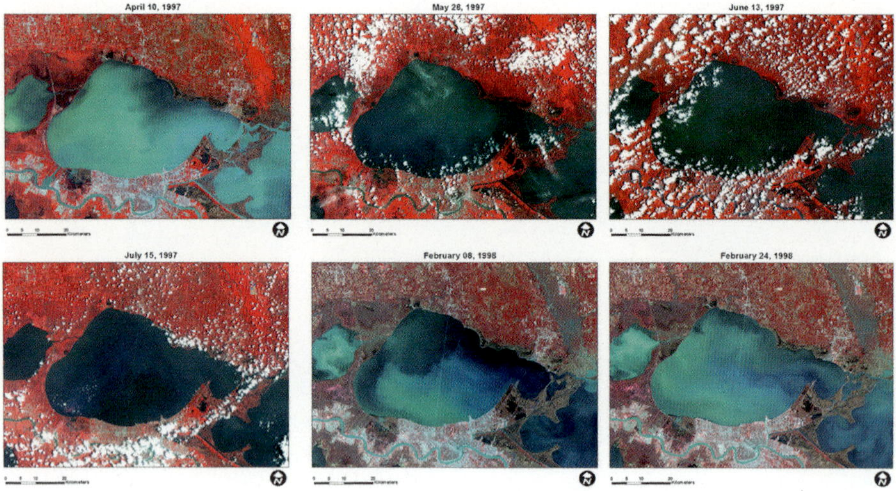

Fig. 12.3 The time series of Landsat 5 TM imagery used. The imagery are shown as 4, 3, 2 false color composite

Table 12.1 List of satellite imagery (Landsat 5 TM) used and their use

Date of image acquisition	Data type	Water quality parameter estimated
April 10, 1997	VNIR	Suspended sediments
February 08, 1998	VNIR + SWIR	Salinity
February 24, 1998	VNIR + SWIR	Salinity
May 28, 1997	VNIR	Chlorophyll-*a*
June 13, 1997	VNIR	Chlorophyll-*a*
July 15, 1997	VNIR	Chlorophyll-*a*

subtracted to calculate the true surface reflectance values. The articles that were consulted to do this calculation include Chander et al. (2009), Song et al. (2001), Chander and Markham (2003) and Teillet et al. (2001). Table 12.1 shows the list of the imagery obtained and their corresponding use for this study.

12.5 Modeling Sediment Transport

In 1997, the BCS was opened for a period of over 31 days of flood release from 3/17 to 4/18. The total volume of sediment-laden water that entered Lake Pontchartrain was approximately 1.18×10^{10} m^3, or twice the volume of the lake (Turner et al. 1999). The total amount of sediment entered the lake was about 9.1 million tons, more than ten times as much as the normal yearly sediment loads of the lake. The suspended sediment concentration (SSC) at the spillway gate was about 240 mg/l (Manheim and Hayes 2002). The calibrated CCHE2D model was applied to simulate the lake flow fields and corresponding sediment transport process during the BCS opening in 1997. The simulation results were validated using Landsat 5 TM imagery derived SSC estimation data over the entire lake as observed on April 10, 1997.

Suspended sediment concentration has been estimated and mapped successfully using remote sensing for the last three decades. Different approaches and algorithms were developed over time for SSC estimation/mapping using optical satellite data. The available techniques can be categorized in four general groups: (1) simple regression (correlation between single band and in-situ measurements) (e.g., Williams and Grabau 1973), (2) spectral unmixing techniques, (3) Band ratio technique using two and more bands (e.g., Lathrop 1992; Populus et al. 1995), and (4) multiple regressions (using multiple bands and in-situ measurements) (e.g., Binding et al. 2005).

Usually when SSC are high, the backscatter/reflectivity of water is high. There are three matters that dominate the reflectance of inland water, which are yellow substance, suspended sediment and phytoplankton. Yellow substance is a soluble matter, which has no scatter capability, but it has a strong absorption effect on short-wave bands that highly reduce the underwater downwelling irradiance. Therefore, when the absorption of water itself and yellow substance is small, actual suspended sediment information could be obtained (Wang et al. 2003).

Hossain et al. (2010) developed a numerical algorithm to estimate and map SSC in the surface water of river/lake environments based on the Normalized Difference Suspended Sediment Index (NDSSI) calculated from Landsat 7 Enhanced Thematic Mapper Plus (ETM+) imagery and near real time in situ measurements. It has been observed for Landsat TM/ETM imagery that Band 1 (blue band/~0.450–0.515 μm) and Band 4 (near-infrared/~0.750–0.900 μm) are most sensitive to water and water transparency (turbidity). Band 1 and Band 4 usually gives the highest and lowest reflectance values respectively for water. These characteristics have been observed for water with different levels of turbidity. NDSSI calculates the ratio between the reflectance variation from the blue and near infra-red bands and the total reflectance of these two bands (Eq. 12.1). It provides a normalized value ranges from −1 to +1 where higher values indicate the presence of more clear water and lower values indicate the presence of more turbid water or land. Hossain et al. (2010) correlated the NDSSI values (calculated from 16 Landsat 7 ETM+ imagery) to near real time in-situ measurements of SSC to estimate the SSC quantitatively in the Mississippi River. Equation 12.2 shows the obtained numerical relationship between NDSSI and measured SSC.

$$NDSSI = \frac{\rho_B - \rho_{NIR}}{\rho_B + \rho_{NIR}} \tag{12.1}$$

$$SSC = 18.69 \times NDSSI^{-3.399} \tag{12.2}$$

Where, $\rho_{B,}$ and $\rho_{NIR,}$ are the reflectance values of Landsat 5/7 TM/ETM+ Band 1, and Band 4 respectively.

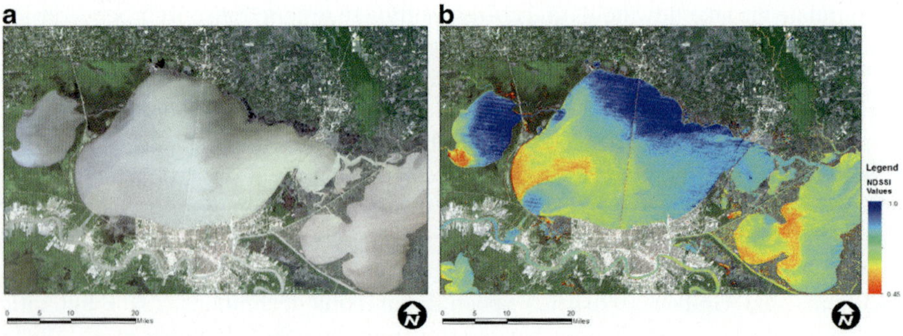

Fig. 12.4 (**a**) True color display of Landsat 5 TM imagery acquired over Lake Pontchartrain on April 10, 1997, (**b**) corresponding NDSSI imagery

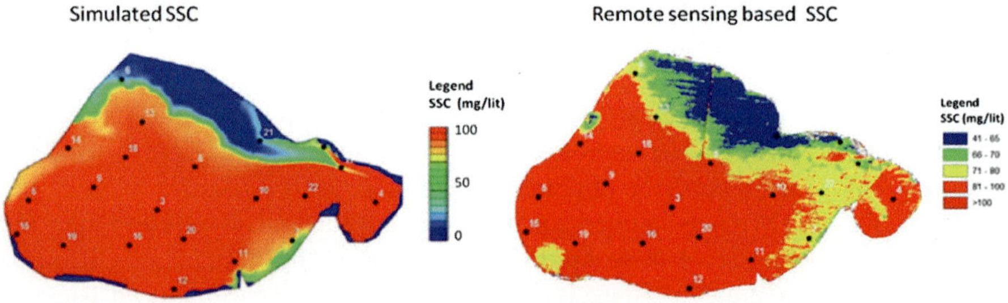

Fig. 12.5 Qualitative comparison (by visual inspection) between simulated SS concentration (by CCHE2D) and remote sensing derived SS concentration estimation (By NDSSI)

NDSSI (Eq. 12.1) was calculated using the Landsat 5 TM imagery acquired over Lake Pontchartrain on April 10, 1997 (Fig. 12.3) to map the relative variation of SSC in the lake surface water (Fig. 12.4) after the BCS opening event. The quantitative estimation of SSC in the lake water was obtained by applying Eq. 12.2 on the NDSSI image data. This data provides SSC estimation in the surface water for the entire lake at 30 m spatial resolution. The obtained SSC estimation was used to validate the numerically computed (simulated) SSC in Lake Pontchartrain by CCHE2D. Figure 12.5 shows the qualitative comparison (by visual inspection) between simulated SSC (by CCHE2D) and remote sensing derived SSC estimation (By NDSSI). Figure 12.6 shows the quantitative comparison between simulated SSC (by CCHE2D) and remote sensing derived SSC estimation (By NDSSI).

The simulated SSC were found in good agreement with satellite observed SSC estimation. The transport processes of SSC in the lake were reproduced by the numerical model. The simulated results and satellite imagery revealed that a large amount of sediment discharged into the lake, moved eastward along the south shore and gradually expanded northward, eventually affecting the entire lake after 1 month of diversion.

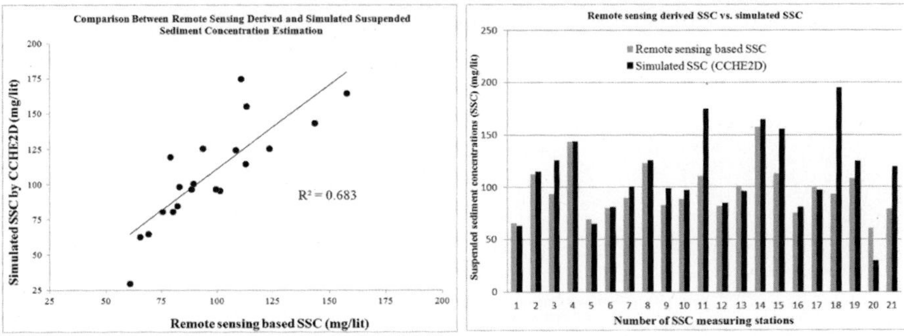

Fig. 12.6 Quantitative comparison between simulated SSC (by CCHE2D) and remote sensing derived SSC estimation (By NDSSI)

12.6 Modeling Salinity Distribution

A large amount of fresh water and sediment were discharged from the Mississippi River into Lake Pontchartrain for 1 month of diversion during the BCS opening for flood release event in 1997. Results obtained from numerical modeling show that the flow pattern during this period was dominated by current instead of circulation. The salinity in the lake decreased significantly since large amount of fresh water discharged into the lake. At Rigolets and Chef Menteur tidal inlets, the salinity reduced from 4.5 ppt to less than 1 ppt (Department of Natural Resources 1997). The water temperature of the lake was also decreased due to the colder water from the Mississippi River. These changes caused negative impacts on oyster beds and fishery nursery grounds in the lake. It was observed that in response to the dramatic changes of the habitat (salinity, temperature and water surface elevation in the lake), some shrimp species, particularly brown shrimp, adjusted their places to stay or migrated away entirely (Penland et al. 2002). Most likely it may take a long time for the fishery resources to recover from the flood release event. The decrease of salinity in the lake also caused negative impact on the growth of submersed aquatic vegetation (SAV).

The recovery of salinity in Lake Pontchartrain from the BCS opening event is governed by the limited water exchanges from three tidal passes (Rigolets, Chef Menteur and IHNC). Because tidal passes are narrow, the salinity recovery could be very slow. Therefore, to improve our understanding of the salinity recovery process for management purpose, the numerical model was applied to simulate the salinity distribution in Lake Pontchartrain during and after the BCS opening for flood release event.

A period from March 17, 1997, when the BCS started to release flood water, to the end of the year 1997, was selected for simulating the salinity distribution in Lake Pontchartrain. The hourly salinity boundary conditions were obtained from USGS measured data. After obtaining flow fields, the prediction of salinity distribution can be solved using the numerical model if it is initialized by well representative initial condition data. Remote sensing technology played a very important role in this case by providing satellite observed salinity data for model initialization.

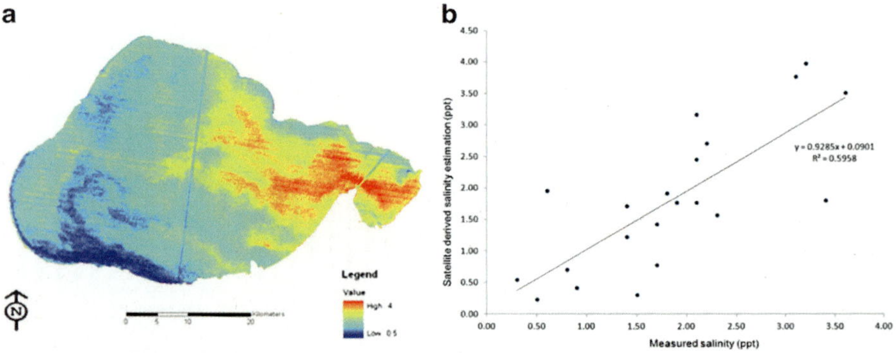

Fig. 12.7 (**a**) Salinity estimated in Lake Pontchartrain using Landsat 5 TM imagery acquired in February 8, 1998. (**b**) Comparison between satellite derived salinity estimation and measured salinity

Large number of studies were conducted to study salinity related estuarine water quality using satellite imagery and digital image processing techniques including the application of Landsat data (Khorram 1982; Baban 1997; Dewidar and Khedr 2001; Lavery et al. 1993). Wang and Xu (2008) recently developed a remote sensing based salinity prediction model to study salinity of Lake Pontchartrain in both normal and extreme conditions using a time series of eight Landsat TM imagery and near real-time salinity measurements in the lake water. Equation 12.3 shows the developed salinity estimation model by Wang and Xu (2008). The results showed that the model achieved a high power in prediction of the lake salinity with a R^2 value of 0.89 and RMSE of validation value of 0.27.

$$\begin{aligned} Salinity =\ & 4.6127 + 106.44 * TM1 + 152.92 * TM2 - 227.46 * TM3 \\ & + 120.47 * TM4 - 139.47 * TM5 \end{aligned} \tag{12.3}$$

Where, TM1, TM2, TM3, TM4, and TM5 represent the surface reflectance values recorded in band 1,2,3,4 and 5 of Landsat 5 TM imagery respectively.

Salinity of Lake Pontchartrain has been estimated for early February 1998 and used as initial condition data for CCHE2D to simulate salinity. Wang and Xu's (2008) numerical model for estuary salinity estimation (Eq. 12.3) was applied on the processed Landsat 5 TM imagery acquired over Lake Pontchartrain on February 08, 1998 to obtain the lake-wise distributed salinity estimation at 30 m spatial resolution. The Landsat 5 TM imagery derived salinity estimated using this method and near real time in situ measurements of salinity were compared and found very good correlation. Figure 12.7a shows the salinity estimated in Lake Pontchartrain using Landsat 5 TM imagery acquired in February 8, 1998. Figure 12.7b shows the comparison between satellite derived salinity estimation and measured salinity.

It would be ideal to initialize the model with the satellite data acquired on March 17, 1997, but it was not possible because there is no satellite imagery available for this date. However, since the pattern of regular salinity distribution of Lake Pontchartrain usually remains similar throughout the season for different years it

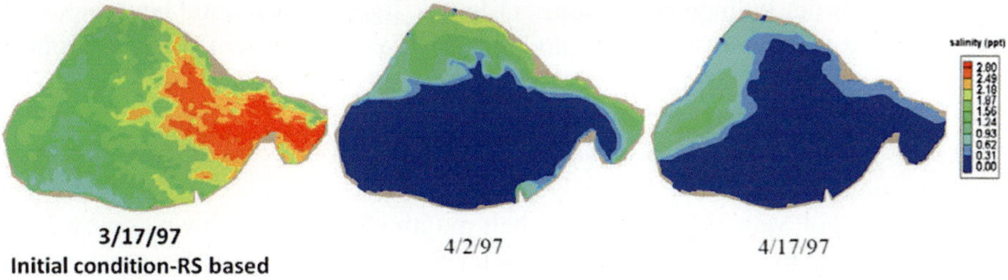

Fig. 12.8 CCHE2D simulated salinity distributions for different time periods

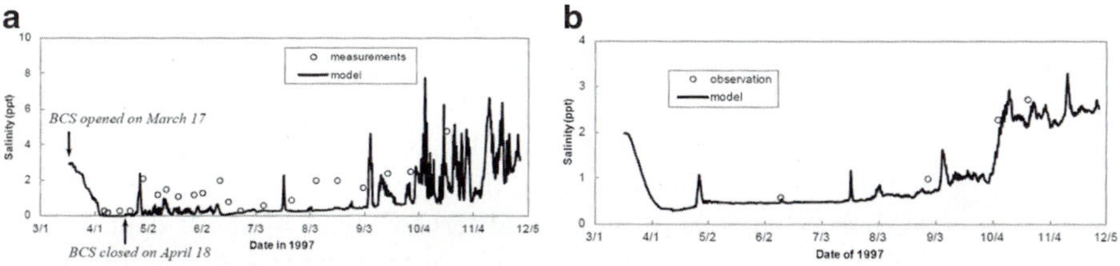

Fig. 12.9 (**a**) Simulated and observed salinity at salt station during and after BCS opening; (**b**) Simulated and observed whole lake-averaged salinity during and after BCS opening

was considered safe to assume that the obtained salinity pattern observed on early February of 1998 would be similar to that of March 17, 1997. Figure 12.8 shows the CCHE2D simulated salinity distributions for different time periods.

To evaluate the model's prediction capability, simulated results were compared with the observed time series of salinity distributions at the salt station (Fig. 12.9a) provided by Gulf Engineers and Consultants (1998). The trend of salinity obtained from the numerical model is generally in good agreement with field measurements. Some peaks were consistent with the higher salinity at the inlet boundary also. To further understand the salinity recovery processes, the time series of whole lake-averaged salinity were obtained based on the numerical results and compared with the available observed data (GEC 1998). The salinity recovery trend predicted by numerical model generally agrees well with observed data (Fig. 12.9b). It was observed that the lake salinity could be reduced as low as 0.28 ppt due to the spillway flood release. After the spillway was closed, the lake salinity slowly recovered. In the middle of June 1997, the lake salinity was in the range of 20–30 % of the value before the spillway was opened. In October 1997, about 6 months after the spillway was closed, the lake salinity recovered to pre-spillway-opening salinity level.

12.7 Prediction of Algal Bloom

The nutrient level in the Mississippi River is usually much higher than that in Lake Pontchartrain. When flood water is released, the river water with higher nutrient concentrations discharges into the lake, and affects the growth of algae. The algal growth rate is determined by the nutrient levels, light intensity and temperature. It was reported that due to the BCS flood release event the nutrient concentrations in the lake increased about 20 times, which would greatly increase the algae growth rate. However, since the SSC increases a lot due to the flood release, algal growth in the lake was restricted and there was no algal bloom observed in the lake during the BCS opening.

By the end of May, about one and a half month after the spillway was closed, the SSC recovered to normal. Due to high levels of nutrients and temperature, the growth rate of algae increased and led to an explosion of blue-green algae (cyanobacteria). The peak of the algal bloom observed. The blooms produced high levels of heptatoxins, which were measured during the peak of the blooms with traces persisting into the fall (Dortch et al. 1998).

The water quality model was applied to simulate the algae biomass distribution after the flood release event from May 28, 1997 to July 02, 1997. In this period, wind and tide were the most important forcing for flow circulations. The flow fields in Lake Pontchartrain were simulated using the measured wind and water surface boundary conditions. Since there was no observed algae biomass data, remote sensing and digital image processing techniques were used to initialize the model and validate the results as well.

Chlorophyll-*a* concentration in Lake Pontchartrain was estimated for Late May 1997 and used as initial condition data for CCHE2D simulation of Chlorophyll-*a* concentration. Chlorophyll-*a* concentration has also been estimated for mid June 1997 to validate the CCHE2D simulation results.

Han and Jordan (2005) developed an algorithm for estimating chlorophyll-*a* concentration in Pensacola Bay using Landsat 7 ETM+ data. As one of 39 estuaries located in the Gulf of Mexico, Pensacola Bay is impacted largely by rivers like Lake Pontchartrain. The techniques used were band ratioing and regression modeling. For the regression models, logarithmically transformed chlorophyll-*a* concentration was used as the dependent variable. Single bands, band ratios and logarithmically transformed band ratios were the independent variables. However, the ratio of ETM+1/ETM+3 was found to be the most effective in estimating chlorophyll-*a* concentration. This model (as shown in Eqs. 12.4 and 12.5) was applied on the processed Landsat 5 TM imagery acquired over Lake Pontchartrain on May 28, 1997 and June 13, 1997 to generate chlorophyll-*a* concentration data for CCHE2D WQ model initialization and validation respectively.

$$\log(chl - a) = y_0 + a * \left(\log b_j / \log b_k\right) \tag{12.4}$$

$$\log(chl - a) = -9.5126 + 12.8315 * (\log TM1/\log TM3) \tag{12.5}$$

Where, TM1 and TM3 represent the surface reflectance values recorded in band 1 and 3 of Landsat 5 TM imagery respectively.

Figure 12.10a shows the remote sensing based chlorophyll-*a* concentration estimation in Lake Pontchartrain as observed on May 28, 1997. Figure 12.10b

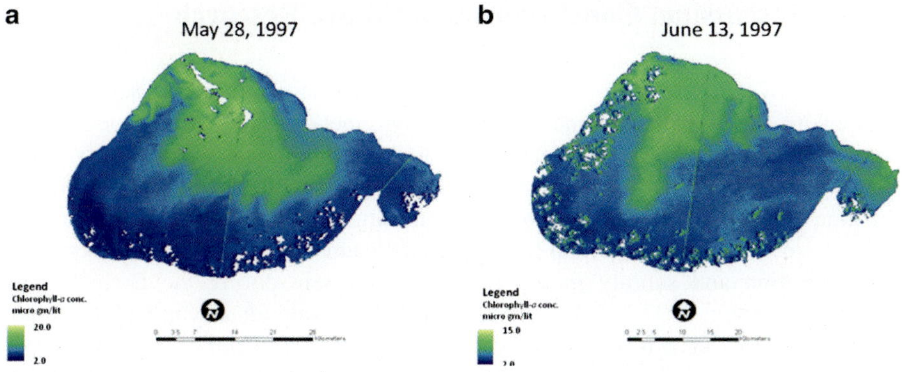

Fig. 12.10 (**a**) Remote sensing based chlorophyll-*a* concentration estimation in Lake Pontchartrain as observed on May 28, 1997; (**b**) Remote sensing based chlorophyll-*a* concentration estimation in Lake Pontchartrain as observed on June 13, 1997.

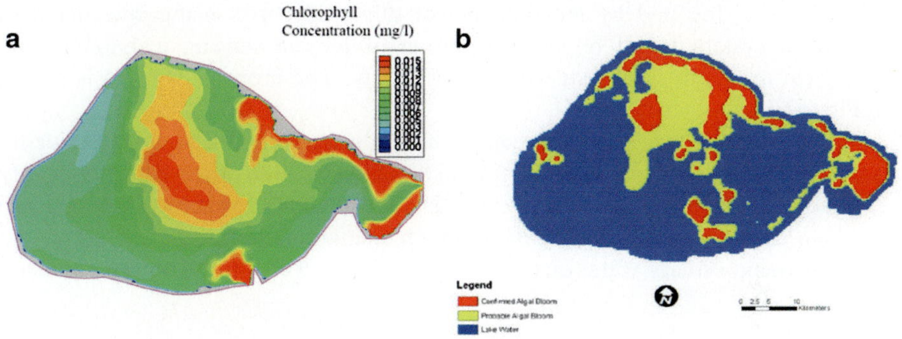

Fig. 12.11 (**a**) Model predicted (simulated) chlorophyll-*a* concentration in Lake Pontchartrain for June 13, 1997; (**b**) Probable algal blooms in Lake Pontchartrain as interpreted from the satellite observed chlorophyll-*a* concentration estimation data for 6/13/97

shows the remote sensing based chlorophyll-*a* concentration estimation in Lake Pontchartrain as observed on June 13, 1997.

Figure 12.11a shows the model predicted (simulated) chlorophyll-*a* concentration in Lake Pontchartrain for June 13, 1997. Figure 12.11b shows the probable algal blooms in Lake Pontchartrain as interpreted from the satellite observed chlorophyll-*a* concentration estimation data for 6/13/97. The visual comparison between Fig. 12.11a, b indicates that modeled chlorophyll-*a* concentration distribution has good agreement with the algal bloom interpreted from satellite observation. It can be observed that algal blooms occurred in a large area of the lake. The blooms caused decreases of dissolved oxygen in the lake, and fish kills occurred in some places in June and July of 1997.

12.8 Discussion, Conclusions and Future Research

The goal of this research was to advance the application of remote sensing technology to study the estuarine water quality. Specific focus was given on the integration of remote sensing technology with the numerical water quality model to map and monitor estuarine water quality at high spatial and temporal resolution. A time series of Landsat 5 TM imagery and digital image processing techniques were used to estimate and map suspended sediment concentrations, chlorophyll-*a* concentration and salinity in the Lake Pontchartrain during the Bonnet Carré Spillway event in 1997. CCHE2D WQ, a numerical water quality model developed at NCCHE was used to model the dynamics of the lake water quality during and after the spill way opening event. The Landsat imagery derived water quality data were used to initialize, calibrate and validate the numerical model. Although both remote sensing and numerical model based methods have weaknesses when used together, they can become a powerful tool to study water quality in estuary. The results obtained in this research clearly demonstrate this capability. This research is still evolving. Inspired by the outcome of this study, it is being considered to develop a system based on the integration of remote sensing technology and numerical model to study estuarine water quality. The proposed system is shown in Fig. 12.12.

This system is expected to be capable of (1) generating different water quality data from multi-sensor satellite platforms, (2) using satellite observed water quality estimation to initialize, calibrate and validate the numerical water quality model to generate estuarine water quality data at high spatial and temporal resolution, and (3) visualizing estuary water quality dynamics in the web based GIS environment.

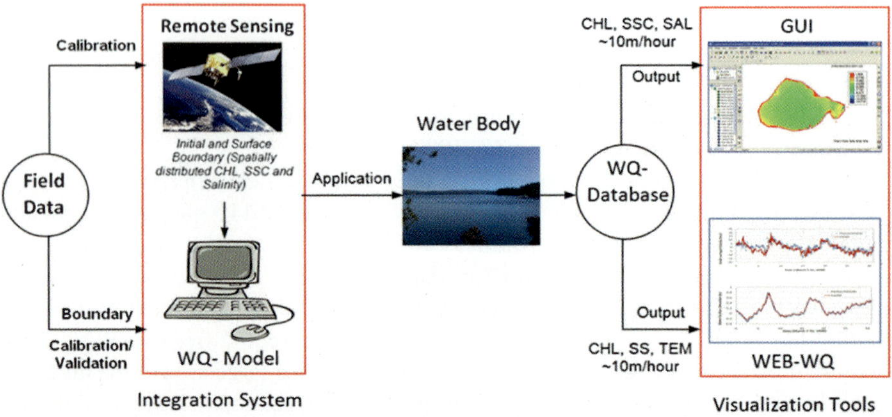

Fig. 12.12 Estuarine water quality monitoring system based on the integration of remote sensing technology and numerical model

Acknowledgements This research was funded by the United States Department of Homeland Security and was sponsored by the Southeast Region Research Initiative (SERRI) at the Department of Energy's Oak Ridge National Laboratory. Thanks are due to the United States Geological Survey (USGS) and the National Aeronautics and Space Administration (NASA) for giving the opportunity to obtain all the satellite imagery used in this research at free of charge.

References

Baban SMJ (1997) Environmental monitoring of estuaries; estimating and mapping various environmental indicators in Breydon Water Estuary, UK, using Landsat TM imagery. Estuar Coast Shelf Sci 44:589–598

Binding CE, Bowers DG, Mitchelson-Jacob EG (2005) Estimating suspended sediment concentrations from ocean color measurements in moderately turbid waters; the impact of variable particle scattering properties. Remote Sens Environ 94:373–383

Caraco NF (1995) Influence of human populations on P transfers to aquatic systems: a regional scale study using large rivers. In: Tiessen H (ed) Phosphorus in the global environment. SCOPE 54. Wiley, New York, pp 235–247

Chander G, Markham BL (2003) Revised Landsat 5 TM radiometric calibration procedures, and post calibration dynamic ranges. IEEE Trans Geosci Remote Sens 41:2674–2677

Chander G, Markham B, Helder DL (2009) Summary of current radiometric calibration coefficients for Landsat MSS, TM, ETM+, and EO-1 ALI sensors. Remote Sens Environ 113:893–903

Chao X, Jia Y, Shields FD, Wang SSY, Cooper CMM (2008) Three-dimensional numerical modeling of cohesive sediment transport and wind wave impact in a shallow Oxbow lake. Adv Water Resour 31:1004–1014

Cracknell AP, Newcombe SK, Black AF, Kirby NE (2001) The ADMAP (algal bloom detection, monitoring and prediction) concerted action. Int J Remote Sens 22:205–247

Dekker AG, Malthus TJ, Hoogenboom HJ (1995) The remote sensing of inland water quality. In: Plummer SE, Danson FM (eds) Advances in environmental remote sensing. Wiley, New York, pp 123–142

Dekker AG, Vos RJ, Peters SWM (2001) Comparison of remote sensing data, model results and in situ data for total suspended matter (TSM) in the southern Frisian lakes. Sci Total Environ 268:197–214

Department of Natural Resources (1997) Bonnet Carré spillway opening. Hydrographic data report, final report

Dewidar K, Khedr A (2001) Water quality assessment with simultaneous Landsat-5 TM at Manzala Lagoon, Egypt. Hydrobiologia 457:49–58

Dortch Q, Peterson J, Turner R (1998) Algal bloom resulting from the opening of the Bonnet Carré spillway in 1997. Fourth bi-annual basics of the basin symposium. United States Geological Survey, pp 28–29

Glasgow HB, Burkholder JM, Reed RE, Lewitus AJ, Kleinman JE (2004) Real-time remote monitoring of water quality: a review of current applications, and advancements in sensor, telemetry, and computing technologies. J Exp Mar Biol Ecol 300:409–448

Gulf Engineers and Consultants (GEC) (1998) Biological and recreational monitoring of the impacts of 1997 opening of the Bonnet Carré spillway Southeastern Louisiana. Final report to U.S. Army Corps of Engineers, New Orleans, Louisiana

Han L, Jordan KJ (2005) Estimating and mapping chlorophyll-*a* concentration in Pensacola Bay, Florida using Landsat ETM+ data. Int J Remote Sens 26(23):5245–5254

Härmä P, Vepsäläinen J, Hannonen T, Pyhälahti T, Kämäri J, Kallio K (2001) Detection of water quality using simulated satellite data and semi empirical algorithms in Finland. Sci Total Environ 268:107–121

Hellweger FL, Schlosser P, Lall U, Weissel JK (2004) Use of satellite imagery for water quality studies in New York Harbor. Estuar Coast Shelf Sci 61:437–448

Hossain A, Jia Y, Chao X (2010) Development of remote sensing based index for estimating/mapping suspended sediment concentration in river and lake environments. In: Proceedings of 8th international symposium on ECOHYDRAULICS (ISE 2010), September 12–16, 2010, COEX, Seoul, Republic of Korea, paper no. 0435, pp 578–585

Hu C, Chen Z, Clayton TD, Swarzenski P, Brock JC, Muller-Karger FE (2004) Assessment of estuarine water-quality indicators using MODIS medium- resolution bands: initial results from Tampa Bay, FL. Remote Sens Environ 93:423–441

Jia Y, Wang SSY (1999) Numerical model for channel flow and morphological change studies. J Hydraul Eng 125(9):924–933

Jia Y, Wang SYY, Xu Y (2002) Validation and application of a 2D model to channels with complex geometry. Int J Comput Eng Sci 3(1):57–71

Khorram S (1982) Remote-sensing of salinity in the San-Francisco Bay Delta. Remote Sens Environ 12:15–22

Lathrop RG (1992) Landsat Thematic Mapper monitoring of turbid inland water quality. Photogramm Eng Remote Sens 58(4):465–470

Lavery P, Pattiaratchi C, Wyllie A, Hick P (1993) Water quality monitoring in estuarine waters using the Landsat Thematic Mapper. Remote Sens Environ 46:268–280

Lindell T, Pierson D, Premazzi G, Zilioli, E. (eds) (1999) Manual for monitoring European lakes using remote sensing techniques. EUR report, vol. 18665. Office for Official Publications of the European Communities (EN), Luxembourg

Manheim FT, Hayes L (2002) Sediment database and geochemical assessment of Lake Pontchartrain Basin. In: Manheim FT, Hayes L (eds) Lake Pontchartrain Basin: bottom sediments and related environmental resources: U.S. Geological Survey professional paper 1634

McCorquodale JA, Georgiou I, Chilmakuri C, Martinez M, Englande AJ (2005) Lake hydrodynamics and recreational activities in the south shore of Lake Pontchartrain, Louisiana, NOAA technical report

Mobley CD (1994) Light and water—radiative transfer in natural waters. Academic, San Diego, 592 pp

National Research Council, C (2000) Clean coastal waters—understanding and reducing the effects of nutrient pollution. National Academy Press, Washington, DC

Penland S, Beall A, Kindinger J (2002) Environmental atlas of the lake Pontchartrain Basin. USGS open file report 02-206

Pettinger LR (1971) Field data collection—an essential element in remote sensing applications. In: Proceedings of the international workshop on Earth resources survey systems, Washington, DC, pp 49–64

Populus J, Hastuti W, Martin JL (1995) Remote sensing as a tool for diagnosis of water quality in Indonesian Seas. Ocean Coast Manag 27(3):197–215

Song C, Woodcock CE, Seto KC, Lenney MP, Macomber SA (2001) Classification and change detection using Landsat TM data: when and how to correct atmospheric effects? Remote Sens Environ 75:230–244

Teillet PM, Dudelzak AE, Pultz TJ, McNairn H, Chichagov A (2001) A framework for in-situ sensor measurement assimilation in remote sensing. In: Proceedings of the 23rd Canadian symposium on remote sensing, Que'bec City, Que'bec, 21–24 August, pp 111–118

Teillet PM, Gauthier RP, Chichagov A, Fedosejevs G (2002) Towards integrated Earth sensing: advanced technologies for in situ sensing in the context of Earth observation. Can J Remote Sens 28:713–718

Turner RF, Dortch Q, Rabalais NN (1999) Effects of the 1997 Bonnet Carré opening on nutrients and phytoplankton in lake Pontchartrain. Final report submitted to the Lake Pontchartrain Basin Foundation, 117 p

United States Army Corps of Engineers (USACE) (2011) Bonnet Carré spillway overview, spillway pace. http://www.mvn.usace.army.mil/bcarre/2011operation.asp

Vermote E, Tanré D, Deuzé JL, Herman M, Morcrette JJ (1997) Second simulation of satellite signal in the solar spectrum (6S), 6S user guide (v. 2) and 6S code (v. 4.1), July 1997
Wang F, Xu YJ (2008) Development and application of a remote sensing-based salinity prediction model for a large estuarine lake in the US Gulf of Mexico coast. J Hydrol 360:184–194
Wang X, Wang Q, Wu Q (2003) Estimating suspended sediment concentration in coastal water of Minjiang River using remote sensing images. J Remote Sens 7(1):54–57
Williams AN, Grabau WE (1973) Sediment concentration mapping in tidal estuaries. In: Third Earth resources technology satellite-1 Sym. NASA SP-351, pp 1347–1386
World Resources Institute (2003) A guide to the global environment—environmental change and human health. The World Resources Institute, The United Nations Environmental Programme, The United Nations Development Program, and The World Bank, New York, pp 1–369
Zhang Y, Jia Y (2009) CCHE2D mesh generator and user's manual, technical report no. NCCHE-TR-2009-1, University of Mississippi

Part III
Advances in Coastal Modeling Using Field Data, Remote Sensing, GIS, and Numerical Simulations

Chapter 13
Developments in Salt Marsh Topography Analysis Using Airborne Infrared Photography

Francisco Andrade, Jackson Blanton, M. Adelaide Ferreira, and Julie Amft

Abstract Salt marshes occur throughout the extra-tropical regions of the world, along low energy shores, where a surplus of fine sediment is available, and they have been estimated as some of the most valuable ecosystems on Earth. They exhibit complex topographies primarily modelled by hydrodynamics which, in turn, determine geobiophysical processes and the pattern of occurring communities. This makes accurate salt marsh topography a prerequisite for the understanding of their function and structure. Only recently, have remote-sensing techniques become widely available to obtain high-resolution topographic data in an environment otherwise extremely arduous to access. Still, extraction of bare-earth surface remains difficult and especially problematic in areas of dense vegetation. LiDAR data, although widely in use still isn't readily available worldwide and requires intensive post-processing and validation.

A detailed digital elevation model (DEM) of the Duplin River (Georgia, Southeastern USA) was constructed with a 1 m^2 resolution. The model was created by the classification and analysis of a time-series of 7 IR (infrared) aerial photographic mosaics taken at 1 h intervals from low- to high-water during a rising tide. The technique is based on the premise that flooded areas can be objectively recognized through image analysis and that the water surface is horizontal throughout the system, thereby defining a reference level at any given time. We focus on the description of the method, and results from its use in a large intertidal area. We also discuss the advantages of the method and its shortcomings when applied to vegetated intertidal areas, and propose further developments and applications.

F. Andrade (✉)
Centro de Oceanografia – Laboratório Marítimo da Guia, Faculdade de Ciências da Universidade de Lisboa, Av. Nª.Srª. do Cabo, 939, Cascais 2750-374, Portugal
e-mail: faandrade@fc.ul.pt

J. Blanton • J. Amft
Skidaway Institute of Oceanography, 10 Ocean Science Circle, Savannah, GA 31411, USA

M.A. Ferreira
Laboratório Marítimo da Guia, Instituto do Mar, Av. Nª Srª do Cabo, 939, Cascais 2750-374, Portugal

C.W. Finkl and C. Makowski (eds.), *Remote Sensing and Modeling: Advances in Coastal and Marine Resources*, Coastal Research Library 9,
DOI 10.1007/978-3-319-06326-3_13,

13.1 Introduction

Salt marshes are transition zones between terrestrial and freshwater habitats, and the sea (Zhao et al. 2004). They occur over wide latitudinal ranges, throughout the extra-tropical regions of the world (mainly above 30°N and below 30°S), in a variety of climate conditions, in intertidal meso- to euhaline sheltered areas with low wave energy and a surplus of fine sediments, and are dominated by halophytic vegetation (Carter 1988; Adam 1990; Allen 2000; Montané and Torres 2006). Salt marshes provide a number of essential ecosystem services vital for mankind, such as shoreline protection from the effects of storms, or removal of excess nutrients and pollutants from the water mass by salt marsh vegetation (Adam 1990). They provide habitats for spawning and nursery grounds, as well as food for a number of animal species, many of them of commercial interest. Moreover, the regulation of the fluxes of water, nutrients, particles and organisms between land, rivers, and the ocean, provided by salt marshes (Zhao et al. 2004), not forgetting their aesthetical value, demand a careful assessment of these systems' functioning. Although they were for long considered wastelands (e.g. Chapman and Roberts 2004), Costanza et al. (1997) have estimated tidal marshes to be among the most valuable ecosystems on Earth, and valued their ecosystem services and functions at US \$9,990 $ha^{-1}year^{-1}$.

Salt marshes exhibit complex topographies of gently sloping flats cut by intricate networks of creeks. These areas are modelled primarily by hydrodynamics but also by biological and ecological factors (Mason et al. 2005). Topography, in turn, conditions and defines the vertical zonation of the salt marsh and its mudflat communities, by determining their immersion/emersion rhythms. Accurate topographic data is paramount for, i.a.: (i) a thorough understanding of the geobiophysical processes taking place in salt marshes, namely the water volumes stored in intertidal areas and released into the tidal channels by the tide (Fagherazzi et al. 2008), and the corresponding material exchanges; (ii) explaining the distribution and variability of the salt-marsh communities, namely halophytic vegetation (Hladik and Alber 2012) and; (iii) the modelling of these environments.

Only recently have remote-sensing techniques become widely available to obtain high-resolution topographic data in these environments, which have been otherwise extremely arduous to access from the ground (Montané and Torres 2006; Hladik and Alber 2012). Topographic techniques of use in intertidal areas are wide-ranging. Besides conventional surveying, they include Real-Time Kinematic (RTK) GPS, bathymetric sounding, LiDAR (Light Detection And Ranging a.k.a. airborne scanning laser altimetry, or airborne Laser Terrain Mapping), stereophotogrammetry and radar stereopairs, airborne or spaceborne SAR interferometry, and the waterline method (Neuenschwander et al. 2000; Mason et al. 2001; Won et al. 2003). Traditional topographic techniques are logistically difficult to use in intertidal areas, expensive and only allow infrequent sampling. Airborne stereophotogrammetry is impaired by the difficulty "of obtaining reference points in flat featureless areas and is extremely expensive." (Mason et al. 1999).

The waterline method, a relatively fast, accurate, and inexpensive method uses the sea level as an altimeter. Throughout the tidal cycle, the waterline (the land-water boundary) moves back and forth. Stacking these land-water boundaries interpreted from images taken at regular time intervals, preferably during rising tide (Sindern and Kathage 1966 in Koopmans and Wang 1995) and relating these boundaries to a measured water level from the time of each image acquisition, the topography of the area between maximum high- and minimum low-water may be compiled (Koopmans and Wang 1995).

The waterline method was first proposed and tested using aerial photographs in the 1960s (Zee 1981 in Koopmans and Wang 1995). A number of studies have used the waterline method with satellite imagery (ERS-1 SAR scenes, Thematic Mapper, Landsat, SPOT and IRS-1C, among others) to build digital elevation models of the intertidal zone over large unvegetated areas, namely beaches and estuarine flats (Mason et al. 1995, 1997a, b, 1998, 1999, 2001; Lohani and Mason 1999; Won et al. 2003), with water level estimated using hydrodynamic models. Errors in shoreline position result from imagery resolution and the process of image registration (Mason et al. 1997a).

In vegetated areas with much more complex elevation patterns, such as salt marshes, extraction of actual topography – bare-earth surface – remains a difficult problem for all sensors, including LiDAR, and is especially problematic in areas of dense vegetation (Ramsey 1995; Neuenschwander et al. 2000; Montané and Torres 2006; Schmid et al. 2011). Nonetheless, LiDAR data are being used to produce high resolution topographical information for the construction of digital elevation models (DEMs) of salt marshes (Neuenschwander et al. 2000; Mason et al. 2005; Montané and Torres 2006; Hladik and Alber 2012) although it requires intensive post-processing and validation (Schmid et al. 2011).

A detailed digital elevation model (DEM) of the Duplin river (Georgia, Southeastern USA) with a 1 m^2 resolution, was constructed through the classification and analysis of a time-series of 7 IR (infrared) aerial photography mosaics taken at 1 h intervals from low- to high-water during a rising tide.

The rationale behind the method is that: (i) flooded areas can be objectively recognized through image analysis, namely by resorting to near infrared radiation, very strongly absorbed at the water surface and; (ii) water surface is horizontal throughout the system and can therefore be used as a reference level. This paper focuses on the development and results of this method, discusses its advantages and shortcomings when applied to vegetated intertidal areas, and proposes future developments.

13.2 Study Area

The Duplin River is a 12.5 km long tidal inlet that borders Sapelo Island to the west and is surrounded by a large salt marsh (Fig. 13.1). The tidal channel experiences a semidiurnal tidal regime typical of the southeastern U.S. Atlantic coast, with tidal ranges between 2.0 and 3.0 m (Blanton et al. 2002). The river opens into Doboy

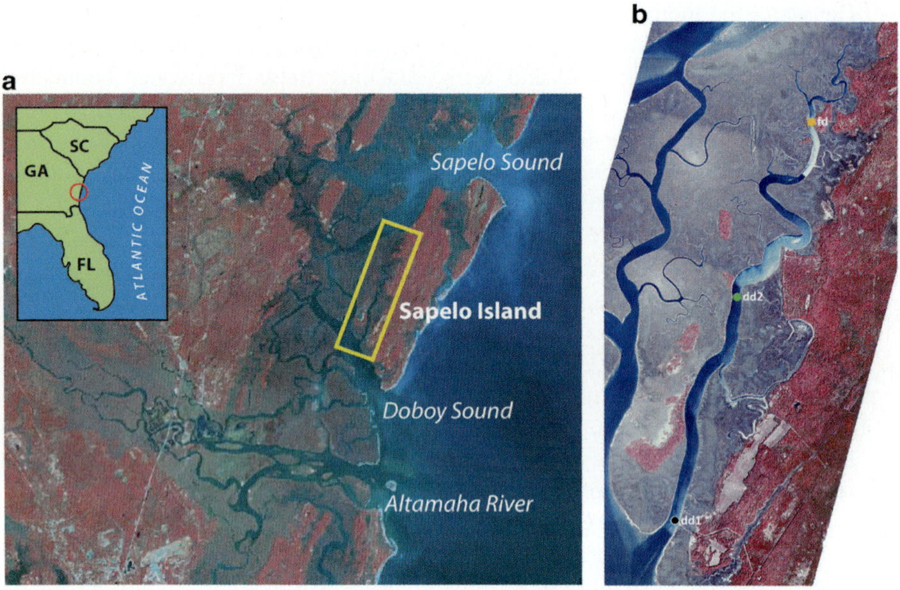

Fig. 13.1 (**a**) Location of the study area: the *yellow box* defines the region covered by the aerial images. (**b**) Duplin river: Location of the tide pressure gauges (color figure online)

Sound. Freshwater sources are rainfall and island groundwater runoff, and the estuary is essentially pristine. Salinity at the river entrance is largely determined by seasonal changes in discharge of the Altamaha River into Doboy Sound, and water movement is controlled by tidal currents (Blanton et al. 2006). The area lies within the Georgia Coastal Ecosystems Long Term Ecological Research domain and is a National Estuarine Sanctuary.

13.3 Method

13.3.1 Waterline Method/Construction of the DEM

13.3.1.1 Aerial Survey

An aerial survey was conducted on April 15, 2004, consisting of a series of seven passes over the Duplin River at a 1 h interval, from low-water (LW) at 12:30 to high-water (HW) at 18:30. The aircraft mission was conducted by Spectrum North Carolina, Inc (presently Geofiny Technologies) using a WILD 15/4 UAG-S camera loaded with Kodak Professional Infrared EIR film and flown at a 1:24,000 scale (1″ = 2,000 ft).

For each pass, a total of eight false color IR images were obtained. Of these, five consecutive photos (Photos 2–6) were selected to cover the Duplin intertidal area, the three bands in each image (BGR) corresponding to green ~550 nm, red ~650 nm and near-infrared ~700–850 nm radiation reflectances (Eastman Kodak Company 2005).

13.3.1.2 Image Processing

The selected photos were digitized at 24 bit/pixel with a 600-dpi resolution to ensure a ground resolution better than 1 m^2 and rectified to NAD83 UTM17 with a 1 m × 1 m final ground resolution. The reference used was the orthophotomap supplied by the image provider, corresponding to the rectified mosaic of the low-water images. The rectification process used a second degree polynomial adjustment and a nearest neighbour resampling in order not to alter the digital numbers (DNs) in the original images (Eastman 2006). This procedure, with an average RMS of 0.9, allowed for the exact superimposition of the selected images throughout the Duplin River domain and yielded seven mosaics of five images across the 6-h time interval, ordered from T1 (LW) to T7 (HW).

The rectification and all subsequent image processing was undertaken using Idrisi Andes (vs. 15.01) software from Clark Labs, Clark University. Each image of the HW mosaic was initially classified using the KMeans classifier. KMeans is an unsupervised clustering technique that partitions an image into K (as defined by the analyst) exclusive clusters and it was initialized with K = 75 centroids randomly distributed over the image, and no class merging. Each pixel in the image is initially assigned to a cluster according to its closest centroid. Cluster centroids are then updated and the process is repeated until the K clusters and their centroids are stable (Eastman 2006). The resulting classes were then manually assigned to flooded (water) or dry soil areas. Although for the majority of the classes produced, and correspondingly, most of the classified area, this process was simple and objective, a number of situations were found where assignment of individual classes had to rely on either terrain knowledge or a best judgement of the information on the original images, including such criteria as feature/morphological continuity. This was achieved through the superimposition of these classes on the false color images.

This is why no assessment of the classification error was carried out since this would result in a circular process, the terrain information available consisting of the images that were subjected to the classification process.

All classes corresponding to flooded areas were reclassified to 1 and all areas of dry soil were reclassified to 0, producing a binary mask that allowed for the exclusion of the HW dry areas.

Superimposing this HW mask (T7) to the corresponding T6 (HW-1 h) mosaic images, only the widest flooded area in the basin was retained. This was the area that was then classified, again using KMeans, now with a random seed for 50 classes (the area to classify being much smaller and less heterogeneous, since terrestrial dry areas had been totally excluded in the first step).

A new binary mask was produced, now excluding the T6 dry area, and the process of classifying flooded and non-flooded areas, of producing a binary mask and superimposing it to the previous (in time) mosaic images was repeated for times

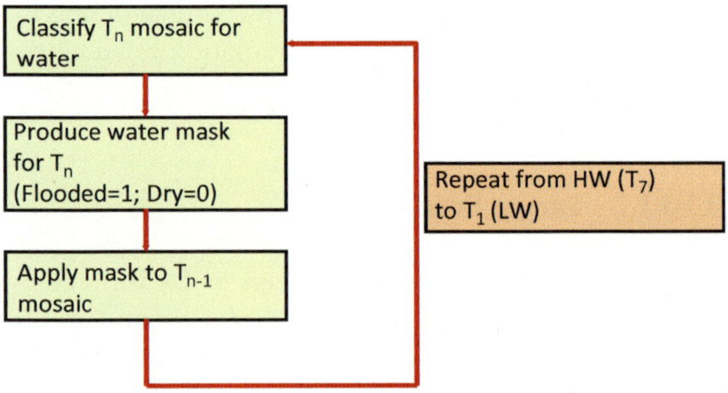

Fig. 13.2 Flow diagram of the method

T5 through T1 (LW). This ensured the total superimposition of successive flooded areas, from HW to LW (Fig. 13.2).

Concatenation of the individual water classifications for each time into LW to HW flooded area mosaics, allowed for the mapping and quantification of the corresponding flooded area inside the overall Duplin drainage basin.

13.3.1.3 DEM Construction

To define the vertical reference water level for each overflight and its corresponding flooded area mosaic (T1 through T7), simultaneous readings were taken from three sub-surface pressure gages deployed along the estuary (Fig. 13.1) and averaged. The relatively short length of the Duplin River assures that the tidal wave approaches a standing wave (Li and O'Donnell 2005) so that LW and HW occur almost simultaneously throughout the whole length of the channel. Water level readings were referenced to local mean-low-water (MLW) level.

The outline of the flooded area for each overflight was then extracted, resulting in a total of 1.7×10^6 points on a 1 m × 1 m X, Y (NAD83 UTM17) grid, each with a corresponding elevation (Z) value (in m above MLW).

Since 2.32 m is the maximum tidal water level found in the vertical reference system used to gauge the different flooded areas, the outline of the overall Duplin basin, as defined by Blanton et al. (2007) as the watershed limit was assumed with that Z value.

These data were used to construct a digital elevation model (DEM) for the whole intertidal domain of the Duplin River inside the watershed limit considered using a universal Krigging interpolation algorithm available in Surfer vs. 10 (Golden Software 2011).

13.3.2 Accuracy Assessment

During 2009, a set of 1,490 X, Y (NAD 83 UTM17) and Z (NAVD88) readings was taken in the salt marsh area of the Duplin basin using a Trimble R6 RTK GPS

Table 13.1 Time and water level as measured at the three gauges used for the seven overflights and corresponding flooded areas

Overflight	Time (hh:mm)	Avg. tidal level (m MLW)	Total flooded area (ha)
T1	12:29–12:31	−0.10	190.54
T2	13.32–13.34	0.09	201.94
T3	14:31–14:33	0.66	238.50
T4	15:29–15:31	1.26	277.94
T5	16:29–16:31	1.8	321.53
T6	17:29–17:31	2.12	385.90
T7	18:26–18:28	2.19	500.47

receiver, with a reported mean vertical error of 0.010 m and mean horizontal error of 0.012 m, both at 68 % confidence level (Hladik and Alber 2012). A total of 736 readings taken within the DEM domain were directly superimposed onto it. Comparison of the corresponding elevation (Z) values ($Z_{DEM} - Z_{RTK}$) for the exact same X and Y coordinates allowed for the assessment of the DEM accuracy according to Maune et al. (2007). For this comparison, DEM Z values (MLW) were first corrected to NAVD88 using the available data from NOAA for nearby St. Simons Island (NOAA 2012). Since local MLW and NAVD88 vertical reference values are 0.582 m and 1.809 m respectively, the corresponding vertical difference of 1.227 m was subtracted from the DEM data. We point out that some caution is warranted in the use of these values since MLW and NAVD88 reference levels are not consistent throughout the geographical range that encompasses Sapelo Island and the Duplin River (cf. NOAA data for Ft. Pulasky, St. Simons and Fernandina Beach, the three closest stations to the Duplin River, with differences between NAVD88 and MLW vertical references of, respectively, 1.170 m, 1.227 m, and 1.108 m).

13.4 Results

On April 15, 2004, low water (LW) level (T1) was −0.10 m MLW and high water (HW) level (T7) was 2.19 m MLW, an unusually large neap tide with a 2.29 m tidal range. Table 13.1 gives the selected tidal levels for each overflight and the corresponding flooded areas inside the DEM domain.

Figure 13.3 shows the hourly results of the classification, and Fig. 13.4 shows the final DEM of the entire study area and two areas of detail. This DEM covers an overall area of 1,200.04 ha, of which 190.54 ha are below the considered LW level of −0.10 m MLW.

In terms of topographic accuracy, the average vertical (Z) difference between the 736 RTK points inside the DEM domain (Fig. 13.5) and the corresponding corrected DEM values over the whole intertidal 1,009.51 ha area was 0.16 m with a standard deviation of 0.18 m and a root mean square error (RMSE) of 0.24 m.

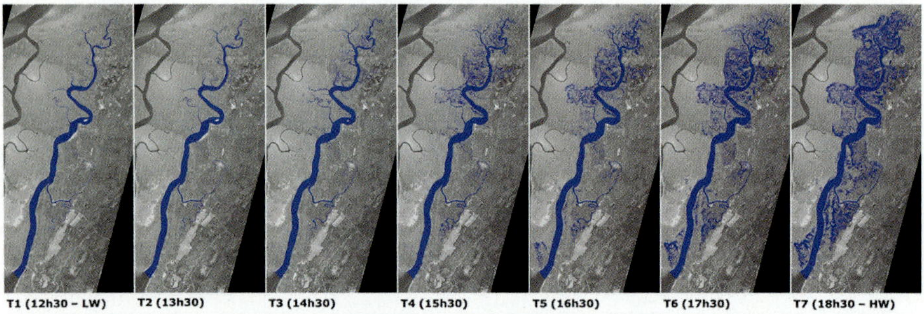

Fig. 13.3 Hourly image mosaics showing the area classified as "water" (*flooded areas*). Note the rapid increase in flooded area in the later stages of flood tide

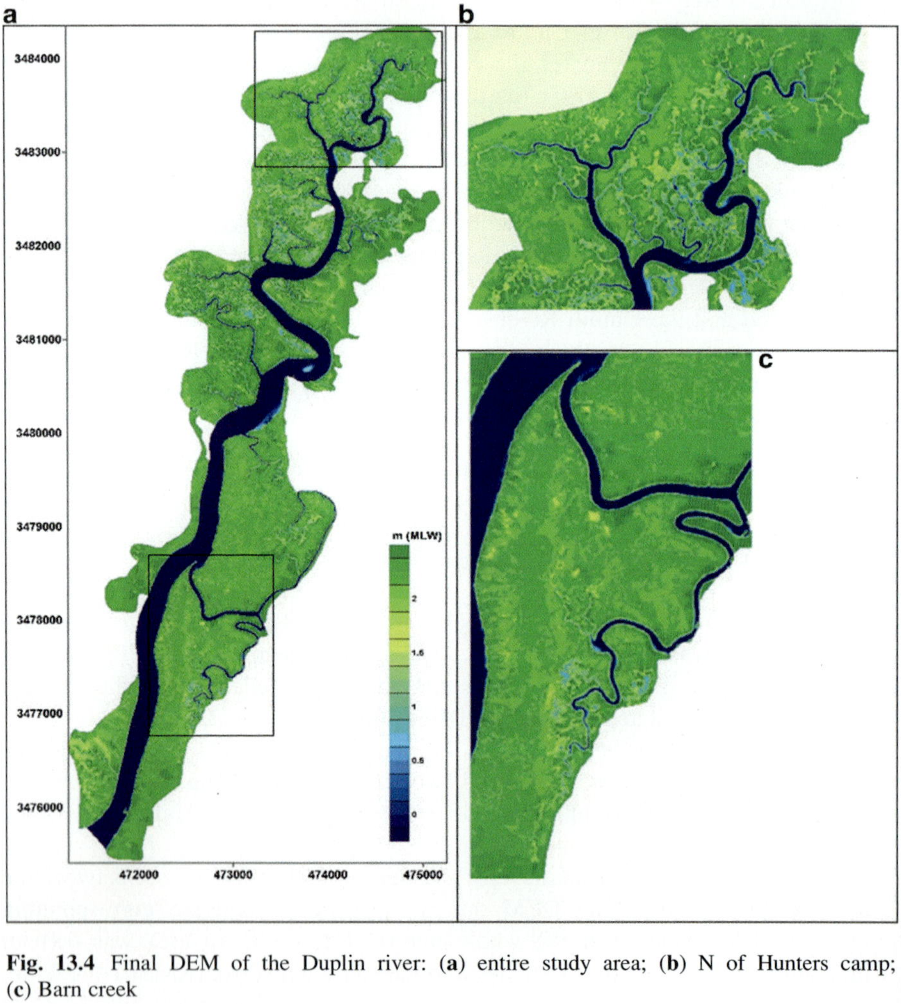

Fig. 13.4 Final DEM of the Duplin river: (**a**) entire study area; (**b**) N of Hunters camp; (**c**) Barn creek

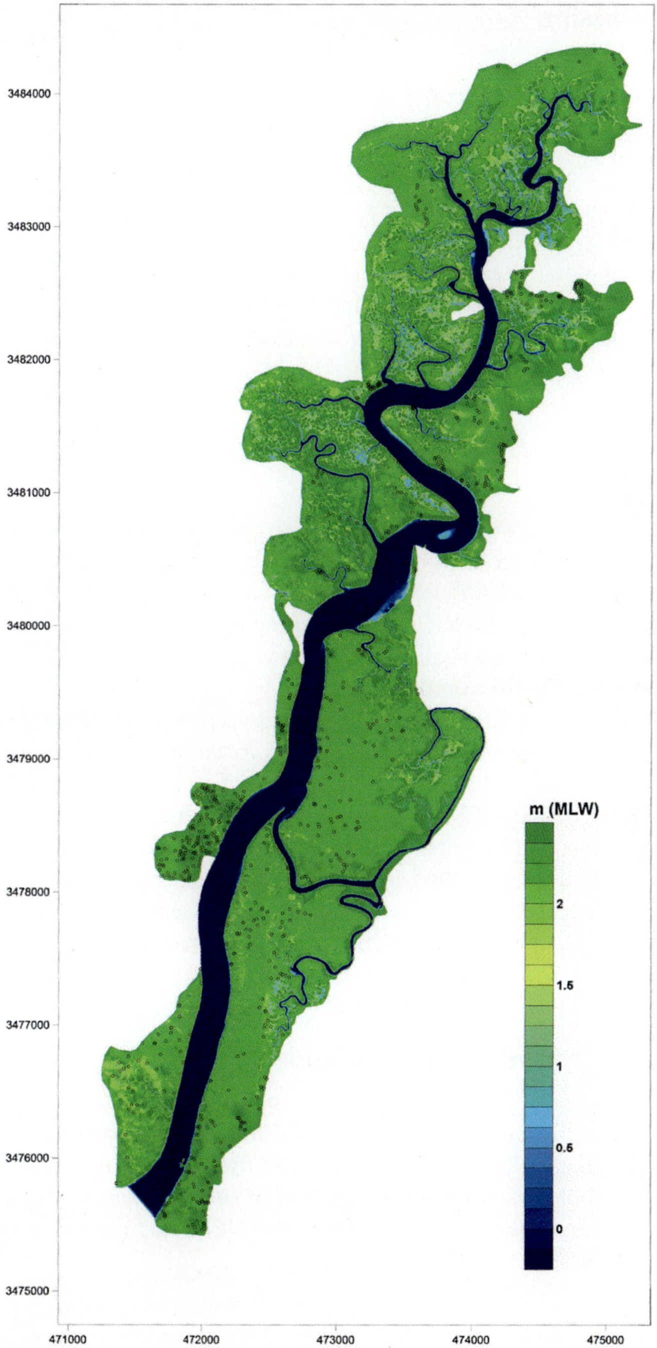

Fig. 13.5 Duplin River DEM showing the 736 RTK GPS points (*black circles*) used in the accuracy assessment

13.5 Discussion

13.5.1 Duplin's Intertidal Marsh Topography and Flooding Patterns

Overall, a very good description of the whole salt-marsh area was achieved, and the DEM captured its fine details (Fig. 13.4).

Variation of flooded area with time (Table 13.1 and Fig. 13.3) shows a rapid increase above the 2 m level suggesting that the water leaves confining channels late in the flood phase.[1]

13.5.2 Duplin's DEM Vertical Accuracy

Five years elapsed between the aerial survey and the field RTK surveys during which data used to gauge the original DEM was obtained. Still, vertical accretion rates found for the marine-influenced marshes surrounding Sapelo Island, with tidal ranges similar to the Duplin River are low, ranging from less than 1 mm $year^{-1}$ for long-term accretion to a maximum of 5.3 mm $year^{-1}$ for short-term accretion (Craft 2007) which validates the comparison.

The average vertical difference of 0.16 m between corresponding DEM and RTK elevation (Z) values may result, at least in part, from the lack of consistency of the vertical reference levels, as mentioned before (cf. 13.3.2. Accuracy assessment). Vertical shifting of the DEM by that 0.16 m difference would make the RMSE converge to the standard deviation value of 0.18 m. This suggests that the surface derived with the present method has a quality comparable to that derived from a LiDAR survey. For the 1,047 nm wavelength LiDAR sensor they used to survey the exact same area, Hladik and Alber (2012) report an absolute RMSE of 0.11 m.

13.5.3 Applications of the Methodology

This same approach was used to generate a DEM for a salt-marsh covered intertidal domain on the Okatee River, SC. This allowed for the construction of hypsometric curves for the whole system and for a number of sub-areas, as well as for studies of its tidal asymmetry (Blanton et al. 2006). The DEM was used in a finite-volume

[1] Animations of the water-level changes in each of the domains shown in Fig. 13.4 throughout the tidal phase from LW to HW can be made available. The rapid increase in surface area, as well as other features, is clearly visualized in these animations.

hydrodynamic model, thus providing for an accurate and morphologically detailed model mesh that included the large intertidal areas, the sinuosity of the river's main channel, and the complex network of side channels (Huang et al. 2008).

The Duplin River intertidal DEM as described herein has been used by the Savannah River National Laboratory (SRNL, Aiken SC, USA) to support its mission to accurately simulate transport and dispersion of radioactive substances in coastal areas (Bollinger et al. 2008). The high resolution of the Duplin DEM provided a testbed for upgrading the hydrodynamic code used by SRNL for other similar environments (Blanton et al. 2009, 2010).

13.5.4 Strengths, Limitations, and Further Potential of the Methodology

The method has a number of strengths shared by more expensive methods. Such strengths include:

- Costs both in terms of image acquisition and processing are relatively low and these will become even lower as the target area size decreases, by resorting to simpler and cheaper platforms and sensors since image obliquity can be readily dealt with (Chandler et al. 2002; Andrade and Ferreira 2011);
- Ground resolutions in the order of 1 m^2 can be easily achieved;
- Final results can be achieved rapidly after imagery acquisition;
- Since images can be acquired from a number of different platforms (e.g., helicopter, balloon, kite) image repetition may allow for high-frequency time analysis of changes in the salt marsh topography;
- Meshes can be created for hydrodynamic tidal models that need to accurately compute currents in areas that alternate between wet and dry;
- Data on marsh areas as a function of depth can provide the information required for accurate calculations of metabolic processes on the marsh surface;
- The hypsometric function describing area and volume as a function of height can be readily derived and used to estimate the properties of tidal transport in any upstream section.

Nonetheless, a number of constraints were identified that may affect the accuracy of results:

- Delineation of each individual waterline relies on the analyst's evaluation of the classification results. Depending on the information available and its quality, namely the original IR photographs, over- or under-classification may occur;
- Vegetation density may mask the underlying flooding of an area and, as such, increase errors in the definition of the water contour;

- Z level definition requires accurate referencing of water level data, as discussed above. Over areas with channels longer than those of the Duplin River, water surface slopes may develop, namely due to the more progressive nature of the tidal wave (Li and O'Donnell 2005);
- Because salt marshes are typically characterized by deeply cut creeks surrounded by flat vegetated platforms, terrain slope can vary strongly over small horizontal extensions. As such, interpolation algorithms used to produce the DEMs must accurately reflect the defined waterlines as much as possible. In areas of abrupt changes in topography, universal Krigging, probably the most common algorithm in use, will tend to exaggerate ditches and ridges translating into an over-estimate of parameters such as the height of the edge of the salt marsh; this is not a limitation of the waterline method itself (i.e., the capacity to discern between water and land/dry and flooded areas) but of the DEM computing process.

In order to ensure the best results, a number of requirements should be met:

- Images should be obtained at a sun angle greater than about 30° above the horizon to achieve proper illumination;
- Atmospheric water content must be low enough so as not to significantly attenuate the ground-reflected infrared radiation reaching the sensor;
- Hydrographic surveys to obtain depth information below the lowest water observed during the overflights are especially important to obtain good depth data to merge with the intertidal areas as determined with the DEM, thus allowing for the accurate determination of the hypsometry of the whole marsh, intertidal and subtidal;
- The temporal frequency of image acquisition should be, at least, in the order of once per hour, and effort should be exercised to achieve a regular time interval.

The method we developed and describe here was applied only during the flood phase of the tide, the corresponding rationale being that this should allow for a negligible number of puddles of water left from the previous ebb stage and no significant sediment swelling due to inundation. However, since tidal currents in channels surrounded by large intertidal areas are usually significantly asymmetric (Dronkers 1986; Blanton and Andrade 2001; Blanton et al. 2002), it is unlikely that the function that defines the temporal growth in water area for the flood phase will be the mirror image of the same function for the ebb stage.

This method has been applied in an area of the globe (SE USA) where flooding during daylight hours only occurs for neap tides. In the case of the North Atlantic Ocean, LW in the SE USA, coincides with HW in the west coast of Europe. Over one-half of the spring-neap period (~7 days), the tidal phase will change so that the number of daylight hours for flooding is maximized closer to spring tide in Europe. This feature could be used with great advantage in such regions since the resulting DEM will cover a larger intertidal area due to a larger tidal range.

13.6 Conclusions

A remote-sensing method was developed that uses a time-series of infrared aerial images taken during the rising tide to delineate and measure the temporal growth of the water surface area over a salt marsh and the corresponding vertical levels, thus producing a digital elevation model (DEM) of that intertidal area.

Results for the c. 1,010 ha of intertidal areas surrounding the Duplin River (SE USA) show that morphology and elevation of such intertidal areas can be resolved to as little as 1 m^2 with a vertical accuracy better than 0.2 m. They also suggest that the method produces good and reliable results, namely in terms of result accuracy vs. effort.

The straightforwardness of the technique, the low costs of imagery and of its analysis, and the presently wider availability and access of IR aerial cameras when compared e.g. to an airborne LiDAR system, suggest that the present method may have significant advantages in a vast range of intertidal environments worldwide.

This also translates into the possibility of using a number of airborne platforms "of occasion" such as helicopters, baloons or kites, further reducing costs and increasing the plasticity and the time and space resolution of the method together with its applicability in a wide range of intertidal environments.

Uncertainties lie mostly with the capacity to detect thin layers of water over the densest marshes. The DEM quality and reliability is also directly influenced by the accuracy of the water level measurements used and its overall vertical position depends on the tide level reference values.

This technique has been applied exclusively to the flood phase of the tide. Similar studies to measure the rate of change of water area and volume during the decrease of water level during the ebb phase of the tidal cycle are likely to yield new insights into the flooding and draining of large intertidal areas.

Acknowledgments We thank Mike Robinson at Skidaway Institute of Oceanography for doing preliminary calculations of water areas, Steven Pennings and Tim Hollibaugh for their support to develop the DEM for the Duplin River. Daniela Di Iorio, Paul McKay and Ken Helm for providing the sub-surface pressure data to relate overflights time to water level. Christine Hladik provided the RTK data used to gauge the DEM. We acknowledge funding from the National Science Foundation (OCE 99-82133), which funded the aerial mission as well as the analyses reported here. We are also grateful to the following agencies that supported the work described in this paper: the Georgia Coastal Zone Management Program (Grant No. RR100-279/9262764), the Department of Energy (Contract No. DE-AC09-96SR18500), and the Luso-American Foundation (FLAD), which funded travelling of the Portuguese co-authors.

References

Adam P (1990) Saltmarsh ecology. Cambridge University Press, Cambridge

Allen JRL (2000) Morphodynamics of Holocene salt marshes: a review sketch from the Atlantic and Southern North Sea coasts of Europe. Quat Sci Rev 19:1155–1231. doi:10.1016/S0277-3791(99)00034-7

Andrade F, Ferreira MA (2011) A method for monitoring shallow eelgrass meadows (Zostera spp.) using terrestrial oblique large-scale photography. Aquat Bot 95:103–109. doi:10.1016/j.aquabot.2011.04.002

Blanton JO, Andrade F (2001) Distortion of tidal currents and the lateral transfer of salt in a shallow coastal plain estuary (O Estuário do Mira, Portugal). Estuaries 24:467–480

Blanton JO, Andrade F, Ferreira MA (2006) The relationship of hydrodynamics to morphology in tidal creek and salt marsh systems of South Carolina and Georgia. In: Kleppel GS, DeVoe MR, Rawson MV (eds) Implications of changing land-use patterns to coastal ecosystems. Springer, New York, pp 93–107

Blanton JO, Andrade F, Ferreira MA, Amft J (2007) A digital elevation model of the Duplin River intertidal area. University of Georgia, Athens, p 9, Final report to the GCE-LTER program

Blanton JO, Lin G, Elston SA (2002) Tidal current asymmetry in shallow estuaries and tidal creeks. Cont Shelf Res 22:1731–1743. doi:10.1016/S0278-4343(02)00035-3

Blanton JO, Garrett AJ, Bollinger JS, Hayes DW, Koffman LD, Amft J (2009) Transport and dispersion of a conservative tracer in coastal waters with large intertidal areas. Estuar Coast 32:573–592. doi:10.1007/s12237-009-9141-4

Blanton JO, Garrett AJ, Bollinger JS, Hayes DW, Koffman LD, Amft J, Moore T (2010) Transport and retention of a conservative tracer in an isolated creek-marsh system. Estuar Coast Shelf Sci 87:333–345. doi:10.1016/j.ecss.2010.01.010

Bollinger J, Garrett A, Koffman L, Hayes D (2008) 3-D hydrodynamic modeling in a geospatial framework. J Map Geogr Libr (Geoscapes) 4:308–318

Carter RWG (1988) Coastal environments: an introduction to the physical, ecological, and cultural systems of coastlines. Academic Press, Suffolk

Chandler J, Ashmore P, Paola C, Gooch M, Varkaris F (2002) Monitoring river-channel change using terrestrial oblique digital imagery and automated digital photogrammetry. Ann Assoc Am Geogr 92:631–644. doi:10.1111/1467-8306.00308

Chapman MG, Roberts DE (2004) Use of seagrass wrack in restoring disturbed Australian saltmarshes. Ecol Manage Restor 5:183–190. doi:10.1111/j.1442-8903.2004.00207.x

Costanza R, D'Arge R, De Groot R, Farber S, Grasso M, Hannon B, Limburg K, Naeem S, O'Neill RV, Paruelo J, Raskin RG, Sutton P, Van Den Belt M (1997) The value of the world's ecosystem services and natural capital. Nature 387(6630):253–260

Craft C (2007) Freshwater input structures soil properties, vertical accretion, and nutrient accumulation of Georgia and U.S. tidal marshes. Limnol Oceanogr 52(3):1220–1230

Dronkers J (1986) Tidal asymmetry and estuarine morphology. Neth J Sea Res 20:117–131

Eastman JR (2006) IDRISI Andes: guide to GIS and image processing. Clark University, Worcester

Eastman Kodak Company (2005) Kodak Ektachrome – professional infrared EIR film. Technical data. Color transparency film. Kodak Publication No. TI-2323

Fagherazzi S, Hannion M, D'Odorico P (2008) Geomorphic structure of tidal hydrodynamics in salt marsh creeks. Water Resour Res 44:W02419. doi:10.1029/2007WR006289

Golden Software (2011) Surfer 10 user's guide: contouring and 3D surface mapping for scientists and engineers. Golden Software, Inc., Colorado

Hladik C, Alber M (2012) Accuracy assessment and correction of a LIDAR-derived salt marsh digital elevation model. Remote Sens Environ 121:224–235. doi:10.1016/j.rse.2012.01.018

Huang H, Chen C, Blanton JO, Andrade F (2008) A numerical study of tidal asymmetry in Okatee Creek, South Carolina. Estuar Coast Shelf Sci 78:190–202. doi:10.1016/j.ecss.2007.11.027

Koopmans BN, Wang Y (1995) Measurement of land-sea transition from ERS-1 SAR at different phases of tidal water. ERSWAD project final report. Netherlands Remote Sensing Board (BCRS)

Li C, O'Donnell J (2005) The effect of channel length on the residual circulation in tidally dominated channels. J Phys Oceanogr 35:1826–1840

Lohani B, Mason DC (1999) Construction of a digital elevation model of the Holderness coast using the waterline method and airborne thematic mapper data. Int J Remote Sens 20(3):593–607

Mason DC, Davenport IJ, Robinson GJ, Flather RA, McCartney BS (1995) Construction of an inter-tidal digital elevation model by the "water-line" method. Geophys Res Lett 22(3):3187–3190

Mason D, Hill D, Davenport I, Flather R, Robinson G (1997a) Improving inter-tidal digital elevation models constructed by the waterline technique. In: Proceedings of the 3rd ERS symposium on space at the service of our environment, Florence, Italy, 17–21 March 1997 (ESA SP-414, 3 vols., May 1997), pp 1079–1082

Mason DC, Davenport IJ, Flather RA (1997b) Interpolation of an intertidal digital elevation model from heighted shorelines: a case study in the Western Wash. Estuar Coast Shelf Sci 45:599–612

Mason DC, Davenport IJ, Flather RA, Gurney C (1998) A digital elevation model of the inter-tidal areas of the Wash, England, produced by the waterline method. Int J Remote Sens 19(8):1455–1460

Mason DC, Amin M, Davenport IJ, Flather RA, Robinson GJ, Smith JA (1999) Measurement of recent intertidal sediment transport in Morecombe Bay by using the waterline method. Estuar Coast Shelf Sci 49:427–456. doi:10.1006/ecss.1999.0508, Interpolation of an intertidal digital elevation model from heighted shorelines: a case study in the Western Wash

Mason DC, Davenport IJ, Flather RA, Gurney C, Robinson GJ, Smith JA (2001) A sensitivity analysis of the waterline method of constructing a digital elevation model for intertidal areas in ERS SAR scene of Eastern England. Estuar Coast Shelf Sci 53:759–778. doi:10.1006/ecss.2000.0789

Mason DC, Marani M, Belluco E, Feola A, Ferrari S, Katzenbeisser R, Lohani B, Menenti M, Paterson DM, Scott TR, Vardy S, Wang C, Wang H-J (2005) LiDAR mapping of tidal marshes for ecogeomorphological modelling in the TIDE project. Eighth international conference on remote sensing for marine and coastal environments, Halifax, Nova Scotia, 17–19 May

Maune DF, Kopp MK, Crawford AC, Zervas CE (2007) Digital elevation model technologies and applications: the DEM User's manual, 2nd edn. American Society for Photogrammetry and Remote Sensing, Bethesda

Montané JM, Torres R (2006) Accuracy assessment of Lidar saltmarsh topographic data using RTK GPS. Photogramm Eng Rem Sens 72(8):1–7

Neuenschwander A, Crawford M, Weed C, Gutierrez R (2000) Extraction of digital elevation models for airborne laser terrain mapping data. In: Proceedings 2000 IEEE, pp 2305–2307

NOAA (2012) NOAA tides and currents: St. Simons Island, GA. http://co-ops.nos.noaa.gov/data_menu.shtml?stn=8677344%20St.simons%20Island,%20GA&type=Datums. Accessed 10 Sept 2012

Ramsey ER (1995) Monitoring flooding in coastal wetlands by using SAR imagery and ground-based measurements. Int J Remote Sens 16:2495–2502

Schmid KA, Hadley BC, Wijekoon N (2011) Vertical accuracy and use of topographic LIDAR data in coastal marshes. J Coast Res 27(6A):116–132. doi:10.2112/JCOASTRES-D-10-00188.1

Won J-S, Na Y-H, Kim S-W (2003) Tidal Flat DEM generation by satellite remote sensing. In: Proceedings 2003 IEEE, pp 2116–2118

Zhao B, Kreuter U, Li B, Ma Z, Chen J, Nakagoshi N (2004) An ecosystem service value assessment of land-use change on Chongming Island, China. Land Use Policy 21:139–148. doi:10.1016%2fj.landusepol.2003.10.003

Chapter 14
Examining Material Transport in Dynamic Coastal Environments: An Integrated Approach Using Field Data, Remote Sensing and Numerical Modeling

Richard L. Miller, Ramón López, Ryan P. Mulligan, Robert E. Reed, Cheng-Chien Liu, Christopher J. Buonassissi, and Matthew M. Brown

Abstract Coastal environments are critical ecological systems and offer vital resources and functions to societies worldwide. As a major interface between terrestrial and ocean environments, coastal water bodies (rivers, estuaries, bays and coastal margins) provide key ecological services and are the major conduit and processors of terrestrially derived particulate and dissolved material as they are transported to the ocean. Consequently, coastal environments have been shown to play a major role in global bio-geochemical cycles and provide critical habitat for a host of marine species. Globally, these important environments are under considerable pressure from high population densities, increasing growth rates and are particularly vulnerable from the effects of projected climate change such as sea level rise and increased storm events. Despite their importance, significant gaps remain in our understanding of how these environments will respond to climate

R.L. Miller (✉)
Department of Geological Sciences, East Carolina University, Greenville, NC, USA

Institute for Coastal Science and Policy, East Carolina University, Greenville, NC, USA
e-mail: millerri@ecu.edu

R. López
Institute for Coastal Science and Policy, East Carolina University, Greenville, NC, USA

R.P. Mulligan
Department of Civil Engineering, Queen's University, Kingston, ON, Canada

R.E. Reed
Center for Applied Aquatic Ecology, North Carolina State University, Raleigh, NC, USA

C.-C. Liu
Department of Earth Sciences, National Cheng Kung University, Tainan, Taiwan R.O.C.

C.J. Buonassissi
Institute for Coastal Science and Policy, Coastal Resource Management PhD Program, East Carolina University, Greenville, NC, USA

M.M. Brown
Department of Geological Sciences, East Carolina University, Greenville, NC, USA

C.W. Finkl and C. Makowski (eds.), *Remote Sensing and Modeling: Advances in Coastal and Marine Resources*, Coastal Research Library 9,
DOI 10.1007/978-3-319-06326-3_14,

change, increasing human population, land use changes, and over exploitation of natural resources. This lack of understanding is due in part to the difficulties in developing effective monitoring and analysis programs using only a single measurement approach that is limited in its spatial and temporal coverage.

We describe an integrated approach based on field measurements, remote sensing and numerical modeling that is being developed to examine the transport of dissolved (colored dissolved organic matter (CDOM), dissolved organic carbon (DOC)) and particulate material (total suspended matter (TSM)) within a complex coastal system, the Albemarle-Pamlico Estuarine System (APES), North Carolina USA. This integrated approach was established to overcome limitations associated with a single measurement and analysis approach. Field measurements and discrete samples are acquired using well-established protocols from small boats, bridges, and from the shore. Remotely sensed data are obtained from several sensors with diverse capabilities including SeaWiFS, MODIS, MERIS, HICO, Landsat and FORMOSAT-2. The numerical model Delft3D is used to simulate freshwater and DOC transport in the estuaries following major rainfall events that lead to high river discharge. Challenges associated with examining the APES using a single vs. an integrated measurement approach along with representative results from a broad suite of measurements are presented. Future advances in technology and refinements in our integrated approach are also considered.

14.1 Introduction

Coastal areas are critical environmental systems worldwide that are highly vulnerable to human-induced and natural change. The highest human population densities and the highest rates of global population growth are in the lowland portion of watersheds that are within 100–200 km inland of the coast (Small and Nicholls 2003). Emigration to coastal towns has shifted expansive urban development toward the coast. For example, in Venezuela 40 % of the population now live within a coastal zone that represents only 2 % of the country's available land (Hinrichsen 1998). In 2010, 123.3 million people in the United States, or 39 % of the nation's population, lived in Coastal Shoreline Counties and 163.8 million people, 52 %, lived in Coastal Watershed Counties (National Oceanic and Atmospheric Administration 2013). While coastal cities are generally prized areas for human habitation and recreation, regions of active shipping commerce, locations of major commercial and recreational fisheries, and vital areas for national defense, it is often the pressures of these human activities that severely threaten the health and availability of resources that these areas provide and that make them such important systems.

In addition to sustaining human habitation and offering various resources, coastal environments provide many key ecological services and play a major role in the global cycling of material such as nitrogen and carbon (Barbier et al. 2011; Bauer and Bianchi 2011; Howarth 2008; Seitzinger et al. 2002). Networks of regional coastal waterways such as rivers, bays and estuaries serve as the major conduit for the mobilization, transport and transformation of terrestrial material to the coastal ocean. Floodplains associated with these networks are the dominant site for alluvial storage of terrestrial materials eroded from the landscape (Walling 1999) and are dynamic process engines that control the biogeochemical transformations and flux of terrestrial materials from landscapes to the ocean and atmosphere. Coastal areas are also vulnerable to climate change that may be expressed locally as accelerated rates of sea level rise (Sallenger et al. 2012) and increased intensity of storms (Elsner et al. 2008). These major natural events can have direct and significant impact on human populations, the availability and use of coastal resources, as well as the transport and cycling of material (Burkholder et al. 2004; Paerl et al. 2001). Despite their importance, significant gaps remain in our understanding of how coastal environments function. More importantly, there is a great need to better understand how coastal environments will respond to climate change, increasing human populations, land-use changes, and increased exploitation of natural resources. These gaps must be addressed with targeted science if coastal ecosystems are to be effectively managed for future generations.

Coastal waters are often complex, dynamic environments where a vast array of coupled biological, chemical, geological, and physical processes occur over multiple time and space scales. This characteristic of coastal waters represents a major challenge in designing a program to adequately monitor and analyze coastal aquatic environments, particularly if one attempts to consider several biologically active components. Field programs consisting of periodic in situ measurements using traditional field instruments and sampling protocols from small boats most often are ineffective in capturing the range and variability of many coastal processes. This approach can frequently under sample in both time and space. In situ instruments deployed from moorings and platforms, while perhaps adequately sampling a variable over the appropriate time scale, will generally represent a major under sampling in space due to a limited number of deployed instruments. Optical remote sensing (i.e., Ocean Color) from space and airborne-based instruments represents a unique approach to gain frequent, synoptic data to address the complex nature of many coastal processes. Numerous investigators have demonstrated the use of remote sensing in coastal systems (see for example, Miller et al. 2005a; Richardson and LeDrew 2006; Wang 2009, and references therein). There are however, several challenges to the use of remote sensing in coastal environments including clouds, effective atmospheric correction schemes, and accurate bio-optical algorithms. The optical properties of coastal waters are particularly complex. The propagation and utilization of light within the water column varies over time and space scales corresponding to changes in concentrations of optically active materials (e.g., phytoplankton, colored dissolved organic matter (CDOM), and suspended particulates). Several components such as CDOM and phytoplankton have overlapping absorption spectra that make the detection of an individual component difficult (Miller et al. 2002). Numerical models have also been developed and applied successfully to an analysis of various coastal processes (see for example, Blaas et al. 2007; Druon et al. 2010; Elias et al. 2012; Green et al. 2008; Mulligan et al. 2008; Wool et al. 2003). Although numerical models can overcome many of the limitations of traditional field sampling and remote sensing, models can be limited by their computational requirements and sufficient appropriate field data for model calibration and validation.

We describe here an approach to investigating material transport in the coastal waters of North Carolina, USA that integrates field measurements, remote sensing and numerical modeling with the goal of better understanding the coupling between terrestrial and ocean processes. This work is presented in the context of the major basic and applied science questions addressed, challenges and benefits of each element of our approach, example results, and future directions.

14.2 Study Site: The Albemarle-Pamlico Estuarine System, North Carolina USA

The Albemarle-Pamlico Estuarine System (APES) of coastal North Carolina (Fig. 14.1a) is the second largest estuary and the largest coastal lagoonal system in the United States with a surface area of ~80,000 km^2 of land and water, a mean depth of ~4.5 m and maximum depth of 7.5 m (Fig. 14.1b). The APES contributes more than $4B annually from fisheries, employment and tourism (NC Division of Marine Fisheries 1995) and serves as an important coastal fisheries nursery in the Southeast/Mid-Atlantic region of the United States.

Major freshwater discharge into the APES is from four rivers (Chowan, Roanoke, Tar-Pamlico, and Neuse) that primarily drain the Piedmont and Coastal-Plain provinces of North Carolina (Giese et al. 1985). The Chowan and Roanoke rivers flow into the Albemarle Sound in the northern reach of the APES while the Tar-Pamlico and Neuse rivers flow into the much larger Pamlico Sound in the south. Land cover in the major river basins is predominantly forest (about 55 %) and agriculture (about 30 %). The Roanoke floodplain contains the largest and least disturbed expanse of bottomland forest on the U.S. Atlantic coastal plain. Urban areas within the Chowan, Roanoke and Tar-Pamlico River Basins comprise less than 6 %. The Neuse River Basin is the largest and most populated watershed draining directly into Pamlico Sound and is experiencing rapid population growth that could increase the total population to about three million by 2050 (NCDENR 2010). Major cities include Rocky Mount and Greenville in the Tar-Pamlico River Basin and Raleigh, Durham, Chapel Hill and New Bern in the Neuse River Basin. Many of the streams, tributaries and rivers within the APES watershed are designated by the US Environmental Protection Agency as 303(d) impaired (Clean Water Act of 1972); chlorophyll *a* (chl *a*) levels in the Neuse River Estuary (NRE) are currently being managed under an EPA mandated TMDL (Total Daily Maximum Load) of total nitrogen.

The Outer Banks barrier island chain seaward of the APES restricts exchange with coastal shelf waters to primarily through three narrow inlets. This restricted flow limits astronomical tides and creates the lagoonal nature of the APES with an

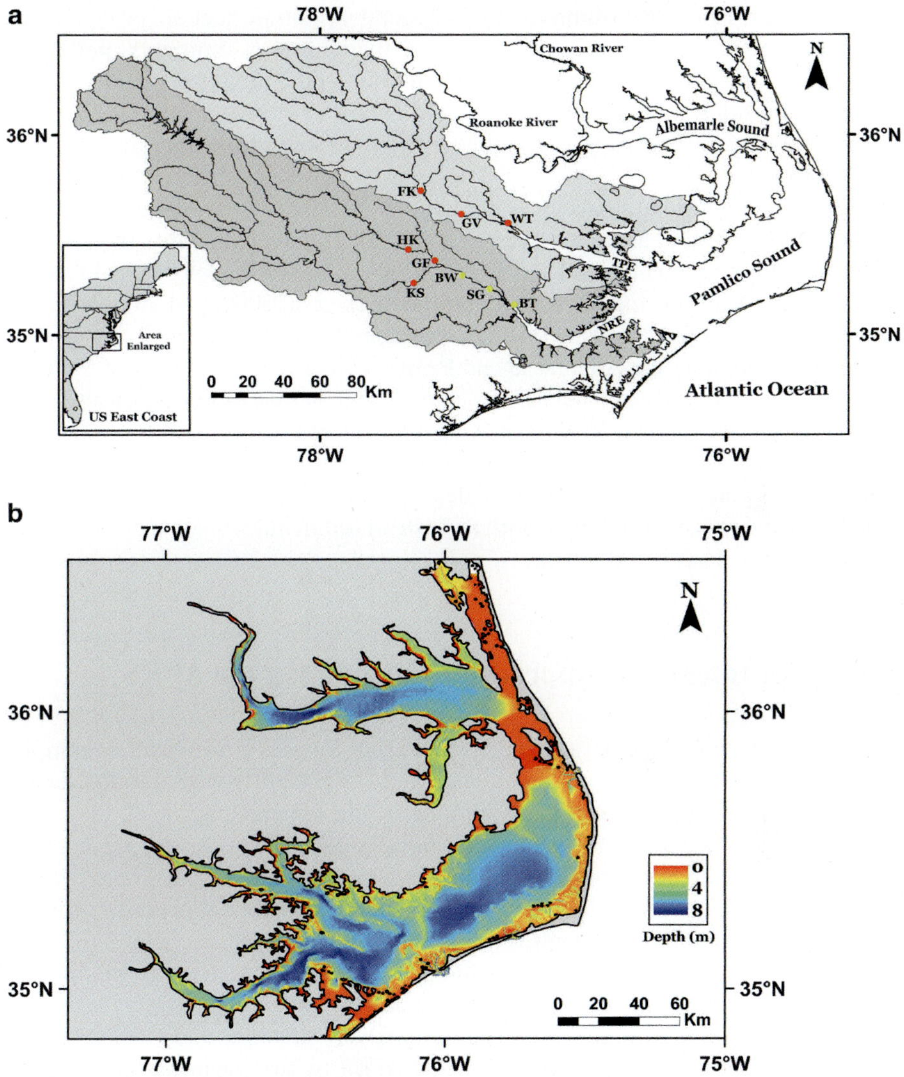

Fig. 14.1 (**a**) The Albemarle-Pamlico Estuarine System (APES) in North Carolina USA showing the Neuse River (*dark grey*) and Tar-Pamlico (*light grey*) drainage basins. Field sample locations are Hookerton (HK), Grifton (GF), Kinston (KS), Barnwell (BW), Spring Garden (SG) and Bridgeton (BT) on the Neuse River and its tributaries, and Falkland (FK), Greenville (GV) and Washington (WT) on the Tar River. Surface water samples are taken near the center of the river channel from a bridge (*red circles*) or from a dock nearshore (*yellow circles*) to examine material transport within the Neuse River Estuary (NRE) and Tar-Pamlico Estuary (TPE). (**b**) Bathymetry of the Albemarle-Pamlico Estuarine System showing the shallow estuaries and broad shoals in both sounds (color figure online)

estimated residence or flushing time of 11 months (Pietrafesa et al. 1986). Hence, the potential impact of terrestrially derived material (e.g., nutrients, pollutants, sediments) transported to the APES is greater due to the prolonged residence time of material within the system. The enclosed nature of the APES makes it an ideal location for examining short- to long-term effects of landscape processes and land-use practices.

Wind forcing primarily drives water circulation in the major sounds and estuaries (e.g., Pietrafesa et al. 1986; Luettich et al. 2002). Given the large surface area, shallow bathymetry with broad shoals and the alignment of the long axis of Pamlico Sound with the prevailing wind direction (Luettich et al. 2000), a dominate feature of the APES is the moderate-to-high concentration of suspended material or turbidity throughout the Albemarle and Pamlico Sounds as frequently observed in remotely sensed images or field observations. This is true following even modest winds (Giffin and Corbett 2003), but is most prevalent following storms such as the frequently occurring winter Nor'easters. However, field data associated with the import, transport and fate of particles (deposition/resuspension) is severely limited and hence the resuspension, transport, and fate of sediments and associated material in the APES is largely unknown.

14.3 An Integrated Approach to Examining the APES

It is clear that the Albemarle-Pamlico Estuarine System is a dynamic, complex environment whose ecological functioning and sustainability is vital to the state of North Carolina and the nation. Additionally, the characteristics of the APES provide a good opportunity to examine how land-ocean processes are coupled, particularly the impact of anthropogenic activities on coastal marine ecosystems, and to gain a better understanding of the transport, transformation and cycling of material in coastal environments. However, the highly dynamic nature and large size of the APES presents major challenges to developing effective monitoring, analysis and management programs. To establish such a program, an integrated approach using multiple measurement and analysis strategies must be employed. We initiated an integrated approach for the APES by first focusing on the Neuse River Basin and Estuary, mainly because several long-term monitoring programs exist and this system has been the focus of numerous investigations.

14.4 Field Sampling and Measurements

A field program was initiated to examine the transport of terrestrially derived dissolved and particulate matter within the NRE following major rain events after the passage of Tropical Storm (TS) Nicole on September 30, 2010. TS Nicole was a short-duration but unusually large asymmetric tropical cyclone that was absorbed into another extratropical storm with large amounts of moisture that brought exceptionally high rainfall (greater than 10 in.) to parts of the Neuse River basin within a 24-h period. Four nearshore stations (HK, BW, SG, BT) were selected to respond quickly to the event (Fig. 14.1a). Surface water samples were collected at the four stations from October 2–13, 2010 following the passage of TS Nicole. For consistency and comparability, the sampling and measurement protocols established during this study are those we continue to apply for all samples collected from the APES. Surface water samples are collected by directly filling acid washed 1 L amber Nalgene bottles or using a 2.2 L horizontal Niskin bottle or a cleaned plastic bucket and then transferred to acid washed 1 L amber Nalgene bottles. All samples are stored on ice and transported to the laboratory where they are stored refrigerated (ca. 4 °C) in the dark until analysis. In the field, chl *a* and turbidity are measured using an AquaFluor handheld fluorometer/turbidimeter (Turner Designs). Salinity is either measured in the field using a Reichert digital refractometer (model AR200) or in the lab using an Orion 4 Star conductivity meter (Thermo Scientific). The program was expanded in 2012 by adding two additional bridge-based stations in the Neuse River Basin (GF and KS) and by locating three new bridge-based stations in the Tar-Pamlico (FK, GV, WT) from the mid-to-lower reach of the system (Fig. 14.1a). Four field campaigns (two each in May and June 2013) were conducted using a small boat to obtain surface water samples at 13 stations along the primary longitudinal axis of the Tar-Pamlico Estuary (TPE) from the head of the estuary to Pamlico Sound. In general, samples are collected at various river stage heights based on a river's hydrograph (Fig. 14.2) to gain perspective on the variability between low-flow and high-flow transport of material within a river basin to the adjacent estuarine waters.

Water quality data are obtained for the NRE from two state supported long-term monitoring programs. Data for the upper-to-mid NRE are collected as part of the Neuse Estuary Monitoring and Research Program (NEMReP) conducted by the Center for Applied Aquatic Ecology, North Carolina State University (http://www.ncsu.edu/wq/RTRM/); data for the mid-to-lower NRE are obtained from the Modeling and Monitoring project (ModMon) field campaigns conducted by the University of North Carolina, Chapel Hill (http://www.unc.edu/ims/neuse/modmon/index.htm). Both monitoring programs collect vertical profiles of temperature, salinity, turbidity, dissolved oxygen and chl *a*. The NEMReP also collects water at three depths for nutrients (Nitrogen, Phosphorus), total organic carbon and dissolved nitrogen, and suspended sediments in addition to discrete surface water samples for analysis within our program. The NEMReP and ModMon stations are shown in Fig. 14.3.

14.5 Laboratory Measurements and Analysis

All discrete surface water samples are filtered and then analyzed for dissolved organic carbon (DOC), total suspended mater (TSM) concentrations and CDOM absorption following NASA standard protocols (Mitchell et al. 2002). The quality of dissolved organic matter (DOM) is determined from fluorescence Excitation-Emission Matrix (EEM) spectroscopy as measured using a FluoroMax-4

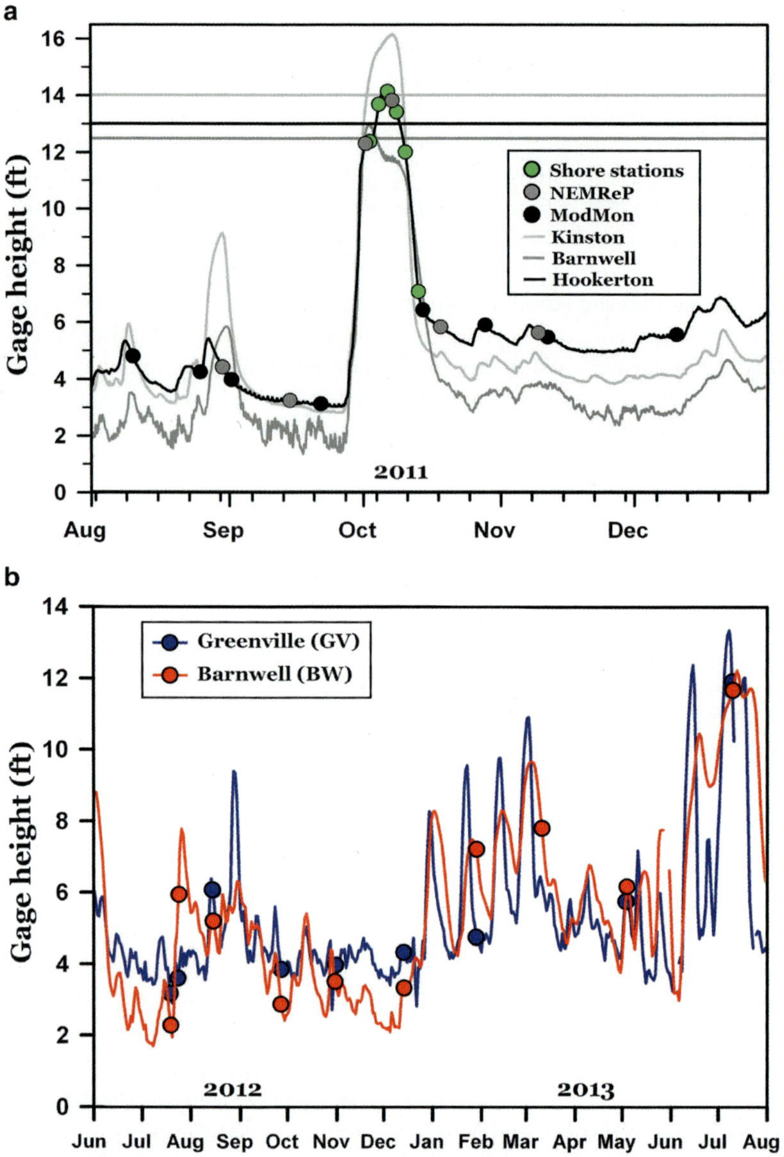

Fig. 14.2 (**a**) Dates when near-surfaces samples were taken to examine the impact of Tropical Storm Nicole on the mobilization and transport of dissolved and particulate material in the Neuse River Estuary. River height at three USGS gage stations in the lower Neuse River basin are shown to indicate localized flooding (*horizontal solid lines* are National Weather Service floodstage) and that river height remained considerably higher than pre-Nicole levels 2 months after the passage of the storm. (**b**) Distribution of field sample dates as a function of river height for a station on the Neuse (Barnwell) and Tar (Greenville) Rivers. The program goal is to sample various stages of river height and its effect on transport processes

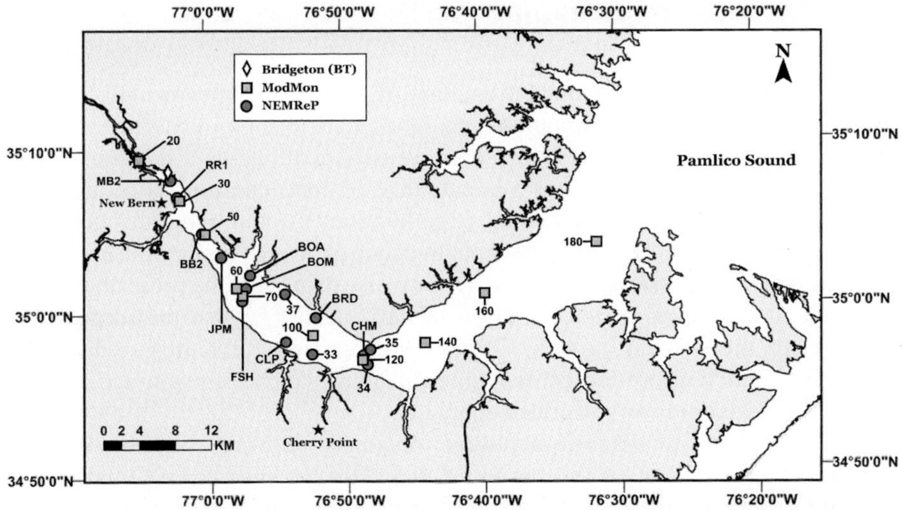

Fig. 14.3 Sample locations for the NEMReP (Neuse Estuary Monitoring and Research Program) and Modeling and Monitoring (ModMon) programs within the Neuse River Estuary. The near-shore station Bridgeton (BT) of our program is shown for reference

Spectrofluorometer (HORIBA Jobin Yvon) following Mitra et al. (2013). Samples are removed from refrigeration and brought to room temperature prior to filtration. Samples are filtered under gentle vacuum (<5 in Hg) through pre-rinsed (Milli-Q water and sample) 0.2 μm Nuclepore polycarbonate filters for CDOM, florescence spectroscopy and TSM samples or through pre-combusted (450 °C for 4 h), pre-rinsed, 0.7 μm Whatman GF/F filters for DOC. The filtrate for CDOM and fluorescence spectroscopy is collected directly into acid-washed, amber glass bottles. The filtrate for DOC samples is collected directly into pre-combusted borosilicate scintillation vials with foil liner caps. DOC samples are acidified with 150 μl of 2 N hydrochloric acid. All filtered samples are stored refrigerated until analysis at room temperature. All measurements are usually made within 2 weeks following filtration.

CDOM UV-Vis absorption spectra are made using a Lambda 850 dual-beam spectrophotometer (PerkinElmer) or a QFT2 (World Precision Instruments, Inc) single-beam fiber optic system according to Miller et al. (2011a). Milli-Q water is used as a reference in both systems. The spectral range (1 nm resolution) used is 250–800 nm for the Lambda 850 spectrophotometer and 187–721 nm for the QFT2.

DOC samples are run on a Shimadzu TOC-V_{CPN} Total Organic Carbon Analyzer following the manufacturers recommended procedures. The concentration (mg/l) of total dissolved nitrogen (TDN) within the DOC prepared sample is provided by the Shimadzu system as well.

TSM is determined gravimetrically by filtering a known volume of sample through pre-weighed 0.2 μm Nuclepore filters. Filters are rinsed following filtration of the sample with a sufficient volume of Milli-Q water to remove salts. The filters are dried in an oven for 6 h at 75 °C then reweighed on a Mettler Toledo X205 high precision balance.

14.6 Example Field Results

Field measurements and laboratory analysis of samples is fundamentally the cornerstone of programs that monitor and investigate environmental systems. To adequately examine material transport in coastal systems, field data is required for the variables of interest over a broad range of concentrations, time and space scales, and following major events (e.g. hurricanes). Our program examining material transport in the APES attempts to acquire these necessary measurements and has collected a considerable volume of diverse data since the program began in 2010. A complete and detailed analysis of these data is beyond the scope of this chapter. In this section as well as the results section for remote sensing and numerical modeling, we present only a few examples of representative data in the context of developing an integrated approach.

As an example of field measurements obtained following a major event, the effect of TS Nicole on CDOM concentration and the transport of DOC and TSM in the lower extent of the Neuse River Basin from the nearshore stations are shown in Fig. 14.4. High CDOM concentrations (Fig. 14.4a, b) were observed during

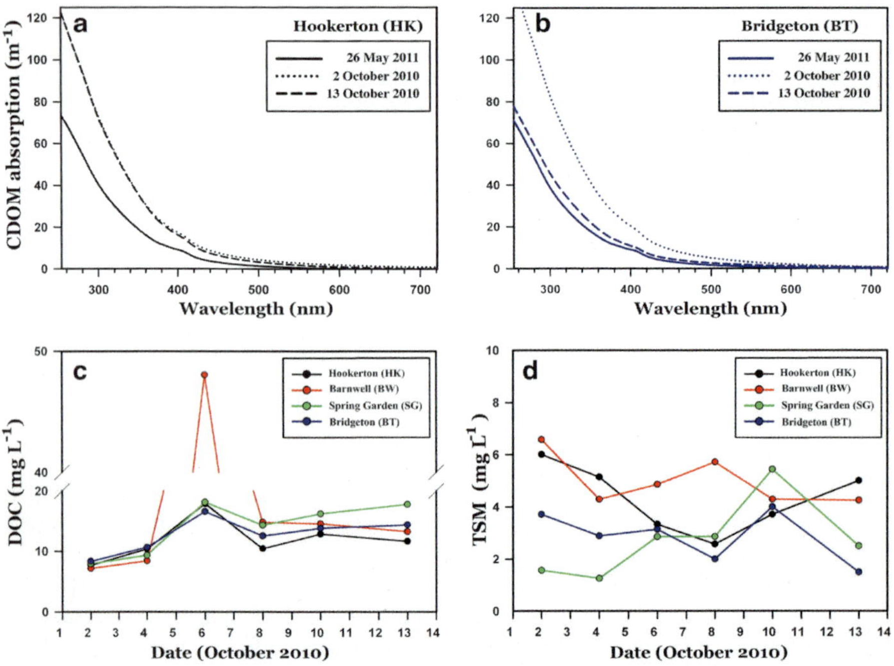

Fig. 14.4 Representative data obtained from samples collected following the passage of TS Nicole (30 September 2010). CDOM absorption spectra (**a, b**) measured at the Hookerton and Bridgeton nearshore during baseline flow (26 May), immediately after (2 October) and 2 weeks following (13 October) TS Nicole. An example of changes in dissolved (**c**) and particulate (**d**) material measured from four nearshore stations on the lower Neuse River

baseline flow conditions (e.g., 26 May 2010) at both the up-river Hookerton ($a_g(443) = 3.8\ m^{-1}$) and lower-river Bridgeton ($a_g(443) = 4.5\ m^{-1}$) stations. These data are typical of the range of CDOM absorption values observed in this system and are generally much higher than values reported for other large river-bay systems such as the Chesapeake Bay (Tzortziou et al. 2007) and Tampa Bay (Chen et al. 2007) but are similar to the range of values observed for the blackwater Suwannee River (Cannizzaro et al. 2013). Immediately following TS Nicole, CDOM absorption at 443 nm more than doubled at each station (8.5 m^{-1} and 9.9 m^{-1} at Hookerton and Bridgeton, respectively). Two weeks following the storm, CDOM concentration at Bridgeton had almost returned to pre-storm values while the concentration at Hookerton remained high ($a_g(443) = 8.5\ m^{-1}$) presumably due to the greater amount of flooding within the Hookerton sub-basin (Fig. 14.2a).

Dissolved organic carbon (Fig. 14.4c) also increased in the Neuse River following TS Nicole and remained high during the sampling period. A significant increase in DOC transport within the NRE following major rain events has been reported by Paerl et al. (2001) following the passage of three hurricanes, Dennis, Floyd, and Irene in September and October 1999 and by Brown et al. (2014) following Hurricane Irene in 2011.

In contrast, TSM (Fig. 14.4d) did not show any apparent trend in either time (days after TS Nicole) or space (up-river, down-river). However, these data may be affected by the flow regime of the river nearshore and may not adequately represent particulate transport in the main river channel. As a result of these findings, we modified our field protocols to include sampling from bridges near the main channel (see Fig. 14.1a).

These results also indicated a need to continue intensive field sampling for a much longer period following a major rain event to adequately capture the full impact of a storm on dissolved material transport within the Neuse River system. For example, to examine CDOM and DOC transport in the NRE following the passage of Hurricane Irene in 2011, Brown (2013) collected frequent field samples at the nearshore stations until 15 days after Hurricane Irene.

The quality of DOM is routinely analyzed using fluorescence spectroscopy. A representative EEM showing the excitation/emission peaks A and C (Coble 1996; Osburn et al. 2012) indicative of terrestrial humic and fulvic substances (peak A) and terrestrial, anthropogenic and agricultural substances (peak C) is shown in Fig. 14.5. To date, our EEMs within the Neuse and Tar Rivers indicate a dominance of terrestrially derived material with an absence or little marine or autochthonous fluorescent substances. Although not shown here, several optical indices including SUVA254 (Weishaar et al. 2003) and absorption spectral slopes and slope ratios (Helms et al. 2008; Inamdar et al. 2011) can provide additional information on the quality and molecular makeup of DOM.

Example results from a typical field sampling design for examining a coastal water body is shown for the TPE in Fig. 14.6. Variation in surface DOC, TSM and salinity along the primary axis of the estuary is shown for the four sample dates in 2013. River height and discharge were at or slightly lower than the median daily average for

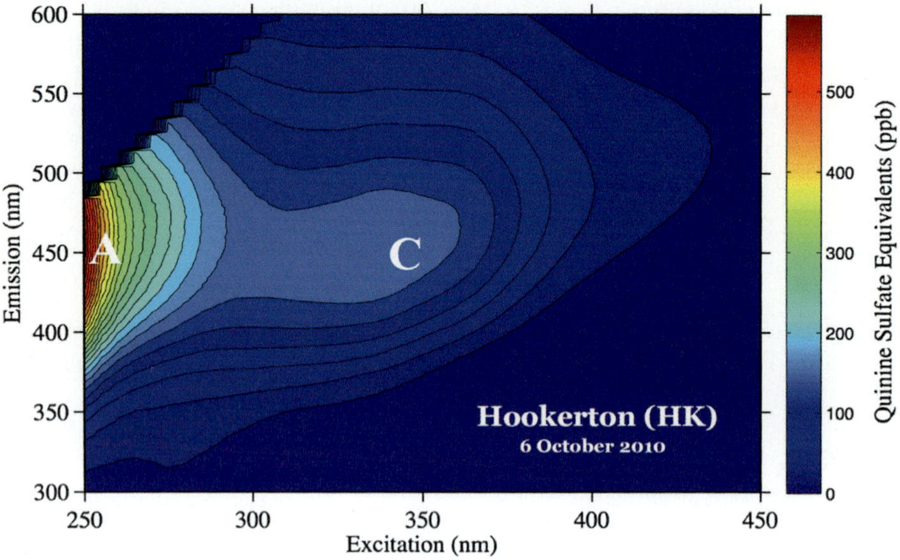

Fig. 14.5 An example EEM (Excitation-Emission Matrix) obtained using fluorescence spectroscopy showing the (**A** and **C**) peaks representing terrestrially-derived humic-like substances

20 years on the first three sample dates (22, 30 May; 12 June, 2013) but was considerably higher on 24 June 2013. As expected, salinity increases from fresh-brackish water near the head of estuary to mid-salinity values at the estuary mouth or opening to Pamlico Sound. The influence of changes in freshwater was restricted to the upper to mid-estuary. Salinity was fairly constant at each location near the opening of the estuary, perhaps due to low freshwater input, wind-driven mixing, and exchange with water from Pamlico Sound. Similar to the Neuse River Estuary, the TPE is primarily characterized by high CDOM and low TSM concentrations. Little variation in TSM was observed throughout most of the estuary with the greatest changes in TSM occurring at the head of the estuary, the main source of freshwater and sediment inputs to the system. In general, DOC concentrations are considerably less than those typically observed within the NRE and show a gradual decrease from the estuary head to the mouth. A common diagnostic approach to examining the dynamics of dissolved substances in coastal waters such as DOC, is to compare the substance concentration as a function of salinity (see for example, Day et al. 2012; Guo et al. 2007) as shown in Fig. 14.7. DOC appears to behave conservatively based on these data with the possible exception of 12 June 2013 where, although the data are more variable than measured on the other sampling dates, there may a DOC source in the mid-estuary region. Osburn et al. (2012) and Brown et al. (2014) reported that DOC behaved conservatively within the NRE following an increase in river discharge (i.e. rapid transport) following Hurricane Irene in 2011. The steeper slope in the change of DOC with salinity observed on 24 June is due to the higher freshwater discharge into the estuary and suggests that DOC is rapidly diluted in the TPE through mixing with lower DOC Pamlico Sound water.

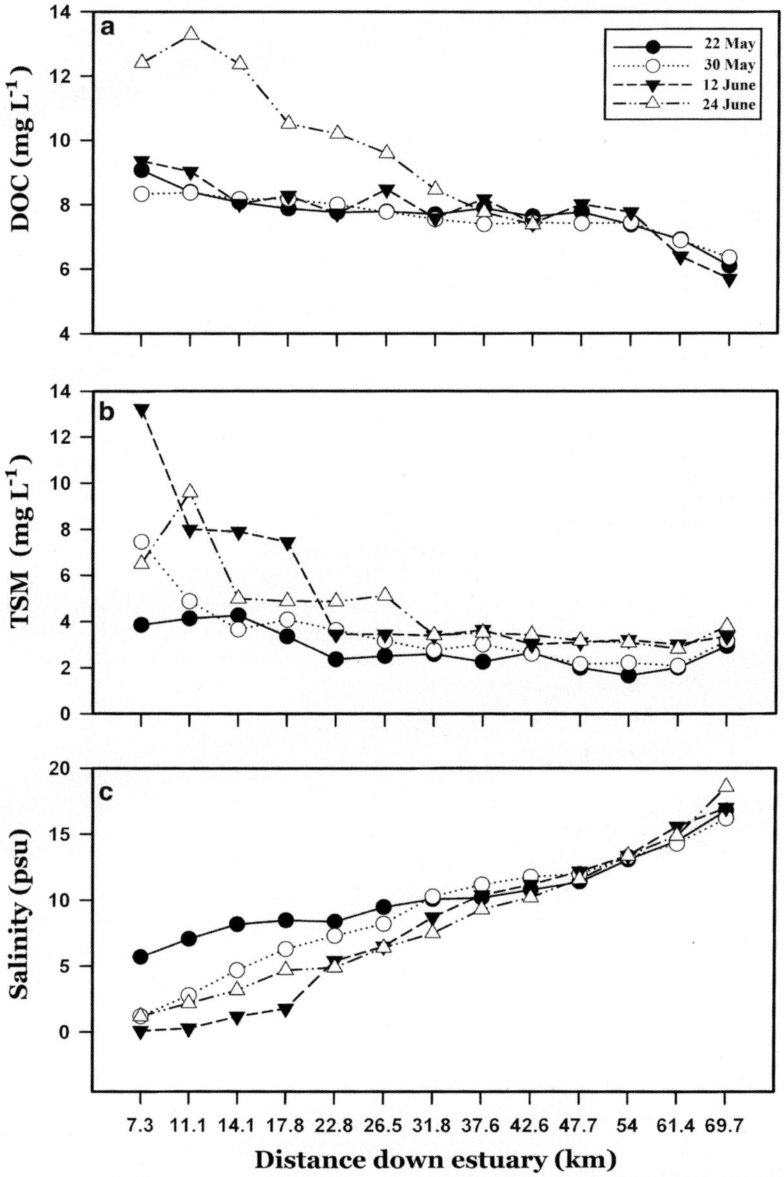

Fig. 14.6 Dissolved organic carbon (**a**), total suspended matter (**b**), and salinity (**c**) measured in discrete surface samples taken along a station transect in the Tar-Pamlico Estuary during four cruises in 2013

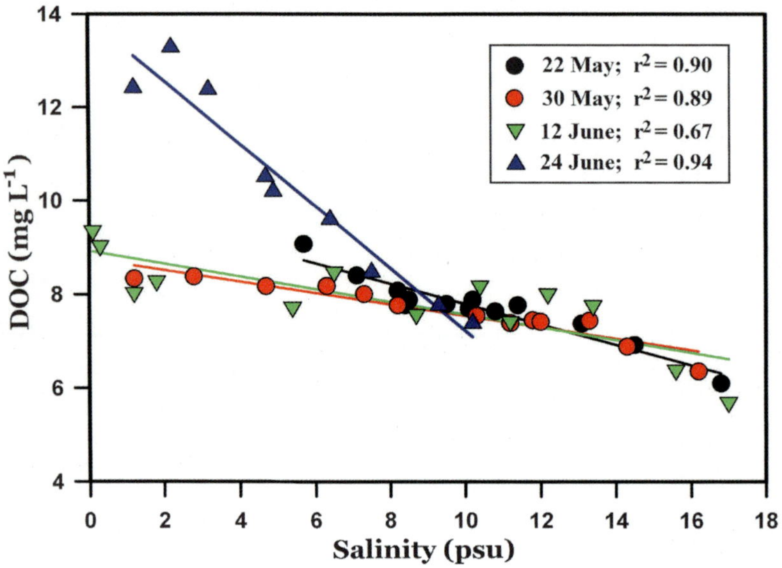

Fig. 14.7 DOC concentration as a function of salinity measured in discrete surface samples showing conservative mixing during the four 2013 cruises in the Tar-Pamlico Estuary. *Solid lines* are the linear least squares regression lines fit to the data for a single cruise (e.g. *solid blue line* corresponds to the 24 June data) (color figure online)

14.7 Remote Sensing

Numerous investigations have firmly established remote sensing as a unique and extremely useful technology for examining a broad range of variables in diverse coastal environments. Meyers and Miller (2005) and Brown et al. (2005) provide a detailed overview of instruments, observations and techniques for airborne and satellite-based remote sensing of coastal aquatic systems, respectively. The unique advantage of remote sensing is that it provides a large-scale synoptic view compared to traditional field sampling. Simply, it is the only technology that can provide a "snapshot" of what is occurring within an environment over a local-to-regional area. This is particularly important when examining dynamic coastal waters where potentially rapidly changing conditions due to many processes such as winds, tides and biological activity can severely bias in situ data taken using traditional field sampling techniques.

The use of satellite ocean color instruments to examine coastal systems worldwide is well documented. For example, Aurin et al. (2013), Del Castillo and Miller (2008), Hu et al. (2004), Miller and McKee (2004), Miller et al. (2005b), Otis et al. (2004) and Zang et al. (2010) used MODIS (MODerate Resolution Imaging Spectroradiometer) images to examine coastal water quality parameters. Representative studies on the application of SeaWiFS (Sea-viewing Wide Field-of-view Sensor) data to coastal waters can be found in Cannizzaro et al. (2013), Cheng et al. (2003), Harding et al. (2005), Hyde et al. (2007), Nezlin et al. (2005), Lopez et al. (2012) and Tomlinson et al. (2004). Coastal investigations using MERIS (MEdium Resolution Imaging Spectrometer) images include Cui et al. (2010), Gons et al. (2008), Kratzer et al. (2008), Sokoletsky et al. (2011) and Xi and Zhang (2011).

Many airborne instruments have also been applied to coastal regions including AVIRIS (Advanced Visible–Infrared Imaging Spectrometer; Lunetta et al. 2009), CASI (Compact Airborne Spectrographic Imager; Casal et al. 2013), CAMS (Calibrated Airborne Multispectral Sensor; Miller and Cruise 1995), and the PHILLS (Portable Imager for Low-Light Spectroscopy; Dierssen et al. 2003). Although the use of airborne instruments is often limited due to deployment logistics and cost, it is important to note that many of these instruments (e.g., AVIRIS, CASI, PHILLS) offer hyperspectral coverage and are therefore extremely useful in developing a range of algorithms and serving as prototypes for existing and planned satellite instruments.

The long lineage of Landsat Thematic Mapper instruments that were designed principally for land applications have also been widely used in coastal aquatic research (see for example, Ekstrand 1992; Khorram et al. 1991; Miller et al. 2011b; Wang 2004). Although Landsat data is not ideal for measuring water quality parameters due to its spectral characteristics (e.g. spectral coverage, signal to noise), it does provide high spatial resolution (30 m) data that is well suited for conducting land cover and land use analyses that are important to investigating the coupling of landscape processes with the transport of terrestrial material to coastal waters.

Our integrated approach attempts to use data from a suite of remote sensing instruments to obtain a range of spatial resolutions, spectral coverage and characteristics, and revisit times necessary to effectively monitor and analyze the dynamic processes of material transport in the APES at the appropriate spatial and temporal scales including the assessment of major events. This capability however, results in numerous challenges not only related to the direct use of remotely sensed data (i.e. acquisition, appropriate software for processing and analysis) but also on the specific requirements for field data to support algorithm development. Currently we obtain and archive all clear sky MODIS data and acquire SeaWiFS, MERIS, HICO (Hyperspectral Imager for the Coastal Ocean), FORMOSAT-2, and Landsat data to support specific investigations. The Hyperspectral Imager for the Coastal Ocean is an imaging spectrometer based on the PHILLS airborne instrument and is flown on the International Space Station. FORMOSAT-2 contains the Remote Sensing Instrument (RSI), a multispectral instrument with a daily revisit period and 8 m resolution that is owned and operated by the National Space Organization (NSPO) of Taiwan (Liu et al. 2007; Miller et al. 2011b).

MODIS, SeaWiFS, MERIS and HICO data are acquired from the NASA Ocean Color Web data portal (http://oceancolor.gsfc.nasa.gov) and processed using the NASA SeaDAS (SeaWiFS Data Analysis System) v6.4, v7.1 software and the ENVI (Exelis) v4.8 image processing software with the atmospheric correction module FLAASH v4.7 (Fast Line-of-sight Atmospheric Analysis of Spectral Hypercubes) and MODIS Conversion Toolkit plugins. L0 or L1 data are obtained for each data type as available. Landsat data are obtained from the USGS Earth Resources Observation and Science (EROS) Center using the USGS Global

Visualization Viewer (GloVis) online search and order tool for satellite data (http://glovis.usgs.gov). FORMOSAT-2 data are obtained directly from NSPO. Landsat and FORMOSAT-2 data are processed using the ENVI software.

14.8 Example Remote Sensing Results

A time series of MODIS images (5, 7, 9 and 15 October) were acquired following the passage of TS Nicole in 2010 to assess the transport of CDOM and DOC from the major rivers of the APES to the Albemarle and Pamlico Sounds and examine suspended sediment dynamics in Pamlico Sound and the adjacent continental shelf. The transport of highly CDOM-laden water (e.g., blackwater) from the Neuse and Tar-Pamlico estuaries to Pamlico Sound can clearly be seen in the MODIS true color images (Fig. 14.8). While these images do not provide a quantitative estimate of CDOM and DOC transport or flux, they do provide a spatial perspective of the change in river and constituent transport to Pamlico Sound over time following a major rain event. Moreover, from these images it appears that a significant amount of CDOM was flushed from the river basins to Pamlico Sound that persisted over 2 weeks after the passage of TS Nicole.

MODIS 250 m data (Bands 1 and 2; 645, 859 nm) from these dates were then processed according to Miller et al. (2011b) to derive estimates of TSM. Briefly, MODIS L0 image files were processed using SeaDAS v6.4 to extract the MODIS 250 m bands. Atmospheric correction was performed with the l2gen module using the single-scattering white aerosol model with 859 nm (MODIS Band 2) as the longest wavelength for aerosol model selection. All L2 mask flags were deselected and normalized water-leaving radiance (nLw, $mW/cm^2/sr/\mu m$) for Bands 1 and 2 were output. Subsequent processing of MODIS images was done in ENVI. Band 1 water-leaving radiance data was converted to TSM concentration (mg/L) using a modified version of the Miller and McKee (2004) TSM algorithm. The resulting time series of TSM 250 m images (Fig. 14.9) clearly demonstrates the dynamic and highly variable nature of suspended matter in the APES. Significant changes in the spatial distribution and concentrations of TSM within the Sounds and offshore occurs over a short period of days. The boundary and distribution of CDOM is also visible in these images as "clearer water" of considerably lower local TSM concentrations. These images also reveal the very large horizontal variability in surface TSM concentrations that is evident in most remotely sensed images of the APES. To date, this variability has been an obstacle to developing a local TSM algorithm using MODIS imagery. For example, two recent cruises near the confluence of the NRE and Pamlico Sound were dedicated to obtain surface samples to develop a regional MODIS TSM algorithm. Although no significant relationship was observed in these data, this analysis suggests that the horizontal variability or surface gradients in TSM resulted in large within pixel (sub-pixel) variability at the 250 m pixel size. Now, consistent with an integrated approach, future field samples to support remote sensing algorithms will be located to incorporate the in situ surface gradients present, the targeted instrument's spatial resolution and its time of data acquisition.

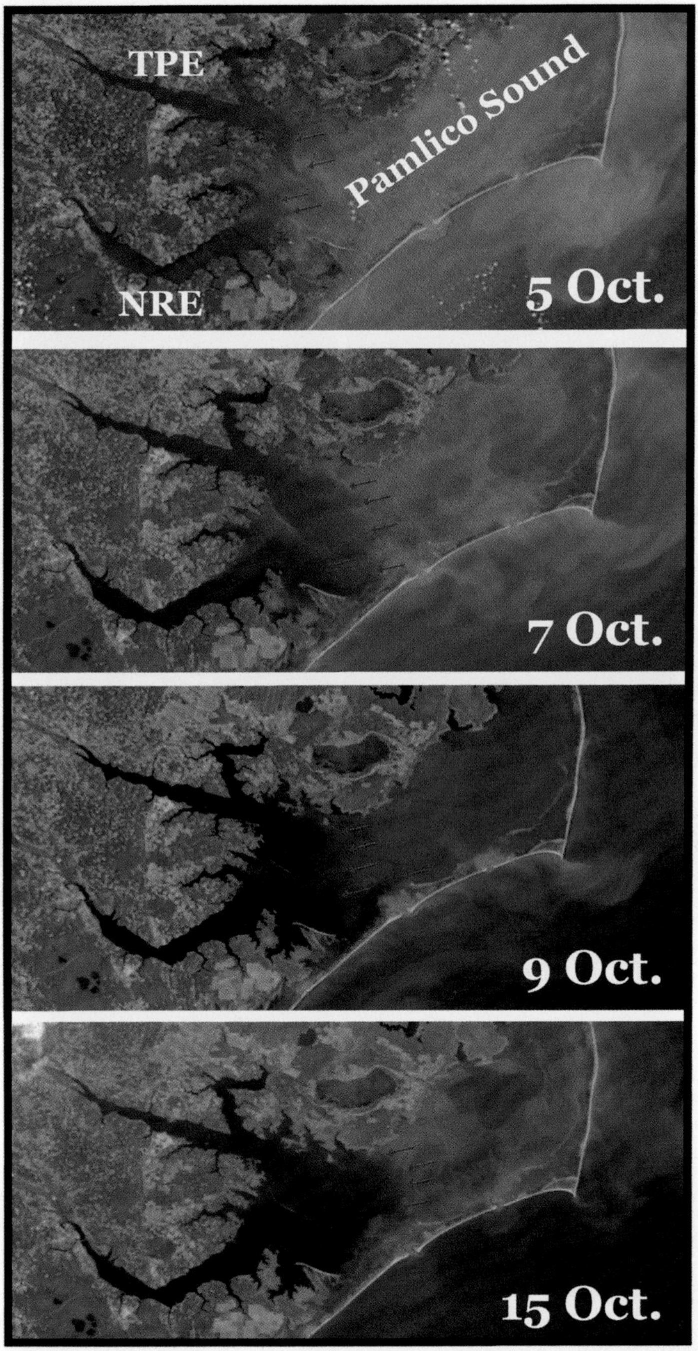

Fig. 14.8 MODIS true-color images of the southern APES showing the progressive discharge of high CDOM waters (blackwater in images) from the Tar-Pamlico Estuary (TPE) and Neuse River Estuary (NRE) into Pamlico Sound following the heavily rainfall from TS Nicole on 30 September 2010. *Red arrows* indicate the "leading edge" of the estuarine CDOM plumes (color figure online)

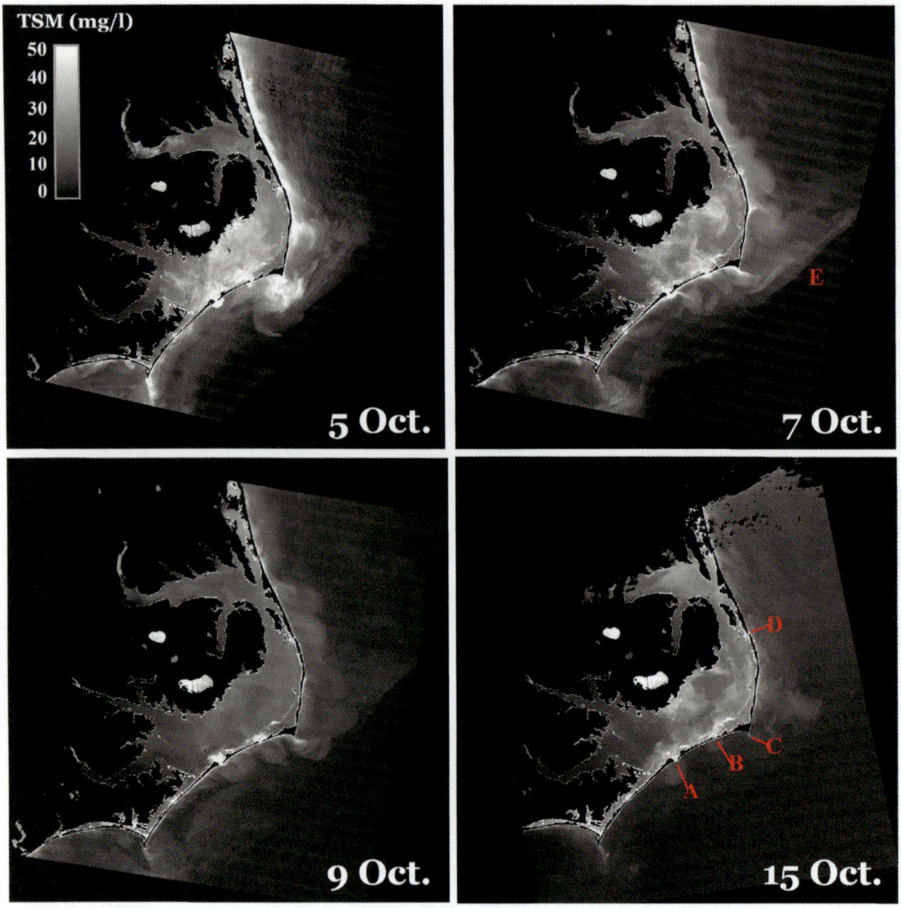

Fig. 14.9 A time series of TSM images showing the large-scale dynamics in suspended particulate matter in the APES and near shore coastal waters derived from MODIS Terra (5–9 October) and MODIS Aqua (15 October). Land and clouds are masked to *black*. Image labels are: **A** = Ocracoke Inlet; **B** = Hatteras Inlet; **C** = Cape Hatteras; **D** = Oregon Inlet; **E** = Edge of Gulf Stream. The 5 October image shows high concentrations of TSM in Pamlico Sound due to wind resuspension with plumes of TSM being transported through the *narrow* inlets

Using a multi-senor approach, Miller et al. (2011b) examined the effect of TSM surface variability in the APES on remote sensing reflectance. Their results indicate a strong relationship between TSM concentrations derived at 8 m (FORMOSAT-2), 30 m (Landsat ETM+) and 250 m (MODIS). Hence, a multi-sensor approach provides a better understanding of the dynamics and distributional patterns of TSM in the APES. Additional efforts are underway to evaluate the coherency in results between other sensors.

As an example, TSM concentrations in the APES derived using MODIS Aqua, MERIS and HICO images collected on 21 October 2009 (Fig. 14.10) were

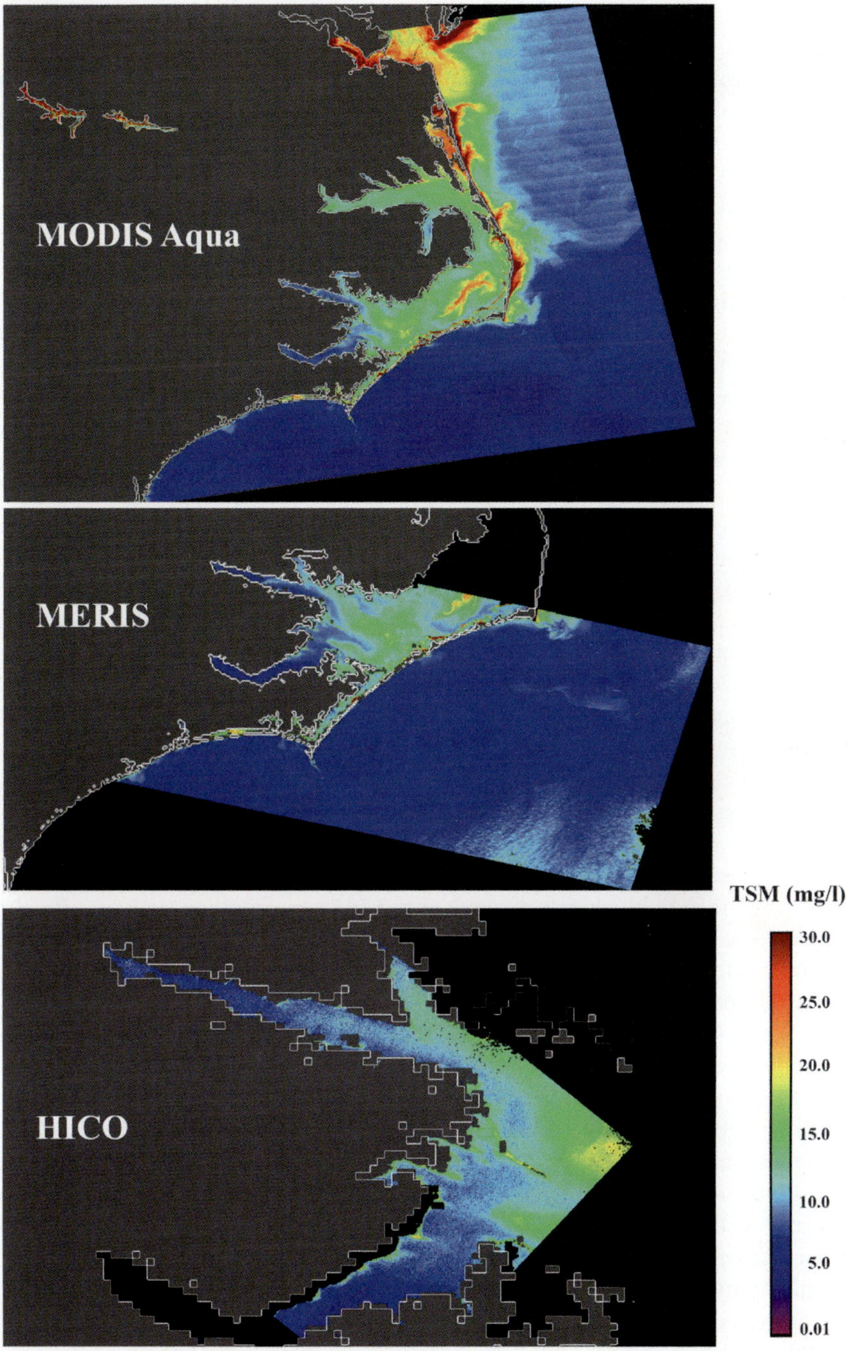

Fig. 14.10 A comparison of TSM concentrations produced using MODIS Aqua, MERIS and HICO data of the APES collected on 21 October 2009. A geographic region (enclosed by 36.90° N 75.12° W and 34.50° N 77.67° W) containing the APES and adjacent drainage basins was subsampled from each image and then reprojected to an equidistant cylindrical projection. Total suspended matter (TSM) was derived from normalized water leaving radiance (nLw) at 645 nm for MODIS and 665 nm for MERIS and HICO using the relationship TSM = nLw * 19.29−0.923. Land and clouds are masked to *dark grey*. All images were scaled to 0.01–30 mg/L

compared following the approach outlined by Miller et al. (2011b). The nominal ground sample distance or spatial resolution for the images is 90 m, 250 m and 300 m for HICO, MODIS and MERIS, respectively. Briefly, MODIS Aqua Level 1a (MODISA), and MERIS and HICO Level 1b data were downloaded from the NASA ocean color web portal (http://oceancolor.gsfc.nasa.gov). MODISA Level 1a granules were processed to Level 1b, retaining the 250 m data (Bands 1 and 2). All sensor Level 1b granules (MODISA 5 min) were processed to Level 2 products (e.g., normalized water leaving radiance and $a_{dg}443$) using SeaDAS v7.0.1 and then reprojected to an equidistant cylindrical projection. Total suspended matter (TSM) was derived from normalized water leaving radiance (nLw) at 645 nm for MODIS and 665 nm for MERIS and HICO using the relationship $TSM = nLw * 19.29 - 0.923$ (Miller et al. 2011c). The TSM concentration of 100 pixels selected at random from an area near the mouth of the NRE imaged by all three sensors were used for this initial inter-sensor comparison (Fig. 14.11). Derived TSM concentrations were significantly correlated ($p < 0.01$) between all three instruments. There is a negative bias (lower values) in MERIS and MODIS TSM concentration when compared to the higher spatial resolution (90 m) HICO data (Fig. 14.11a). In general, the MERIS bias is less than in the MODIS data. The highest linear correlation observed ($r = 0.84$) is between MODIS Aqua and MERIS, although again, the MODIS TSM concentration is about 1–2 mg/L less than the corresponding MERIS TSM estimate (Fig. 14.11b). It is important to note that the spatial resolutions of these instruments, 250 m for MODIS and 300 m for MERIS are closer than the 90 m for HICO. Hence, the variation in TSM estimates may be due at least in large part to sub-pixel variability as related to reasonably high horizontal variation in surface TSM concentrations. These findings are encouraging, however, given the various potential sources of error associated with processing the individual instrument data to TSM concentration such as differences in spectral band pass, spectral signal-to-noise, effectiveness of atmospheric correction and differences in the time of data collection. The time difference in image acquisition between the instruments was 1.35 h for HICO and MERIS, 2.77 h for MERIS and MODIS, and 4.13 h for HICO and MODIS. These time differences, especially between HICO and MODIS Aqua, may be significant depending on the local wind conditions and currents.

As described in the field measurements section, CDOM is a major constituent throughout most of the APES. Therefore, there is a strong need to monitor the large-scale dynamics of CDOM over a time scale that depends on the amount of river discharge, estuarine and sound hydrography, or the time-scale of physical processes associated with a major event. Typically the time scale is a few days to a week to adequately monitor CDOM in the APES. Despite its importance, the measurement of CDOM concentration using remote sensing in the APES is challenging due mainly to the very high CDOM absorption in the blue-green region of the visible spectrum and the fairly narrow spatial dimensions of the estuarine systems. Another significant challenge is the need for an effective atmospheric correction scheme to correct remotely sensed imagery in this coastal region. These challenges and limitations are, however, common to those encountered during other investigations

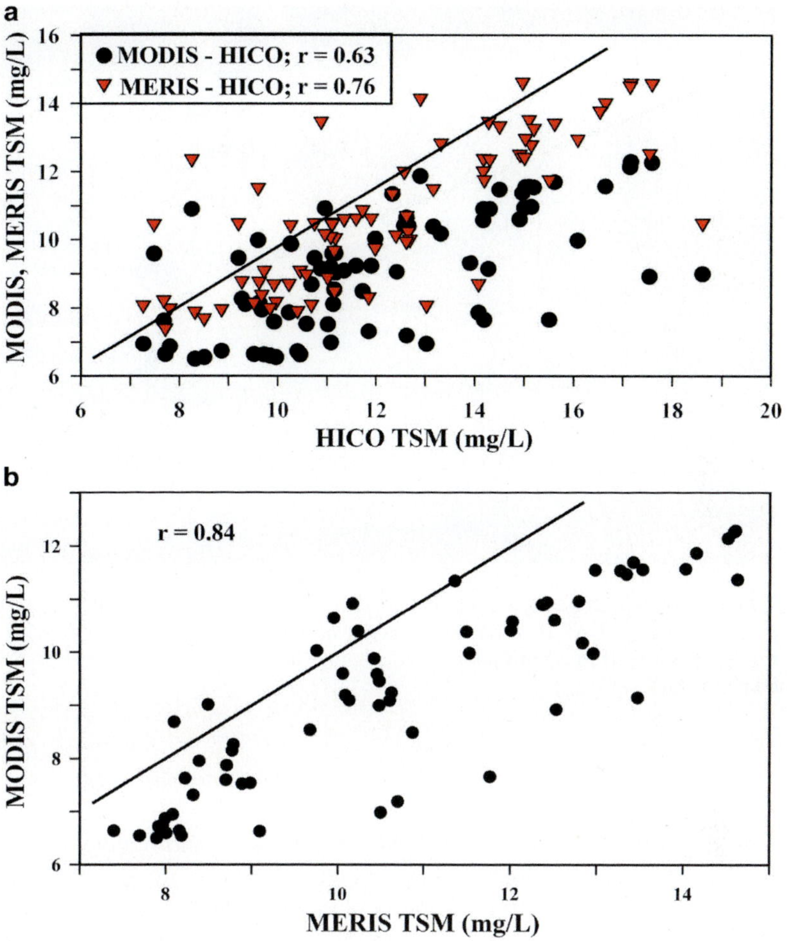

Fig. 14.11 Correlations between TSM estimates from the MODIS Aqua, MERIS and HICO images collected on 21 October 2009. Random pixels (n = 100) from the images were sampled from an area enclosed by 35.26° N 76.27° W and 35.10° N 76.48° W. *Solid lines* are the 1:1 lines

in similar systems. Nevertheless, our program is attempting to gain a better understanding of CDOM dynamics using several ocean color sensors, in particular MODIS Aqua and MERIS. An example of a CDOM image derived using MODIS is shown in Fig. 14.12. Although there is no field data to determine the accuracy of these estimates, the values and distributional gradients within the NRE do not seem correct based on a review of our field measurements. These results also clearly indicate the need for field campaigns dedicated to mapping CDOM distribution in waters with lower CDOM absorption during instrument overpass so that CDOM-based algorithms can be developed within the capabilities of the instruments used and error associated with atmospheric processing.

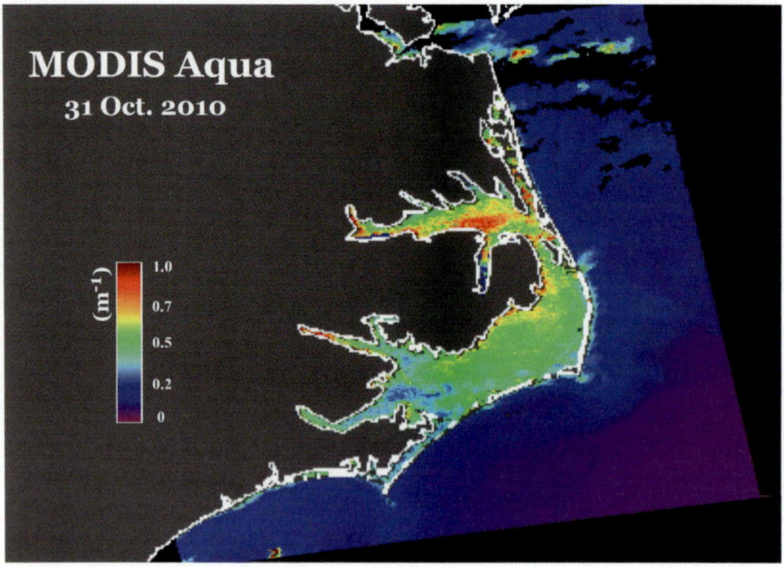

Fig. 14.12 A sample CDOM absorption product (adg(443)) produced using MODIS Aqua data. The QAA method in SeaDAS v6.4 was used to derive absorption values. SeaDAS v7.0.1 was used to reproject the image to an equidistant cylindrical projection. Land is masked *dark grey*. Clouds are masked to *black*

14.9 Modeling

Although the combined use of both field measurements and remote sensing data together has been widely adopted in many investigations of coastal environments, limitations remain in the combined ability of these methods to effectively monitor dynamic, spatially complex coastal systems such as the APES. The cost and logistics of field campaigns and field monitoring stations typically yield only a sparse network of data and it is often very difficult to impossible to sample immediately prior to and after a major event such as a hurricane. Measurements from optical remote sensing only provide near-surface estimates and are frequently limited by clouds. To circumvent at least some of the limitations from field and remote sensing measurements, numerical models that characterize material transport in the APES are being integrated into our program.

Numerical modeling of material transport in coastal environments such as estuaries may be particularly complex due to the large range of important physical and material-related processes. To simulate coastal processes using numerical models, the initial conditions of the system must first be specified. This could be relatively simple or very difficult depending on the horizontal and vertical variability of the system parameters. Data to set the initial conditions may come from various sources including previously published results relevant to a specific site, in situ data, remotely sensed observations, or even results from other numerical models. Calibration and validation data are also critical components in any numerical modeling project that allow the model free parameters to be adjusted such that the simulated conditions best match the observations (calibration) for certain events and to confirm that the numerical predictions are accurate (validation) for other events or over longer time scales.

There are different approaches to numerical modeling and different types of models that can be used to simulate coastal processes. For example, Mulligan et al. (2010) simulated the coastal region of the Mackenzie River Delta in the Arctic Ocean using the Delft3D three-dimensional numerical flow and transport model (Lesser et al. 2004) to examine the processes that are most important in governing the circulation and structure of a thermal plume in the shallow coastal ocean. In the APES, Haase et al. (2012) used the ADCIRC (Luettich et al. 2002) circulation model in combination with a particle-tracking model to simulate the wind-forced hydrodynamics and dispersal of oyster larvae. Mulligan et al. (2014) used the SWAN wave model (Booij et al. 1999) coupled to the Delft3D model to show that under strong wind forcing from hurricanes, wind-generated surface waves and wind-driven storm surge induce changes in water-level that contribute to shoreline flooding. Recent work by Dresback et al. (2013), attempts to couple a suite of numerical models that include the watershed hydrologic model HL-RMS (Koren et al. 2004) and the circulation model ADCIRC to examine the impact of storm surge on flooding. Coupled hydrologic-hydrodynamic models may become important tools for understanding material transport.

Brown et al. (2014) present the results of an initial model to examine DOC transport in the NRE following Hurricane Irene in 2011. Briefly, the Delft3D (Lesser et al. 2004) three-dimensional numerical flow and transport model was used to simulate freshwater and DOC over a 22-day period to assess the impact of rainfall associated with Hurricane Irene on material transport. A high-resolution (~200 × 140 m) model grid with ten vertical layers was constructed in the NRE with a domain extending about 80 km from the head of the NRE near Bridgeton, NC to an open boundary at the intersection of the NRE and Pamlico Sound. Hydrodynamic and transport conditions at the open model boundary were defined using a coarse grid (~350 × 285 m), vertically averaged model that simulated conditions throughout the entire APES. DOC is modeled as a passive tracer within the NRE. Results of simulated DOC concentrations are presented for several selected days within the modeled time period in Figs. 14.13 and 14.14. These results exhibit the simulated transport of DOC along the primary channel axis of the NRE, cross-channel transport and vertical distribution. Modeled salinity is in good agreement with field measurements. As expected, DOC initially propagates downriver in the near-surface freshwater plume and then increases with depth due to vertical mixing processes. Brown (2013) and Brown et al. (2014) provide a more detailed description of the model development, setup, and model results and interpretation. As shown here, a major benefit of employing numerical models is that, given a well-calibrated, accurate model, information on the distribution and transport of material governed by physical processes can be obtained for a spatially detailed, vertically resolved grid over a high temporal time frame that is not constrained by the factors that limit the use of field measurements or remotely sensed images. Upon further validation of this current model with additional field data, the model will be

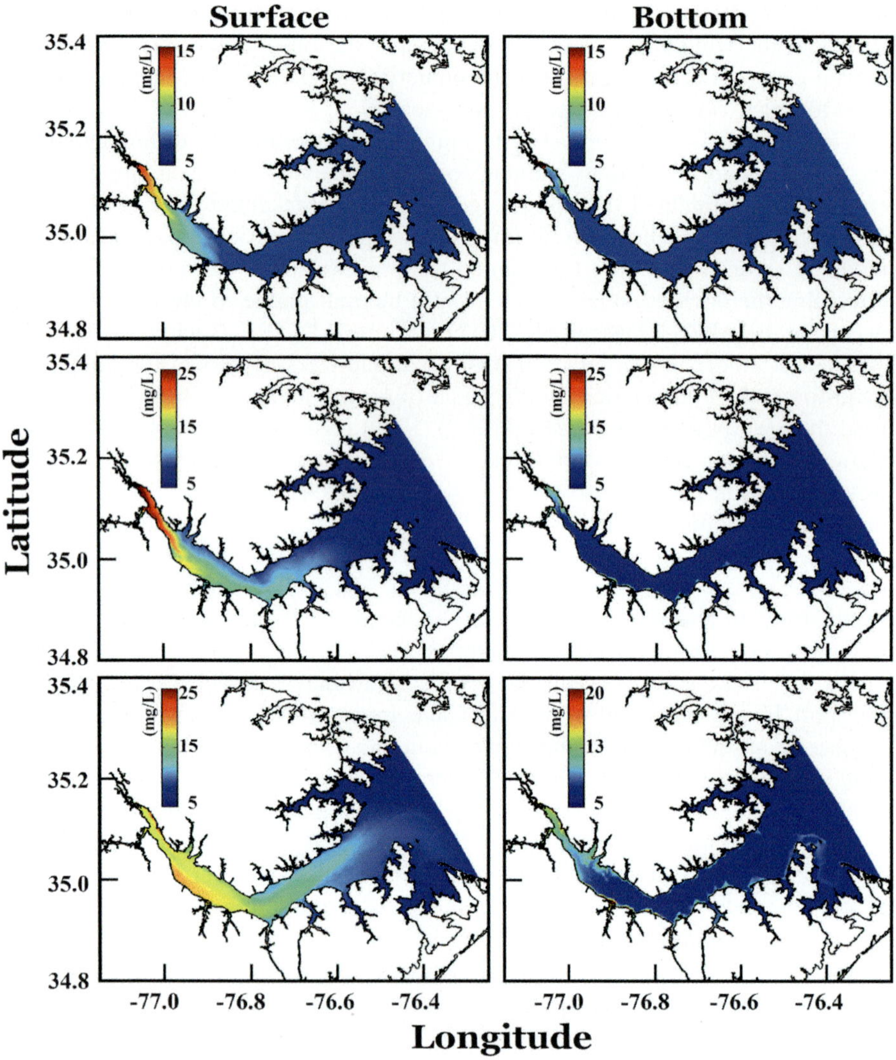

Fig. 14.13 Representative results of DOC concentrations in the surface and bottom layers of the NRE Delft3D model simulated for September 5 (*top panel*), 10 (*middle panel*) and 15 (bottom panel), 2011 following the passage of Hurricane Irene. These maps provide unique information on the spatial and vertical distribution of DOC within the system driven by physical processes. Note the differences in scale between various panels

expanded to include the TPE. Of course, the calibration of a model is directly related to the quantity and quality of the associated field data. Remotely sensed data can also serve as a calibration source for numerical models over a reasonably frequent, high spatial grid, albeit for near-surface values only. This further demonstrates the need for additional development of an integrated approach for analyzing material transport in the APES based on field, remote sensing, and numerical models.

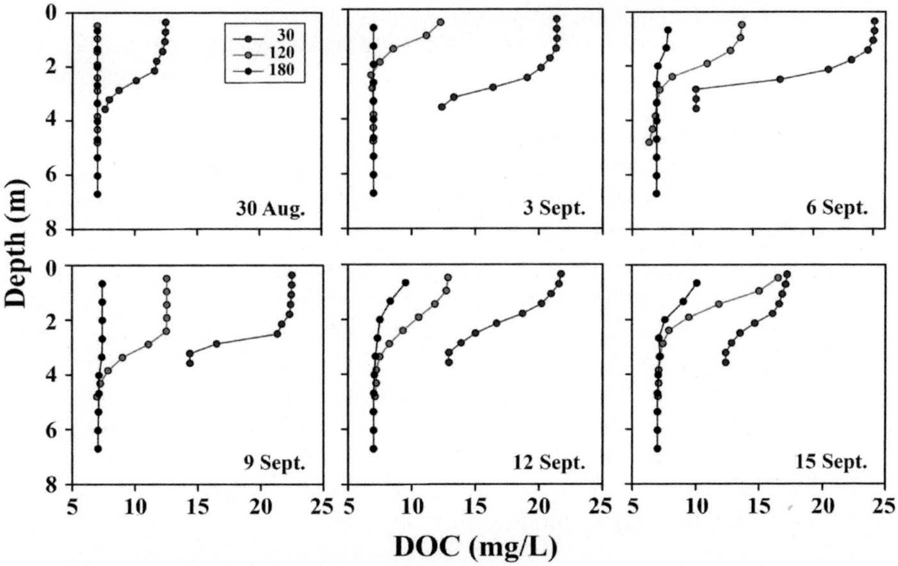

Fig. 14.14 Vertical profiles of model DOC results at an up-estuary, mid-estuary and lower-estuary ModMon station (Fig. 14.2) showing the transport of DOC down estuary as a near-surface plume. Each point represents the DOC concentration within a vertical layer of the model at that location

14.10 Discussion

We presented here a brief overview of an effort to develop an effective program for monitoring material transport in the dynamic and complex environments of the Albemarle-Pamlico Estuarine System. To establish the most comprehensive program, results from an integrated approach using field measurements, remote sensing and numerical models are used. Each of these components has advantages, disadvantages, inherent limitations, current issues and potential advancements that may significantly impact their use in coastal environments.

Accurate field data, without question, should be the cornerstone of any program. Measurements made with instruments in situ or measurements made on discrete field samples in a laboratory are fundamentally a "direct" measurement of some desired aspect of an environmental parameter. Simply, field data are the best measure of what is occurring in nature; however, these data typically represent a small fraction of a system in time and space. Since field data are a direct measurement of the true, current conditions they are also fundamental to the application of indirect technologies such as remote sensing (algorithm development) and numerical modeling (calibration and validation). The major disadvantage of field programs is that the acquisition of traditional field data over a network that is adequate to characterize the most relevant time and space scales of an environmental variable is often costly (perhaps prohibitive) and labor intensive. To offset this disadvantage, a field program must be optimized to support the various elements of the overall program, such as the three components of our integrated approach. Given the findings from an initial integration of remote sensing and numerical modeling, we are now further refining the field program into three components that are dedicated to: (1) gaining a better understanding of field-related processes such as the transport of dissolved material following typical rain events, (2) developing variable-specific (e.g. CDOM, TSM) remote sensing algorithms, and (3) numerical model calibration. Collectively, the major goal of the field program in this approach is to sample the range of concentrations (i.e. gradients) that are observed in the system.

To maximize the amount and use of field data, new field portable sampling technologies will be employed such as the QFT2 (Miller et al. 2011a) and Ultrapath multiple pathlength liquid waveguide (Miller et al. 2002) for particle and CDOM absorption measurements, respectively. To facilitate our efforts related to remote sensing and modeling, continuous vertical profiles of bio-optical parameters will be obtained using in situ flow-through absorption instruments (ac-9, acs, WET Labs, Inc), a backscattering sensor (ECO VSF3, WET Labs, Inc), and CDOM and chl *a* fluorometers with 700 nm scattering meter (BBFL2 ECO Triplet sensor, WET Labs, Inc). The bio-optical profiles will be obtained from a small boat during dedicated cruises and cruises of opportunity with existing programs such as the NEMReP field campaigns in the upper Neuse River Estuary. The absorption measurements will enable accurate estimates of CDOM concentration that can be used to directly characterize CDOM (DOC) dynamics. The backscattering and scattering measurements will be related directly to TSM and particle dynamics. The near-surface measurements will directly support the remote sensing effort while the vertical profiles will directly support the calibration and validation of numerical models. Additional field measurement technologies could contribute to our integrated program as well. For example, underway measurements made during a cruise using a near-surface water intake and flow-through instruments aboard ship can provide a high density of data. This approach however, provides only near-surface measurements and can be problematic due to contaminants and bubbles. Miller et al. (2003) presented an alternate technology where they described a small undulating towed body equipped with a suite of optical and hydrographic instruments designed to collect horizontal and vertical profiles in shallow coastal environments to support remote sensing applications.

This study demonstrated the use of remotely sensed data from different optical sensors for examining temporal and spatial dynamics in dissolved and particulate material in the APES. Recent advancements in computer-based technologies and an increase in international collaborations have lead to an unprecedented use of remote sensing. Data portals now offer an impressive catalog of data from multiple sources that can easily and quickly be downloaded over the Internet using transfer speeds that are available to most users. Highly capable software such as SeaDAS that is available without cost has enabled users from many different disciplines and varying computational experience to conduct sophisticated processing and analysis of data. These capabilities have significantly enabled the local use of remote sensing of various aspects of coastal environments such as the APES. While the general use of remote sensing is widely expanding, its application to coastal environments is not without challenges. For example, the use of remotely sensed images of a specific coastal system might be limited in part due to the size and spatial dimensions of the system, image contamination by persistent clouds or difficulties in developing robust remote sensing algorithms due to complex optical characteristics (Muller-Karger et al. 2013). These are some of the challenges that our program is attempting to overcome through a multi-sensor approach coupled to an improved field program of bio-optical measurements as described above. However, the greatest potential future challenge for the use of remote sensing may be a lack of contemporary data due to a limited number of operational ocean color sensors. With the demise of SeaWiFS in 2010 and MERIS in 2012, only data from the MODIS and the recently launched VIIRS (Visible Infrared Imager Radiometer Suite) instruments are readily available to our APES program with which to examine modern events and processes. However, both MODIS instruments are well beyond their design lifetime and much work remains to demonstrate the utility of VIIRS data for monitoring coastal waters (Hlaing et al. 2013). Effort is now being initiated within this program to compare products generated by MODIS Aqua and VIIRS. Several global ocean color sensors are being planned that should directly support coastal research (National Academy of Sciences 2011) including the NASA PACE (Pre-Aerosol, Clouds, and ocean Ecosystem) mission. NASA is also planning a geostationary mission, GEO-CAPE (GEOstationary Coastal and Air Pollution Events), to provide multiple observations during daylight of the coastal regions of North and South America (Fishman et al. 2012) to examine diurnal changes and impacts of ephemeral events. It is now critically important that the user community of ocean color data, researchers, government resource managers, and private industry become active advocates for the continued and non-interrupted availability of coastal ocean color observations. The sustained availability of remotely sensed data of coastal environments is most important now as coastal regions come under ever increasing pressure from increased population growth and impacts of global climate change.

Numerical models are now commonly used to examine a host of coastal physical processes. Beyond their ability to provide an independent analysis of numerous coastal processes, numerical models can also significantly supplement the data obtained from field measurements and remote sensing. Numerical models are also invaluable tools for understanding coastal processes, predicting system response to forcing conditions, and estimating water and material transport under different event scenarios. Perhaps the most important models still to be developed are those capable of simulating all the relevant and interconnected processes for tracking water masses with complex-time histories of multi-constituent particle plumes and biogeochemical processes for long-term predictions of changes to coastal estuaries. While a numerical model can presumably provide meaningful data to fill the gaps in time and space that exist using either field measurements or remote sensing, an effective model must accurately represent the physics of the system and may require a reasonably large field database. Recent advances in computer hardware have significantly reduced runtime limitations of most models allowing for the increased spatial and temporal resolution necessary for coastal research.

The Albemarle-Pamlico Estuarine System, similar to many coastal systems worldwide, is experiencing significant pressure from human-induced and natural change. These pressures are expected to increase as coastal human populations continue to grow and the direct consequences of global climate change on coastal areas increase. Successful management of these critical, vulnerable environments requires knowledge gained from good scientific programs. Integrating the results from field measurements, remote sensing and numerical models is proving to be a good start to developing an effective, comprehensive approach to examining material transport in the dynamic environments of the APES. This approach is offering data over a broad spectrum of time and space scales available only through the integration of these three methods. The integrated approach requires a multidisciplinary team, a team that not only has the experience and knowledge necessary to implement the technologies inherent in field measurements, remote sensing, and modeling, but also the scientific understanding for data integration and synthesis. This approach has facilitated an increased understanding of key processes within and among the different disciplines. Therefore, the benefits from an integrated approach will hopefully become a model of a cost-effective program to support resource managers to better manage these vital resources and offer a better understanding of the global role that these dynamic systems play in the transport and cycling of material.

Acknowledgements We gratefully acknowledge the contributions of the National Aeronautics and Space Administration Ocean Biology and Biochemistry Program by collecting and maintaining the Ocean Color Data Sets. We thank the Institute of Marine Sciences, University of North Carolina, for sharing their Modeling and Monitoring Project (ModMon) data and to the Center for Applied Aquatic Ecology, North Carolina State University for sharing their Neuse Estuary Monitoring and Research Program (NEMReP) data. We thank the National Space Organization of Taiwan and the Global Earth Observation and Data Analysis Center, National Cheng Kung University, for their help in obtaining and processing FORMOSAT-2 imagery. Mulligan acknowledges funding for travel through a Research Initiation Grant at Queen's University. Partial funding of this program was provided by a grant to Miller from the NC Sea Grant Program.

References

Aurin D, Mannino A, Franz B (2013) Spatially resolving ocean color and sediment dispersion in river plumes, coastal systems, and continental shelf waters. Remote Sens Environ 137:212–225

Barbier EB, Hacker SD, Kennedy C, Koch EW, Stier AC, Silliman BR (2011) The value of estuarine and coastal ecosystem services. Ecol Monogr 81:169–193

Bauer JE, Bianchi TS (2011) Dissolved organic carbon cycling and transformation. In: Wolanski E, McLusky DS (eds) Treatise on estuarine and coastal science. Academic, Waltham, pp 7–67

Blaas M, Dong C, Marchesiello P, McWilliams JC, Stolzenbach KD (2007) Sediment-transport modeling on Southern Californian shelves: a ROMS case study. Cont Shelf Res 27(6):832–853

Booij N, Ris R, Holthuijsen L (1999) A third-generation wave model for coastal regions: 1. Model description and validation. J Geophys Res 104(4):7649–7666

Brown MM (2013) An integrated method for monitoring material transport in a coupled land-estuary system following a dynamic storm event: the Neuse river and Estuary, NC and Hurricane Irene. MS thesis, East Carolina University, 111 pp

Brown CW, Connor LN, Lillibridge JL, Nalli NR, Legeckis RV (2005) An introduction to satellite sensors, observations and techniques. In: Miller RL, Del Castillo CE, McKee BA (eds) Remote sensing of coastal aquatic environments: technologies, techniques and application. Springer, New York, pp 21–50

Brown MM, Mulligan RP, Miller RL (2014) Modeling the transport of freshwater and dissolved organic carbon in the Neuse River Estuary, NC, USA following Hurricane Irene (2011). Estuar Coast Shelf Sci 139:148–158

Burkholder JM et al (2004) Comparative impacts of two major hurricane seasons on the Neuse River and Western Pamlico Sound Ecosystems. Proc Natl Acad Sci 101:9291–9296

Cannizzaro JP, Carlson PR Jr, Yarbro LA, Hu C (2013) Optical variability along a river plume gradient: implications for management and remote sensing. Estuar Coast Shelf Sci 131:149–161

Casal G, Kutser T, Domínguez-Gómez JA, Sánchez-Carnero N, Freire J (2013) Assessment of the hyperspectral sensor CASI-2 for macroalgal discrimination on the Ría de Vigo coast (NW Spain) using field spectroscopy and modelled spectral libraries. Cont Shelf Res 55:129–140

Chen C, Shi P, Zhan H (2003) A local algorithm for estimation of yellow substance (gelbstoff) in coastal waters from SeaWiFS data: Pearl River estuary, China. Int J Remote Sens 24:1171–1176

Chen Z, Hu C, Conmy R, Muller-Karger F (2007) Colored dissolved organic matter in Tampa Bay, Florida. Mar Chem 104:98–109

Coble P (1996) Characterization of marine and terrestrial DOM in seawater using excitation-emission matrix spectroscopy. Mar Chem 51:325–346

Cui T, Zhang J, Groom S, Sun L, Smyth T, Sathyendranath S (2010) Validation of MERIS ocean-color products in the Bohai Sea: a case study for turbid coastal waters. Remote Sens Environ 114:2326–2336

Day J, Kemp M, Yaez-Arancibia A, Crump B (2012) Estuarine ecology, 2nd edn. Wiley-Blackwell, New York, p 568

Del Castillo CE, Miller RL (2008) On the use of ocean color remote sensing to measure the transport of dissolved organic carbon by the Mississippi River Plume. Remote Sens Environ 112:836–844

Dierssen HM, Zimmerman RC, Leathers RA (2003) Ocean color remote sensing of seagrass and bathymetry in the Bahamas Banks by high-resolution airborne imagery. Limnol Oceanogr 48:444–455

Dresback KM, Fleming JG, Blanton BO, Kaiser C, Gourley JJ, Tromble EM, Luettich RA Jr, Kolar RA, Hong Y, Van Cooten S, Vergara HJ, Flamig ZL, Lander HM, Kelleher KE, Nemunaitis-Monroe KL (2013) Skill assessment of a real-time forecast system utilizing a coupled hydrologic and coastal hydrodynamic model during Hurricane Irene (2011). Cont Shelf Res 71:78–94

Druon JN, Mannino A, Signorini S, McClain C, Friedrichs M, Wilkin J, Fennel K (2010) Modeling the dynamics and export of dissolved organic matter in the Northeastern U.S. continental shelf. Estuar Coast Shelf Sci 88(4):488–507

Ekstrand S (1992) Landsat TM based quantification of chlorophyll-a during algae blooms in coastal waters. Int J Remote Sens 13:1913–1926

Elias EPL, Gelfenbaum G, Van der Westhuysen AJ (2012) Validation of a coupled wave-flow model in a high-energy setting: the mouth of the Columbia River. J Geophys Res 117:C09011

Elsner JB, Kossin JP, Jagger TH (2008) The increasing intensity of the strongest tropical cyclones. Nature 455:92–95

Fishman J, Iraci LT, Al-Saadi J, Chance K, Chavez F, Chin M, Coble P, Davis C, DiGiacomo PM, Edwards D, Eldering A, Goes J, Herman J, Hu C, Jacob DJ, Jordan C, Kawa SR, Key R, Liu X, Lohrenz S, Mannino A, Natraj V, Neil D, Neu J, Newchurch M, Pickering K, Salisbury J, Sosik H, Subramaniam A, Tzortziou M, Wang J, Wang M (2012) The United States' next generation of atmospheric composition and coastal ecosystem measurements: NASA's Geostationary Coastal and Air Pollution Events (GEO-CAPE) Mission. Bull Am Meteorol Soc 93:1547–1566

Giese GL, Wilder HB, Parker GG (1985) Hydrology of major estuaries and sounds of North Carolina: U.S. Geological Survey Water-Supply Paper 2221. USGS, 108 p

Giffin D, Corbett DR (2003) Evaluation of sediment dynamics in coastal systems via short-lived radioisotopes. J Mar Syst 42:83–96

Gons HJ, Auer MT, Effler SW (2008) MERIS satellite chlorophyll mapping of oligotrophic and eutrophic waters in the Laurentian Great Lakes. Remote Sens Environ 112:4098–4106

Green RE, Gould RW Jr, Ko DS (2008) Statistical models for sediment/detritus and dissolved absorption coefficients in coastal waters of the northern Gulf of Mexico. Cont Shelf Res 28:1273–1285

Guo W, Stedmon CA, Han Y, Wu F, Yu X, Hu M (2007) The conservative and non-conservative behavior of chromophoric dissolved organic matter in Chinese estuarine waters. Mar Chem 107:357–366

Haase AT, Eggleston DB, Luettich RA Jr, Weaver RJ, Puckett BJ (2012) Estuarine circulation and predicted oyster larval dispersal among a network of reserves. Estuar Coast Shelf Sci 101:33–43

Harding LW, Magnuson A, Mallonee ME (2005) SeaWiFS retrievals of chlorophyll in Chesapeake Bay and the mid-Atlantic bight. Estuar Coast Shelf Sci 62:75–94

Helms JR, Stubbins A, Ritchie JD, Minor EC, Kieber DJ, Mopper K (2008) Absorption spectral slopes and slope ratios as indicators of molecular weight, source, and photobleaching of chromophoric dissolved organic matter. Limnol Oceanogr 53:955–969

Hinrichsen D (1998) Coastal waters of the world: trends, threats and strategies. Island Press, Washington, DC

Hlaing S et al (2013) Evaluation of the VIIRS ocean color monitoring performance in coastal regions. Remote Sens Environ 139:398–414

Howarth RW (2008) Coastal nitrogen pollution: a review of sources and trends globally and regionally. Harmful Algae 8:14–20

Hu C, Chen Z, Clayton T, Swarzenski P, Brock J (2004) Assessment of estuarine water-quality indicators using MODIS medium-resolution bands: Initial results from Tampa Bay, FL. Remote Sens Environ 93:423–441

Hyde KJW, O'reilly JE, Oviatt CA (2007) Validation of SeaWiFS chlorophyll-a in Massachusetts Bay. Cont Shelf Res 27:1677–1691

Inamdar S, Finger N, Singh S, Mitchell M, Levia D (2011) Dissolved organic matter (DOM) concentration and quality in a forested mid-Atlantic watershed, USA. Biogeochemistry. doi:10.1007/s10533-011-9572-4

Khorram S, Cheshire H, Geraci AL, La Rosa G (1991) Water quality mapping of Augusta Bay, Italy from Landsat-TM data. Int J Remote Sens 12(4):803–808. doi:10.1080/01431169108929696

Koren V, Reed S, Smith M, Zhang Z, Seo D-J (2004) Hydrology laboratory research modeling system (HL-RMS) of the US national weather service. J Hydrol 291(3–4):297–318

Kratzer S, Brockmann C, Moore G (2008) Using MERIS full resolution data to monitor coastal waters—a case study from Himmerfjärden, a fjord-like bay in the northwestern Baltic Sea. Remote Sens Environ 112:2284–2300

Lesser GR, Roelvink JA, Van Kester JATM, Stelling GS (2004) Development and validation of a three dimensional morphological model. J Coast Eng 51:883–915

Liu C, Liu J, Lin C, Wu A, Liu S, Shieh C (2007) Image processing of FORMOSAT-2 data for monitoring the South Asia tsunami. Int J Remote Sens 28:3093–3111

López R, Del Castillo CE, Miller RL, Salisbury J, Wisser D (2012) Examining organic carbon transport by the Orinoco river using SeaWiFS imagery. J Geophys Res 117, G03022. doi:10.1029/2012JG001986

Luettich RA Jr, McNinch JE, Paerl HW, Peterson CP, Wells JT, Alperin M, Martens CS, Pinckney JL (2000) Neuse River Estuary modeling and monitoring project stage 1: hydrography and circulation, water column nutrients and productivity, sedimentary processes and benthic-pelagic coupling. Report UNC-WRRI-2000-325B. Water Resources Research Institute of the University of North Carolina, Raleigh

Luettich RA Jr, Carra SD, Reynolds-Fleminga JV, Fulchera CW, Mcninch JE (2002) Semi-diurnal seiching in a shallow, micro-tidal lagoonal estuary. Cont Shelf Res 22:1669–1681

Lunetta RS et al (2009) Measurement of water colour using AVIRIS imagery to assess the potential for an operational monitoring capability in the Pamlico Sound Estuary, USA. Int J Remote Sens 30:3291–3314

Meyers JS, Miller RL (2005) An introduction to satellite sensors, observations and techniques. In: Miller RL, Del Castillo CE, McKee BA (eds) Remote sensing of coastal aquatic environments: technologies, techniques and application. Springer, New York, pp 53–69

Miller RL, Cruise J (1995) Effects of suspended sediments on coral growth: evidence from remote sensing and hydrologic modeling. Remote Sens Environ 53:177–187

Miller RL, Mckee BA (2004) Using MODIS Terra 250 m imagery to map concentrations of total suspended matter in coastal waters. Remote Sens Environ 93:259–266

Miller RL, Belz M, Castillo CD, Trzaska R (2002) Determining CDOM absorption spectra in diverse coastal environments using a multiple pathlength, liquid core waveguide system. Cont Shelf Res 22:1301–1310

Miller RL, Twardowski M, Moore C, Casagrande C (2003) The dolphin: technology to support remote sensing bio-optical algorithm development and applications. Backscatter, Alliance for Marine Remote Sensing, spring, pp 8–12

Miller RL, Del Castillo CE, Chilmakuri C, McCorquodale JA, Georgiou I, McKee BA, D'Sa EJ (2005a) Using multi-temporal MODIS 250 m data to calibrate and validate a sediment transport model for environmental monitoring of coastal waters. MultiTemp'05, Biloxi, MS. 5 pp

Miller RL, Del Castillo CE, Mckee BA (2005b) Remote sensing of coastal aquatic environments: technologies, techniques and applications. Springer, New York, p 347

Miller RL, Buonassissi C, Del Castillo CE, Belz M (2011a) A portable fiber optic system for measuring particle absorption using the quantified filter technique (QFT). Limnol Oceanogr Methods 9:554–564

Miller RL, Liu C-C, Buonassissi CJ, Wu A-M (2011b) A multi-sensor approach to examining the distribution of total suspended matter (TSM) in the Albemarle-Pamlico estuarine system, NC, USA. Remote Sens 3:962–974

Miller RL, Mitra S, Buonassissi CJ, Brown M (2011c) Transport of dissolved organic matter following major rainfall events in a coupled terrestrial – ocean system: the Neuse River, North Carolina USA. 11th International Estuarine Biogeochemisty Symposium. Atlantic Beach, NC, 15–19 May

Mitchell BG, Kahru M, Wieland J, Stramska M (2002) Determination of spectral absorption coefficients of particles, dissolved material, and phytoplankton for discrete water samples. In: Mueller JL, Fargion G, McClain C (eds) Ocean optics protocols for satellite ocean color sensor validation NASA Tech. Man. 2003-211621. Goddard Space Flight Center, Natl. Aeronautics and Space Admin, Greenbelt, pp 39–64

Mitra S, Wozniak SA, Miller RL, Hatcher PG, Buonassissi C, Brown M (2013) Atlantic hurricanes as a source of marine dissolved organic matter onto land. Mar Chem 154:67–76

Muller-Karger F, Roffer M, Walker N, Oliver M, Schofield O, Abbott M, Graber H, Leben R, Goni G (2013) Satellite remote sensing in support of an integrated ocean observing system. Geosci Remote Sens Mag IEEE 1(4):8–18

Mulligan RP, Hay AE, Bowen AJ (2008) Wave-driven circulation in a coastal bay during the landfall of a hurricane. J Geophys Res 113:C05026

Mulligan RP, Perrie W, Solomon S (2010) Dynamics of the Mackenzie River plume on the inner Beaufort Shelf during an open water period in summer. Estuar Coast Shelf Sci 89:214–220

Mulligan RP, Walsh JP, Wadman HM (2014) Storm surge and surface waves in a shallow lagoonal estuary during the crossing of a Hurricane. J Waterw Port Coast Ocean Eng. A5014001, doi: 10.1061/(ASCE)WW.1943-5460.0000260

National Academy of Science (2011) Assessing requirements for sustained ocean color research and operations. The National Academies Press, Washington, DC

NC Division of Marine Fisheries (1995) NC commercial landings. http://www.ncfisheries.net/statistics/comstat/95land.htm. Accessed 3 Jan 2014

NCDENR (2010) North Carolina Department of Environment and Natural Resources (NCDENR), Division of Water Resources. 2010. Neuse River Basin Water Resources Plan, July 2010. NCDENR, Raleigh, NC

Nezlin NP, Digiacomo PM, Stein ED, Ackerman D (2005) Stormwater runoff plumes observed by SeaWiFS radiometer in the Southern California Bight. Remote Sens Environ 98:494–510

NOAA (2013) National Oceanic and Atmospheric Administration. National Coastal Population Report. Population trends from 1970 to 2020. http://stateofthecoast.noaa.gov/features/coastal-population-report.pdf. Accessed 3 Jan 2014

Osburn CL, Handsel LT, Mikan MP, Paerl HW, Montgomery MT (2012) Fluorescence tracking of dissolved and particulate organic matter quality in a river-dominated estuary. Environ Sci Technol 46:8628–8636

Otis D, Carder K, English D, Ivey J (2004) CDOM transport from the Bahamas banks. Coral Reefs 23:152–160

Paerl HW et al (2001) Ecosystem impacts of three sequential hurricanes (Dennis, Floyd, and Irene) on the United States' largest lagoonal estuary, Pamlico Sound, NC. Proc Natl Acad Sci 98:5655–5660

Pietrafesa LJ, Janowitz GS, Chao T-Y, Weisberg RH, Askari F, Noble E (1986) The physical oceanography of Pamlico Sound. University of North Carolina Sea Grant Publication UNC-SG-WP-86-5, 126 pp

Richardson LL, Ledrew EF (2006) Remote sensing of aquatic coastal ecosystem processes. Springer, New York, p 324

Sallenger AH, Doran KS, Howd PA (2012) Hotspot of accelerated sea-level rise on the Atlantic coast of North America. Nat Clim Chang 2:884–888

Seitzinger SP, Kroeze C, Bouwman AF, Caraco N, Dentener F, Styles RV (2002) Global patterns of dissolved inorganic and particulate nitrogen inputs to coastal systems: Recent conditions and future projections. Estuaries 25:640–655

Small C, Nicholls RJ (2003) A global analysis of human settlement in coastal zones. J Coast Res 19:584–599

Sokoletsky LG, Lunetta RS, Wetz MS, Paerl HW (2011) MERIS retrieval of water quality components in the turbid Albemarle-Pamlico Sound Estuary, USA. Remote Sens 3:684–707

Tomlinson MC, Stumpf RP, Ransibrahmanakul V, Truby EW, Kirkpatrick GJ, Pederson BA, Vargo GA, Heil CA (2004) Evaluation of the use of SeaWiFS imagery for detecting Karenia brevis harmful algal blooms in the eastern Gulf of Mexico. Remote Sens Environ 91(3–4):293–303

Tzortziou M, Osburn C, Neale P (2007) Photobleaching of dissolved organic material from a tidal marsh-estuarine system of the Chesapeake Bay. Photochem Photobiol 83:782–792

Walling DE (1999) Linking land use, erosion and sediment yields in river basins. Hydrobiologia 410:233–240

Wang Y (2004) Using Landsat 7 TM data acquired days after a flood event to delineate the maximum flood extent on a coastal floodplain. Int J Remote Sens 25:959–974

Wang Y (2009) Remote sensing of coastal environments. CRC Press, Boca Raton, 457 pp

Weishaar JL, Aiken GR, Bergamaschi BA, Fram MS, Fugii R, Mopper K (2003) Evaluation of specific ultraviolet absorbance as an indicator of the chemical composition and reactivity of dissolved organic carbon. Environ Sci Technol 37:4702–4708

Wool TA, Davie SR, Rodriquez HN (2003) Development of three-dimensional hydrodynamic and water quality models to support total maximum daily load decision process for the Neuse River Estuary, North Carolina. J Water Resour Plan Manag 129:295–306

Xi H, Zhang Y (2011) Total suspended matter observation in the Pearl River estuary from in situ and MERIS data. Environ Monit Assess 177:563–574

Zhang M, Tang J, Dong Q, Song Q, Ding J (2010) Retrieval of total suspended matter concentration in the Yellow and East China Seas from MODIS imagery. Remote Sens Environ 114:392–403

Chapter 15
Simulation Management Systems Developed by the Northern Gulf Coastal Hazards Collaboratory (NG-CHC): An Overview of Cyberinfrastructure to Support the Coastal Modeling Community in the Gulf of Mexico

Robert R. Twilley, Steve Brandt, Darlene Breaux, John Cartwright, Jim Chen, Greg Easson, Patrick Fitzpatrick, Kenneth Fridley, Sara Graves, Sandra Harper, Carola Kaiser, Alexander Maestre, Manil Maskey, William H. McAnally, John McCorquodale, Ehab Meselhe, Tina Miller-Way, Kyeong Park, Joao Pereira, Thomas Richardson, Jian Tao, Amelia Ward, Jerry Wiggert, and Derek Williamson

Abstract Given the significance of natural and built assets of the Gulf of Mexico region, the three states of Alabama, Louisiana, and Mississippi, leveraged their unique partnerships, proximity, and significant prior investments in cyberinfrastructure (CI) to develop the Northern Gulf Coastal Hazards Collaboratory (NG-CHC). This collaboratory was established to catalyze collaborative research

---

R.R. Twilley (✉)
Department of Oceanography and Coastal Sciences, School of the Coast and Environment, Louisiana State University, Baton Rouge, LA 70803, USA
e-mail: rtwilley@lsu.edu

S. Brandt • J. Tao
Center for Computation and Technology, Louisiana State University, 231 Johnston Hall, Baton Rouge, LA 70803, USA

D. Breaux
Department of Civil Engineering, University of Louisiana at Lafayette, P.O. Box 42291, Lafayette, LA 70504, USA

J. Cartwright • W.H. McAnally
Geosystems Research Institute, Mississippi State University, Box 9652, Mississippi State, MS 39762, USA

J. Chen
Center for Computation and Technology, Louisiana State University, 231 Johnston Hall, Baton Rouge, LA 70803, USA

Department of Civil and Environmental Engineering, Louisiana State University, 3418D Patrick F. Taylor Hall, Baton Rouge, LA 70803, USA

C.W. Finkl and C. Makowski (eds.), *Remote Sensing and Modeling: Advances in Coastal and Marine Resources*, Coastal Research Library 9,
DOI 10.1007/978-3-319-06326-3_15,

via enhanced CI to reduce the regions vulnerability to natural and human disasters by facilitating high performance modeling to test hypotheses focused on engineering design, coastal system response, and risk management of coastal hazards. The objective of the NG-CHC is to advance research and inspire collaboration through highly available innovation-enabling CI, with a particular focus on geosciences and engineering from the watershed to the coast. An integrated CI capable of simulating all relevant interacting processes is needed to implement a system that captures the

G. Easson
Department of Geology and Geological Engineering, University of Mississippi, 118E Carrier Hall, University, Oxford, MS 38677, USA

P. Fitzpatrick
Northern Gulf Institute, Mississippi State University Science and Technology Center, Stennis Space Center, 1021 Balch Blvd, Mississippi State, MS 39529, USA

K. Fridley • A. Maestre • D. Williamson
Department of Civil, Construction and Environmental Engineering, University of Alabama, Box 870205, Tuscaloosa, AL 35487-0205, USA

S. Graves • S. Harper • M. Maskey
Information Technology and Systems Center, University of Alabama in Huntsville, Technology Hall S339, Huntsville, AL 35899, USA

C. Kaiser
Department of Oceanography and Coastal Sciences, School of the Coast and Environment, Louisiana State University, Baton Rouge, LA 70803, USA

Center for Computation and Technology, Louisiana State University, 231 Johnston Hall, Baton Rouge, LA 70803, USA

J. McCorquodale • J. Pereira
Department of Civil and Environmental Engineering, and Pontchartrain Institute for Environmental Sciences, University of New Orleans, 2000 Lakeshore Drive, New Orleans, LA 70148, USA

E. Meselhe
Department of Civil Engineering, University of Louisiana at Lafayette, P.O. Box 42291, Lafayette, LA 70504, USA

The Water Institute of the Gulf, One American Place, 301 N. Main St., Suite 2000, Baton Rouge, LA 70825, USA

T. Miller-Way • K. Park
Department of Marine Sciences, University of South Alabama, Dauphin Island Sea Lab, 101 Bienville Blvd., Dauphin Island, AL 36528, USA

T. Richardson
Jackson State University, P.O. Box 17249, Jackson, MS 39217, USA

A. Ward
Center for Freshwater Studies, University of Alabama, Box 870206, Tuscaloosa, AL 35487-0206, USA

J. Wiggert
Department of Marine Sciences, University of Southern Mississippi, Stennis Space Center, 1020 Balch Boulevard, Hattiesburg, MS 39529-9904, USA

dynamic nature of coastal surface processes. The NG-CHC has implemented CI to locate appropriate data and computational resources, create necessary workflows associated with different simulation demands, and provide visualization tools for analysis of results. Three simulation management systems, SIMULOCEAN, SULIS, and ASGS, were implemented, each with a defined suite of hypotheses and institutional participants to run collaborator experiments. The NG-CHC focused on developing suites of CI tools centered on handling the functional needs of each simulation management system in a collaborative environment. The NG-CHC also developed curriculum units, computer games and simulations to extend the knowledge of coastal hazards to students from middle school to college. Education and outreach activities were developed to increase public understanding and support for sustainable coastal practices. The elements of the CI tool box within NG-CHC describe generic tools needed to promote a 'collaborative modeling environment' in other coastal systems.

15.1 Introduction

Coastal margins are important to the sustainability of economically important coastal fisheries, to strategic and economically important energy activities around the world, and as critical transportation centers associated with port activities (Adger et al. 2005; Crossett 2004; Day et al. 2007). They are regions of increased risks to human and natural resource infrastructure due to environmental drivers associated with climate change, sea level rise, hurricane frequency, and land-surface dynamics (Crossett 2004; Crowell et al. 2013; Cutter et al. 2008). These risks are amplified by significant human populations living in coastal regions, where approximately 8.5 million people live in census blocks that border ocean or are contained within the 1 % annual chance of a coastal flood (Crossett 2004; Crowell et al. 2013). Eight of the ten largest cities in the world and 44 % of the world's seven billion population are located within 150 km of the ocean (Vörösmarty et al. 2009). As a result, coastal hazards and public health are major concerns to agencies responsible for the safety of coastal regions (Adger et al. 2005; Hallegatte et al. 2013; Laska et al. 2005; Norris et al. 2008; Syvitski and Saito 2007). Coastal floodplains and rivers, the coastal ocean, and the underlying geology form a complex hydrodynamic, geomorphic, and ecological system that substantially impacts coastal inhabitants and natural resources (NSF 2009; Norris et al. 2008; Opperman et al. 2009; Pinter 2005). Understanding and predicting the behavior of coastal systems, under prevailing and extreme environmental conditions, is an important problem in many countries around the world, particularly those with deltaic coasts (Nicholls et al. 2011; Sherrieb et al. 2010; Vörösmarty et al. 2009).

The northern coast of the Gulf of Mexico is a region where weather and storm disturbance plays a significant role in public safety (Lin et al. 2012; NRC 2006), and are expensive entities to the Gulf of Mexico's offshore energy industry (Fischetti 2001). Every day, the Gulf of Mexico production facilities extract, process, and transport approximately three million equivalent barrels of oil (27 % of U.S. production) and natural gas (25 % of U.S. production), with at least 12.5 % going to state and federal royalties (NRC 2006). The oil and gas infrastructure in the Gulf of Mexico is substantial, with 6,400 producing wells, 4,000 active platforms, and 29,000 miles of pipelines. Hurricanes pose a considerable threat to production operations, forcing evacuation of offshore drilling rigs and temporarily ceasing production. For example, during hurricane Katrina more than 700 platforms were evacuated reducing the oil production to less than 10 % (Feltus cited by Cruz and Krausmann 2008). The Northern Gulf Coast also constitutes one of the most culturally and ecologically rich coastal ecosystems in the world (Laska et al. 2005). Within this region are found 25 % of the nation's coastal wetlands and 40 % of all salt marshes in the contiguous 48 states (Day et al. 2007). These ecosystems are a buffer against storm damage and a nursery and foraging area for fish and crustaceans. Louisiana ranks first among all states in the commercial harvest of menhaden, oysters, and crabs and is a major producer of shrimp (NRC 2006). Overall commercial fisheries in the Northern Gulf Coast states account for over 50 % of the nation's total fish catch. Long-term trends in the estuarine fish community in the northern Gulf of Mexico indicate that during the period 1992–2004 fish communities increased in abundance, species richness, and trophic balance. However, it has been considered that moderate increases in nutrient loads, habitat loss, and climate change could cause a "tipping point" with a rapid decay in fish community structures, number of species, and abundance (Jordan et al. 2010; Rabalais et al. 2002).

The Northern Gulf Coast provides an ideal setting for developing new insights into modeling coastal environmental responses to dynamic external forcings associated with sea level and watershed processes (Dietrich et al. 2010; Galloway et al. 2009; Twilley 2007; Westerink et al. 2008). Thus, the Northern Gulf Coast is a region of dynamic land-water interfaces – with extensive coastal wetlands, barrier islands and river networks that challenge a static approach to modeling and analytical approaches used to design scenarios commonly employed by regional planning and risk management (Paola et al. 2011; Twilley et al. 2008). Finally, coastal regions around the world will experience increased risks from climate change and sea level rise (Day et al. 2008; Vörösmarty et al. 2009). These processes and human development scenarios of the Northern Gulf Coast represent research opportunities to develop solutions to living and working in healthy coastal landscapes nation- and worldwide.

Given the significance of natural and built assets of the Gulf Coast region, and their vulnerability to natural and human disasters, the Northern Gulf Coastal Hazards Collaboratory (NG-CHC) was established to accomplish the following goals: (1) enhance the research competitiveness of the region, (2) advance economic opportunities by reducing risks to coastal communities, and (3) catalyze collaborative research via enhanced cyberinfrastructure (CI) that facilitates high-performance modeling to test hypotheses focused on engineering design, coastal system response, and risk management of coastal hazards. The three states in the consortium, Alabama, Louisiana, and Mississippi, leveraged their unique partnerships, proximity, and significant prior investments in CI to advance science and engineering of coastal hazards across the region (Fig. 15.1). One of the grand

Fig. 15.1 Description of the Northern Gulf Coastal Hazards Collaboratory showing the location of universities involved and some of the topics of the simulation management systems developed by the consortium

challenges for earth system science is to characterize dynamic environmental processes at appropriate space and time scales with integrated observation networks and simulation models (NSF 2009). Such capabilities are critical to societal needs for reduction of risks to built, human and natural environments (Galloway et al. 2009). The observational and data storage systems located at university, government and private industry in the Northern Gulf Coast have increased capacity due to recent major investments in CI. However, this region lacks the necessary regional scale CI to integrate these data inventories into information and knowledge to improve our ability to simulate and forecast dynamics of water-land interactions in coastal and watershed environments. The strategic plan of NG-CHC was to develop integrated CI for a research and education environment to promote regional capability in simulating coastal hazards by enhancing the linkages between modeling and data sources in a multidisciplinary environment that couples geoscience, engineering, geoinformatics, computational science, and STEM education (Fig. 15.2). Coastal hazards represent generic environmental, engineering and social problems worldwide in which human and natural dynamics are strongly and inherently coupled (Galloway et al. 2009). Thus CI developed in the NG-CHC has national and international implications to living and working in coastal environments. This overview describes the strategies developed within the NG-CHC to

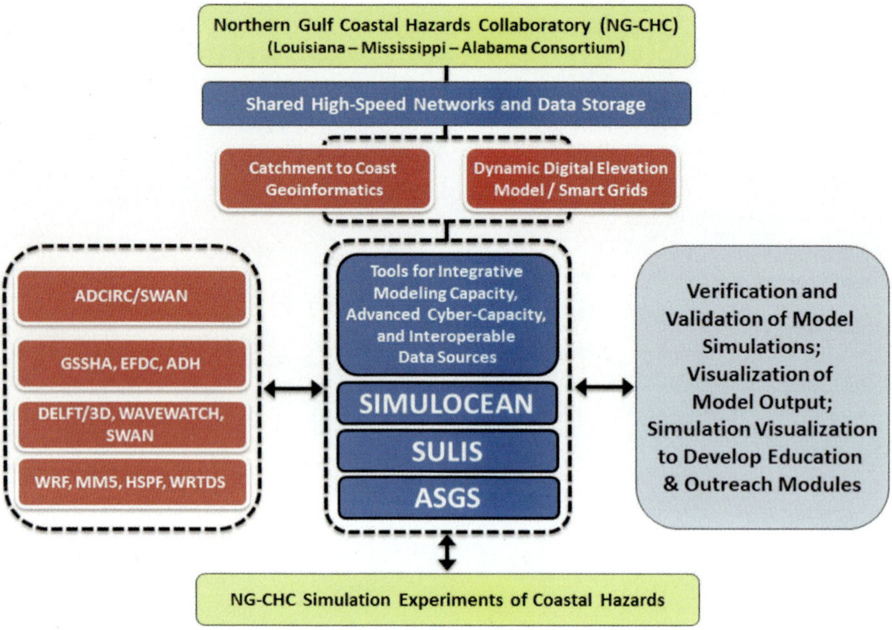

Fig. 15.2 Conceptual diagram of the simulation management systems in the Northern Gulf Coastal Hazard Collaboratory including geoinformatics with vision of developing specific cyberinfrastructure to evaluate coastal risks

define a more integrated research environment by interpreting landscape patterns, forecasting hydrodynamics, and applying critical thinking and problem solving techniques in 'simulation system designs'.

15.2 Simulation Management Systems

There are three types of simulation management systems developed to drive the NG-CHC CI development (Fig. 15.2). The strategy in selecting these three systems was to include: (1) modeling systems that are defined for one geographic region that can be expanded to other areas of need in the northern Gulf, as noted in the Introduction; (2) modeling platforms that capture the diversity of processes from the catchment to the coast for more robust simulations of ecosystem and geosciences processes related to coastal hazards specific to the northern Gulf; (3) simulations that couple hazard mitigation using ecosystem restoration with estimates of risk reduction. These simulation management systems were also designed to provide guidance to coastal hazard preparedness and response (disaster science and management), and economic development. The scientific goal of these simulation management systems was to develop the mathematical, computational, and engineering framework for modeling the geosciences, including watershed science, oceanography, meteorology, ecology, and geomorphology under a changing climate. The need for such a framework was driven by numerous applications that will be considered by each of the simulation experiments on coastal hazard planning, preparedness, and response. By pursuing each approach, and linking these collective efforts within the NG-CHC, our strategy was to build incremental capacity with strategic CI to promote discovery in coastal hazards using the Northern Gulf Coast landscape as a laboratory.

There is no single model that addresses all the coastal hazards described in the Introduction, so a collection or a "community" of models connected by cyberinfrastructure is a basic requirement of the Collaboratory. Further, the diversity of hazard management needs makes single management system far too ambitious for a single project and much too limiting for the Collaboratory. The NG-CHC focus was to produce three simulation management systems including the following: SIMULOCEAN, SULIS, and ASGS (Fig. 15.2) that connect with each other and with other systems (e.g., the Corps of Engineers eCoastal system, USACE 2013). These interconnected 'systems of systems' represents the major accomplishments of the NG-CHC in providing a modeling framework whereby scientists and engineers from across the institutions of the northern Gulf of Mexico Coast could participate in discovery on critical issues of coastal hazards.

SIMULOCEAN is a web-based deployment and visualization framework for coastal modeling (Fig. 15.3), available at http://simulocean.org. SIMULOCEAN collects observational data, schedules modeling codes for execution, manages data transfer, and visualizes both observational and numerical results. SIMULOCEAN can also provide direct validation and verification for models, and automatically generate high quality technical reports. The SIMULOCEAN framework leverages the Advanced Message Queuing Protocol (AMQP) in the background to collect and dispatch various tasks necessary for simulation and data management. The backend of SIMUOCEAN is written with the Python scripting language and built on top of a MySQL database. The front end is written in HTML, CSS and JavaScript (Table 15.1).

There are five components of SIMULOCEAN to achieve the functionality described above. (1) A module is implemented in SIMULOCEAN to support job management and data transfer (system workflow implementation). A user can create, control, and manage simulations running on remote supercomputers via a web-based interface (Fig. 15.3). (2) A GIS-based interface was implemented to support observational data retrieval and archiving in SIMULOCEAN. (3) FVCOM, ADCIRC, and Delft3D have been compiled, tested and implemented in SIMULOCEAN, in addition to SWAN and CaFunwave (Table 15.2 identifies all model acronyms used in this review). Users in the NG-CHC can now configure and run simulations using those three additional models and retrieve the output data following the simulations. (4) A user's manual and step-by-step tutorial were added to make it easy for users to get started with SIMULOCEAN. Online documents were also developed to provide guidance. A video tutorial of how to use SIMULOCEAN to run Delft3D models is available at http://www.youtube.com/watch?v=LfwaYjDbjWE. (5) Collaborating with the University of Alabama in Huntsville (UAH) the metadata for each simulation executed on SIMULOCEAN is now automatically injected into the NG-CHC metadata server to facilitate searching and analysis of simulation output data.

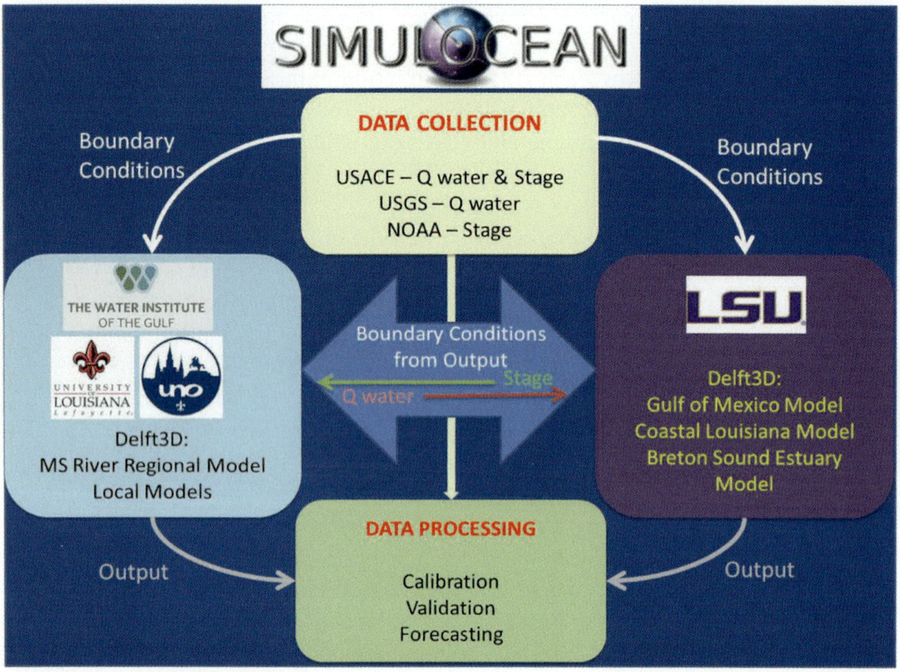

Fig. 15.3 Example of how a SIMULOCEAN user can create, control, and manage simulations running on remote supercomputers with ease using the web-based interface. *Upper panel* describes the basic flow of information; *lower panel* shows the web-based interface that the user utilizes to submit jobs for simulation

Table 15.1 Data sources used to develop merged product and test extraction toolset

Data source	Name	Date completed	Spatial resolution	Horizontal datum	Vertical datum
USGS NED	NED 1/9	N/A	1/9 arc-second	NAD83	NAVD88
USGS NED	NED 1/3	N/A	1/3 arc-second	NAD83	NAVD88
USGS NED	NED 1	N/A	1 arc-second	NAD83	NAVD88
NOAA VDatum	Mobile	11/30/2009	1/3 arc-second	NAD83	NAVD88
NOAA VDatum	New Orleans	04/30/2010	1/3 arc-second	NAD83	NAVD88
NOAA VDatum	Northern Gulf Coast	12/31/2010	1 arc-second	NAD83	NAVD88

SULIS is a manager's assessment system with a public interface and supports two functioning sub-sections of NG-CHC: a section utilizing models to simulate water, sediment, and constituent flow from watershed to coastal bay to Gulf, and an informatics services section that translates model results for use by resource managers (Fig. 15.4). The modeling section simulates river, watershed and gulf

Table 15.2 Acronyms of the models used by the Northern Gulf Coastal Hazards Collaboratory

Model	Name
ADCIRC	Advanced circulation model for oceanic, coastal and estuarine waters
SWAN	Simulating waves nearshore
GSSHA	Gridded surface/subsurface hydrologic analysis
EFDC	Environmental fluid dynamics code
ADH	Adaptive hydraulics modeling
DELFT3D	3D modeling suite to investigate hydrodynamics, sediment transport and morphology
WRF	Weather research and forecasting model
MM5	PSU/NCAR Mesoscale model
HSPF	Hydrological simulation program fortran
WRTDS	Weighted regressions in time, discharge, and season
FVCOM	The unstructured grid finite volume coastal ocean model

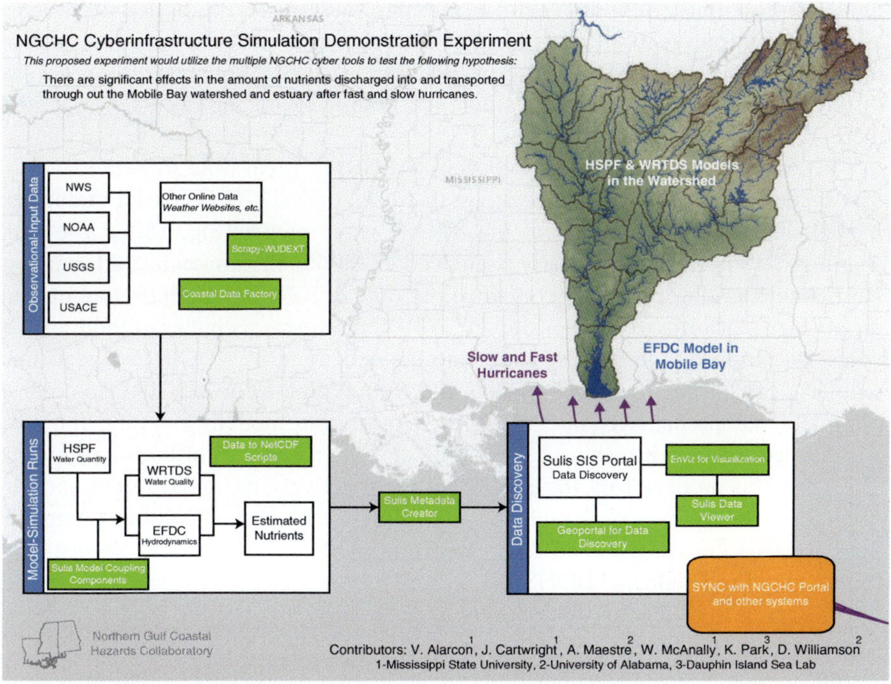

Fig. 15.4 SULIS is a simulation management system with public interface, developed by MSU that functions in two groups: the modeling group and the informatics services (SIS) group

processes based upon hurricane surge data, constituent loads in the watershed and basin, and rainfall. For example, the coupled EFDC model is able to simultaneously run multiple modeled systems, exchanging inflow, water surface, and salinity data among all model instances in the coupled system. This approach was used to allow surge from an ADCIRC model operated within SIMULOCEAN to drive a Mobile

Bay surge model, which propagated the surge into a smaller embayment, Week's Bay. The data transfer system allows data to be passed among models instantly, to achieve more complex model goals such as storm-induced hypoxia.

An open source metadata and data discovery system (built on the Esri Geoportal) serves as the foundation for SULIS Informatics Services (SIS). The SIS capability provides tools for managers to use in assessing environmental hazard risks. The custom install can handle complex model data for exploration, visualization and analysis. The data within the Geoportal can be associated with specific projects (NG-CHC, Mobile Bay RSM, etc.) and their component products, model data, and observed data can be tied to a specific sub-project, allowing easier exploration and discovery of data. The Inference Engine is an analysis component of SULIS, which includes a form of Reduced Order Modeling (ROM) that uses complex, physics-based (high-fidelity) computational model outputs to predict new results with substantially less computational effort than additional model simulations. Support vector regression and spline regression techniques allow predictions to be made in a fraction of the time needed by the original models. Tests using EFDC and ADCIRC model simulations involving several different variables show that the Inference Engine ROM represents an effective method of extending model results.

The ADCIRC Surge Guidance System (ASGS) is a real time software automation system for coastal ocean modeling that uses the ADCIRC coastal circulation model with optional coupling to the SWAN wave model (Fig. 15.5). The open source ASGS project was initially conceived in 2006 in the immediate aftermath of Hurricane Katrina. It is a self-contained package of software modules that operate on any size platform, from a desktop PC to HPC platforms with tens of thousands of cores. Its purpose is to reliably automate the execution of ADCIRC in real time, regularly consuming meteorology data and formulating it for ADCIRC in order to provide a comprehensive, integrated, high resolution modeling system for wind and storm surge conditions associated with specific advisories of hurricane events. The guidance produced by the ASGS has been actively used and relied upon by agencies at the state and federal levels, including the US Army Corps of Engineers, the US Coast Guard, the Louisiana Governor's Office of Homeland Security and Emergency Preparedness (GOHSEP), as well as emergency managers and weather forecasters at the state and local levels in Louisiana, Mississippi, Alabama, North Carolina and Texas.

Storm forecasts demand accuracy to build user confidence, timeliness to be relevant, and robustness to be a stable and reliable product. Predictions also require modeling results to be effective and user-friendly to emergency managers, using visually appealing products that convey detailed results sufficient to evaluate flooding risks. The NG-CHC simulation research environment developed an interactive website known as Coastal Emergency Risks Assessment (CERA), available at http://coastalemergency.org, which presents ADCIRC (and SWAN) results in an environment to serve disaster management needs (Fig. 15.5). CERA has been designed to be an intuitive and easy to understand tool for the scientific community, emergency managers, and decision makers. The CERA website generates data

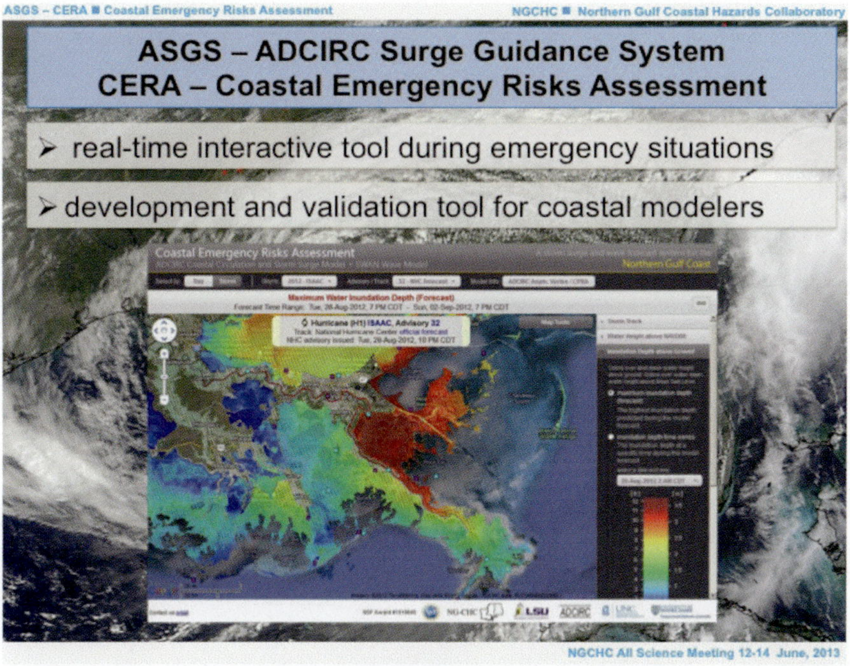

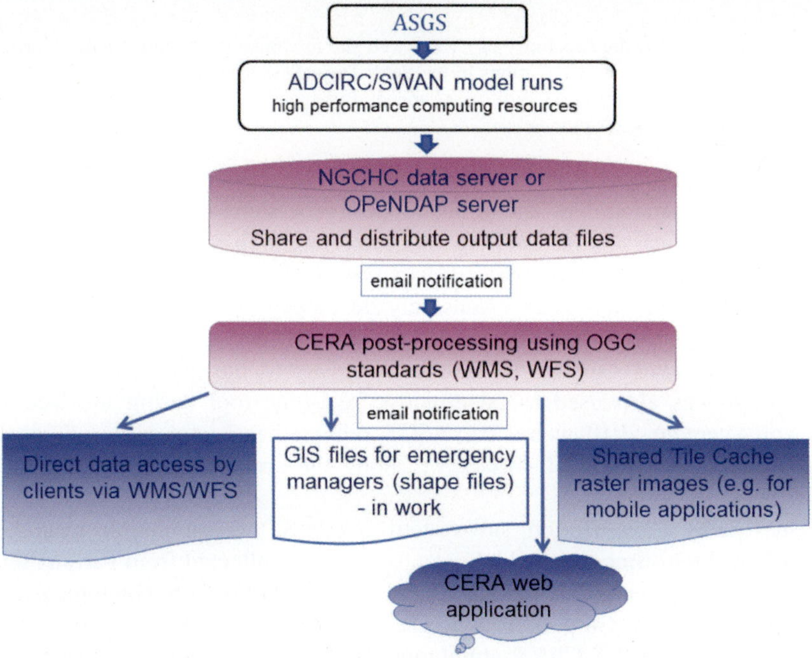

Fig. 15.5 Coastal Emergency Risks Assessment (CERA) has provided operational advisory services related to impending hurricane events (**a**), including a diverse group of clients during Hurricane Isaac during 2012. ASGS system was modified to compare ISAAC hindcasts with different grid, wind speed, and speed conditions tested to evaluate the effects on maximum water inundation

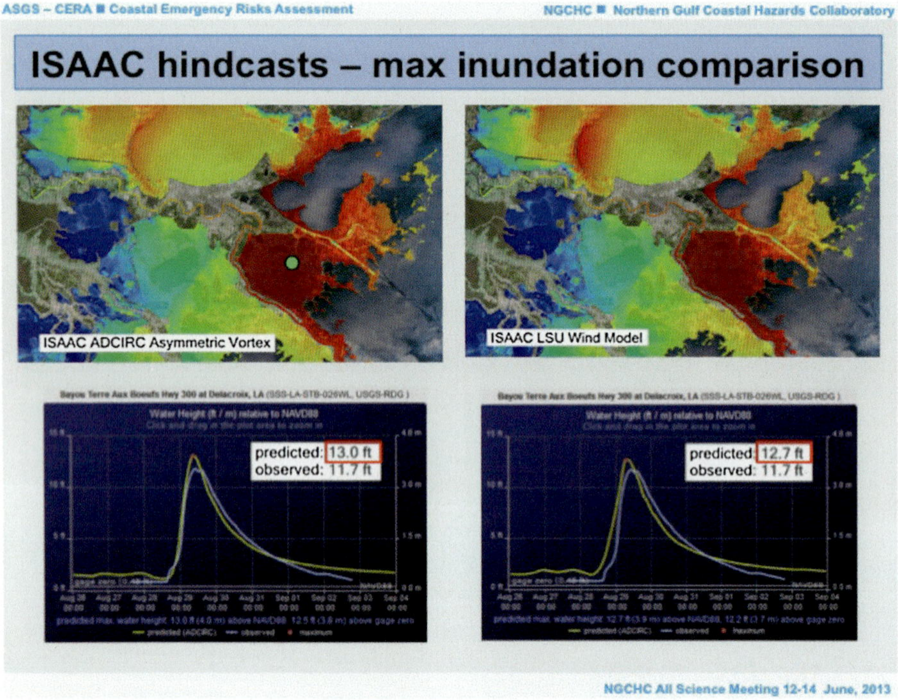

Fig. 15.6 Example of the capability of CERA web site to compare simulated water levels using ASGS compared to actual observations in the field. The ASGS system was modified to compare ISAAC hindcasts with different grid, wind speed, and speed conditions tested to evaluate the effects on maximum water inundation

visualization products directly from ASGS using a stable, reliable, and robust workflow implemented by an effective post-processing system. This workflow delivers huge data sets to hundreds of users via a web mapping application and enables timely and accurate data distribution to the wider disaster management community. CERA has been extensively developed using feedback from emergency managers and federal agencies over the last 8 years taking advantage of real-time hurricane events such as Hurricanes Rita, Gustav, Ike, and Isaac in the Gulf of Mexico. It was also used as an effective real-time tool during the Deepwater Horizon event in 2010 where the ASGS provided daily forecasts of nearshore and inshore penetration of surface oil along the northern Gulf Coast (Dietrich et al. 2012).

The CERA visualization sub-system overlays the simulation results with consolidated real-time data, which is automatically collected from various sensors distributed along the Northern Gulf Coast such as NOAA, U.S. Geological Survey (USGS), US Army Corp of Engineers, and National Hurricane Center (NHC) (Fig. 15.6). The ASGS/CERA simulation management system uses several standards and technologies like the usage of Open Geospatial Consortium (OGC) standards to ensure interoperable data exchange. The interface also allows visualization of multiple runs based on different track and wind projections, facilitating probability projections and sensitivity studies. For example, CERA presented results of different ADCIRC forecasts for Hurricane Isaac (2012) from Louisiana State University and Mississippi State University, based on simulation runs supported on each institution's high-performance computing systems.

15.3 Geoinformatics

To support the application of these three simulation management systems, NG-CHC also invested in CI systems that could enhance the geoinformatics used in each of the modeling grids. A key objective of the Dynamic Digital Elevation Models (DDEM) was to form digital grids that capture the elevation of natural and built infrastructure of the region and change the way the diverse environmental systems interact (Fig. 15.2). Dynamics in both elevation models and models of environmental systems (e.g. sea level rise, subsidence, and construction of civil infrastructure) produces complex behavior of water distribution in the northern Gulf Coast. The majority of digital elevation and bathymetry data are created and housed by federal and state agencies in the northern Gulf Coast. Limited, specialized data are available from selected county governments in the region and is usually the result of flood mapping and other countywide (parish) projects. Primary sources of data are the U.S. Geological Survey (USGS) and the National Geophysical Data Center (NGDC) for land surface and bathymetry, respectively.

The Coastal Relief Model (CRM, V. 4) released by the NGDC in 2001 was chosen as the basis for the DDEM. The CRM encompasses the Northern Gulf of Mexico (NGOM) from Pensacola, FL to Sabine Pass, TX with a three-arc second resolution. The comprehensive documentation that accompanies the CRM provides transparent understanding of the bathymetric surveys that have been included. Another appealing aspect of the CRM that led to its adoption as initial framework is that it represents the first high-quality, high-resolution DEM covering the U.S. coastal zone.

The bathymetric component of the merged DDEM (BDDEM) consists of two stages (Fig. 15.7). The first stage incorporated five NGDC-produced digital elevation models into the original CRM consisting of one arcsec resolution image from Northern Gulf Coast and one-third arcsec resolution images from Biloxi, Mobile, SE Louisiana and New Orleans (Table 15.3). These datasets were incorporated into the merged DDEM, with vertical datum established as NAVD88 and one-third arcsec resolution with bathymetric data sources current through 2001. The second stage BDDEMs were systematically updated by incorporating bathymetric survey data released by NOAA-NOS (National Ocean Service) over the period 2002–2011. The 2002 BDDEM used the first stage product as its basis, and subsequent years used the previous year as its basis (i.e., BDDEM 2003 uses BDDEM 2002 etc.). For the three member states of the NG-CHC, the unincorporated single beam and multi-beam bathymetric surveys through 2011 included 19 surveys for Alabama, 12 for Mississippi, and 80 for Louisiana, as well as three surveys from western Florida (Fig. 15.7).

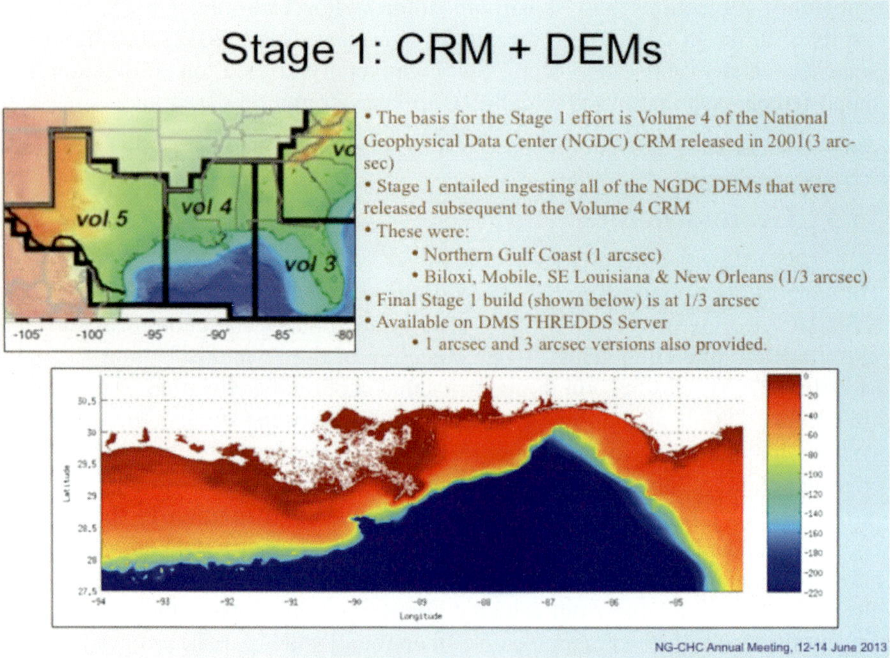

Fig. 15.7 Example of the first stage in adding bathymetric data to the dynamic digital elevation model (DDEM) strategy of NG-CHC

Table 15.3 Specifications for the NGDC high resolution DEM in the northern Gulf of Mexico

Name	Date completed	Spatial resolution	Horizontal datum	Vertical datum
Biloxi	03/29/2007	1/3 arc-second	WGS84	MHW
Mobile	11/30/2009	1/3 arc-second	NAD83	NAVD88 and MHW
Northern Gulf Coast	12/31/2010	1 arc-second	WGS84	NAVD88 and MHW
New Orleans	04/30/2010	1/3 arc-second	NAD83	NAVD88, MHW and MLLW
Southern Louisiana	12/31/2010	1/3 arc-second	WGS84	NAVD88 and MHW

The workflow applied for incorporating these NOS surveys by year employed open source software tools consisting of the Generic Mapping Tools (GMT, Wessel and Smith 1998), MB-System (Caress and Chayes 1996) and NOAA's Vertical Datum Transformation Tool (VDatum, http://vdatum.noaa.gov/welcome.html). The workflow consisted of: (1) translating the NOS survey to NAVD88 and gridding them to a resolution of one-third arcsec; (2) merging and annealing the gridded NOS survey within a surrounding subset extracted from the BDDEM basis; and (3) placing the updated bathymetry field subset back into the basis BDDEM at one-third arcsec. These software tools and workflow were streamlined to produce an open source toolkit that will allow either further updating of the BDDEM, or application of these tools to researchers in other coastal regions.

In addition to developing the first two stages of BDDEM products on the University of Southern Mississippi THREDDS server, NG-CHC developed a web-based Graphical User Interface (GUI) extraction module that facilitates extraction of bathymetry, using the second stage updates (2011) of three-arcsec resolution as the prototype. A BDDEM website was established (http://bddem.ngchc.org) that leverages the data identification and extraction capabilities developed for SIMULOCEAN. The BDDEM website is currently under active development as a resource designed to facilitate extraction of BDDEM fields to match a user's specific interest.

15.4 Collaborative Environment

From the above descriptions it can be seen that each of the three simulation management systems serves a somewhat different constituency ranging from science to engineering to management and functions from hazard prevention planning to emergency operations to ecosystem restoration. They share a common toolkit (Table 15.4) and the "connective tissue" of the cyberinfrastructure.

15.4.1 Cybertools of NG-CHC

The NG-CHC focused on developing a suite of CI tools (Table 15.4) centered on handling the needs of the simulation management systems in an integrated view (Nicolson et al. 2002). CI tools were used as a method for rapid prototyping all NG-CHC modeling efforts. These tools assisted researchers in identifying critical limitation that each simulation management system will face during its refinement and generation of future prototypes. Conducting the three collaboratory experiments in parallel, facilitated conditions where tools developed by one team enhanced the capabilities of the other NG-CHC teams, thus fostering collaboration. In addition, the group found that it was critical to engage a combination of modelers and non-modelers in the process of developing CI tools as well as educational aids/tools. This was in part accomplished by regularly scheduled discussions during bi-weekly teleconferences. The CI tools and services developed by the NG-CHC collaborators focused on functional attributes and services as follows: metadata extraction; cataloging and discovery; large file data transfers; workflow services; data access, formatting, processing, storage and visualization; and educational tools (Table 15.4). These CI tools provided scientists and educators with a reusable suite of standards-based tools and services to enable effective science collaboration, as well as improved access to high-performance computing (HPC) systems. Applications of the cybertools listed in Table 15.4 are described in the "Collaboratory Experiments" Sect. 15.5 below.

Table 15.4 Description of the various tools developed to support simulation management systems in NG-CHC

Tool title	Descriptions
Metadata extractions, cataloging, and discovery	
Sulis metadata creator <beta>	Web-based metadata template and creator that generates xml metadata records
NG-CHC Metadata driven visualization	Metadata driven visualization capability has been integrated into the NGCHC metadata catalog so that users can visualize distributed data within the portal
NG-CHC Metadata catalog	Developed a highly adaptable information catalog that is also deployable within collaborative environments. The catalog provides users with the ability to search, browse, and discover the information at different levels of granularity (Collection and Data Granule). Also developed a deployable visualization tool that consumes the catalog information. Packaged together, the catalog and the supporting visualization tool provide a suite of solutions to distributed coastal data management
Large files data transfer	
Globus online	Installed GlobusOnline server to facilitate staging and transfer of large data files during simulation experiments using the GridFTP protocol to facilitate data transfers. The team is working on automatically capturing the metadata for the transferred data and publishing it to the metadata catalog, allowing researchers to discover the large simulation experiment data via the portal or other tools and use Globus Connect to download
Workflow services	
Visual workflow system	A visual workflow system to assist the DDEM group with streamlining bathymetry data generation. The workflow will be triggered when new NOAA survey files become available
Science workflow deployment suite	Developed an apache airavata based science workflow management tool that can be used for constructing, deploying and executing workflows. This tool allows any application to be wrapped as a service. These services are deployed in cloud or computing clusters. Deployed mining workflows at the NSF XSEDE resources successfully. This tool abstracts various XSEDE job related functionalities, hence submitting and monitoring such jobs via workflow composer (XBaya) are seamless

(continued)

Table 15.4 (continued)

Tool title	Descriptions
Data access, formatting, processing, storage and visualization services	
GeoTIFF to NetCDF4 tool	MATLAB based scripts to extract header information from GeoTIFF, input GeoTIFF, transpose, and write NetCDF4 using low-level NetCDF MATLAB functions
ADCIRC to ESRI shape file scripts	C++ programs that take ADCIRC model data as an input and allows either the entire data or a selected geographic region of the data to be converted to an ESRI shape file
SULIS portal viewer	Custom viewer based on the Esri javascript API for searching and visualizing data from the Sulis Informatics Services Portal
SULIS severe weather viewer	Custom viewer based on Esri javascript API for viewing and comparing model forecast data for specific storms. Different forecasts can be compared side by side. Additional layers can be loaded to show additional data such as contextual information
Weather Underground Web Data Extraction Tool (WUDEXT)	This new tool extracts information stored in tables located at the bottom of any weather underground webpage. Such data, when used with caution, can be a useful supplement to traditional stations. The tool is able to extract the climate data from multiple days, months, and even years. All the extracted information is stored in a single text file
Coastal datafactory	The Coastal Data Factory (CDF) is a web-based data grabbing and preprocessing toolkit for gathering, refining, storing and plotting coastal data. CDF provides a data warehouse with an on-demand data grabbing engine that allows users to easily fetch data from multiple data sources or/and local data center and check data availability via web-based graphical user interface
Data to NetCDF scripts	C++ programs take ADCIRC, ADH, or EFDC model data and converts the data to a NetCDF format to be processed by the Sulis viewer. This provides a global maximum layer that will show the highest predicted values at each location of the ADCIRC grid
ASGS -NHC text advisories to ATCF format	Perl script to parse text advisories (shtml) from the National Hurricane Center. Converts them to ATCF format for use within ADCIRC. Writes a forecast track (fst) file by default
ASGS -ADCIRC ASCII to NetCDF script	A windows executable that allows the conversion of ADCIRC ascii files to NetCdf as used for the ASGS workflow

(continued)

Table 15.4 (continued)

Tool title	Descriptions
EnVis	Enhancements to scientific visualization package for complex model data. Envis can handle both unstructured and structured data. It can also handle rendering using level-of-detail and tiling which makes it appropriate to use within Esri map layers. It is used in the Sulis Geoportal to handle interactive visualizations
Coastal events driven data delivery and visualization	Event-Driven Data Delivery (ED3), aggregation and processing automates the access and processing of data for situational awareness in a hazardous event that could potentially become a disaster. ED3 is being integrated with decision support systems to state agencies in Alabama, and others to help define the capabilities that will best suit the end users for a variety of events. In NG-CHC, ED3 can be used to trigger data collection and subsequent workflow during tropical storm events
FloodViz	Visualization tool that calculates the inundation of land based on the river levels over time provided from HEC-RAS over the DEM, geo-referenced imagery, and ESRI shape files. The batch mode capability of this tool will be used to output contour lines, which can be used by the Sulis viewer. Once this is in place, FloodViz may also be expanded to handle other model data as inputs
Educational tools	
Research-based interactive computer simulation to model response of floodplains to precipitation events	An interactive research-based computer simulation and educational tool for floodplain analysis
Online storm surge game	An interactive web-based user interface for a simple one dimensional storm surge model

15.4.2 NG-CHC Portal

The NG-CHC collaborative environment enables close interaction among coastal scientists, educators and computer scientists. Minimizing the complexities of sharing, discovering and using data and information can enable the scientific process and stimulate discovery, by allowing users to concentrate on science. Based on a freely available Content Management System (CMS), the extensible environment provides a portal with modules specifically developed to support science research, analysis, and visualization for distributed coastal data management. Users can organize, discover, share, and use information about data, models, tools and other

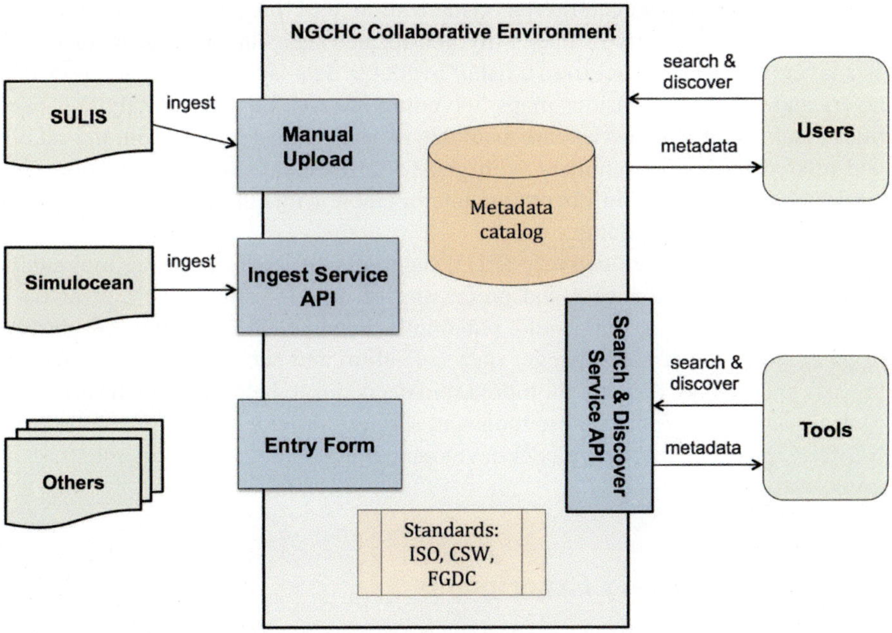

Fig. 15.8 The information architecture of the NG-CHC Collaborative Environment Metadata Catalog shows how the standards-based catalog can be updated manually with ingest scripts, via an automatic ingest service, or with a simple user entry form. The catalog can be searched and accessed by either users or tools to discover and consume data of interest

resources; discuss project activities and results; view publications, presentations and other documents; and track the history of project activities. Information can be categorized and aggregated by content type and group audience. Integrated tools and services include collaboration support; standards-based metadata extraction, cataloging and discovery; large file data transfers; workflow services; data access, formatting, processing, storage and visualization. The NG-CHC portal includes both public and private areas, as well as an education and outreach area with project information, educational tools and learning modules, to increase public knowledge and understanding.

The NG-CHC collaborative environment hosts a Federal Geographic Data Committee (FGDC)-based metadata catalog, providing users with a central location to search, browse, and discover information at different levels of granularity. A variety of metadata ingestion tools are supported, including ingest scripts that plug into a modeling workflow, a file upload interface, a manual form-based interface, and a THREDDS metadata harvester, to aggregate information and catalogs from a variety of coastal information systems. Both SULIS and SIMULOCEAN publish metadata to the NG-CHC metadata catalog in support of simulation experiments, as depicted in the information architecture (Fig. 15.8). Standards-based publication of catalog entries allows for existing tools to search, discover, and consume the information.

An integrated metadata-driven visualization tool provides custom mapping capabilities for a variety of data. Any geo-located data (image) can be displayed as a layer. Users can select from a list of available data layers or add their own data layers and overlay on various maps, including 2D and 3D maps. Analytic capabilities (such as feature queries) are available over displayed layers. Both the catalog and related tools are compliant with geospatial metadata and visualization standards, making this cyberinfrastructure interoperable with other standards-compliant geospatial tools and systems.

An event-driven data delivery (ED3), aggregation, and processing tool can be used to automate the access and processing of data for situational awareness in coastal hazard events that could potentially become disasters. For instance, a tropical storm event could trigger data collection and subsequent workflows that process and prepare the data for more rapid decisions to help mitigate the effects of coastal hazards. Together, these tools and services provide a suite of solutions for NG-CHC project activities, model development and simulation experiments.

15.5 Collaboratory Experiments

The following three experiments were developed during the 3-year NG-CHC project. These experiments were contributions from the group to develop interdisciplinary models that inform coastal and inland water resources planners and managers. This exercise initiated an open, highly participatory, and coordinated collaboration among the universities involved, and it is in line with the goal of developing tools that promote general agreements and a shared vision of future users (Loucks 2012).

15.5.1 River Management and Ecosystem Restoration

Ecosystem restoration of the northern Gulf coast requires an accurate prediction of sediment transport and landscape changes caused by tropical cyclones and winter storms (Chen et al. 2007, 2008; Zhao and Chen 2008). To meet the needs, NG-CHC adopted the open-source software package Delft3D as a key component of SIMULOCEAN (Figs. 15.3 and 15.9). Nested, curvilinear grids ranging from basin-scale domains similar to the ADCIRC mesh to local-scale domains that cover Breton Sound and Terrebonne Bay, coastal basins of Mississippi River delta, have been developed. Simulations of Hurricanes Katrina (2005) and Gustav (2008) as well as Tropical Storm Lee (2011) using the nested Delft3D model, and comparisons with field observations have been carried out. It is found that the domain decomposition function (nesting with different resolution) in Delft3D is particular useful to capture storm surge in this complex coastal region.

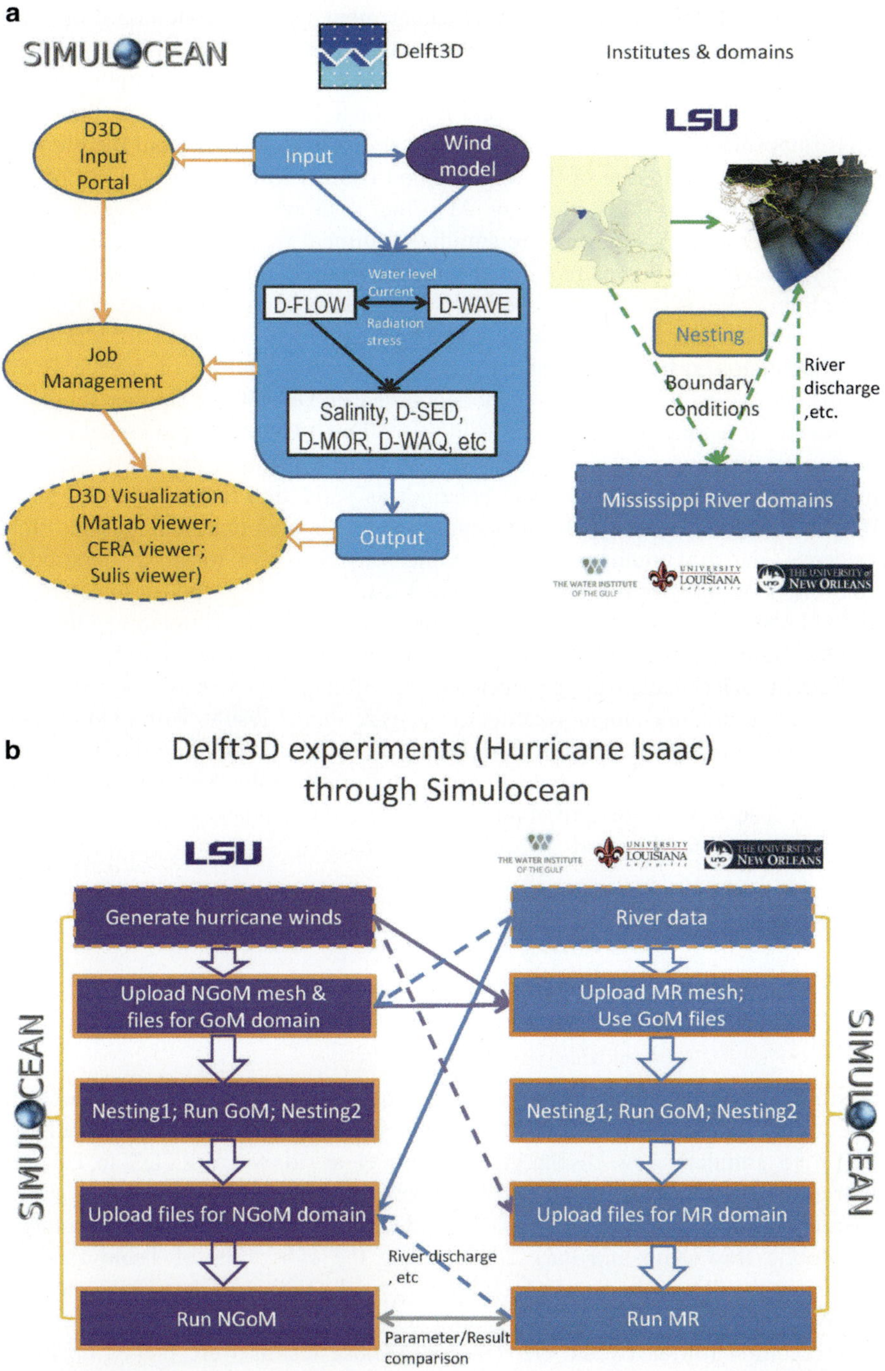

Fig. 15.9 SIMULOCEAN was used to perform a suite of experiments on Lower Mississippi River using following models: (**a**) a regional Delft3-d model for the Lower Mississippi River; (**b**) a 3-d FVCOM surge model for the Pontchartrain Estuary

SIMULOCEAN was also used as a management system to simulate storm surge movement up the Mississippi River during extreme events (Fig. 15.9). The following models were validated and applied to evaluate interaction of coastal restoration features along the Mississippi River. A regional Delft3D model for the Lower Mississippi River captures storm surge and sediment transport from the mouth of the river upstream to one of the largest flood control structures just north of New Orleans, the Bonnet Carré Spillway. At a finer scale resolution of river dynamics, two Delft3D models were used to simulate sediment transport in segments of the river upstream and downstream of the Bonnet Carré Spillway. The validation of these finer scale models during operation of this flood control structure, during high flood events, were used to simulate anticipated sediment dynamics for a proposed river diversion structure planned for location in the vicinity of Myrtle Grove, Louisiana. Along with these river models based on Delft3D, is another simulation approach of estuarine processes using a 3-D FVCOM surge model for the Pontchartrain Estuary. This model was also operated within the SIMULOCEAN management system to test model behavior as validation technique using results from Hurricane Isaac (2011). The estuary model showed that, for the same storm category, the maximum surge heights are very sensitive to storm duration, as controlled by the storm's forward speed. Finally, a rapid response dynamic 1-D (HECRAS) was designed to compute storm surges as far upstream as Old River Control Structure, nearly 485 km upstream from mouth of Mississippi River. These numerical experiments were conducted using validated models as described, using the simulation management systems to compare model results with field observations. The local high-resolution models have been used to develop sediment water delivery ratios (SWR) for river diversion structures with deep and shallow sill designs. The SWRs are important in assessing the efficiency of a diversion in building land (Meselhe et al. 2012).

15.5.2 Mobile Bay Experiment: Processes from Catchment to the Coast

A simulation experiment was conducted using SULIS to develop a protocol for integrating four models for the Mobile Bay system (Figs. 15.4 and 15.10), including HSPF to simulate flow discharge (Alarcon et al. 2009), WRTDS to calculate nutrient loads (Maestre et al. 2012; Hirsch et al. 2010), ADCIRC to simulate storm surges on the shelf, and EFDC to forecast circulation and nutrient transport in Mobile Bay (Kim and Park 2012), using the cybertools of Table 15.4. For example, the Watershed model HSPF imported landscape data from elevation database using the metadata catalog, computed rainfall runoff which was converted to a common format by the GeoTiff tool. Those results were picked up by the WRTDS model to calculate nutrient loads to the bay. All simulations during the experiment were logged into the database by the Metadata Creator. Surge data

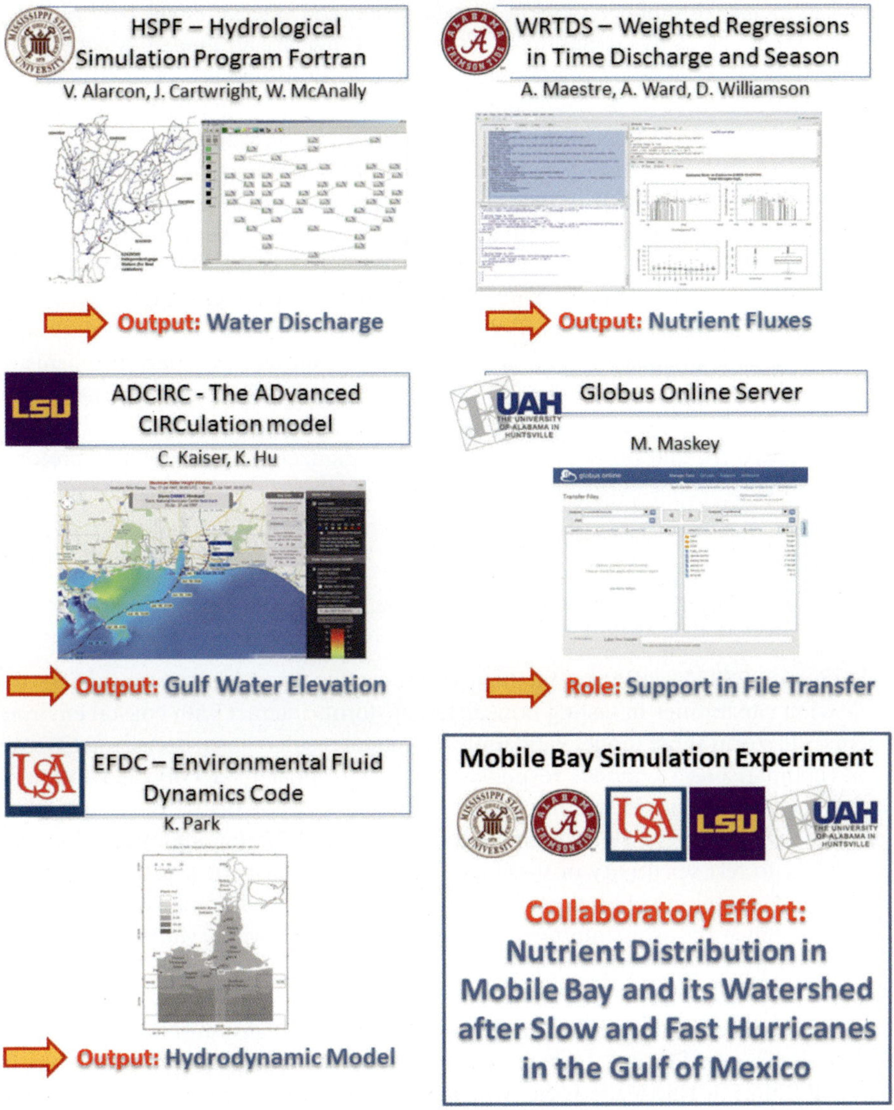

Fig. 15.10 Joint effort from NG-CHC universities to estimate nutrient transport in Mobile Bay and its watershed as part of the loose coupling experiments

from ADCIRC were supplied the same way and all inputs merged for use in the water quality simulation of Mobile Bay and subsequent visualization and analysis by the cybertools. The impact of the protocol is to allow coastal scientists and engineers to model the interactive effects of hurricane storm surge and watershed processes in Mobile Bay and resource managers to examine the results for planning and operations requirements. This is a new capability provided from the collaboratory to the principal disciplines of coastal engineering and science along with resource management across the spectrum of physical, biological and human systems. This capability can be used by practitioners to investigate a number of environmental phenomena that may have significant social and economic impacts (e.g., hypoxia in coastal areas directly affects fisheries, recreation, and land values).

While much research exists in freshwater studies regarding transport of nutrients through watersheds, efforts to investigate the dynamic role of the watershed on the receiving bay during and after a hurricane, as well as testing hurricanes with different intensity, are lacking. The protocol developed in the Mobile Bay Simulation Experiment provided an avenue for evaluating watershed impacts on coastal ecosystems during these events. For example with this protocol one can ask how different characteristics of storms would affect nutrient transport mechanisms and thus impact the distribution of land-derived materials along a coastal area. The Environmental Protection Agency (EPA), The National Marine Fisheries Service (NMFS), and other state and federal organizations can use this protocol to better manage coastal resources.

15.5.3 Forecasting Hurricane Storm Surge

ASGS and CERA software facilitates hindcasting exercises by providing the following capabilities in testing how different storms interact with coastal environments: (1) storm surge model validation; (2) model result comparisons; and (3) real-time comparisons of model results with observational data collected during a hurricane. For example, during Hurricane Isaac (2012) operational runs were provided using different grid configurations and wind forcing algorithms by LSU and MSU to test sensitivity of these parameters to forecasts of coastal inundation. LSU used a high-resolution grid configured for the Louisiana coast including updated elevations for improved levee systems following Hurricane Katrina (produced by Coastal Protection and Restoration Authority for the 2012 Coastal Master Plan, known as the CPRA grid). While this grid contains updated information for Louisiana coast, it has lower resolution for the eastern Northern Gulf in Mississippi and Alabama. In contrast, MSU used the S08 grid for simulations of ADCIRC during Hurricane Issac, which more accurately represents elevations of the Mississippi and Alabama coastlines, but does not capture the detail of coastal Louisiana. Developing software that can quickly visualize simulation results from different grid systems is advantageous since even small errors in hurricane track forecasts can result in large surge differences along the coast of these three states. In addition, MSU and LSU could focus their supercomputing capability to local impacts, but at the same time emergency managers could compare relative estimates for respective coasts in the northern Gulf region. Both institutions used shell scripts for preprocessing and post-processing data. LSU used ASGS for preprocessing while MSU's was in-house. Both relied on software tools such as Google Earth, OpenDAP, GIS Shapefile generation, perl, netcdf, FORTRAN, and

GMT. This helped to improve forecasting methods, validate ADCIRC model grids, and enhance modeling research by the ADCIRC community.

Both LSU and MSU downloaded intensity, track, and storm size parameters in the National Hurricane Center forecast statements, but incorporated these parameters in different wind forcing schemes to generate a forecast ensemble. LSU used the "asymmetric vortex" method discussed by Mattocks and Forbes (2008), while MSU used a variant of the Holland scheme, which included storm translation and radius of 34-knot winds implicitly in the wind shape parameter. CERA provided graphics for both simulations comparing results to real-time observations and hindcast analysis for Isaac. For instance, different wind models were compared to the official wind model (asymmetric vortex). Maps were created to identify locations of significant differences in coastal inundation predicted between the two simulation efforts. In addition, USGS rapid deployment stations were posted on the CERA website to compare observed and predicted values of the two grids and to compare the different wind models (Figs. 15.5 and 15.6).

15.6 Education and Outreach

The focus of the NG-CHC has been to create a collaborative environment that allows the integration of models from the catchment to the coast. While products have included tools that facilitate this integration, the output of these efforts has been focused on scientific visualizations. Thus, the focus of the education effort has been to introduce students to scientific visualizations, models and modeling through a lens of coastal hazards.

The NG-CHC developed a Coastal Hazard Curriculum Unit for middle and high school students consisting of four lesson modules (Fig. 15.11). The curriculum unit is currently based on visualizations of hurricane storm surge using the REAL tool. The REAL (Rapid Estimates of Approaching Landfall) tool uses pre-computed, high resolution model data to generate fast, accurate, time dependent and operation estimates of local storm surge. As a unit, these modules help students learn about accessing data online, the use of visualizations in scientific research, how to use Google Earth (GE), hurricanes and the importance of storm surge as a coastal hazard. While initially available only in paper and electronic form (during testing), this unit has being converted to an online tool. The curriculum unit is openly available at http://stormsurge.disl.org.

Formative assessments of curriculum unit (Fig. 15.11, upper panel) were conducted on the coastal hazard curriculum unit at 3 K-12 educator workshops ($n = 58$ participants total: Teachers Exploring Coastal Hazards, MSU, 6/2012; Climate Change in the Gulf of Mexico, DISL, 7/2012; iSTEM meeting, Huntsville, AL 2/2013; Climate Change in the Gulf of Mexico, DISL 6/2013). The GE module was added as a result of need indicated by formative evaluations.

Student evaluations of the curriculum unit were conducted in the spring and summer of 2013 ($n = 110$; 70 % 11th graders; Auburn High School marine science

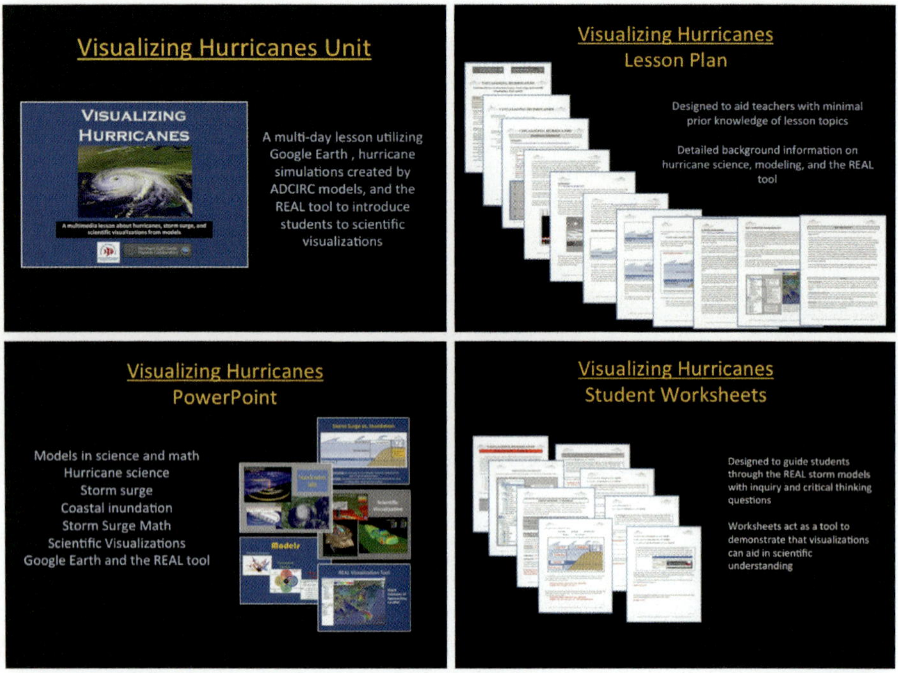

Fig. 15.11 Coastal Hazard Curriculum Unit for middle and high school students entitled 'Visualizing Hurricanes'. The *upper left panel* and *bottom two panels* define some of the materials developed as part of the curriculum unit including lesson plans, power point presentations, and student worksheets

club; Baker High School marine biology classes; Discovery Hall Programs Summer High School program) (Fig. 15.11, lower panel). Student evaluations indicated that the lesson plans were interesting (75 %), clear and easy to understand (75 %) and resulted in "improved knowledge of storm surge and inundation" (92 %). Interestingly, 72 % indicated they would be more likely to pay attention to evacuation notices about storm surge as a result of these lessons. Lastly, 81 % agreed or strongly agreed with the statement "using visualizations helps me to better understand scientific information".

This interactive simulation approach to the exploration of hurricanes and associated storm surge has been developed into a storm surge 'game' suitable for K-12 students and the public. The 'game' allows the user to explore the impacts of storm surge at the local scale by selecting both storm intensity and storm landfall location along the northern gulf coast. Model simulations then display maximum surge elevation or maximum surge inundation as color superimposed on a map or satellite view. Included is the ability to display hydrographs showing water levels at tide gauge locations over time. Users can zoom in and out displaying a smaller or larger geographic area. The 'game' can be accessed at http://stormsurge.disl.org.

An interactive research-based computer simulation and educational tool for floodplain analysis (Fig. 15.12, lower panel) simulated the response of upland

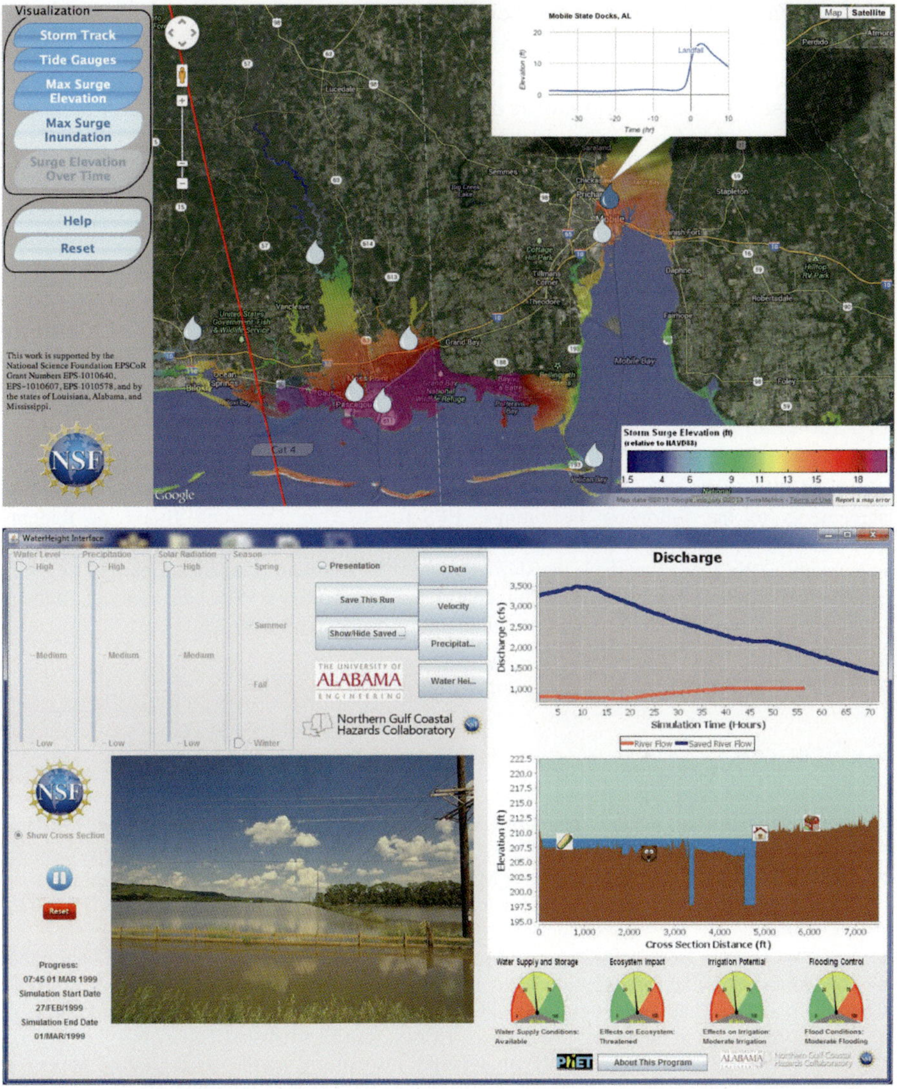

Fig. 15.12 Screen clips of interactive educational tools developed by NG-CHC. The online storm surge 'game' (*upper panel*) allows users to explore the impacts of storm intensity and landfall location on surge elevations and inundations on specific geographic locations using a Google Earth interface. The tool allows the simulation of hypothetical storm tracks (*red line*), the display of location-specific hydrograph plots, and the ability to zoom in geographically. The floodplain computer simulation and educational tool (*lower panel*) allows users to set initial conditions of water elevation, precipitation, and solar radiation in specific seasons, and observe the changes in discharge, velocity, and water stage in a floodplain located in the Sipsey River (Alabama) (color figure online)

flood plains to extreme event precipitation. The simulation is based on similar efforts developed by the PhET project of the University of Colorado to model real-life science and engineering phenomena. Variables including water level, solar radiation, season, and storm intensity can be adjusted by the user to assess the relative impacts of different scenarios. Target audiences for the educational tool are K-12 to college students. The simulation allows students to choose the initial conditions of a section of the Sipsey River (located near Tuscaloosa, Alabama) and observe the potential impacts in the floodplain. The simulation is initiated from historical records, while the user observes how the discharge, stage, and velocity of the river changes. The simulations are based on hourly records collected at the Elrod station managed by the USGS. The first draft version of the simulation has been posted as an executable Java executable file in the NG-CHC portal. An initial testing of the tool was conducted by students at the University of Alabama.

Acknowledgements This research was supported by NSF EPSCoR Program with cooperative agreements to the following institutions: EPS-1010640 to Louisiana Board of Regents; EPS-1010578 to Mississippi State University; and EPS-1010607 to The University of Alabama in Huntsville. We appreciate the leadership of Michael M. Khonsari (Louisiana Board of Regents), Sandra H. Harpole (Mississippi State University) and Sara J. Graves (The University of Alabama in Huntsville) for guidance through this collaborative research program.

References

Adger WN, Hughes TP, Folke C, Carpenter SR, Rockstrom J (2005) Social-economical resilience to coastal disasters. Science 309:1036–1039

Alarcon V, McAnally W, DiazRamirez J, Martin J, Cartwright J (2009) A hydrological model of the mobile river watershed, southeastern USA. AIP Conf Proc 1148:64. doi:10.1063/1.3225392

Caress DW, Chayes DN (1996) Improved processing of Hydrosweep DS multibeam data on the R/V Maurice Ewing. Mar Geophys Res 18:631–650. doi:10.1007/bf00313878

Chen Q, Wang L, Zhao H, Douglass SL (2007) Predictions of storm surges and wind waves on coastal highways in hurricane-prone areas. J Coast Res 23:1304–1317

Chen Q, Wang L, Tawes R (2008) Hydrodynamic response of northeastern Gulf of Mexico to hurricanes. Estuar Coasts 31(6):1098–1116. doi:10.1007/s12237-008-9089-9

Crossett KM (2004) Population trends along the coastal United States: 1980–2008/ Kristen M, Crossett T, Culliton P, Wiley T, Goodspeed (2004) NOAA. U.S. Dept. of Commerce, National Oceanic and Atmospheric Administration, National Ocean Service, Management and Budget Office, Special Project, Silver Spring

Crowell M, Westcott J, Phelps S, Mahoney T, Coulton K, Bellomo D (2013) Estimating the United States population at risk from coastal flood-related hazards. In: Coastal hazards. Springer, Dordrecht, pp 151–183

Cruz AM, Krausmann E (2008) Damage to offshore oil and gas facilities following hurricanes Katrina and Rita: an overview. J Loss Prev Process Ind 2(6):620–626

Cutter SL, Barnes L, Berry M, Burton C, Evans E, Tate E, Webb J (2008) A place-based model for understanding community resilience to natural disasters. Glob Environ Chang 18:598–606

Day JW Jr, Boesch DF, Clairain EJ, Kemp GP, Laska SB, Mitsch WJ, Orth K, Mashriqui H, Reed DR, Shabman L, Simenstad CA, Streever BJ, Twilley RR, Watson CC, Wells JT, Whigham DF (2007) Restoration of the Mississippi delta: lessons from hurricanes Katrina and Rita. Science 315:1679–1684

Day JW, Christian RR, Boesch DM, Yáñez-Arancibia A, Morris J, Twilley RR, Naylor L, Schaffner L, Stevenson C (2008) Consequences of climate change on the ecogeomorphology of coastal wetlands. Estuar Coasts 31:477–491

Dietrich JC, Bunya S, Westerink JJ, Ebersole BA, Smith JM, Atkinson JH, Jensen R, Resio DT, Luettich RA, Dawson C, Cardone VJ, Cox AT, Powell MD, Westerink HJ, Roberts HJ (2010) A high-resolution coupled riverine flow, tide, wind, wind wave and storm surge model for southern Louisiana and Mississippi: part ii – synoptic description and analysis of hurricanes Katrina and Rita. Mon Weather Rev 138(2):378

Dietrich JC, Trahan CJ, Howard MT, Fleming JG, Weaver RJ, Tanaka S, Yu L, Luettich RA Jr, Dawson CN, Westerink JJ, Wells G, Luf A, Vega K, Kubach A, Dresback RL, Kolar KM, Kaiser C, Twilley RR, (2012) Surface trajectories of oil transport along the Northern Coastline of the Gulf of Mexico. Cont Shelf Res. http://dx.doi.org/10.1016/j.csr.2012.03.015

Fischetti M (2001) Drowning New Orleans. Sci Am, October, 77–85

Galloway G, Boesch D, Twilley RR (2009) Restoring and protecting coastal Louisiana. Issues Sci Technol Winter 2009:29–38

Hallegatte S, Green C, Nicholls RJ, Corfee MJ (2013) Future flood losses in major coastal cities. Nat Clim Chang 3:802–806. doi:10.1038/NCLIMATE1979

Hirsch RM, Moyer DL, Archfield SA (2010) Weighted Regressions on Time, Discharge, and Season (WRTDS), with an application to Chesapeake Bay river inputs. JAWRA J Am Water Resour Assoc 46:857–880. doi:10.1111/j.17521688.2010.00482.x

Jordan SJ, Lewis MA, Harwell LM, Goodman LR (2010) Summer fish communities in northern Gulf estuaries: indices of ecological condition. Ecol Indic 10(2):504–515. doi:10.1016/j.ecolind.2009.09.003

Kim C, Park K (2012) A modeling study of water and salt exchange for a micro-tidal, stratified northern Gulf of Mexico estuary. J Mar Syst 96–97:103–115. doi:10.1016/j.jmarsys.2012.02.008

Laska S, Woodell G, Hagelman R, Gramling R, Teets Farris M (2005) At risk: the human, community and infrastructure resources of coastal Louisiana. J Coast Res 44:90–111

Lin N, Emanuel K, Oppenheimer M, Vanmarcke E (2012) Hurricane surge and global warming: a physically-based risk assessment. Nat Clim Chang 2:462

Loucks DP (2012) Water resource management modeling in 2050. Toward a sustainable water future visions for 2050. American Society of Civil Engineers

Maestre A, Williamson D, Ward A (2012) Nutrient fluxes in rivers of the mobile – Alabama River system using WRTDS. 2012 Alabama water resources conference. Orange Beach, Alabama

Mattocks C, Forbes C (2008) A real-time, event-triggered storm surge forecasting system for the state of North Carolina. Ocean Model 25:95–119

Meselhe EA, Georgiou I, Allison MA, McCorquodale JA (2012) Numerical modeling of hydrodynamics and sediment transport in lower Mississippi at a proposed delta building diversion. J Hydrol 472–473:340–354

National Research Council (2006) Drawing Louisiana's new map: addressing land loss in Coastal Louisiana Committee on the restoration and protection of coastal Louisiana. National Academy Press, Washington, DC

National Science Foundation (2009) The GeoVision report, NSF Advisory Committee for Geosciences, 44 pp

Nicolson CR, Starfield AM, Kofinas GP, Kruse JA (2002) Ten heuristics for interdisciplinary modeling projects. Ecosystems 5(4):376–384

Nicholls RJ, Woodroffe CD, Burkett V, Hay J, Wong PP, Nurse L (2011) 12.14 – Scenarios for coastal vulnerability assessment, reference module in earth systems and environmental sciences. In: Wolanski E, McLusky D (Editors-in-Chief) Treatise on estuarine and coastal science. Waltham, pp 289–303. doi:10.1016/B978-0-12-374711-2.01217-1

Norris FH, Stevens SP, Pfefferbaum B, Wyche KF, Pfefferbaum RL (2008) Community resilience as a metaphor, theory, set of capacities, and strategy for disaster readiness. Am J Community Psychol 41:127–150
Opperman JJ, Galloway GE, Fargione J, Mount JF, Richter BD, Secchi S (2009) Sustainable floodplains through large-scale reconnection to rivers. Science 326:1487–1488
Paola C, Twilley RR, Edmonds DA, Kim W, Mohrig D, Parker G, Viparelli E, Voller VR (2011) Natural processes in delta restoration: application to the Mississippi Delta. Ann Rev Mar Sci 3 (1):67–91
Pinter N (2005) One step forward, two steps back on U.S. floodplains. Science 308:207–208
Rabalais NN, Turner RE, Scavia D (2002) Beyond science into policy: Gulf of Mexico hypoxia and the Mississippi River. Bio-Science 52:129–147
Sherrieb K, Norris F, Galea S (2010) Measuring capacities for community resilience. Soc Indic Res 99(2):227–247
Syvitski J, Saito Y (2007) Morphodynamics of deltas under the influence of humans. Glob Planet Chang 57:261–282
Twilley RR (2007) Comprehensive systems analysis and planning to secure water and coastal resources: lessons learned from hurricanes Katrina and Rita. Natl Wetl Newsl 29:33–35, Environmental Law Institute. Washington D.C., USA
Twilley RR, Couvillion BR, Hossain I, Kaiser C, Owens AB, Steyer GD, Visser JM (2008) Coastal Louisiana ecosystem assessment and restoration (CLEAR) program: the role of ecosystem forecasting in evaluating restoration planning in the Mississippi River deltaic plain. In: McLaughlin KD (ed) Mitigating impacts of natural hazards on fishery ecosystems. American Fisheries Society, symposium 64, Bethesda, Maryland, pp 29–46
USACE (2013) eCoastal enterprise GIS for coastal engineering and science, U.S. Army Corps of Engineers. http://ecoastal.sam.usace.army.mil/. Accessed Dec 2013
Vörösmarty CJ, Syvitski J et al (2009) Battling to save the world's river deltas. Bull At Sci 65 (2):31
Wessel P, Smith WHF (1998) New, improved version of generic mapping tools released. EOS Trans Am Geophys Union 79:579
Westerink JJ Jr, Luettich RA, Feyen JC, Atkinson JH, Dawson C, Powell MD, Dunion JP, Roberts HJ, Kubatko EJ, Pourtaheri H (2008) A basin-to channel-scale unstructured grid hurricane storm surge model as implemented for southern Louisiana. Mon Weather Rev 136:833
Zhao H, Chen Q (2008) Characteristics of extreme meteorological forcing and water levels in Mobile Bay, Alabama. Estuar Coasts 31(4):704–718. doi:10.1007/s12237-008-9062-7

Chapter 16
Advancement of Technology for Detecting Shoreline Changes in East Coast of India and Comparison with Prototype Behavior

Ramasamy Manivanan

Abstract Developments in the coastal area have significant impact on the adjacent shorelines. Mathematical modeling provides a useful tool for predicting such changes in shorelines in advance. Processing and analysis of satellite imageries of coastal area enables us to estimate and monitor the shoreline changes, which is otherwise extremely difficult, time consuming and costly by field surveying. In this paper the shoreline changes obtained by mathematical modelling and by image processing technique are compared by applying these techniques to shoreline adjacent to Ennore region. The study indicated that the cross-shore and longshore impact predicted by mathematical model and satellite information match satisfactorily. Thus the satellite information is useful for calibrating the mathematical model which can be further used for predictive purposes.

16.1 Introduction

The coastline is an interface between the land and sea where winds, waves, tides and currents attack the land. The land responds to this attack by dissipating the energy of the sea and changing the shape and alignment of the coastline continuously. Waves play a prominent role in the nearshore processes. The action of waves is the principal cause for changes in the shorelines.

Human interventions into the natural processes taking place in the coastal zone have an impact on the shorelines. These are construction of ports and coastal structures, dredging, mining etc. In order to anticipate the impact and suggest remedial measures to minimize the adverse effects, it is desirable to predict the

R. Manivanan (✉)
Mathematical Modeling for Coastal Engineering (MMCE), Central Water and Power Research Station, Khadakwasla, Pune 411 024, India
e-mail: vananrmani@rediffmail.com

C.W. Finkl and C. Makowski (eds.), *Remote Sensing and Modeling: Advances in Coastal and Marine Resources*, Coastal Research Library 9,
DOI 10.1007/978-3-319-06326-3_16, © Springer International Publishing Switzerland 2014

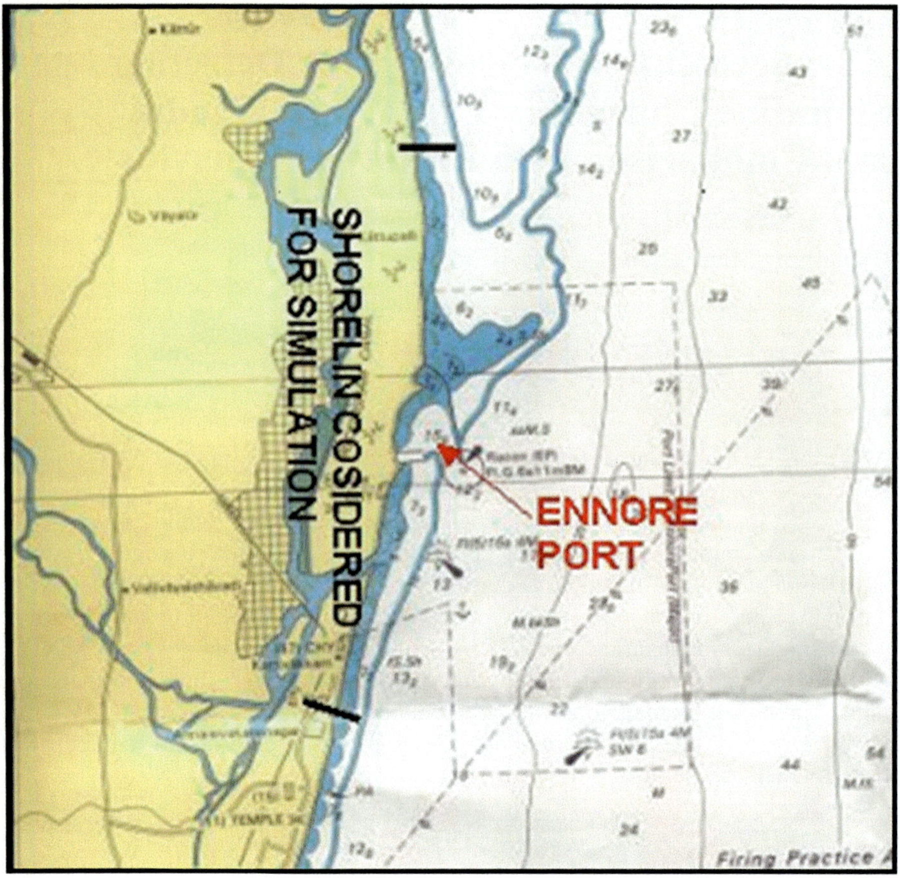

Fig. 16.1 Location of Ennore region

changes in the shorelines before undertaking the developments. Mathematical modeling provides a useful tool for predicting such changes in the shorelines.

Satellite remote sensing technique is a very useful tool for studies regarding coastline changes. It can provide synoptic view in the form of satellite imageries of coastal area at different time instants. Processing and analysis of these imageries enables us to estimate, monitor and understand the shoreline changes, which is otherwise extremely difficult, time consuming and costly by field surveying.

This study is carried out to demonstrate the usefulness of the satellite derived information for mathematical modelling of the shoreline changes. The coastal site around Ennore region is considered for this study (Fig. 16.1). Shoreline changes due to the development of the port are simulated by mathematical modelling techniques and compared with the shoreline changes derived from remote sensing/image processing techniques. The methodology and findings of this study is described below.

16.2 Development of Port at Ennore

It is located on the east coast of India about 20 km north of Chennai port. It was constructed in 1999 to meet the increasing demand of cargo. It is in operation since 2001 with coal as major cargo. The port is sheltered by northern breakwaters of 3.2 km length and southern breakwaters of 1.1 km length with entrance from southeast direction. Since the region is located in highly sensitive zone of littoral drift, the development of the port was expected to cause significant impact on the coastline. After the construction of the port, the southern shoreline has witnessed accretion due to the southern breakwater and also due to presence of Ennore creek while the northern coastline has experienced severe erosion.

16.3 Site Conditions

For simulation of littoral drift and shoreline changes, data on bathymetry, tidal levels, currents, waves and sediments at the site are required. The site at Ennore is near Chennai where the tidal levels MHWS, MHWN, MSL, MLWN and MLWS are 1.1 m, 0.8 m, 0.6 m, 0.4 m and 0.1 m respectively. Tidal currents are unidirectional, northward during SW and non-monsoon period and southward during NE monsoon period with average magnitude of the order of 0.1–0.2 m/s. The major data required for simulation of shoreline changes is the nearshore wave climate at the site.

Instrumentally observed wave data at the site over a period of several years, if available, are best suited for design purposes. As such data were not available, the ship observed deep water wave data reported by India Meteorological Department (IMD) covering a period of about 30 years for quadrant between latitude 10–150°N and longitude 80–85°E were analysed to obtain seasonal and annual wave climates in the offshore region of Ennore. It is seen from the annual wave climate (Fig. 16.2a) that the predominant wave direction is from SW and NW quadrants with maximum wave height of the order of 4.5 m. This wave climate in the deep sea was transformed to get the nearshore wave climate at Ennore using a mathematical model which is described subsequently. The orientation of the coastline in this region of Ennore is approximately in North-South direction.

Littoral drift rates near the site at Ennore were computed by various researchers. Chandramohan et al. (1990) estimated northward drift of 1.027×10^6 m^3 and southward drift of 0.683×10^6 m^3 with net drift towards north to be 0.344×10^6 m^3 along Madras coast. Natesan and Subramanian (1994) indicated that along the coastal stretches of Grid I (Lat. 9–12 N, Long. 80–83 E) and Grid II (12–15 N, 81–84 E), a net transport of 0.35–0.4 million m^3 of sand towards north occurs. Indomer (2005) estimated northward drift of 0.977313×10^6 m^3 and southward drift of 0.512757×10^6 m^3 with net drift towards north to be 0.464556×10^6 m^3 along Minjur coast. Shoreline studies were carried out by ICMAM (2007) in which the net transport towards north at Ennore coast was assumed to be 0.35×10^6 m^3.

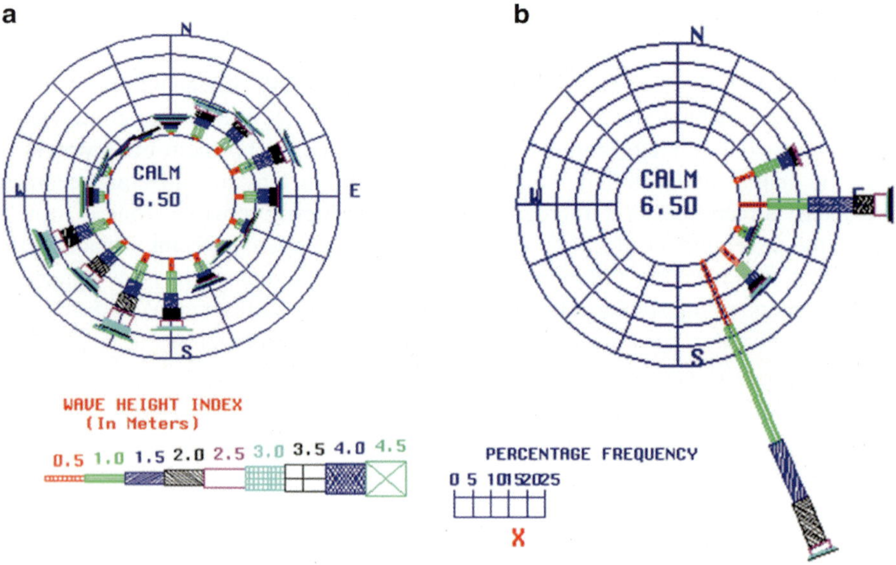

Fig. 16.2 Rose diagram for wave heights for annual period. (**a**) Offshore. (**b**) Inshore

From the above discussions it is seen that there is considerable variation in the estimation of littoral drift in the region due to the uncertainties involved. However, in general, the annual rate of northward and southward transport at the Chennai coastline are of the order of 0.6 and 0.1 million m^3 respectively. Hence the average northward and southward transports for Ennore coastline are considered to be 0.60 and 0.1 million m^3 so that net and gross transports are 0.5 and 0.7 million m^3.

16.4 Transformation of Waves

The wave climate prevailing at the site is one of the important parameters for estimating littoral drift and shoreline changes which is derived by wave transformation from deep to shallow waters using OUTRAY (1989) model. This model takes the deep water wave height, period and direction as input and computes the wave height, period and direction at the inshore point of interest. The wave height and period are input to the model by specifying the spectral distribution of wave energy at the offshore boundary. The offshore wave climate described earlier was transformed to the inshore location near Ennore in 12 m contour to obtain the seasonal and annual inshore wave climates. It is seen that during SW monsoon (June–Sept.), the waves approach predominantly from SSE direction with 90 % occurrence. During NE monsoon (Oct.–Jan) the predominant wave directions are East, and SSE at the nearshore location with percentages 48 % and 29 % respectively. In non-monsoon period (Feb. to May), the predominant wave directions are East, ESE, SE and SSE with percentages 23 %, 17 %, 15 % and 40 % respectively. The maximum wave

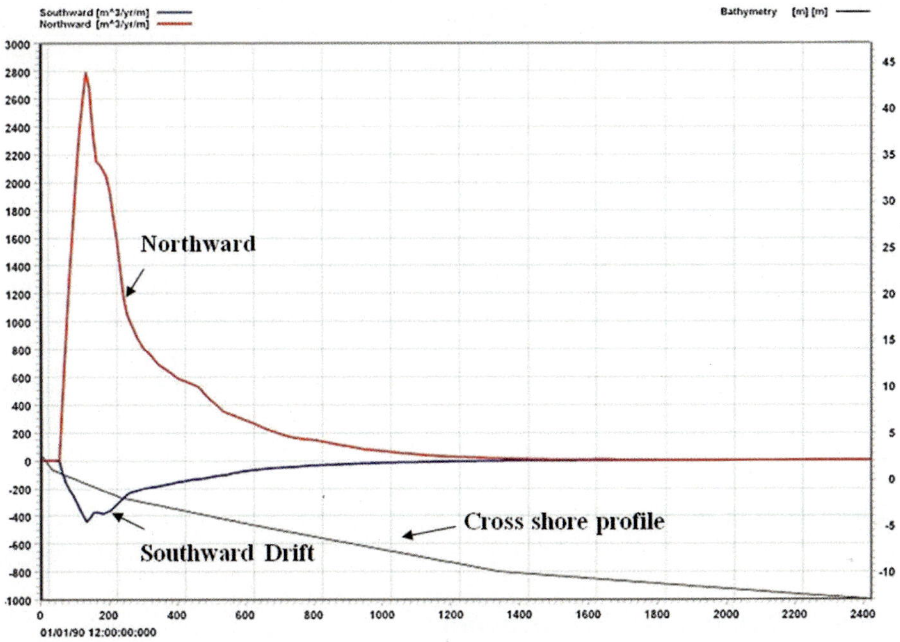

Fig. 16.3 Distribution of littoral drift during annual period

height of 3.5 m is seen to occur. The annual frequency distribution of waves at the inshore location near Ennore is shown as rose diagram in Fig. 16.3b which indicates that the waves approach predominantly from East and SSE directions with percentages 25 % and 48 % respectively and waves up to 4.0 m are seen.

16.5 Simulation of Littoral Drift and Shoreline Changes

For these studies, LITPACK (2000) software was used to simulate the longshore transport and impact of the construction of Ennore port. Computation of longshore sediment transport is based on the local wave, current and sediment characteristics. The sediment transport model takes into account time-varying distribution of both suspended load and bed load in combined wave and current motion including the effect of wave breaking, when relevant. The model gives a deterministic description of cross-shore distribution of longshore sediment transport for an arbitrary, non-uniform bathymetry and sediment profile. The seasonal/annual sediment budget is found by the contribution of transport from each of the wave incidents occurring during the season/year. For simulation of shoreline evolution, the following continuity equation for sediment volumes is numerically solved.

$$\frac{\partial y_c}{\partial t} = -\frac{1}{h_{act}(x)}\frac{\partial Q(x)}{\partial x} + \frac{Q_{sou}(x)}{h_{act}(x)\Delta x} \tag{16.1}$$

Table 16.1 Seasonal and annual transport rates (million cu m)

Profile	Season	Northward	Southward	Net	Gross
South profile	SW monsoon	−0.34	0.00	−0.34	0.34
	NE monsoon	−0.11	0.07	−0.04	0.18
	Non-monsoon	−0.16	0.02	−0.14	0.18
	Annual	−0.61	0.09	−0.52	0.70
North profile	SW monsoon	−0.29	0.00	−0.29	0.29
	NE monsoon	−0.13	0.08	−0.05	0.21
	Non-monsoon	−0.15	0.02	−0.13	0.17
	Annual	−0.57	0.10	−0.47	0.67

Note : +ve sign indicates southward transport, −ve sign indicates northward transport

where, y_c = distance from the base line to the coastline, h_{act} = height of the active cross-shore profile, Q = longshore transport of sediment expressed in volume, x = longshore position, Δx = longshore discretisation step, Q_{sou} = source/sink term expressed in volume. Thus the term Q_{sou} can be conveniently used to include sand bypassing arrangement.

16.5.1 Littoral Drift Distribution

The LITPACK (2000) model assumes the depth contours parallel to the coast and longshore transport is computed along a representative cross-shore profile. The longshore transport was computed for 2.5 km long bed profiles normal to shore with depth varying from +2.9 m to −15 m. The profiles were divided into 250 grid points with grid size of 10 m. For these studies, besides the inshore wave climate and bathymetry along the cross-shore profile, grain size distribution over the profiles is also required. At the site grain size was observed to be of the order of 0.2 mm Hence grain size D50 over the profiles was assumed to be 0.2 mm.

Since the general orientation of the coastline at Ennore is North-South, depending on the wave direction with respect to the shoreline, the littoral drift will be directed towards north or south along the shoreline. The model was run for the two cross-shore profiles for seasonal and annual inshore wave climates. The model was calibrated for the net and gross annual transport of 0.70 million cum and 0.50 million cum and the seasonal/annual northward, southward, net and gross transport quantities computed are given in Table 16.1.

The distributions of seasonal and annual transports over the profile are shown in Fig. 16.3 which indicate that the transport is mainly confined within 400 m from the shore. The maximum transport occurs at about 50–70 m from the shore.

16.5.2 Shoreline Changes

For the shoreline evolution model, shoreline of 14.4 km length extending 6 km towards north and 6 km on the south of Ennore region was considered. This was

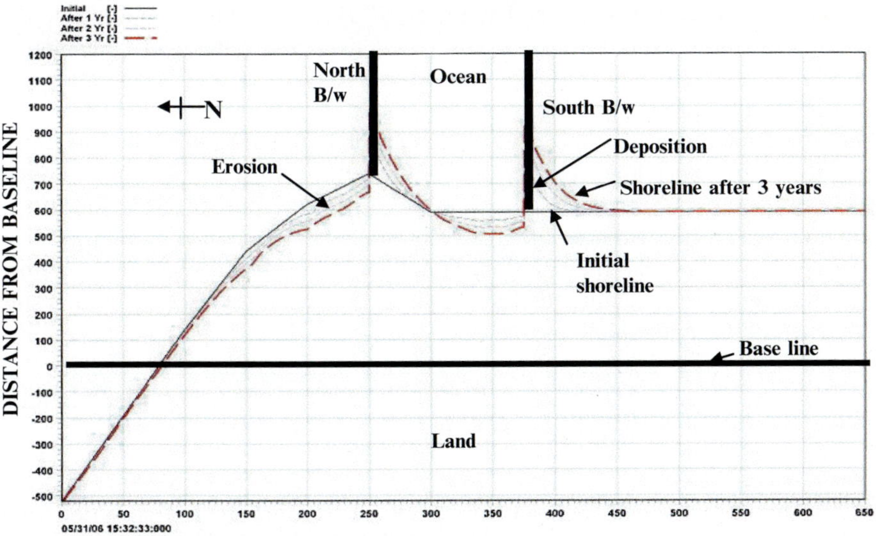

Fig. 16.4 Shoreline changes after 1, 2, 3 years

divided into 721 grid points with grid size of 20 m. The port with 3.2 km long north breakwater and 1.1 km long south breakwater was considered in the model. The effective blocking lengths normal to shoreline of the north and south breakwaters are considered to be 1,500 m and 1,000 m. With these conditions the shoreline evolution model is run with the schematic layout of the breakwaters for 3 years period. The shorelines evolved after each year is shown in Fig. 16.4.

As the net transport is towards north, the coastline on the south side of the south breakwater has advanced and the coastline on the north side of the north breakwater has receded. The maximum cross-shore advancement on the south side is about 300 m after 3 years and its longshore accretion effect on south side of south breakwater is felt for about 1.5–2 km. The maximum cross-shore recession on the north side is about 100 m after 3 years and its longshore erosion effect on north side of north breakwater is felt for about 2 km.

16.6 Shoreline Changes by Image Processing

Remote sensing satellite records the features of earth's environment in the form of spectral response without physically coming in contact with the earth's surface. The electro-magnetic energy in the discrete bands of different wavelengths of electro-magnetic spectrum reflected or radiated by different objects enables identification of different objects, their locations, spatial distribution and helps in obtaining their properties. The advantage of remote sensing satellites is their capability to map regions including coastal, oceanic and land features like wetlands, intertidal zones etc., at regular interval of time. The temporal data obtained from remote sensing satellite can effectively be used to assess the dynamic changes in the shorelines by delineation of water and land. The infrared band in the spectral range of 0.77–0.86 μm is found to be suitable for demarcation of shoreline as the contrast between land and water is very sharp. Multi-date digital satellite images can be compared to detect any changes in the shoreline using digital satellite image processing software. While comparing two imageries of different dates, it is important to ensure that both the imageries are at similar tidal condition.

In order to evaluate the shoreline, satellite data sets of two dates covering a period of 3 years were selected. The details of the imageries selected for this study are listed in the following table.

Sl. No.	Satellite	Sensor	Date	Path	Row	Spatial resolution (m)
1.	IRS-1D	LISS-III	12.02.2001	102	064	23.8
2.	IRS-1D	LISS-III	17.04.2004	102	064	23.8

The imageries are in digital form having visible and infrared spectral bands. Subset imageries bounded by latitude 13.2795 and longitude 80.3574 were used for the analysis of shoreline changes. The images of shorelines of 2001 and 2004 near Ennore region as viewed by LISS-III sensor of IRS-1D satellite are compared.

16.7 Comparison of Shoreline Changes Obtained by Mathematical Model and Image Processing

The shoreline changes obtained by mathematical modelling and image processing are shown in Fig. 16.5 for comparison. It can be seen that the cross-shore and longshore shoreline changes obtained by the model and image processing match

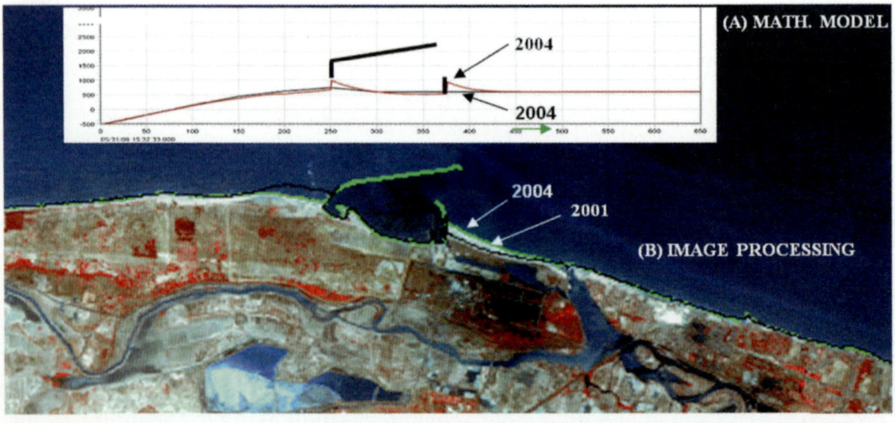

Fig. 16.5 Comparison of shoreline changes obtained by mathematical model

satisfactorily. With further refinement in the model results by tuning model parameters and analysis of additional imageries, the matching of the two results can be improved.

16.8 Conclusion

Human interventions into the natural processes taking place in the coastal zone such as construction of ports has significant impact on the shorelines. In order to assess the effect of such developments and suggest remedial measures in anticipation, it is desirable to predict the shoreline changes in advance. Mathematical modeling provides a useful tool for predicting such changes in advance. Processing and analysis of satellite imageries of coastal area enables us to estimate and monitor the shoreline changes, which is otherwise extremely difficult, time consuming and costly by field surveying. Satellite remote sensing is a very useful technique for studies regarding coastline changes. Processing and analysis of satellite imageries of coastal area enables us to estimate and monitor the shoreline changes, which is otherwise extremely difficult, time consuming and costly by field surveying.

Shoreline changes due to the development of Ennore port are simulated by mathematical modelling technique and compared with the shoreline changes derived from remote sensing/image processing technique. The study indicated that the cross-shore and longshore impacts predicted by mathematical model and satellite information match satisfactorily. Thus the satellite information is useful for calibrating the mathematical model which can be further used for predictive purposes. With further refinement in the model results by tuning model parameters and analysis of additional imageries, the model could be improved.

Acknowledgement Authors are thankful to Shri. S. Govindan Director, Central Water and Power Research Station, Pune for kind consent for publishing this paper.

References

Chandramohan P, Nayak BU, Raju VS (1990) Longshore sediment model for south Indian and Shri Lankan coasts. J Waterw Port Coast Ocean Eng Proc ASCE 116:408–424

ICMAM (2007) Shoreline management plan for Ennore coast, Tamil Nadu

Indomer Coastal Hydraulics (P) Ltd (2005) Rapid marine EIA for CWDL desalination plant at Minjur

LITPACK (2000) An integrated modelling system for littoral processes and coastal kinetics. Danish Hydraulic Institute.

Natesan U, Subramanian SP (1994) Study of transport processes along Tamil Nadu Coast. In: Proceedings of the Indian National Conference on Harbour and Ocean Engineering (INCHOE-94), Vol. 1, pp D-103–D-108

OUTRAY (1989) Refraction model-training and user manual, hydraulics research, Wallingford, UK

Chapter 17
Advances in Remote Sensing of Coastal Wetlands: LiDAR, SAR, and Object-Oriented Case Studies from North Carolina

Thomas R. Allen

Abstract Coastal wetlands provide crucial ecosystem services to society including pollution filtration, fish and wildlife nursery and habitat, storm surge mitigation, and sinks for atmospheric carbon. Uncertainty of wetland responses to sea-level rise is a pervasive concern in coastal science and management. Advances in Light Detection and Ranging (LiDAR), space-borne Synthetic Aperture Radar (SAR), and multi-sensor and object-oriented image analysis techniques are poised to aid the inventorying, monitoring and management of wetlands to an even greater extent. This chapter reviews developments and coastal wetland applications of these state of the art remote sensing data and techniques and evaluates the utility of these products for management of coastal reserves in case studies. Following concise review of the literature on remote sensing technological and image processing advances, case studies from North Carolina coastal wetlands are presented: (1) multidate SAR and LiDAR imagery for regional salt marsh mapping in Cedar Island National Wildlife Refuge, (2) object-based image analysis (OBIA) for transitional marshes and *Phragmites australis* inventory in Alligator River National Wildlife Refuge, and (3) very fine resolution barrier island mapping using multisensor and multidate imagery and OBIA techniques in the Rachel Carson Coastal Reserve. Drawing upon these developments and case studies, insights for practical applications are evaluated to further even wider application to coastal management.

T.R. Allen (✉)
Department of Geography, Planning, & Environment, East Carolina University, Greenville, NC 27858, USA
e-mail: allenth@ecu.edu

C.W. Finkl and C. Makowski (eds.), *Remote Sensing and Modeling: Advances in Coastal and Marine Resources*, Coastal Research Library 9,
DOI 10.1007/978-3-319-06326-3_17, © Springer International Publishing Switzerland 2014

17.1 Introduction

Remote sensing technology continues its rapid advances in applications to coastal ecosystems research and management. Over the last decade the diversity of sensors and their applications has expanded notably, with new utility and operational uses ranging from disasters to environmental restoration. A number of reviews and books have inventoried these myriad developments. Klemas (2009) inventories the wide breadth of state-of-the-art remote sensing applications for coastal waters, watersheds, and ecosystems. In particular, the variety of sensors and technologies have continued their expansion in the field, ranging from space-borne to aerial platforms, UAVs, and improving thermal infrared (TIR) and light detection and ranging (LiDAR) sensors. Klemas (2013a) reviews several technical advances for remote sensing of wetland biomass and environmental change detection. Notable research progress includes the integration of multi-sensor data, regional applications and the continuity of Landsat, spectral mixture methods, and increasing capabilities for mapping mangroves, submerged aquatic vegetation (SAV) and marshes. Canopy height and volume, in particular, are now achievable measures that encourage applications requiring above ground biomass (ABG) and its dynamics (Klemas 2013b.) Monitoring wetland changes, whether biomass or areal habitat coverage, is important for management, yet comparisons of wetland losses and gains have been shown to be disparate between airborne surveys and satellite data (Carle 2011). Ramsey et al. (2012) also highlight the capabilities and limitations of L- and C-band SAR and Landsat Thematic Mapper (TM) for monitoring coastal wetlands impacted by flooding. Their findings note a sensitivity of the SAR backscatter to damage after storms yet also potential confounding factors such as look angle and wind roughening that may be overcome with more frequent observations. Increasing observations and standardized monitoring strategies could improve our understanding of the causes of marsh dieback related to water-logging and salinity. Isolating environmental variability in remotely sensed water level monitoring could also improve the use of remote sensing for a variety of process studies, such as storm surges, riverine flooding, and sediment transport (Allen et al. 2011) and tidal inundation (Kearney et al. 2009).

17.1.1 Coastal Wetland Biomass and Topography

Coastal marshes are among the most productive as well as vulnerable ecosystems. Studies in vulnerable deltaic and trailing edge continental margins are now relating their biomass derived to spectral indices such as the Normalized Difference Vegetation Index (NDVI) in order to predict responses to sea-level rise point and reveal thresholds in their inundation depth and frequency (Couvillion and Beck 2013; Couvillion et al. 2013). Wetland elevation also can be a functionally useful addition to broad-scale mapping efforts, such as the use of Digital Elevation Models (DEMs) and topographic derivatives to differentiate and classify forested wetlands and improve the accuracy and efficiency of their monitoring (Lang et al. 2013). In addition, although coastal marshes are prone to a large relative error for their low-lying location even with LiDAR (Schmid et al. 2011), the potential to apply post-process corrections can improve their accuracy (Hladik and Alber 2012) and aid confidence in model forecasting of marsh responses to tidal inundation and future sea-level rise (Kearney and Rogers 2010).

Elevation data in coastal wetlands is crucial for accurate mapping of species patterns and zonal delineation, especially for forecasting salt marsh responses to sea level rise. Noting that classification based on optical imagery alone fail to accurately discriminate between salt-marsh vegetation species, DeHouck et al. (2012) explored the use of both high-resolution SAR data and high resolution optical imagery for mapping intertidal coastal areas. Results showed little improvement in the detection of salt marsh vegetation. Furthermore, the radiometric variability inherent to SAR data (speckle) introduced noise in the classification results (DeHouck et al. 2012).

Accurate elevation mapping is also essential, as elevation is an important determinant of the extent of tidal flooding and, consequently, species distribution and patterns in marshes (Hladik and Alber 2012). The elevation distribution of *Spartina alterniflora* in relation to mean sea level is a key indicator of marsh stability (Morris et al. 2002). Due to the relatively flat topography of salt marshes, elevation differences of less than 10 cm can affect overall marsh extent and species patterns (Hladik and Alber 2012). Thus, there is a need for accurate elevation mapping in salt marshes to identify vulnerable habitat and predict how marshes will respond to perturbations that might alter plant distributions, such as sea level rise (Hladik and Alber 2012). In recent years, LiDAR has been used to ascertain ground elevation and vegetation structure for developing digital elevation models (DEMs). LiDAR has been used to determine elevation in tidal areas with vertical resolution as fine as 16 cm in North Carolina (Poulter and Halpin 2008). However, an accuracy assessment of LiDAR models in salt marshes showed that they overestimate marsh ground elevations, with mean errors increasing with increased vegetation density and plant height (Hladik and Alber 2012). These errors may result from inadequate horizontal and vertical sensor resolution that, given the small topographic variability in salt marshes, makes it difficult to detect meaningful differences in elevation. Another source of error may be the inability of LiDAR to differentiate between vegetation types with heights less than 2 m, which describes much of salt marsh vegetation (Hladik and Alber 2012).

17.1.2 SAR and LiDAR Data

Hydrologic dynamics of wetlands compound the difficulties faced in their mapping and monitoring. Temporal complexity of wetlands is evidenced in both optical signatures and sub-canopy structure (e.g., Kasischke and Borgeau-Chavez 1997). Spurred by operational needs for monitoring, the research community has increasingly been challenged to develop algorithms that are robust and portable, especially for areas of study that are vast or difficult to access. Some approaches have sought to reduce the problems of classification, such as incorporating hydrology directly into the classification products in hydrogeomorphic classification (Hamilton et al. 2007). Advanced image classification techniques such as spectral unmixing to discern fractional composition of wetlands undergoing change (e.g., Rogers and Kearney 2004) and hydrologic simulation modeling have also provided additional input data for classification. For instance, floodplain inundation models have been incorporated into classification (Townsend and Walsh 1998; Wang 2004) and such data integration has an analogous history of success at improving forest classification in complex montane environments (c.f., Franklin 1991).

Synthetic Aperture Radar (SAR) and Light Detection and Ranging (LiDAR) sensors have rapidly developed and found application for coastal mapping and monitoring. C-band SAR imagery such as ERS-1 has been applied in the Florida Everglades for detecting hydroperiod changes across a region (Borgeau-Chavez et al. 2005). Historic problems with signal-to-noise and speckle have found solutions with improved engineering or post-processing filters. Kushwaha et al. (2000), for instance, found several filters could reduce noise and speckle reduction in time series ERS-1 SAR imagery. Multispectral processing, such as Intensity-hue-saturation (HIS) analysis has also been applied to SAR imagery over time, including ERS-1 improvements for wetland discrimination when combined with Indian Remote Sensing Satellite (IRSS). Forested wetlands, a challenging vegetation assemblage for multi-spectral or single-source remote sensing imagery, have also been examined and found to be improved in image classifications with SAR imagery. A multi-temporal analysis of ERS-2, ENVISAT/ASAR, and Landsat Enhanced Thematic Mapper Plus (ETM+) data by Lang et al. (2008) found principal components analysis (PCA) could delineate hydroperiod-related classes of forested wetlands in the Mid-Atlantic, USA. Time series remote sensing and change detection have also been advancing, with seasonal, inter-annual, and phenologic changes of terrestrial land cover explored in multi-date imagery (Lunetta and Evlidge 1988). For example, multisensor SAR imagery over a decadal timespan could detect coastal wetland shoreline change between JERS-1 and Advanced Land Observation Satellite (ALOS) Phased-Array L-band SAR (PALSAR) imagery in the Pamlico Sound, North Carolina (NC), USA (Wang and Allen 2008).

17.1.3 Fine Spatial Resolution and Object-Based Image Analysis

With increasing spatial resolutions, alternatives to pixel-based classification are being developed. A major drawback of pixel-based classification is that a pixel's spatial extent may not match the extent of the land cover feature of interest (Aplin and Smith 2008). This results in mixed pixels, whereby a pixel represents more than a single type of land cover. High resolution images support several scales within their images. Object-based image analysis (OBIA), on the other hand, uses spatially explicit information to derive objects that are made up of several pixels. OBIA builds on older segmentation, edge-detection, feature extraction and classification concepts that have been used in remote sensing image analysis for decades, but were not used extensively in geospatial applications until the past decade (Blaschke 2010). Segments are regions which are generated by one or more criteria of homogeneity in one or more dimensions of a feature space (Blaschke 2010). Segments have more spectral information than single pixels; more importantly, they have more spatial information (e.g., distances, neighborhood, topology) that is crucial to OBIA methods. Grouping pixels into image objects mitigates the "salt and pepper effect" that occurs with misclassification of pixels caused by shadow, mixed species, and differences in reflectance values (Kelly et al. 2011). Some object-based image classification techniques attempt to replicate traditional pixel-based classification techniques using the spatial scale of the object instead of the pixel. For instance, the maximum likelihood classification algorithm has been used for object-based classification, either (i) by classifying objects directly (e.g., by comparing each object's group of pixels as a whole against the training classes, or (ii) by first classifying pixels individually and then grouping these to populate each object (Aplin and Smith 2008).

The efficacy of OBIA techniques has been demonstrated in several studies. Platt and Rapoza (2008) compared results from a maximum likelihood classification with results from OBIA and found that the combination of segmentation into image objects, use of the nearest neighbor classifier, and the integration of expert knowledge yielded substantially improved classification accuracy compared to a per-pixel method. Kamal and Phinn (2011) compared pixel-based and object-based mapping techniques using spectral angle mapper (SAM) and linear spectral unmixing (LSU) for the pixel-based approaches, and multi-scale segmentation for the OBIA approach. They found that OBIA mapping produced the most accurate results (overall accuracy 76 %, Kappa 0.67). Weih and Riggan (2010) obtained similar results comparing three methodologies (object-based, and supervised and unsupervised pixel-based classifications) to determine if an object-based analysis of remotely sensed imagery would produce a statistically more accurate LULC classification than a pixel-based analysis applied to the same imagery. Results showed that when merging high-spatial resolution aerial imagery with medium-spatial resolution satellite imagery, the object-based classification outperformed both supervised and unsupervised pixel-based methods.

OBIA has also illustrated robustness for multi-scale mapping, expanding beyond its initial focus on ultra-high resolution imagery to extend to landscape mapping. Phinn et al. (2012) demonstrate the multi-scale mapping for coral reef benthic habitats and geomorphology, highlighting the value of hierarchical classification. Using high-resolution color- and near-infrared imagery, Kelly et al. (2011) note that care should be taken in multi-scale analyses for marsh vegetation undergoing restoration, as pattern metrics such as patch statistics can be sensitive to scale. Kim et al. (2011) also find that multi-scale OBIA classification produces higher map accuracies in marshes. Nonetheless, some coastal ecosystem zones, such as fringing mangroves, may present a challenge to even ultra-high resolution imagery. Heumann (2011) achieved accuracy exceeding 94 % for mangroves in the Galapagos but describes limitations for OBIA and WorldView-2 imagery of mixed zonation or sparse vegetation. Timm and McGarigal (2012) document the classification of 1 m resolution coastal dune and marsh vegetation at Cape Cod, crediting success to a supervised random forest classification algorithm. These studies collectively affirm that very fine spatial resolution coastal marshes are able to be accurately mapped, and moreover that advanced image segmentation and classification algorithms are able to efficiently handle landscape extents. A prevailing challenge cited remains the standardization of imagery, acquisition and environmental factors, and the portability of OBIA parameters, rule sets, and resulting variation in accuracy.

17.1.4 Rapid Coastal Response

In addition to further disentangling water level effects on the spectral and derived structure characteristics of wetlands, greater attention is being given to measurement of climate change and natural as well as technological hazards and actual disasters. Chambers et al. (2007) using spectral mixture analysis and MODIS data to assess carbon stocks in coastal ecosystems, noting the effect of Hurricane Katrina in forest damage. Wang et al. (2010) also evaluate damage data from Hurricane Katrina in coastal forests. In a study comparing Landsat and MODIS, Chen et al. (2013) observe that MODIS 8-day composite imagery offer a feasible surrogate for daily observing of flood inundation. With careful attention to LiDAR DEM error and inundation modeling, Yoon and Shim (2013) have demonstrated an operational storm surge hazard mapping in South Korea that combines hydrodynamic models corroborated by field observations of inundation. Dresback et al. (2013) further assessed the quality of coupled hydrologic and hydrodynamic models for coastal terrain in North Carolina mapped primarily from LiDAR. Demonstrating remote sensing's relevance to emergency management to coastal disasters, Ramsey et al. (2011) assess the utility of unmanned aerial vehicle SAR (UAVSAR) for mapping oil spill in marshes from the Deepwater Horizon disaster in the Gulf of Mexico and environmental response to hurricane storm surges (Ramsey et al. 2012). Rodrigues and Souza-Filho (2013) underscore the utility of Landsat for regional environmental sensitivity to oil spills in a study on the Amazonian coastal plain. Among these case studies, there is a shared commonality rapid and systematic acquisition for highly time-dependent environmental response, factors are critical to time-dependent emergency management. Coastal remote sensing thus offers valuable information at multiple spatial and temporal scales to augment modeling and forecasting systems, a complementarity that is gaining traction in the field of emergency management and visualization of coastal impacts and situational awareness (Akbar et al. 2013).

17.2 Coastal Ecosystem Classification and Change

17.2.1 Wetland Hydrogeomorphic Classification

Remote sensing, field, and environmental modeling data stored and analyzed within the geographic information system (GIS) have played a strong and increasing role in wetland mapping, such as the provision of floodplain boundaries and frequency of inundation simulations (Coe 2000; Wang et al. 2002). Hydrogeomorphic conditions dominate the environmental gradients of riverine and coastal wetlands, affecting their biodiversity patterns (Hupp 1988) and zonation that can be discerned via remote sensing and environmental simulations of hydrology. Digital shoreline data, mapped via field, boat, or aerial imagery, are frequently used to mask aquatic and terrestrial zones for simplification of coastal remote sensing problems. Multi-sensor data can also be valuable, such as the utilization of imagery with near infrared reflectance data to rapidly map shorelines and develop mask layers for land-water separation. In addition, LiDAR has emerged to provide water levels useful to hydrogeomorphic classes (Hofel et al. 2009). Several comprehensive remote sensing products have demonstrated useful insights from well-thought feasibility, prototyping, and accuracy assessments, including the NOAA Coastal Change Analysis Program (C-CAP) (Jensen et al. 1993), CoastWatch (Dobson and Bright 1991; Klemas et al. 1993), and USGS North American Landscape Characterization (NALC) archive (Loveland et al. 1991).

Assessment of the value of potential data combinations, fusion, multi-date and alternative classification algorithms has a pronounced research agenda in remote sensing that furthers thematic detail and hydrgeomorphic characterization. Evaluating alternative combinations of data is given substantial methodological attention. South et al. (2004), for example, analyzed several supervised classification methods to ascertain their performance for classifying agriculture tillage practices. In addition, the use of multiple classifiers in ensemble approach has shown promise by comparative accuracy assessments (Dai and Liu 2010). Multi-sensor or comparative results of classifications by individual sensors have also been rigorously scrutinized, such as individual SAR versus optical Landsat TM forested wetland change mapping in the Roanoke River basin, North Carolina (Townsend and Walsh 1998). These studies further prompt the current investigation to compare what possible accuracy could be efficiently achieved by a robust technique combining multi-date SAR imagery with increasingly available LiDAR bare earth digital elevation models (DEMs) and surface height of vegetation canopy. Given a profusion of methodological papers on accuracy assessment, there is some pause or concern for cross-comparisons. Kappa statistics, in particular, have become a nearly ubiquitous measurement for quantitatively comparing alternative methods of classification. However, some new critical work has revealed the limitations of this family of indices (Pontius and Millones 2011).

Nonetheless, wetland mapping at the landscape scale using remotely sensed data has been deemed efficient and accurate with proven and evolving classification methods. Traditional wetland mapping methods often rely on optical imagery and manual photo interpretation or classification. The integration of multi-platform and multi frequency remotely sensed data can provide superior data for mapping wetlands. Data from multiple sources has the capacity to capture greater variation in surface features, such as elevation and vegetation structure (Corcoran et al. 2013), and thus has the potential to increase both classification accuracy and confidence. Given the wealth of remotely sensed and ancillary data, a robust wetland classification method applied to large geographic areas needs to be computationally fast, require no assumptions about data distribution, handle nonlinearity in relations between input variables, and be capable of using numeric and categorical data. In addition, the assessment of results will be improved if the classification method identifies outliers in the training data, provides rankings of the importance of the input variables, and produces internal estimates of error and confidence of the output classification. Decision tree classifiers that have been used in land cover mapping for years meet most of these requirements (Pal and Mather 2003), including the See5 classifier (RuleQuest Research). These approaches at a regional or national scale have yet to find a secure place in federal coastal resource management and monitoring programs. In lieu of this, landscape scale, synoptic remote sensing for wider land use/land cover mapping with sensors such as Landsat in CCAP or orthophoto inventories and field mapping (e.g., NWI) are likely to continue. The improved multispectral and radiometric resolution of Landsat 8 may itself play a stronger role in monitoring and invite the integration of previously noted LiDAR, SAR, and OBIA techniques, at least at the fine-scale.

17.2.2 Salt Marsh Zonation and Classification

Salt marshes are among the most productive ecosystems on earth, providing benefits to coastal communities by filtering pollutants from surface waters, buffering storm energies, and providing nursery and habitat for coastal wildlife and fisheries. These functions also provide economic benefits to coastal communities that provide services and supplies for recreational hunting and fishing. Salt marshes also provide highly effective sinks for atmospheric carbon dioxide, and efforts are emerging to use salt marsh preservation or restoration in carbon offset programs (Chmura 2013). Salt marshes are threatened by increasing rates of sea level rise. In many areas, the vegetation responsible for marsh soil accretion may not survive increased flooding periods, resulting in submergence of the marsh in its present location or inability to restore a marsh at its previous elevation.

The distribution of salt marsh communities is governed by the magnitude, frequency, and duration of tidal inundation (Donelly and Bertness 2001). Salt marshes in southeastern North Carolina typically comprise three parts – high marsh, low marsh, and creekbank marsh. Creekbank marsh and low marsh are flooded twice a day. High marsh forms where trapped sediment has built up over time; while flooded periodically it may be dry for several days at a time. Different species of vegetation, each important to the provision of ecosystem services, inhabit the different zones. Low marsh and creekbank marsh are dominated by smooth cordgrass (*Spartina alterniflora*), an important producer in the estuarine food web. High marsh plants include spikegrass (*Distichlis spicata*), Black needlerush (*Juncus roemerianus*), salt meadow cordgrass (*Spartina patens*), sea ox-eye and glasswort (*Salicornia species*) (Hladik and Alber 2012). Vegetation at this level provides much of the pollutant filtering and storm surge protection services. In addition, salt marsh ecosystems have relatively high rates of carbon sequestration. Carbon burial rates within a salt marsh may be affected by variability in hydroperiod, salinity, and suspended sediment supply (McLeod et al. 2011). The alteration of the demarcations among species is an indicator of sea level and environmental changes within a salt marsh (Lee et al. 2012). Changes in distribution patterns can alter erosion and accretion rates in marshes, increasing their vulnerability to sea level rise. Therefore, conservation of marsh communities at each level is critical to the continued provision of ecosystem services.

17.2.3 Invasive Phragmites

On the eastern seaboard of the USA, *Phragmites australis* is often considered an invasive marsh grass, encroaching along estuarine shorelines and creeks in colonies, but this claim is contested and not yet thoroughly proven (Saltonstall 2002). Regardless of *Phragmites australis* region of exotic versus indigenous origin, it is frequently treated as invasive due to its robust colonial growth – growing up to six meters tall. With a dearth of herbivores to naturally control the spread, the plants are prone to contagiously spread along shorelines, tributaries, and along ditches in ruderal corridors. Efforts to monitor its distribution and manage, even eradicate it locally, are fraught with sparse management resources spread across extensive geography, especially in eastern North Carolina. Dedicated aerial campaigns to inventory and monitor its spread are thus limited, and alternatives to map and track the spread are sought using more efficient satellite-based data. For instance, *Phragmites* is known to exhibit differential morphologies (basal area, height, and density) along riverine to estuarine gradients (Zhang et al. 2003) and biogeographic setting (Meyerson et al. 2009). Classification accuracy results may thus depend on regional stratification of morphology (salt water, riverine, or brackish) or latitudinal phenology and require adaptation of the techniques. SAR, LiDAR, and multispectral imagery, including Landsat 8, hold strong promise for satellite-based monitoring of *Phragmites*.

17.3 Case Studies

17.3.1 Case Study: Multi-sensor and Multi-date SAR for Emergent Marsh Mapping in Cedar Island NWR

A principal challenge to operational wetland monitoring is maintaining the currency of maps for a landscape ecosystems that are spatially dynamic. Salt marshes in low-lying tidal locations such as Cedar Island National Wildlife Refuge (NWR), North Carolina, are disposed to dynamics from persistent as well as episodic storm erosion, dieback from long-term sea-level rise and associated tidal inundation, anthropogenic change such as diking, dredging or filling, and invasion from *Phragmites australis*. Some estuarine marshes in North Carolina have erosion rates exceeding 10 m per year (Wang and Allen 2008.) Traditional aerial photointerpretation, such as the US Fish and Wildlife Service National Wetland Inventory (NWI) utilizes protocols that could be enhanced with the sensors and technological advances previously noted. Cedar Island National Wildlife Refuge is an exemplary location for pilot testing methods for rapid and regional wetland inventorying at a screening or root-level of the NWI hierarchy. Classification of estuarine emergent marshes, shrub-scrub, forested wetlands and Phragmites on an extensive and efficient basis could save the NWI program time and guide decisions for map updates. A pilot study was undertaken to apply image classification of a multi-date, multi-sensor image stack to inventory marshes and identify Phragmites (if present.) The area includes an estuarine barrier island and pine forest hammocks amidst dune ridges adjoining extensive *Juncus* marshes and fringing *Spartina alterniflora*. The marshes are situated very low in the tidal frame and undergoing transition and dieback from increasing inundation in the area of the *Thorofare* and Core Sound. Four ALOS PALSAR images (HH and HV polarization) were obtained, transformed to db power, and geometrically co-registered for April 30, June 15 and September 15, 2010. These images were initially masked using a water mask digitized from orthophotos. A 6 m resolution LiDAR DEM was integrated within the data stack for unsupervised classification. Details of the classification method are described in Allen et al. (2013) in a parallel study of the Alligator River National Wildlife Refuge.

Figure 17.1 depicts the multidate PALSAR HH bands and the stark visual contrast in backscatter between extensive *Juncus sp.*, fringing *Spartina sp.* and wrack, and the stronger backscatter of shrub-scrub and forest vegetation. The classified image (Fig. 17.2) easily differentiates these among four basic classes in a rapid, albeit traditional ISODATA cluster analysis. Twelve clusters were aggregated to produce these four basic information classes. Unfortunately, the pilot study was unable to evaluate the potential to map *Phragmites sp.* at Cedar Island, which was only found outside the refuge. A subsequent study in Alligator River, however, more rigorously evaluated the classification potential for *Phragmites*, also using an OBIA approach. Nonetheless, this demonstration of a simple image classification requiring minimal pre-processing (limited to image acquisition and georectification checks) and an elementary classifier provides land managers accurate, timely, and cost-effective information on major ecosystem states across an extensive area.

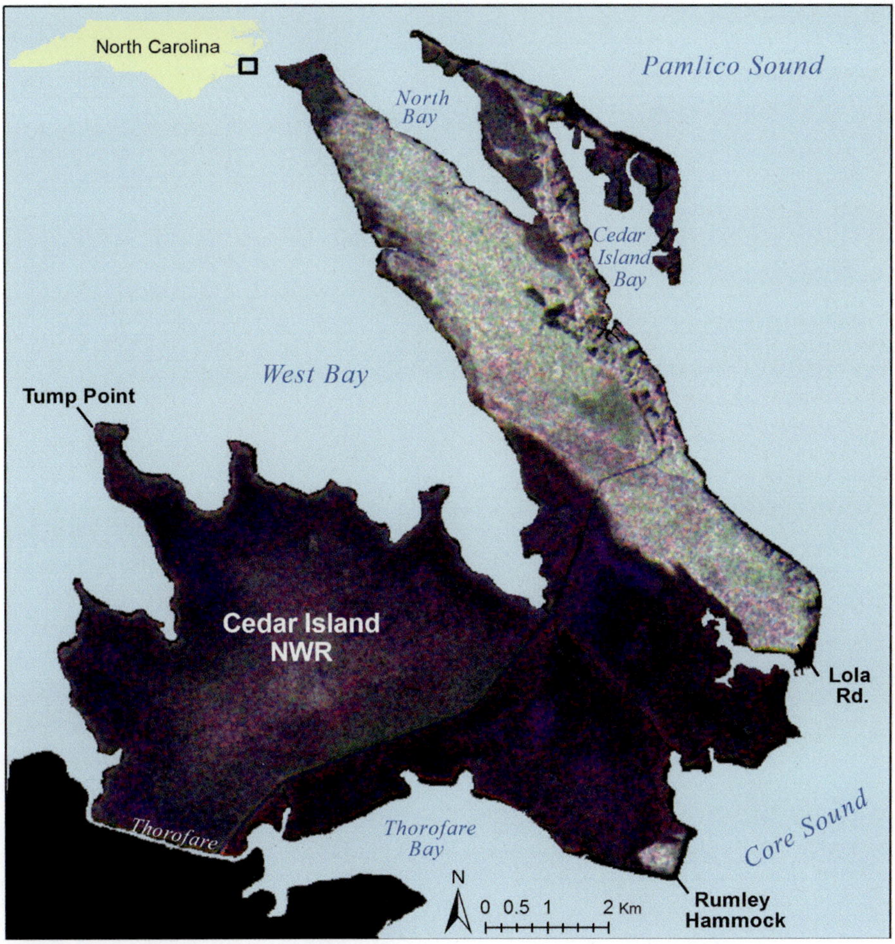

Fig. 17.1 ALOS PALSAR multi-date composite image of Cedar Island NWR, North Carolina, showing RGB composite HH polarization on April 30, June 15, and September 15, 2010, respectively

17.3.2 Case Study: OBIA for Transitional Marshes and Phragmites in Spencer Creek, Alligator River NWR

A second case study sought to further explore the potential to map multiple marsh communities, including expansive areas of *Phragmites australis*. The latter community has been apt to establish along estuarine fringes and in expanding, patchy colonies throughout the Pamlico and Currituck Sounds. Within the Alligator River National Wildlife Refuge and others in the northern Outer Banks area, *Phragmites*

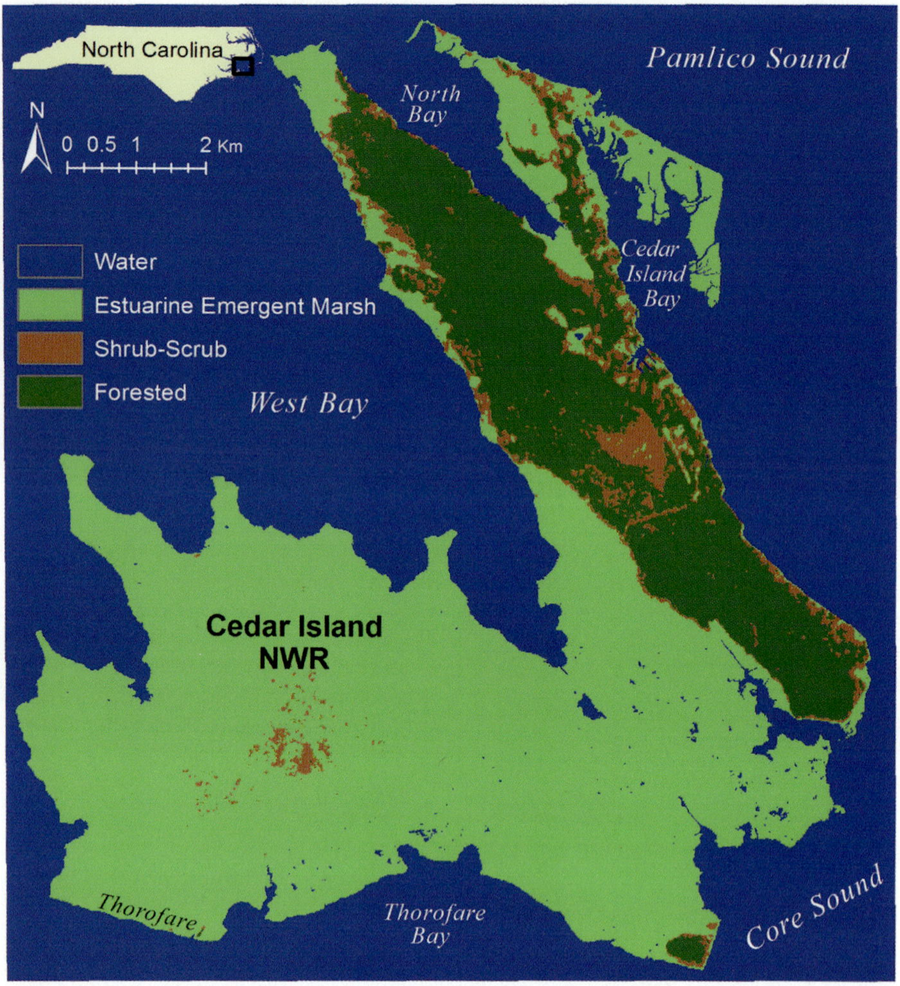

Fig. 17.2 Classified ALOS PALSAR of Cedar Island National Wildlife Refuge, NC

has displaced low marshes and threatened the habitats of associated migratory birds and other endemic species, requiring refuge staff and FWS personnel to exert management and control to mitigate its spread. One area in particular, from Point Peter Road north to Spencer Creek on the outer Pamlico Peninsula has witness rapid estuarine hotspot erosion and *Phragmites* colonization. Strong interest there by the FWS and NGOs such as The Nature Conservancy could be served by remotely sensed monitoring and inventory.

Similar to the Cedar Island NWR case study, ALOS PALSAR were acquired and stacked for image classification. In addition, a LIDAR DEM was obtained, including bare earth and vegetation canopy height layers. For classification, Erdas OBJECTIVE software was used to iteratively segment then classify vector

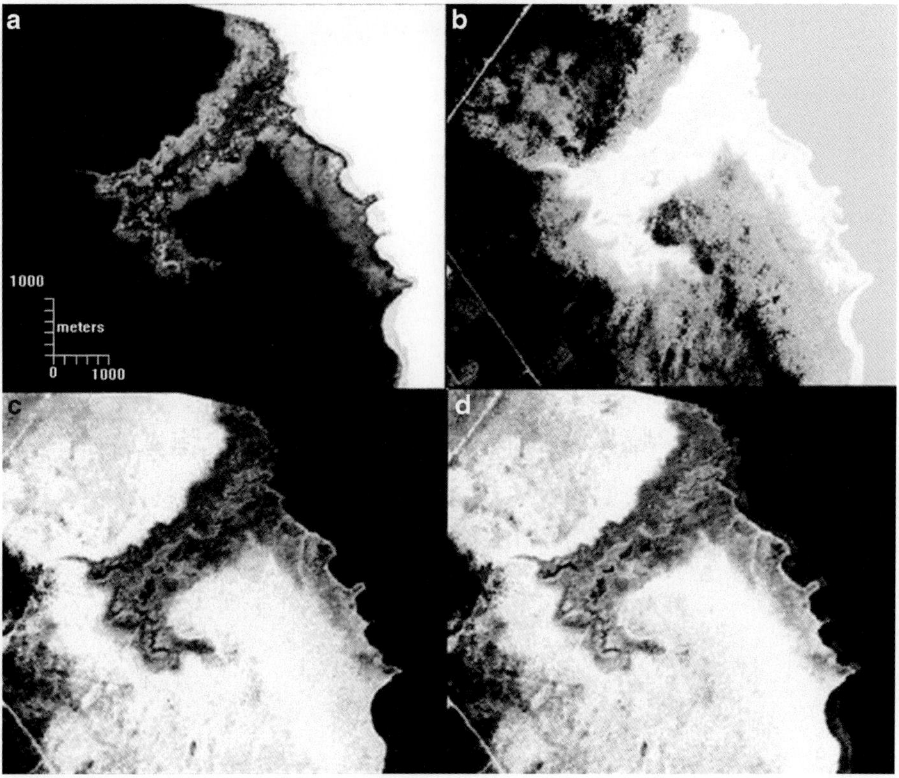

Fig. 17.3 Pixel spectral feature probability values derived from multidate ALOS PALSAR for selected coastal vegetation of Spencer Creek, Alligator River National Wildlife Refuge, NC. (**a**) Water, (**b**) *phragmites*, (**c**) shrub-scrub, (**d**) marsh

polygons of homogeneous spectral and topographic conditions in the area. Figure 17.3 depicts the results from one step of the process, the derivation of pixel spectral feature probabilities based on training sites for marsh, *Phragmites*, open water, and shrub-scrub.

One of three intensive study areas explored by Allen et al. (2011) and assessed in detailed in Allen et al. (2013), the resulting image classification (Fig. 17.4) achieved class accuracies exceeding 85 %. In comparison, accuracies for wetlands ranged from 80 to 82 % with per pixel classification (Allen et al. 2013). The map illustrates a process of rapidly expanding *Phragmites* along the shoreline and within the Spencer Creek drainage that presents a severe management challenge for native habitats.

One pitfall of discrete, state-based wetland classification in such a dynamic estuarine system is the inability of capture gradual shifts in ecosystem state and health. Towards this, the normalized differences (NDX) composites and spectral mixture analysis techniques proposed by Rogers and Kearney (2004) composed of Normalized Difference Vegetation Index (NDVI), Normalized Difference Water Index (NDWI), and the Normalized Difference Soil Index (NDSI). These

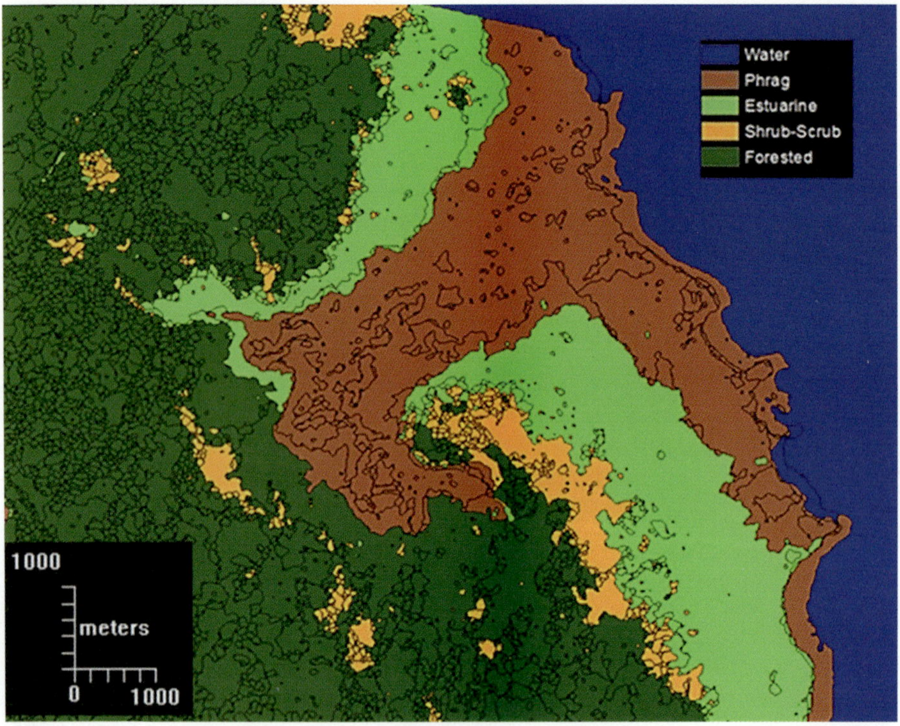

Fig. 17.4 Classified segments from multidate ALOS PALSAR for Spencer Creek, Alligator River NWR, NC

multispectral normalized differences afford the potential to compare and track changes across fixed indices through time (as compared to scene-dependent data compression approaches such as PCA) and also to fix axes for spectral unmixing (e.g., fractional pixel components of soil, water, and marsh vegetation). Figure 17.5 illustrates an exploratory application of the NDX composite indices for the Alligator River NWR and adjoining Roanoke Island and Outer Banks. The indices in composite RGB display strongly differentiate the Phragmites, emergent salt marsh, shrub-scrub, and swamp forests in the Alligator NWR as well as the southern marshes of Roanoke Island (near Wanchese), emergent post-fire marshes near Stumpy Point, and recovering Atlantic White Cedar peatland swamp forests in the Buckridge Coastal Reserve. Although exploratory in our case study, future work using the NDX image composites might also investigate the sensitivity of the multi-date SAR and LiDAR to biophysical and seasonal parameters pertinent to wetland monitoring, including classification, migration, colonization, punctuated events (e.g., storm erosion and marsh wrack deposition) and *Phragmites* condition. With the advantage of temporal stability of the indices, there is also a strong multitemporal change detection application, such as has found a rich body of literature exploiting the Tasseled Cap Transformation (Crist and Cicone 1984) and the inclusion of multi-date imagery for improved wetland classification accuracy (Allen et al. 2012).

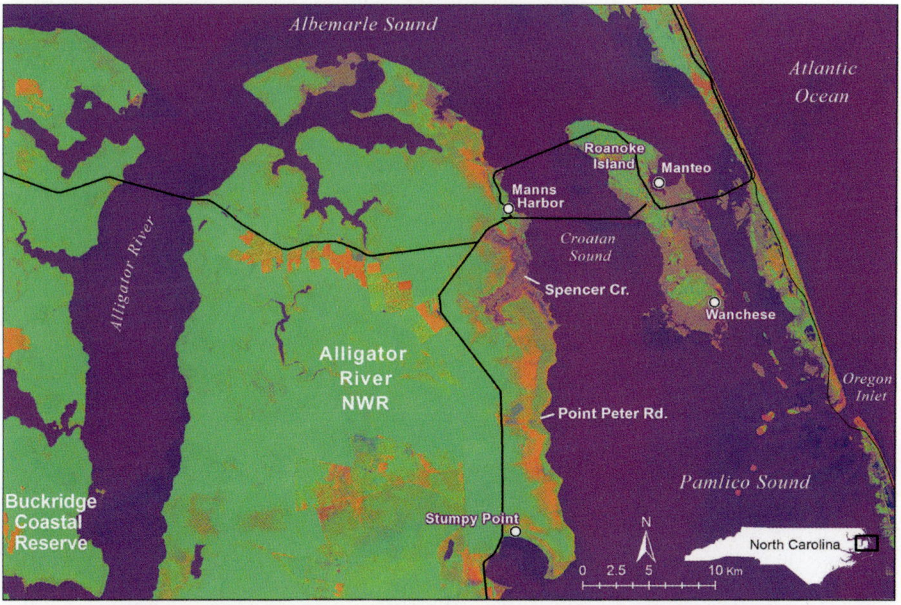

Fig. 17.5 Landsat 8 OLI Normalized Difference Transformation (NDX) soil-water-vegetation composite for the northern Pamlico, Croatan, and Roanoke Sound, North Carolina. 321 RGB composite (NDSI, NDVI, NDWI, respectively)

17.3.3 Barrier Island Vegetation Mapping with OBIA Techniques, Rachel Carson Coastal Reserve

The goal of this case study was to develop remote sensing protocols for mapping a dynamic saltmarsh and associated communities using a combination of LiDAR, aerial digital orthophotography, and satellite imagery and OBIA classification methods. The resulting ecogeomorphic map would incorporate predominantly vegetative or geomorphic units previously mapped and monitored by the NC Coastal Reserve and NC Division of Coastal Management (NC Department of Environment and Natural Resources.) This scheme required utilizing an ultra-high spatial resolution (previous work has traditionally used orthophoto digitization and high-resolution field-based GPS surveys) as well as including the wide range of classes from terrestrial maritime forests and shrub-scrub through wetlands (supratidal to intertidal) as well as subtidal flats and water courses dynamically changing in the Bird Shoal area of the reserve near Beaufort Inlet and Shackelford Banks, an oceanfront barrier island.

This research could further the contribution of remote sensing for scientific research and resource management. A systematic image processing-based approach and new sensor technology have the potential to reduce mapping error owing to misinterpretation and map generalization. This research uses recent technological advances to design a methodology aimed at producing highly

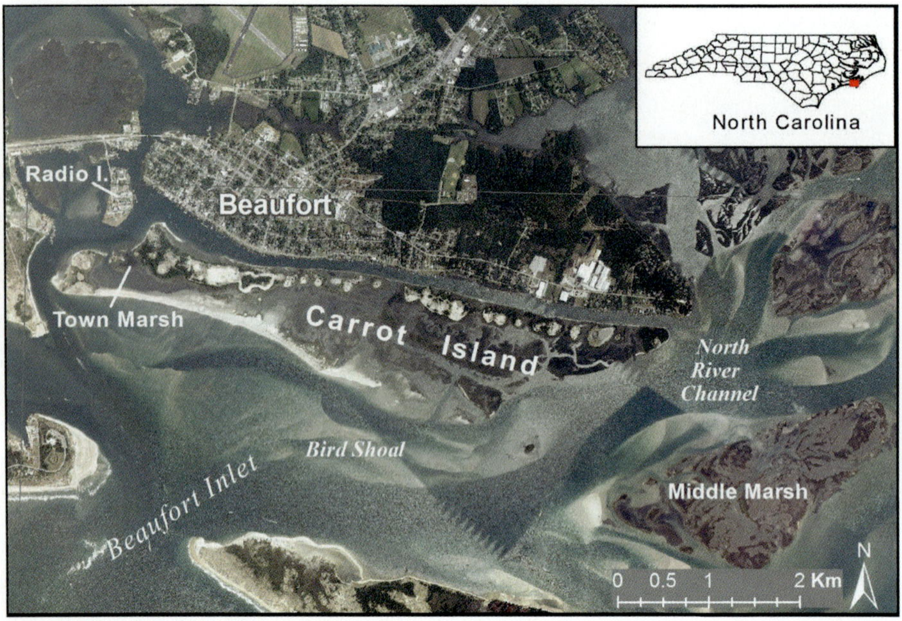

Fig. 17.6 Carrot Island study area within the Rachel Carson Reserve, NC Coastal Reserve, North Carolina, April 2010 color orthophoto

accurate, cost-effective classification and elevation maps and for resource managers, academics, and researchers. Accurate classification and elevation information is essential to informed policy- and decision-making with respect to coastal wetland vulnerability.

The Rachel Carson component of the NCNERR is located near the mouth of the Newport River in southern Carteret County, North Carolina, directly across Taylor's Creek from the historic town of Beaufort (Fig. 17.6). The site is approximately 10.6 km^2 and comprises several small islands (Carrot, Town Marsh, Bird Shoal, Horse Island, and Middle Marshes). The State port at Morehead City is located three miles to the west. Rachel Carson is bounded to the north by Taylor's Creek and the city of Beaufort, to the east by Back Sound, to the south by the Cape Lookout National Seashore, and to the west by Piver's and Radio Islands (Site Profile). The site is approximately 10.6 km^2 and comprises several small islands (Fig. 17.7). The Middle Marshes were not included in the mapping project, as adequate LiDAR data was not available for that portion of the Reserve. The 2010 true-color orthophoto (Fig. 17.7a) depicts the diversity of terrestrial wetlands, highly modified dune and sand flats, and intertidal marshes spread across the reserve. In addition, broad expanses of subtidal flats and sporadic submerged aquatic vegetation (SAV) may sometimes be found. The variable depth and surface roughness owing to wave fetch are evident, as are numerous sand bars and shoals. The historic Town of Beaufort and adjoining developed land are excluded from the mapping.

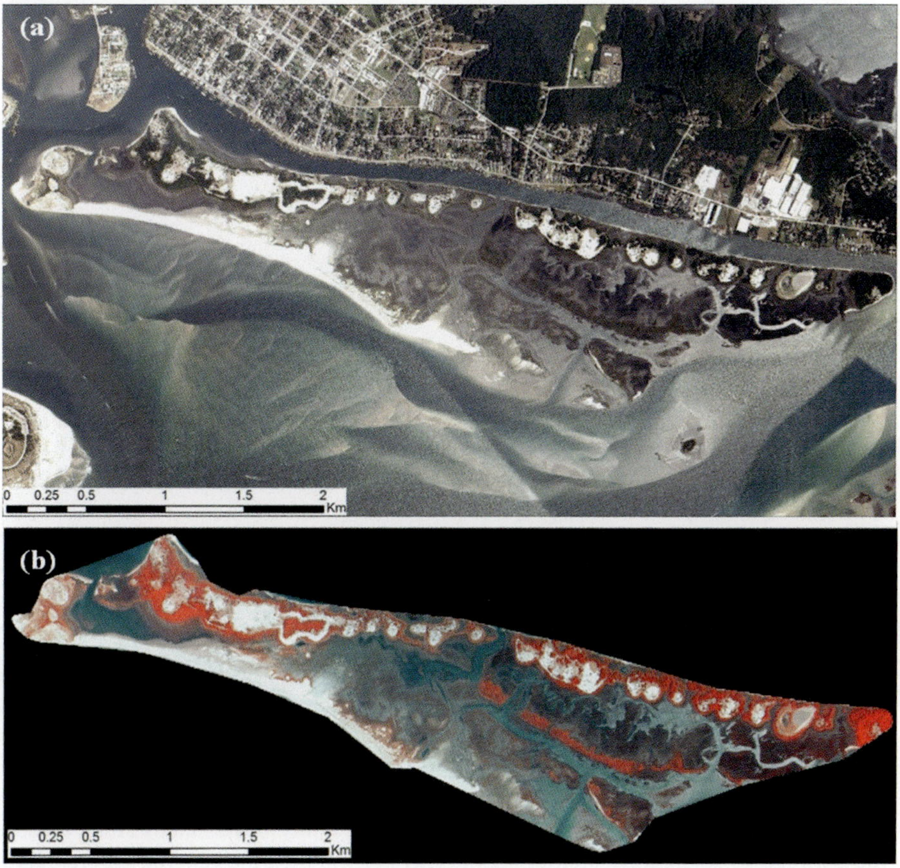

Fig. 17.7 Rachel Carson Reserve true color orthophoto mosaic (**a**) and color-infrared mosaic (**b**) excluding Middle Marsh March 2010

As with the multisensory stack in case studies at Cedar Island and Alligator River NWR, a multidate stack of ALOS PALSAR imagery were acquired and preprocessed for Rachel Carson Reserve. The multiple images visually depict subtle changes in cover type cover and structure over the year 2010 (Fig. 17.8). These data were stacked with bare earth DEMs and the four-band CIR composite (Fig. 17.7) and input to Erdas OBJECTIVE module for segmentation.

Derivative statistics for each layer included up to 76 different variables for each spectral and topographic characteristic, in addition to patch shape and areal indices. A sample of segments were used for training a classification and regression tree (CART) (See5 software) to generate a pruned and winnowed set of classification rules. Twenty parsimonious rules were generated from the CART analysis and used to classify the segments in an ArcGIS Python script. More than 900 segments of the 36,000 generated were randomly selected for accuracy assessment. Initial results

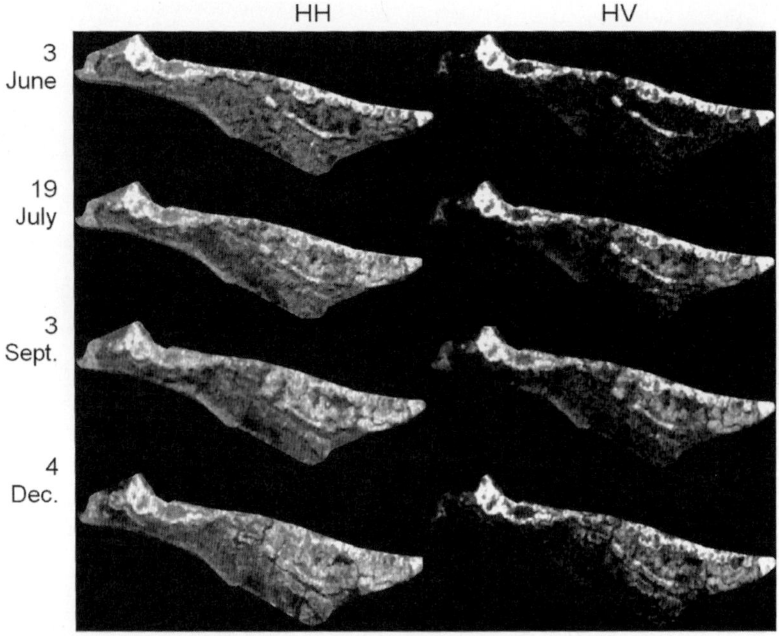

Fig. 17.8 Multi-date ALOS PALSAR 2010 imagery subset and masked for Rachel Carson Reserve, North Carolina Estuarine Research Reserve

indicate a very high classification accuracy with overall sample accuracy of 88 % for eleven classes. If the shrub-scrub classes and sandy beach/sandy dune classes are aggregated to create a smaller set of nine classes, then the accuracy increases to 95 % overall. This level of accuracy presents a strong case for expanded application of OBIA and multi-sensor high resolution mapping of coastal environments (Fig. 17.9).

17.3.4 Case Study Synthesis

The overall results from the case studies presented highlights the robustness of remote sensing across a diversity of multi-sensor, multi-date, and alternative classification algorithms for coastal wetlands. Two general insights are noteworthy. First, it is increasingly feasible to classify finer thematic detail *and* spatial resolution of coastal wetlands with improved data and algorithms. The potential to map discrete classes of coastal wetlands has been long-established, although now it is possible to accurately identify and classify wetland communities that have historically been challenging. In these case studies, those difficult to map and monitor classes include forested wetlands and *Phragmites*. Secondly, there is

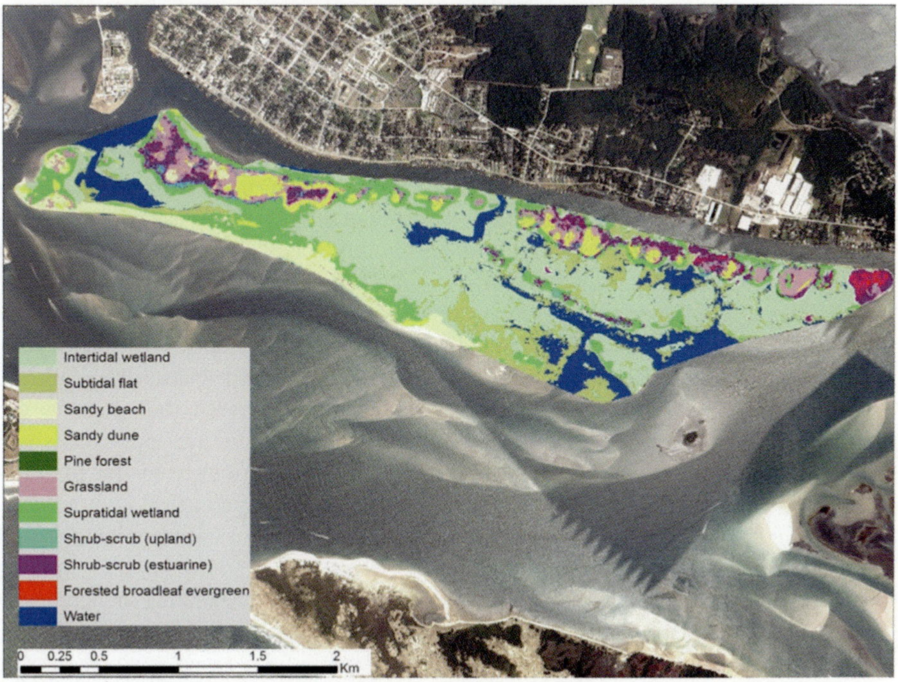

Fig. 17.9 Resulting image classification for 11 island vegetative and surficial categories

expanding opportunities to incorporate LiDAR and SAR imagery for their capabilities to characterize structural and continuous variation of wetland characteristics, such as biomass, elevation, density, and canopy structure. LiDAR data in the case studies often disentangled marshes that are spectrally similar but isolated elevationally. Similarly, SAR imagery captured the differential and seasonal changes of wetland canopies, particularly shrub-scrub, forests, and *Phragmites* In the fine-scale study of Carrot Island, both LiDAR and SAR variables formed the majority of the CART rule sets, despite the ultra-high resolution and CIR bands in the orthophotography. Finally, Landsat 8 is illustrated in the NDX composites for the Alligator NWR case study. Although significantly coarser than the other remotely sensed data applied in this chapter, Landsat 8 offers extremely improved radiometric and spectral characteristics. In conjunction with its systematic and low-cost image acquisition and the continuity of preceding Landsat sensors, the potential for monitoring and change detection with Landsat is robust, indeed. As an assemblage of remote sensing systems, these data offer new opportunities for screening-level inventorying, hotspot and anomaly detection (e.g., erosion events, *Phragmites* colonization, land use/cover change or loss), and investigation of coastal ecological processes as never before (e.g., marsh dieback, ecotone switching, and biological invasions.)

17.4 Conclusions

This chapter sought to assess developments in the state of the art of remote sensing of coastal wetlands through the lens of selected case studies. Throughout the investigations described, the incorporation of exotic or novel imagery was often deemed experimental or risky among cooperating coastal resource managers. The familiarity of traditional optical imagery, albeit digital and perhaps multispectral, was a comfort zone that end-users of remote sensing products were favorable toward, yet they also recognized the potential efficiency, scalability, and repeatability of satellite sensors. King and Green (2009) remark on the continuing, automatic assumption by managers and non-remote sensing specialists that aerial photography is the natural and most desirable form of remotely sensed information for dynamic coasts. They cite the need for better education and awareness and applied research that matches the pertinent issue to the scale of imagery prescribed. While there are applications and especially ultra-fine resolution that aerial or even UAV drones may be the only solution, Since humans to a great degree can intuitively interpret aerial photography, efforts to educate sectors of end-users on the wider variety of data available should demonstrate, guide, and evaluate the degree to which users can incorporate evolving data into their products and workflows. One possibility, organizing imagery and derivative map products hierarchically, could allow variable resolution and "drilling down" to the appropriate detail and accuracy desirable for a given application. The structure of such a hierarchy can be quite readily developed with modern GIS and decision support systems. Indeed, a variety of coastal web atlases have begun doing that, such as the NOAA Digital Coast (NOAA 2014). While both satellite and airborne remote sensing are essential for coastal applications, the increasing resolution and volume of data calls for more automation and object-based techniques (Kelly and Tuxen 2009) and exploitation of innovative multi-temporal and multi-sensor data.

References

Akbar M, Aliabadi S, Patel R, Watts M (2013) A fully automated and integrated multi-scale forecasting scheme for emergency preparedness. Environ Model Software 39:24–38

Allen TR, Wang Y, Gore B, Swords J, Newcomb D (2011) Coastal Wetland mapping Using Time Series SAR Imagery and LiDAR: Alligator River National Wildlife Refuge, North Carolina. In: Proceedings, Pecora 18 symposium, Herndon, Virginia, 14–17 November

Allen YC, Couvillion BR, Barras JA (2012) Using multitemporal remote sensing imagery and inundation measures to improve land change estimates in coastal wetlands. Estuar Coasts 35:190–200

Allen TR, Wang Y, Gore B (2013) Coastal wetland mapping combining multi-date SAR and LiDAR in North Carolina. Geocarto Int 28(7):616–631

Aplin P, Smith GM (2008) Advances in object-based image classification. Int Arch Photogramm Remote Sens Spat Inf Sci vol XXXVII(B7):725–728, Beijing

Blaschke T (2010) Object based image analysis for remote sensing. ISPRS J Photogramm Remote Sens 65:2–16. doi:10.1016/j.isprsjprs.2009.06.004

Borgeau-Chavez LL, Smith KB, Brunzell SM, Kasischke ES, Romanowicz EA, Richardson CJ (2005) Remote sensing of regional inundation patterns and hydroperiod in the Greater Everglades using synthetic aperture radar. Wetlands 25(1):176–191

Carle MV (2011) Estimating wetland losses and gains in coastal North Carolina: 1994–2001. Wetlands 31:1275–1285

Chambers JQ, Fisher JI, Zeng H, Chapman EL, Baker DB, Hurtt GC (2007) Hurricane Katrina's carbon footprint on U.S. Gulf Coast forests. Science 318(5853):1107. doi:10.1126/science.1148913, 16

Chen Y, Huang C, Ticehurst C, Merrin L, Thew P (2013) An evaluation of MODIS daily and 8-day composite products for floodplain and wetland inundation mapping. Wetlands 33:823–835

Chmura GL (2013) What do we need to assess the sustainability of the tidal salt marsh carbon sink? Oceans Coast Manag 83:25–31. doi:10.1016/j.ocecoaman.2011.09.006

Coe MT (2000) Modeling terrestrial hydrologic systems at the continental scale: testing the accuracy of an atmospheric GCM. J Climate 13:686–704

Corcoran JM, Knight JF, Gallant AL (2013) Influence of multi-source and multi-temporal remotely sensed and ancillary data on the accuracy of random forest classification of wetlands in Northern Minnesota. Remote Sens 2013(5):3212–3238. doi:10.3390/rs5073212

Couvillion BR, Beck H (2013) Marsh collapse thresholds for coastal Louisiana estimated using elevation and vegetation index data. J Coast Res 63:58–67

Couvillion BR, Steyer GD, Wang H, Beck HJ, Rybczyk JM (2013) Forecasting the effects of coastal protection and restoration projects on wetland morphology in coastal Louisiana under multiple environmental uncertainty scenarios. J Coast Res 67:29–50

Crist EP, Cicone RC (1984) A physically-based transformation of thematic mapper data – the TM tasseled cap. IEEE Trans Geosci Remote Sens GE-22:256–263

Dai L, LIU C (2010) Multiple classifier combination for land cover classification of remote sensing image. In: Proceedings, 2nd International Conference on Information Science and Engineering (ICISE), Beijing, Institute of Geographic Sciences and Natural Resources Research, pp 3835–3839

Dehouck A, Lafon V, Baghdadi N, Marieu V (2012) Use of optical and radar data in synergy for mapping intertidal flats and coastal salt-marshes (Arcachon Lagoon, France). Geosci Remote Sens Symp (IGARSS) 2011, IEEE International. Available http://www.halcyon.com/pub/journals/21ps03-vidmar. Last accessed 1 May 2011

Dobson JD, Bright EA (1991) CoastWatch – detecting change in coastal wetlands. Geo Info Syst 1:36–40

Donelly JP, Bertness MD (2001) Rapid shoreward encroachment of salt marsh cordgrass in response to accelerated sea-level rise. Proc Natl Acad Sci 98(25):14218–14223. doi:10.1073/pnas.251209298

Dresback KM, Fleming JG, Blanton BO, Kaiser C, Gourley JJ, Tromble EM, Luettich RA Jr, Kolar RL, Hong Y, van Cooten A, Vergara HJ, Flamig ZL, Lander HM, Kelleher KE, Nemunaitis-Monroe KL (2013) Skill assessment of a real-time forecast system utilizing a coupled hydrologic and coastal hydrodynamic model during Hurricane Irene (2011). Cont Shelf Res 71:78–94

Franklin SE (1991) Satellite remote sensing of mountain geomorphic surfaces. Can J Remote Sens 17(3):218–229

Hamilton SK, Kellndorfer J, Lehner B, Tobler M (2007) Remote sensing of floodplain geomorphology as a surrogate for biodiversity in a tropical river system (Madre de Dios, Peru). Geomorphology 89:23–38

Heumann BW (2011) An object-based classification of mangroves using a hybrid decision tree – support vector machine approach. Remote Sens 3:2440–2460

Hladik C, Alber M (2012) Accuracy assessment and correction of a LIDAR-derived salt marsh digital elevation model. Remote Sens Environ 121:224–235

Hofel B, Vetter M, Pfeiffer N, Mandlburger G, Stotter J (2009) Water surface mapping from airborne laser scanning using signal intensity and elevation data. Earth Surf Proc Land 34 (12):1635–1649

Hupp CR (1988) Plant ecological aspects of flood geomorphology and paleoflood history. In: Baker VR, Kochel RC, Patton PC (eds) Flood geomorphology. Wiley, New York, pp 335–356

Jensen JR, Cowen D, Althausen JD, Narumalani S, Weatherbee O (1993) An evaluation of CoastWatch change detection protocol in South Carolina. Photogramm Eng Remote Sens 59 (6):1039–1046

Kamal M, Phinn S (2011) Hyperspectral data for mangrove species mapping: a comparison of pixel-based and object-based approach. Remote Sens 3:2222–2242. doi:10.3390/rs3102222

Kasischke ES, Bourgeau-Chavez LL (1997) Monitoring south Florida wetlands using ERS-1 SAR imagery. Photogramm Eng Remote Sens 33:281–291

Kearney MS, Rogers AS (2010) Forecasting sites of future coastal marsh loss using topographical relationships and logistic regression. Wetlands Ecol Manag 18(4):449–461

Kearney MS, Stutzer D, Turpie K, Stevenson JC (2009) The effects of tidal inundation on the reflectance characteristics of coastal marsh vegetation. J Coast Res 25(6):1177–1186

Kelly M, Tuxen KA (2009) Remote sensing support for tidal wetland vegetation research and management. In: Yang X (ed) Remote sensing and geospatial technologies for coastal ecosystem assessment and management. Lecture notes in geoinformation and cartography. Springer, Heidelberg, pp 341–363

Kelly M, Tuxen KA, Stralberg D (2011) Mapping changes to vegetation pattern in a restoring wetland: finding pattern metrics that are consistent across spatial scale and time. Ecol Indic 11:263–273. doi:10.1016/j.ecolind.2010.05.003

Kim M, Warner TA, Madden M, Atkinson DS (2011) Multi-scale GEOBIA with very high resolution digital aerial imagery: scale, texture and image objects. Int J Remote Sens 32(10):2825–2850

King SD, Green DR (2009) Matching issue to utility: an hierarchical store of remotely sensed imagery for coastal zone management. In: Green DR, King SD (eds) Coastal and marine geo-information systems. Coastal systems and continental margins. Kluwer, Dordrecht, pp 474–486

Klemas V (2009) Sensors and techniques for observing coastal systems. In: Yang X (ed) Remote sensing and geospatial technologies for coastal ecosystem assessment and management. Lecture notes in geoinformation and cartography. Springer, Heidelberg, pp 17–43

Klemas V (2013a) Remote sensing of coastal wetland biomass: an overview. J Coast Res 29(5):1016–1028

Klemas V (2013b) Remote sensing of emergent and submerged wetlands: an overview. Int J Remote Sens 34(18):6286–6320

Klemas VV, Dobson JE, Ferguson RL, Haddad KD (1993) A coastal land cover classification system for the NOAA CoastWatch change analysis project. J Coast Res 9(3):862–872

Kushwaha SPS, Dwivedi RS, Rao BRM (2000) Evaluation of various digital image processing techniques for detection of coastal wetlands using ERS-1 SAR data. Int J Remote Sens 21(3):565–579

Lang MW, Kasischke ES, Prince SD, Pittman KW (2008) Assessment of C-band synthetic aperture radar for mapping and monitoring coastal plain forested wetlands in the Mid-Atlantic region, U.S.A. Remote Sens Environ 112:4120–4130

Lang M, Mccarty G, Oesterling R, Yeo I (2013) Topographic metrics for improved mapping of forested wetlands. Wetlands 33:141–155

Lee Y, Park J, Choi J, Oh Y, Choi J, Oh HY, Won J (2012) Potential uses of TerraSAR-X for mapping herbaceous halophytes over salt marsh and tidal flats. Estuar Coast Shelf Sci 115:366–376

Loveland TR, Merchant JW, Ohlen DO, Brown JF (1991) Development of a land-cover characteristics database for the conterminous U.S. Photogramm Eng Remote Sens 57(11):1453–1463

Lunetta RS, Evlidge CD (eds) (1988) Remote sensing change detection: environmental monitoring methods and applications. Taylor and Francis, London

Mcleod E, Chmura GL, Bouillon S, Salm R, Bjor M, Duarte CM, Lovelock CF, Schlesinger WH, Silliman BR (2011) A blueprint for blue carbon: toward an improved understanding of the role of vegetated coastal habitats in sequestering CO_2. Front Ecol Environ 9(10):552–560. doi:10.1890/110004

Meyerson LA, Saltonstall K, Chambers RM (2009) Phragmites australis in eastern North America: a historical and ecological perspective. In: Silliman BRE, Grosholz M, Bertness D (eds) Salt marshes under global siege. University of California Press, Berkley, pp 57–82

Morris JT, Sundareshwar PV, Nietch CT, Kjerfve B, Cahoon DR (2002) Responses of coastal wetlands to rising sea level. Ecology 83:2869–2877

Pal M, Mather P (2003) An assessment of the effectiveness of decision tree methods for land cover classification. Remote Sens Environ 86:554–565

Phinn SR, Roelfsema CM, Mumby PJ (2012) Multi-scale, object-based image analysis for mapping geomorphic and ecological zones on coral reefs. Int J Remote Sens 33(12):3768–3797

Platt RV, Rapoza L (2008) An evaluation of an object-oriented paradigm for land use/land cover classification. Prof Geogr 60(1):87–100

Pontius RG, Millones M (2011) Death to Kappa: birth of quantity disagreement and allocation disagreement for accuracy assessment. Int J Remote Sens 32(15):4407–4429

Poulter B, Halpin PN (2008) Raster modelling of coastal flooding from sea-level rise. Int J Geogr Info Sci 22(2):167–182. doi:10.1080/13658810701371858

Ramsey E III, Rangoonwala A, Suzuoki Y, Jones C (2011) Oil detection in a coastal marsh with polarimetric Synthetic Aperture Radar (SAR). Remote Sens 3(12):2630–2662

Ramsey E III, Werle D, Suzuoki Y, Rangoonwala A, Lu Z (2012) Limitations and potential of satellite imagery to monitor environmental response to coastal flooding. J Coast Res 28 (2):457–476

Rodrigues SWP, Souza-Filho P (2013) Mapping of environmental sensitivity index to oil spill from Landsat TM images: a study case on the Amazon coastal plain. Rev Bras Geofis 30 (4):533–543

Rogers AS, Kearney MS (2004) Reducing signature variability in unmixing coastal marsh Thematic Mapper scenes using spectral indices. Int J Remote Sens 25:2317–2335

Saltonstall K (2002) Cryptic invasion by a non-native genotype of the common reed, Phragmites australis, into North America. Proc Natl Acad Sci 99(4):2445–2449

Schmid KA, Hadley BC, Wijekoon N (2011) Vertical accuracy and use of topographic LIDAR data in coastal marshes. J Coast Res 27(6):116–132

South S, Qi J, Lusch DP (2004) Optimal classification methods for mapping agricultural tillage practices. Remote Sens Environ 91:90–97

Timm BC, Mcgarigal K (2012) Fine-scale remotely-sensed cover mapping of coastal dune and salt marsh ecosystems at Cape Cod National Seashore using random forests. Remote Sens Environ 127:106–117

Townsend PA, Walsh SJ (1998) Modeling floodplain inundation using an integrated GIS with radar and optical remote sensing. Geomorphology 21:295–312

Wang Y (2004) Seasonal change in the extent of inundation on floodplains detected by JERS-1 synthetic aperture radar data. Int J Remote Sens 25(13):2497–2508

Wang Y, Allen TR (2008) Estuarine shoreline change detection using Japanese ALOS PALSAR HH and JERS-1 L-HH SAR data in the Albemarle-Pamlico Sounds, North Carolina, USA. Int J Remote Sens 29(15):4429–4442

Wang Y, Colby JD, Mulcahy KA (2002) An efficient method for mapping flood extent in a coastal floodplain using Landsat TM and DEM data. Int J Remote Sens 23(18):3681–3696

Wang W, Qu JJ, Hao X, Liu Y, Stanturf JA (2010) Post-hurricane forest damage assessment using satellite remote sensing. Agric For Meteorol 150(1):122–132

Weih Jr RS, Riggan ND (2010) Object-based classification vs. pixel-based classification: comparative [sic] importance of multi-resolution imagery. Int Arch Photogramm Remote Sens Spat

Inf Sci, Vol. XXXVIII-4/C7. Available at http://www.isprs.org/proceedings/xxxviii/4-c7/pdf/Weih_81.pdf. Last accessed 23 Apr 2013

Yoon J, Shim J (2013) Estimation of storm surge inundation and hazard mapping for the southern coast of Korea. J Coast Res 65:856–861

Zhang SP, Wang RQ, Zhang Z (2003) Study on morphological variation of Phragmites australis in the Yellow River downstream wetland. Chin J Plant Ecol 27(1):78–85

Part IV
Advances in the Management of Coastal Resources Using Remote Sensing Data and GIS

Chapter 18
Numerical Modelling and Satellite Remote Sensing as Tools for Research and Management of Marine Fishery Resources

Grinson George

Abstract This chapter outlines the use of modelled and satellite remote sensing (SRS) data in supporting the research, technology-development and management of marine fishery resources. The value of such methods used in marine fisheries management is emphasized. State of art techniques in fisheries resource management utilizing numerical models, and SRS, separately and jointly, are described here. Numerical models are useful for studying fish and other aquatic invertebrate larval transport. SRS data are used to locate fish stocks, locate areas of reef stress and delineate areas of high productivity in the wake of cyclone paths. Coupling SRS with models helps to manage fishery resources on an ecosystem scale, generate potential fishing zones (PFZ), forecast ocean state (OSF), detect meso-scale features such as eddies and track cyclones threatening coastal resources. Modelled, SRS and *in situ* data sets in combination can be used in the estimation of potential fishery resources in the exclusive economic zone (EEZ), which in turn can help in fishing fleet management. Hence, there is a huge scope for application of numerical modelling and SRS in marine fisheries research and management.

18.1 Introduction

Data collection in an oceanic environment is tedious and expensive. So as to enable wide use of *in situ* data, various organizations are hosting their databases on World Wide Web (Table 18.1). But often marine fisheries research and management lack *in situ* environmental time series data (Platt et al. 2007). Both modelled and Satellite Remote Sensing (SRS) data validated for time and space can be used to fill such gaps. Implementation of complex numerical models is frustrated by lack of

---

G. George (✉)
Fishery Resources Assessment Division, Central Marine Fisheries Research Institute, Ernakulam North P.O., Kochi, P.B.No. 1603 682018, India
e-mail: grinson@cmfri.org.in; grinsengeorge@gmail.com

C.W. Finkl and C. Makowski (eds.), *Remote Sensing and Modeling: Advances in Coastal and Marine Resources*, Coastal Research Library 9,
DOI 10.1007/978-3-319-06326-3_18,

Table 18.1 Various organizations maintaining world wide web databases on fish resources

Organization	Web address
Central Marine Fisheries Research Institute	http://cmfri.org.in/annual-data.html
Commission for the Conservation of Southern Bluefin Tuna (CCSBT)	http://www.ccsbt.org/
Conservation of Arctic Flora and Fauna (CAFF)	http://www.grida.no/prog/polar/caff
Convention on the Conservation of Antarctic Marine Living Resources (CCAMLR)	http://www.ccamlr.org/
Food and Agriculture Organization of the United Nations (FAO)	http://www.fao.org/
Indian National Centre for Ocean Information Services (INCOIS)	http://odis.incois.gov.in/
Indian Ocean Tuna Commission (IOTC)	http://www.iotc.org/English/index.php
Inter-American Tropical Tuna Commission (IATTC)	http://www.iattc.org/
International Commission for the Conservation of Atlantic Tunas (ICCAT)	http://www.iccat.int/
International Council for the Exploration of the Seas (ICES)	http://www.ices.dk/
International Pacific Halibut Commission (IPHC)	http://www.iphc.washington.edu/
International Whaling Commission (IWC)	http://www.iwcoffice.org/
North Atlantic Salmon Conservation Organization (NASCO)	http://www.nasco.int/
North Pacific Anadromous Fish Commission (NPAFC)	http://www.npafc.org/
North Pacific Marine Science Organization (PICES)	http://www.pices.int/
North-East Atlantic Fisheries Commission (NEAFC)	http://www.neafc.org/
Northwest Atlantic Fisheries Organization (NAFO)	http://www.nafo.int/
Secretariat of the Pacific Community (SPC)	http://www.spc.int/
Western and Central Pacific Fisheries Commission (WCPFC)	http://www.wcpfc.int/

data inputs whereas simple models ignore some complexities in the marine ecosystem. As a result model outputs have not been analysed to the extent they deserved. In case of SRS, algorithms for data retrieval vary spatially and temporally depending on the nature of optical constituents present, especially in coastal waters. But modelled and SRS data can permit at least qualitative inferences when we address some of the major unresolved questions in fisheries biology. With the advent of improved computing facilities and SRS, last two decades have seen increased activity in both ecosystem modelling and ocean biology from space (Chassot et al. 2011). The results are used for operational and applied marine fisheries research (Stuart et al. 2011). This chapter discusses some of the applications and illustrates them with particular case studies.

18.2 Numerical Models and Their Potential Application to Marine Fish and Invertebrate Larval Transport

The benefit of numerical modelling is that the necessary state variables can be simulated at each grid point of the study domain. Advanced techniques such as finite element mesh and curvilinear contours have made modelling of irregular coastline easier and decreased the complexities involved in formulating the numerical equations.

Biological processes such as fish larval transport can be modelled based on a clear understanding of the physics of a water body. Knowledge of local hydrodynamics is a pre-requisite to modelling coastal processes, given that physical drivers such as tides and currents control them. There is a major role of diffusion and related physical processes in dispersal and recruitment of marine populations (Okubo 1994). Tidal flows can move larvae passively in peak tidal velocities (Levin 1990; Gross et al. 1992). Physical processes influence the distribution of larval fish on a variety of scales, ranging from few meters to thousands of kilometers (Bruce et al. 2001; Hare et al. 2002). There are few larval transport studies in the coastal waters in particular regions (Moser and Smith 1993; Oliver and Shelton 1993; Grothues and Cowen 1999; Hare et al. 2001). Two related studies are discussed in the following section.

18.2.1 Fish Larval Transport Modelling as an Example of Bio-physical Processes

The basic idea in fish larval transport studies is to characterise the passive movement of larvae during the planktonic larval duration (PLD) phase of the species studied. During the pelagic larval phase, the larvae may be dispersed or retained in passive response to physical forcing (Cowen and Sponaugle 2009). It is a phase that larvae are considered as "poorswimmers" (Leis et al. 2006) because the hydrodynamic (HD) forcing on them exceeds their swimming ability. There are various HD models to provide the spatial and temporal current patterns. Digitized bathymetry maps are used for defining the study domain. Inputs such as tide and wind are given in the model as the major physical forcings driving the current. Simulation will produce the HD variables as output at every grid point for the time interval required. The currents generated in these models can be validated using observed data at certain grid points to ascertain the model accuracy.

This HD input, along with the physical forcings, is applied to larval transport models to deduce the dispersion pattern of larvae. Simulation results provide information on the retention of larvae (as concentration of larvae) at every grid point (Cowen and Sponaugle 2009). A common strategy employed in this kind of model is to predict the maximum likelihood of retention of investigated species based on habitat attributes (Guisan and Zimmermann 2000; Moisen et al. 2006; Elith and Graham 2009). During breeding seasons, the areas of maximum likelihood retention of larvae can be demarcated as nursery grounds or marine protected areas for aiding larval survival.

Numerical modelling of fish egg dispersion at the Patos Lagoon estuary in Brazil was carried out by Martins et al. (2007). A similar study combining observational data with a two-dimensional numerical model product has been carried out to determine the fate of fish eggs released in a semi enclosed basin (Grinson et al. 2011a). Fish eggs were treated as passive particles in the model, and were released from probable spawning sites identified during exploratory surveys. Areas with retention of larvae above 30 % were demarcated as nursery areas. Model simulation of eggs from different spawning sites showed varying dispersal patterns. About 80 % of the particles were retained in the basin for all the three seasons studied. Complete retention of particles occurred in the southern part of study domain. A small quantity is flushed out at the northern boundary. This could be the reason for more fish larvae in the southern part. Trawler catches from scientific surveys show a higher abundance of adult fish in the southern part compared to the north. Marine protected areas (MPA) in this study region are demarcated based on ecologically significant factors such as the presence of corals and mangroves. The areas of maximum likelihood of retention of fish larvae differentiated by the numerical model reasonably corraborated with the MPA (Grinson et al. 2011a). Therefore, this sort of simulation can also help as a decision support tool in rightly demarcating the MPA.

Similar experiment was carried out along an open coastal region (Grinson 2011). HD model results were validated with measurements (Fig. 18.1a–c). The modelled and predicted tidal levels showed very good match with no phase difference. The modelled u and v-velocities agree well with the measured current velocities. The larval tracks have a uniform pattern of dispersal southwards and they aggregate near the southern boundary during all the seasons (Fig. 18.1). We can infer that the hydrodynamics of the region promote occurrence of fish assemblages in the southern part of study domain. This is consistent with earlier field studies from this region (Rivonker et al. 1990) which have shown that the fishing grounds are spread over the southern part of the region without any seasonal change in the pattern.

The scale and predictability of fish larval dispersion or retention remain unknown, largely due to measurement difficulties. Utilization of high-resolution biophysical model allows multiple releases of virtual eggs, thus making each individual simulation equivalent to observations of numerous dispersal events.

18.2.2 Modelling Larval Transport and Settling Areas in Case of Bio-fouling Organisms from Ballast Waters

Biofouling by barnacles poses a major threat to fishing, navigation, tourism and port-related activities. The life cycle of barnacles includes a pelagic larval phase. Therefore, the methodology used for modelling dispersal of larval fish can be applied for larval barnacles. The scale of dispersal depends on movement of the water masses, larval behaviour and duration of pelagic stages (Table 18.2). This is characteristic of oceanic diffusion with a positive correlation between time and spread (Pineda 2000).

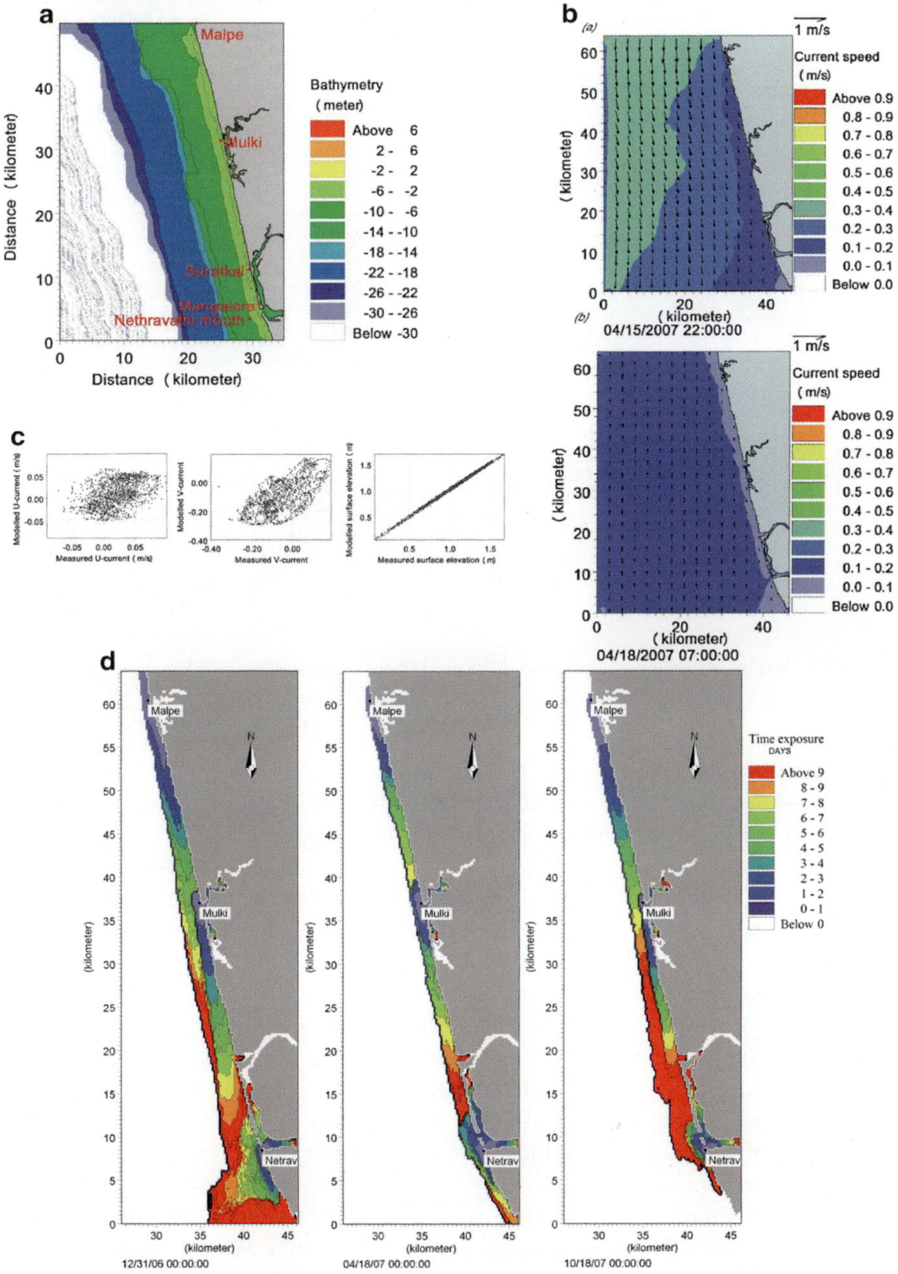

Fig. 18.1 Mangalore open coastal waters (**a**), HD modelling (**b**), HD validation (**c**) and fish larval transport indicating their aggregation towards southern boundary of the domain (**d**), which is a traditional fishing ground

Table 18.2 Distance traversed by barnacles for different pelagic larval duration scales as calculated by Pineda (2000)

Pelagic larval duration	Distance traversed
Few hours	100–1,000 m
1–2 days	1 km
7–14 days	10–20 km
10–15 days	20–30 km
1–2 months	50–100 km
1 year	1,000 km

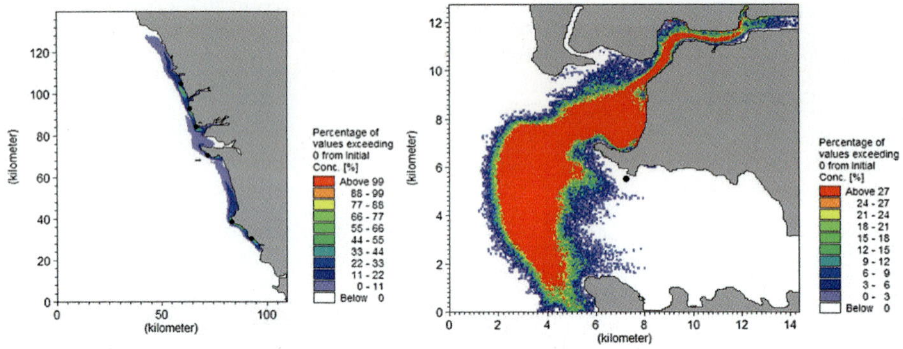

Fig. 18.2 Barnacle larval transport modelling carried out in an open coast of Goa and in Mandovi-Zuari estuary

An application was worked out to estimate the extent of larval spread of barnacles from ballast water sources in different seasons. In the case-study discussed below, the larval dispersal and retention of barnacles in an estuarine mouth was examined (Gaonkar et al. 2012; Grinson et al. 2013a). There is a spatial and temporal variation in the larval dispersal pattern from different spawning sites. Northern spawning sites of the study domain contributed larvae to the estuary during December and May. But, with the reversing wind and current, southern spawning sites contributed larvae to the estuary in October. Larval transport is southward during December and May, reflecting the predominant wind direction. In October the larvae spawned at northern sites were transported to the north and as consequence are lost from the estuary. Larvae spawned at remaining sites in October might be transported either to the north or the south, depending on the timing of the reversal in winds. The net transport in October is usually to the north (Fig. 18.2).

Numerical simulations of larval dispersal have limitations arising from the simplicity of the model in its biological and ecological aspects. Rates of settling, natural mortality, predation and reproductive output may vary with season and with location. For example, higher temperatures can lead to early metamorphosis of larvae, leading to earlier settlement than at lower temperatures. Heavy rainfall result in lower salinity and increased land runoff leading to turbidity which increases larval mortality. Therefore, the results of the simulations can be interpreted as maximum dispersal

Table 18.3 Geo-physical databases useful to fishery scientists

Website	Parameters
http://218.248.0.134:8080/OCMWebSCAT/html/controller.jsp	SST, Chl-a, WS and WD
http://cersat.ifremer.fr/data/products/latest-products	WS and WD
http://oceandata.sci.gsfc.nasa.gov/readme.html	SST, Chl-a, PP
http://odis.incois.gov.in/index.php/remote-sensing-data	Aerosol, Chl-a, fog, humidity, SST, Kd-490 and wind
http://podaac.jpl.nasa.gov/CoreMeasurements	SST, SSS, OCC, SSH, SSHA, WS and WD
http://www.aviso.oceanobs.com/en/data/products.html	SSH, Wind and wave
http://www.eumetsat.int/website/home/Data/Products/Ocean/index.html	SST, Wind
http://www.src.com/mm5/MM5_Main_Page.html	WS and WD

OCC ocean currents and circulation, *SSHA* sea surface height anomaly; Kd-490- diffuse attenuation coefficient of down-welling irradiance at 490 nm

and survival rates. However, this study did provide some basis for the observed variations in larval abundance, settlement and recruitment of barnacles which represent the major biofouling agent from ballast waters.

18.3 Geo-physical Datasets from SRS in the Context of Marine Fisheries Research and Management

SRS datasets are often used in empirical or semi-analytical validated models, either to extrapolate regional datasets in space or to generate derived geo-physical products. A simple example for this can be the summation of thermal signals from different wavelengths for generation of SST. In a similar way, some of the most useful and relevant environmental properties in fisheries research such as sea surface salinity (SSS), WS and WD, sea surface height (SSH), chlorophyll-a (Chl-a) and Chl-a derived primary production (PP) are available online as processed and unprocessed geo-physical datasets (Table 18.3). These datasets can be used to advantage in various fisheries research and management programmes. A few such case studies are illustrated in this section.

18.3.1 SRS Chlorophyll Data Providing Cues on Fish Stock Variability

Variations between years in the seasonal cycle of SRS Chl-a have been implicated in fluctuations in fish stock variability (Platt et al. 2003). In this section we describe the results of an analysis of Chl-a with Indian oil sardine in the coastal waters of

India. Fishing effort in the coastal waters of India changed little in the period 1998–2006, with 238,772 fishing craft in 2005 (CMFRI 2005) in comparison with 239,000 craft in 1997 (Sathiadhas 2006). Thus, the variability in sardine landings during the study period, despite steady fishing effort, indicates that other factors such as environment or food to the sardines are involved. A correlation analysis between available environmental factors (SST, sea bottom temperature, surface salinity, surface dissolved oxygen, bottom dissolved oxygen, pH, nutrients, chlorophyll, zooplankton, rainfall, multivariate El Niño Southern Oscillation index, coastal upwelling index, and derived SST) and sardine catch from the study area emphasised the high significance of chlorophyll compared with other environmental factors in explaining the variability in sardine catch (Krishnakumar and Bhat 2008). Using their fine branchial apparatus, sardines feed predominantly on phytoplankton and zooplankton. In a given area, Chl-a is a good index of the food availability to sardines. Summer surface Chl-a from the study area lies in the range 0.1–5 mg/m^3, and can be very high, from 5 to 10 mg/m^3, during bloom periods (Raghavan et al. 2006). Given the wide dynamic range of chlorophyll concentration in the coastal waters of southwest India and the dominant role of chlorophyll as a determinant of variability in sardine stocks, it seems likely that much will be gained in studying this link in detail.

Algal bloom in the study area often occurs during upwelling. Upwelling in the waters of the southwest coast of India (5–15°N latitude) and the variability in local physical parameters drives changes in the chlorophyll concentration (Smitha et al. 2008). Physical processes affect not only the magnitude of the plankton biomass, but also its species composition (Huntsman et al. 1981), which may in turn affect larval fish feeding and survival (Lasker 1975; Simpson 1987). According to the Hjort-Cushing match-mismatch hypothesis (Hjort 1914; Cushing 1974, 1990), the survival rate of fish larvae is a function of the match between timing of hatching of eggs and initiation of spring phytoplankton bloom. The advent of SRS provides information at the appropriate temporal and spatial scales for testing this hypothesis (Platt et al. 2007). With SRS, it is possible to characterize the spring bloom objectively based on the timing of initiation, amplitude and duration. The statistical moments of all of these properties, and their inter-annual variation, can be calculated and the results used to analyze the effect of ecosystem fluctuations on exploited fish stocks (Platt et al. 2003).

The case study presented below deals with the interannual variability of Indian oil sardine (*Sardinella longiceps*) stock in the southwest coastal waters of India and its relationship with the phytoplankton bloom characteristics computed from SRS, with a view to explain larval survival and interannual variability at the synoptic scale (Grinson et al. 2012).

The life cycle of sardines includes an active breeding season from May to September. This coincides with the high chlorophyll concentration seen during May to September every year (Fig. 18.3). Thus, we find a probable connection between the life history of sardines and phytoplankton bloom dynamics. This supports the finding that the fish itself times its breeding and adjusts its migration

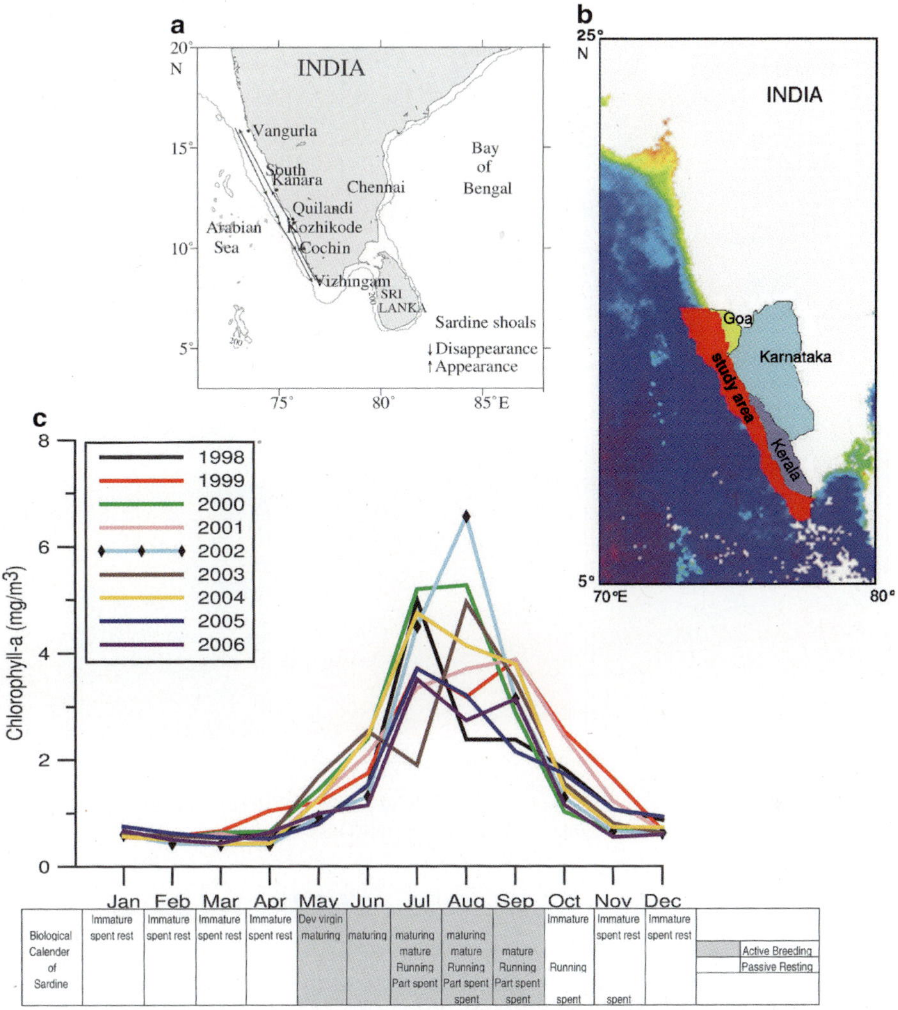

Fig. 18.3 Indian oil sardine migration (**a**), and SeaWiFS satellite chlorophyll (**b**) peaks utilized for explaining the trophic link between the upwelling bloom biology (**c**) and sardine interannual variability along the south west coastal waters of India (Source: Grinson et al. 2012)

to exploit the productive southwest monsoon period. In this study, magnitude of the bloom during initiation month is selected for characterization of bloom, which naturally falls in the month of May every year. May is the most critical month for sardines because both bloom initiation and the beginning of sardines' active breeding phase occur during this month. A delay in the initiation of bloom in the area results in a delay in the onset of suitable conditions for survival of sardine larvae (Grinson et al. 2012).

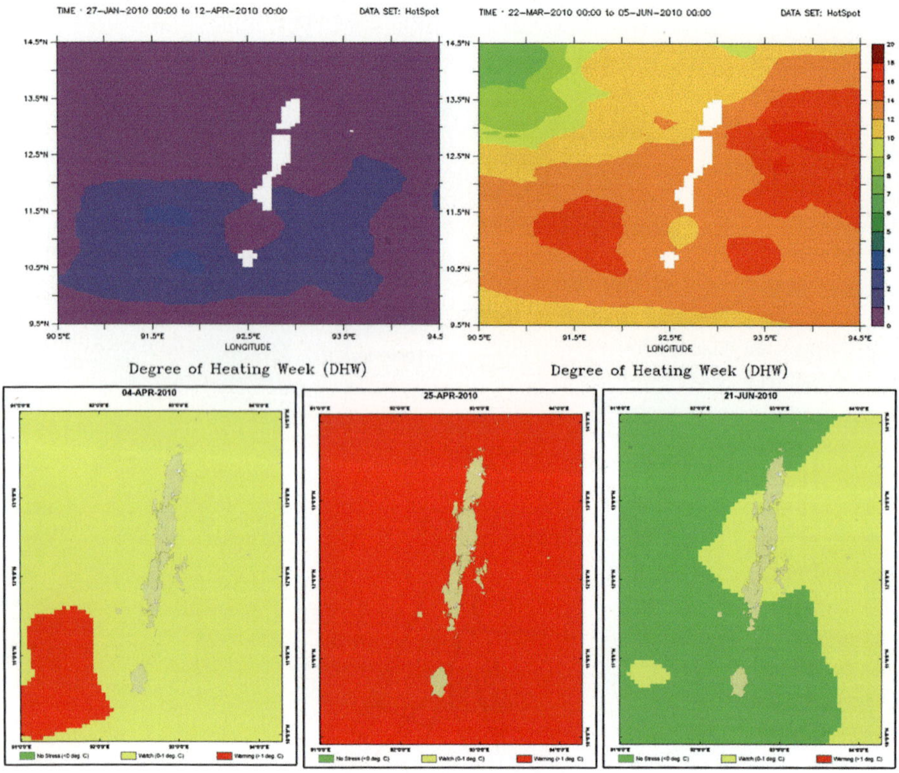

Fig. 18.4 HS and DHW generated during the bleaching events in 2010 in Andaman Sea indicating the stress levels reefs faced (provided as advisory by INCOIS)

18.3.2 Reef Health Advisories Using SRS Derived SST

Globally, there are several instances of mass coral bleaching incidents leading to heavy reef mortality (Krishnan et al. 2011). The application of SRS provides synoptic views of the global oceans in near-real-time for monitoring the reef areas (Liu et al. 2003; Bahuguna et al. 2008; Mahendra et al. 2010). SST during night time is an important parameter for assessment of the thermal conditions inducing the bleaching. SRS provides SST information during day and night routinely, facilitating the development of a coral reef bleaching warning system to generate early warning advisories/bulletins in near real time. The estimation of monthly maximum mean using night time SST climatology retrieved using NOAA, AVHRR is used for generating reef health advisories to eliminate the effect of solar glare and reduce the variation in SST caused by the heating during day time. Threshold hotspot (HS) and daily heating week (DHW) values for a region are calculated the advisory (Mohanty et al. 2013). Depending on the intensity of HS and DHW there can be advisories such as ‘no stress’; ‘watch’; ‘warning’ and ‘alert levels-I & II’ which progressively indicate the severity of a potential bleaching event. Based on this study INCOIS offers reef stress advisories (Fig. 18.4) to alert the reef managers to take appropriate measures to reduce the damage caused to reefs during bleaching events.

18.3.3 SRS Data for Cyclone Tracks Creating Productive Fishing Grounds

Even though cyclones are devastating, there are some positive effects of cyclones on the fishery. Study of the effect of tropical cyclones on biological processes has gained momentum in the recent past. In thermally-stratified coastal waters, cyclones trigger the breaking up of nutrient-depleted surface waters and bring in nutrient-rich sub-surface waters inducing sudden algal blooms and enhancing the regional scale PP. The effect of physical forcings on PP, its variation and associated hydrography in the southwestern Bay of Bengal during the southwest monsoon (July) and post-cyclone period (November) of 1999 was studied by Madhu et al. 2002. In the postcyclone period, the combined effects of well-mixed coastal waters and freshwater injection from the land runoff associated with the cyclone brought nutrients to the mixed layer, which enhanced PP. Potentially, such enhancement of PP results in improving the regional fishery. But cyclone tracks alone will not provide the information on enhanced PP. SRS is able to detect the environmental changes caused by tropical cyclones. Geo-physical data sets from SRS are useful in such studies for indicating possible productive fishing grounds after a lag following the cyclone (Rao et al. 2006).

18.3.4 Demarcation of Ecological Provinces in Support of an Ecosystem Approach to Fisheries Management

Globally, the ecosystem approach to fisheries management (EAFM) is preferred as a basis for sustainable management of fish stock (Garcia et al. 2003b). In this context, it is useful to have a spatial structure for global oceans defined on the basis of ecological provinces rather than geo-political considerations. There are various approaches for classifying the global oceans into ecological provinces (Ekman 1953; Margalef 1961; Yentsch and Garside 1986; Cushing 1989; Fanning 1992; Sathyendranath et al. 1995). The classification by Longhurst et al. 1995 is the most comprehensive, identifying some 50 biogeochemical provinces globally (Longhurst et al. 1995). Some other methodologies require huge data sets for demarcating ecological provinces (Hooker et al. 2000; Li et al. 2004; Alvain et al. 2005; Sherman et al. 2011). But there is lack of *in situ* data to support these approaches. As oceanic realms are dynamic, there are logistic issues in sampling.

Consequently, SRS data are very useful to classification protocols. PP derived from SRS can be a very useful input as PP provinces subsume many oceanographic forcing mechanisms on synoptic scales (Platt and Sathyendranath 2008). These ecological provinces are useful in fisheries management as the physical processes and the ecosystems in each province support characteristic fisheries different from those in nearby provinces (Stuart et al. 2011). Beyond static partitioning, there is a further goal for dynamic bio-geography at regional scales that would incorporate complexities of a dynamic marine environment and their effect on the phytoplankton. SRS will be an invaluable source of inputs in case of such partitioning. Changes in spatial extent of the ecological provinces arising from temporal variations in physical forcing can be captured in a SRS climatology of ocean colour.

18.4 Coupling Modelled and SRS Data for Effective Fishery Management

So far in this chapter, we have discussed the usage of environmental data sets from models and SRS for various aspects in fisheries research and management. But lack of environmental time series data sets pointed to the need for more data. Coupled with SRS, numerical modelling is an alternative tool to generate environmental and biological datasets, which can help to mitigate problems arising from data gaps. Some relevant case studies are described below.

18.4.1 Trophic Modelling Using SRS Data as an Ecosystem Approach to Fisheries Management

Trophic levels in the marine ecosystem are similar to those in terrestrial systems starting with primary producers and ending in scavengers. But, the trophic structure in marine systems is web like, rather than a linear food chain. Fishing often alters the ecosystem structure. Trophic webs will respond differently to fishing depending on whether the target species is a predator or prey species. Single-species fish stock-assessment models ignore food web interactions. Ecosystem based fish stock assessment is offered as another option. EAFM models often resort to SRS-based PP as an input for forcing at the base of the food web to investigate energy transfers and biomass in an ecosystem without fishing, from lower to upper trophic levels (Chassot et al. 2011).

18.4.2 Generating Potential Fishing Zones (PFZ) and Their Dissemination Along with Ocean State Forecasts (OSF)

Identification of PFZ involves an understanding of oceanic processes and interaction of hydro-biological parameters (Desai et al. 2000). The forage base and the physical gradients of temperature and Chl-a help the predatory fish to locate their prey and the same cues are used by fishermen. A number of studies have examined the use of SRS as an aid to locate more productive fishing areas (Waluda et al. 2001). Indian Remote Sensing Satellite P4 Ocean Colour Monitor (IRS P4 OCM) derived chlorophyll concentration and National Oceanographic Aerospace Administration Advanced Very High Resolution Radiometer (NOAA AVHRR) derived SST images have been used to characterise the relationship between the biological and physical variables in coastal waters and it was observed that chlorophyll concentration and SST were inversely correlated with each other (Solanki et al. 1998). The relationship between these two parameters was estimated by a clustering technique called ARNONE (NCAER 2010) and the matching features were selected for generating integrated PFZ forecasts from the composite images on the basis of latitude and longitude (Solanki et al. 2005; NCAER 2010).

Validation of studies of PFZ forecasts have shown that the forecast may lead to substantial increase in fish catch (Solanki et al. 2001, 2003; Nayak et al. 2003). PFZ forecasts in near-real time indicating the likely availability of fish stocks for the next 2 days are disseminated in the Indian EEZ by INCOIS (Fig. 18.5) to about 225 nodes for operational use (Nayak et al. 2003). A significant increase in total catch by following PFZ forecasts has been documented from ANI (Grinson et al. 2011b, 2013b).

18.4.3 Detection of Meso-scale Features Such as Eddies and Fronts That May Indicate Productive Fishing Grounds

Oceanographic features such as eddies, currents and meanders are pervasive features in the world's oceans. These conspicuous hydrographic features influence the horizontal and vertical distributions of the chemical (e.g. nutrients), physical (e.g. SST) and biological (e.g. Chl-a) properties in pelagic systems (Yoder et al. 1981; Seki et al. 2001). Eddies have been found to be localized regions of higher PP leading to aggregation and development of forage species base communities. The presence of mesoscale eddies and their detection by the fishing fleet is an important factor in fishery performance, leading to increased catch per effort for most pelagic species (Laurs and Lynn 1977; Laurs et al. 1984). The influence of mesoscale processes at fronts, such as the formation of rings,

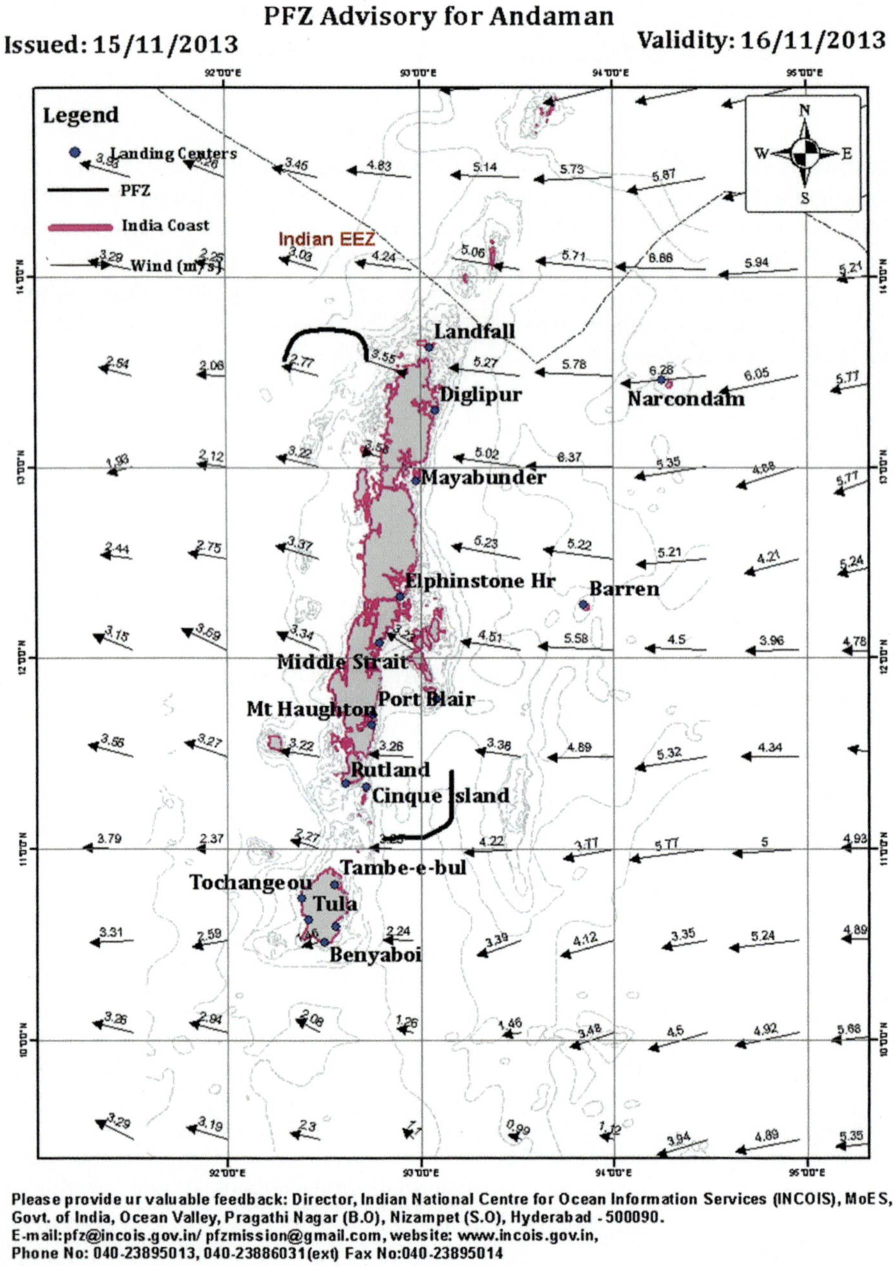

Fig. 18.5 An INCOIS PFZ imagery for Andaman group of islands

meanders and streamers arising or breaking off from these dynamic current systems, has also been shown to be important in shaping the distribution of pelagic fish and shellfish (Waluda et al. 2001). Studies linking the physical oceanographic processes with fish have been carried out around the major boundary currents and related mesoscale processes, such as in the fishing grounds associated with Kuroshio frontal regions (Yokouchi et al. 2000), mesoscale eddies and pelagic fisheries off Hawaiian waters (Seki et al. 2001), upwelling and longline fishery of Portuguese waters (Santos et al. 2006), Atlantic tuna and Gulf of Mexico circulation (Block et al. 2005), oceanographic conditions of spawning grounds of bluefin tuna in the NE Indian ocean (Matsuura et al. 1997), bluefin and frigate tuna spawning along the Balaeric archipelago (Garcia et al. 2003a) and tuna exploitation near the mesoscale processes near the Sechelles (Fonteneau et al. 2006).

The chlorophyll-SST based advisories depend on the surface manifestation by algal blooms and thermal fronts which result from eddies and upwelling. Using altimetry data however, one would be able to follow the evolution of feature from inception to maturation and dissipation with time. There is a time lag between physical upwelling of nutrients to the ocean surface and development of phytoplankton blooms, and subsequently the aggregation of planktivorous and piscivorous fish. Altimetry data helps to identify the fish-aggregating meso-scale features from the outset giving valuable time to forecast and exploit the consequences. Difficulties in getting cloud free imageries sometimes limits the scope of this approach. Altimetry data, especially the SSH have been useful to study the physical oceanography and mesoscale circulation. Advances in SRS altimetry are making it possible to extend the information to the coastal areas where the fishermen are most active. Inputs from the altimetry data on the mesoscale features can be used to augment the PFZ advisories and also provide data during cloud cover.

18.4.4 Forecasting Cyclones and Ocean State to Reduce Impacts on Coastal Fisher Folk and Resources

Apart from elucidating the areas of likelihood of fish/shellfish distributions during the PLD phase, the wind models used for generating wind inputs in simulation of physical process can be utilized for studying cyclone tracks. Fisheries is one of the sectors with high occupational hazard. The extent of direct mortality caused by storms at local or regional scales is severe (Gardner et al. 2005; Done 1992). OSF derived as products of numerical models are provided as input to fishermen to mitigate this risk. OSF provides wave and swell height as well as period, WS as well as wind direction (WD), Tsunami and rough sea warnings and coastal current details. To ensure safe navigation and operations at sea, and to forewarn the fishermen community, INCOIS started the OSF service in 2005 by issuing forecasts 7 days in advance and at three hourly intervals, with daily updates. Fishermen utilize these forecasts to guide their daily operational activities and to ensure safe

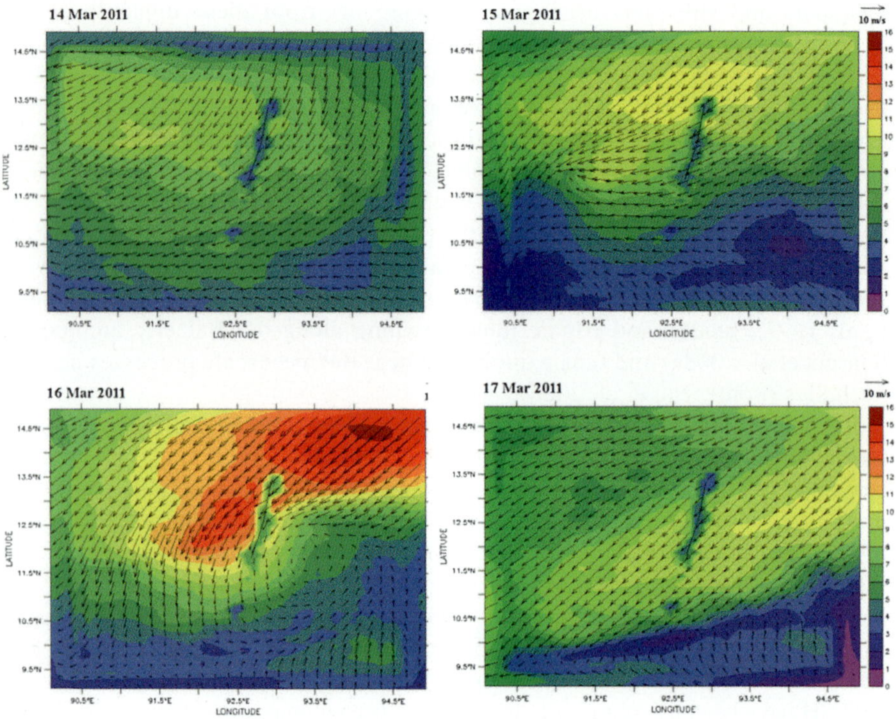

Fig. 18.6 Tropical cyclone wind climatology modelled using WRF for studying the reef damages in Andaman Sea

navigation. Though international agencies such as National Centres for Environmental Prediction (NCEP), USA and European Centre for Medium-Range Weather Forecasts (ECMWF) and UK issue sea state forecasts based on models such as WAVEWATCH III and WAM, these forecasts are for the open ocean. The INCOIS model provide accurate location-specific forecast in the coastal waters using high resolution local bathymetry, and tuning them using observed wave measurements. Real-time and on-line validation of the forecast products is disseminated through various means by INCOIS (Nair et al. 2013).

Cyclones also render coastal resources vulnerable. The ecological effects of cyclones on coral reefs have been reviewed by Harmelin-Vivien (1994). Tropical storms cause severe damage to the reefs; their impacts include the removal of reef matrix, scouring and fragmentation (Rogers and Fish 1991; Done 1992), deposition of loosened material onto beaches above sea level or transporting it into deeper sub-reef environments (Done 1992). The reefs in Andaman and Nicobar Islands (ANI) suffered severe damage following a tropical cyclone in the Bay of Bengal off Myanmar coast during 13–17 March 2011 (Krishnan et. al. 2013). The investigation exposed the vulnerability of the reefs to oceanographic features which generally remain unnoticed unless they directly affect the life or the property of coastal inhabitants. The wind tracks of cyclone were generated using weather research and forecasting (WRF) models (Fig. 18.6) which clearly indicated the passage of cyclone where reefs suffered damage.

18.4.5 Estimation of Potential Fishery Resources of an Exclusive Economic Zone (EEZ) for Fishing Fleet Management

Global marine fish production increased from less than 20 million tons per year in early 1950s to average around 90 million tons per year during the last decade. If the unreported and discarded catches are also taken into account, the global catches will be around 120 million tons per year (Dirk and Pauly 2005). The general trend in shortfall from traditional fishing grounds in the EEZ's of developed countries is compensated by the increasing exploitation of resources in developing countries (Pauly and Watson 2003). The United Nations Convention on the Law of the Sea (UNCLOS) bestows the coastal states with the right to exploitation and responsibility for management of fishery resources of their EEZs. Observations are of paramount importance for managing the resources, and there is a need to establish acccurate catch data collection systems. Fish captured are considered to reflect fish abundance in coastal waters. From marine fish catch data, we can estimate the potential harvestable fish by plotting the catch effort curve, and estimate the maximum sustainable yield (MSY). But, mere post-mortem analysis of landed fish may lead to imperfect estimates as fish catch data without geotags of catching locations may not provide samples representative of the stock in the sea. Therefore, an estimate of harvestable fish based on *in situ* water productivity, taking into account the tropho-dynamics in the EEZ may afford very useful complimentary information.

Chlorophyll, which is an index of algal biomass (ML^{-3}) present in a water column (L) is a prerequisite for primary production and subsequent fish production ($ML^{-2}\,T^{-1}$) which is the annual rate of production of fish biomass per unit area of sea bed. The importance of the potential link between PP and fish was understood decades ago (Ryther 1969), but the advent of SRS Chl-a and modelled PP data sets now available on global and meso-scale prompted policy planners to utilize this for estimation of fishery potential in the EEZ. Past studies relied on *in situ* datasets resulting from different sampling and processing methods and were generally characterized by low spatiotemporal sampling coverage. SRS Chl-a data are now basic to cross-trophic-level analyses of ecosystem production, structure, and function because of the easy and free availability of a wide-ranging, high resolution, and consistent sampling framework (Platt et al. 2007) at a reliable accuracy.

18.5 Overview

This chapter has reviewed some recent advances in numerical modelling, SRS and their coupled applications in marine fisheries research and management. Globally, there is considerable potential for phytoplankton data to serve in future as a proxy for estimation of fish biomass. Usually, investigations in fisheries biology lack environmental time series or other biological data sets (Longhurst and Wooster 1990; Madhupratap et al. 1994). Data from SRS, numerical modelling and their combinations can lead to robust conclusions. Numerical simulation is an attempt to supplement the existing decision support systems. The biological relevance of numerical modelling and SRS at present is restricted to a few operational activities in fisheries. But the issues are complex and there are many uncertainties. The role of coastal processes in fish and shell fish production and dynamics deserves further investigation with improved time series sampling of early life stages in marine organisms. But in the years to come, numerical modelling, SRS and their coupled usage will provide major technological advances supporting operational fisheries research and management.

Acknowledgement This chapter was written during the fellowship offered by NAIP, Indian Council of Agricultural Research for my PML visit with Dr. Trevor Platt as my mentor. My previous research associations with Dr. Vethamony at NIO, Goa, Dr. Dam Roy and Dr. Krishnan at CARI, Port Blair, Dr. T. Srinivasakumar and Dr. R.S. Mahendra at INCOIS and A. Anand at RRSC, Nagpur are reflected in this chapter. I acknowledge the support from Dr. B. Meenakumari, DDG (Fisheries), ICAR, India and Dr. A. Gopalakrishnan, Director, CMFRI, India and the ChloRIFFS project which sponsored this work.

References

Alvain S, Moulin C, Dandonneau Y, Bre'on FM (2005) Remote sensing of phytoplankton groups in case 1 waters from global SeaWiFS imagery. Deep Sea Res I: Oceanogr Res Pap 52:1989–2004

Bahuguna A, Nayak S, Roy D (2008) Impact of the tsunami and earthquake of 26th December 2004 on the vital coastal ecosystems of the Andaman and Nicobar Islands assessed using RESOURCESAT AWiFS data. Int J Appl Earth Observ Geoinf 10(2):29–237

Block BA, Steven LH, Andreas W, Boustany A, Michael J, Stokesbury W, Farwell CJ, Weng KC, Dewar H, Williams TD (2005) Electronic tagging and population structure of Atlantic blue fin tuna. Nature 434:1121–1127

Bruce BD, Evans K, Sutton CA, Young JW, Furlani DM (2001) Influence of mesoscale oceanographic processes on larval distribution and stock structure in jackass morwong Nemadactylus macropterus: Cheilodactylidae. ICES J Mar Sci 58:1072–1080

Chassot E, Bonhommeau S, Reygondeau G, Nieto K, Polovina JJ, Huret M, Dulvy NK, Demarcq H (2011) Satellite remote sensing for an ecosystem approach to fisheries management. ICES J Mar Sci: J Cons 68(4):651–666

CMFRI Marine fisheries census (2005) volume part I. Central Marine Fisheries Research Institute, Cochin, 114 pp

Cowen RK, Sponaugle S (2009) Larval dispersal and marine population connectivity. Ann Rev Mar Sci 1:443–466

Cushing DHA (1974) The natural regulation of fish populations. In: Jones FRH (ed) Sea fisheries research, vol 2. Elek Sci, London, pp 399–412

Cushing DH (1989) A difference in structure between ecosystems in strongly stratified waters and in those that are only weakly stratified. J Plankton Res 11:1–13

Cushing DH (1990) Plankton production and year–class strength in fish populations: an update of the match–mismatch hypothesis. Adv Mar Biol 26:249–294

Desai PS, Honne Gowda H, Kasturirangan K (2000) Ocean research in India: perspective from space. Curr Sci 78(3):268–278

Dirk Z, Pauly D (2005) Good news, bad news: global fisheries discards are declining, but so are total catches. Fish Fish 6(2):156–159

Done TJ (1992) Effects of tropical cyclone waves on ecological and geomorphological structures on the Great Barrier Reef. Cont Shelf Res 12:859–887

Ekman S (1953) Zoogeography of the sea. Sidgwick and Jackson, London, 417 pp

Elith J, Graham C (2009) Do they? How do they? Why do they differ? On finding reasons for differing performances of species distribution models. Ecography 32:66–77

Fanning K (1992) Nutrient provinces in the sea – concentration ratios, reaction rate ratios, and ideal covariation. J Geophys Res 97:5693–5712

Fonteneau A, Lucas V, Delgado A, Demarcq H (2006) Meso-scale exploitation of a major tuna concentration in the Indian Ocean. Document- IOTC-2006-WPTT-24, 16 pp

Gaonkar CA, Samiksha SV, Grinson G, Aboobacker VM, Vethamony P, Anil AC (2012) Numerical simulations of barnacle larval dispersion coupled with field observations on larval abundance, settlement and recruitment in a tropical monsoon influenced coastal marine environment. J Mar Syst 94:218–231

García A, Alemany F, Velez-Belchí P, Rodríguez JM, López Jurado JL, González Pola C, de la Serna JM (2003a) Bluefin and frigate tuna spawning off the Balearic archipelago in the environmental conditions observed during the 2002 spawning season. Col Vol Sci Pap ICCAT 55(3):1261–1270

Garcia SM, Zerbi A, Do Chi T, Lasserre G (2003b) The ecosystem approach to fisheries. Issues, terminology, principles, institutional foundations, implementation and outlook. FAO fisheries technical paper, 443, 81 pp

Gardner TA, Gill IMCA, Grant A, Watkinson AR (2005) Hurricanes and Caribbean coral reefs: impacts, recovery patterns, and role in long-term decline. Ecology 86(1):174–184

Grinson G (2011) Fish larval transport in the coastal waters through ecological modelling. Ph.D. thesis, Goa University- National Institute of Oceanography, Goa, India, 165 pp

Grinson G, Vethamony P, Sudheesh K, Babu MT (2011a) Fish larval transport in a macro-tidal regime: Gulf of Kachchh, west coast of India. Fish Res 110(1):160–169

Grinson G, Krishnan P, Kamal-Sarma KR, Goutham-Bharathi MP, Kaliyamoorthy M, Krishnamurthy V, Srinivasa Kumar T (2011b) Integrated Potential fishing Zone (IPFZ) forecasts: a promising information and communication technology tool for promotion of green fishing in the islands. Indian J Agric Econ 66(3):513–519

Grinson G, Bharathiamma M, Mini R, Srinivasa K, Vethamony P, Babu MT, Xivanand V (2012) Remotely sensed chlorophyll: a putative trophic link for explaining variability in Indian oil sardine stocks. J Coast Res 28(1A):105–113

Grinson G, Dattesh Desai V, Chetan Gaonkar A, Aboobacker VM, Vethamony P, Arga Chandrashekar A (2013a) Barnacle larval transport in the Mandovi–Zuari estuarine system, central west coast of India. J Oceanogr 69(4):451–466

Grinson G, Krishnan P, Roy SD, Sarma K, Goutham Bharathi MP, Kaliyamoorthy M, Krishnamurthy V, Srinivasa Kumar T (2013b) Validation of Potential Fishing Zone (PFZ) forecasts from Andaman and Nicobar Islands. Fish Technol 50:1–5

Gross TF, Werner FE, Eckman JE (1992) Numerical modelling of larval settlement in turbulent bottom boundary layers. J Mar Res 50:611–642

Grothues TM, Cowen R (1999) Larval fish assemblages and water mass history in a major faunal transition zone. Cont Shelf Res 19:1171–1198

Guisan A, Zimmermann NE (2000) Predictive habitat distribution models in ecology. Ecol Model 135:147–186

Hare JA, Fahay MP, Cowen RK (2001) Spring time icthyoplankton of the slope region off the north-eastern United States of America: larval assemblages, relation to hydrography and implications for larval transport. Fish Oceanogr 10(2):1774–1789

Hare JA, Churchill JH, Cowen RK, Berger TJ, Cornillion PC, Dragos P, Glenn SM, Giovoni JJ, Lee TN (2002) Routes and travel rates of larval fish transport from the southeast to the northeast United States continental shelf. Limnol Oceanogr 47(6):1774–1789

Harmelin-Vivien M (1994) The effects of storms and cyclones on coral reefs: a review. J Coast Res 12:211–231
Hjort J (1914) Fluctuations in the great fisheries of Northern Europe viewed in the light of biological research. Rapports et Procs-Verbaux des Runions du Conseil Permanent International pour l'Exploration de la Mer 20:1–228
Hooker SB, Rees NW, Aiken J (2000) An objective methodology for identifying oceanic provinces. Prog Oceanogr 45:313–338
Huntsman SA, Brink KH, Barber RT, Blasco D (1981) The role of circulation and stability in controlling the relative abundance of dino-flagellites and diatoms over the Peru Shelf. In: Richards FA (ed) Coastal upwelling. American Geophysical Union, Washington, DC, pp 357–365
Krishnakumar PK, Bhat GS (2008) Seasonal and interannual variations of oceanographic conditions off Mangalore coast (Karnataka, India) during 1995–2004 and their influences on pelagic fishery. Fish Oceanogr 17:45–60
Krishnan P, Dam-Roy S, Grinson G, Srivastava RC, Anand A, Murugesan S et al (2011) Elevated sea surface temperature during May 2010 induces mass bleaching of corals in the Andaman. Curr Sci 100(1):111–117
Krishnan P, Grinson George, Vikas N, Titus-Immanuel T, Goutham-Bharathi MP, Anand A, Vinod Kumar K, Senthil Kumar S (2013) Tropical storm off Myanmar coast sweeps reefs in Ritchie's Archipelago, Andaman. Environ Monit Assess 185(6):5327–5338
Lasker R (1975) Field criteria for survival of anchovy larvae: the relation between inshore chlorophyll maximum layers and successful first feeding. Fish Bull 73:453–462
Laurs RM, Lynn RJ (1977) Seasonal migration of North Pacific albacore, *Thunnus alalunga*, into North American coastal water: distribution, relative abundance, and association with transition zone waters. Fish Bull US 75:795–822
Laurs RM, Fiedler PC, Montgomery DR (1984) Albacore tuna catch distributions relative to environmental features observed from satellites. Deep Sea Res 31:1085–1099
Leis JM, Hay AC, Trnski T (2006) In situ ontogeny of behaviour in pelagic larvae of three temperate, marine, demersal fishes. Mar Biol 148:655–669
Levin LA (1990) A review of methods for labelling and tracking marine invertebrate larvae. Ophelia 32:115–144
Li WKW, Head EJH, Harrison WG (2004) Macroecological limits of heterotrophic bacterial abundance in the ocean. Deep Sea Res I: Oceanogr Res Pap 51:1529–1540
Liu G, Strong AE, Skirving W (2003) Remote sensing of sea surface temperature during 2002 Barrier Reef coral bleaching. EOS 84(15):137–144
Longhurst AR, Wooster WS (1990) Abundance of oil sardine Sardinella longiceps and upwelling on the southwest coast of India. Can J Fish Aquat Sci 47:2407–2419
Longhurst AR, Sathyendranath S, Platt T, Caverhill C (1995) An estimate of global primary production in the ocean from satellite radiometer data. J Plankton Res 17:1245–1271
Madhu NV, Maheswaran PA, Jyothibabu R, Sunil V, Revichandran C, Balasubramanian T, Gopalakrishnan TC, Nair KKC (2002) Enhanced biological production off Chennai triggered by October 1999 super cyclone (Orissa). Curr Sci 82(12):1472–1479
Madhupratap M, Shetye SR, Nair KNVS (1994) Oil sardine and Indian mackerel: their fishery, problems and coastal oceanography. Curr Sci 66s:340–348
Mahendra RS, Bisoyi H, Prakash CM, Velloth S, Sinivasa Kumar T, Nayak S (2010) Applications of the multi-spectral satellite data from IRS-P6 LISS-III and IRS-P4 OCM to decipher submerged coral beds around Andaman Islands. Int J Earth Sci Eng 3(5):626–631
Margalef R (1961) Correlations entre certains caracte 'res synthe' tiques des populations de phytoplancton. Hydrobiologia 18:155–164
Martins IM, Dias JM, Fernandes EH, Muelbert JH (2007) Numerical modelling of fish eggs dispersion at the Patos Lagoon estuary–Brazil. J Mar Syst 68:537–555
Matsuura H, Sugimoto T, Nakai M, Tsuji S (1997) Oceanographic conditions near the spawning ground of southern bluefin tuna; Northeastern Indian Ocean. J Oceanogr 53:421–434

Mohanty PC, Mahendra RS, Bisoyi H, Tummula SK, Grinson G, Nayak S, Sahu BK (2013) Assessment of the coral bleaching during 2005 to decipher the thermal stress in the coral environs of the Andaman Island using remote sensing. Eur J Remote Sens 46:417–430
Moisen GG, Freeman EA, Blackard JA, Frescino TS, Zimmermann NE, Edwards TC Jr (2006) Predicting tree species presence and basal area in Utah: a comparison of stochastic gradient boosting, generalized additive models, and tree-based methods. Ecol Model 199:176–187
Moser HG, Smith PE (1993) Larval fish assemblages and oceanic boundaries. Bull Mar Sci 53:283–289
Nair TB, Sirisha P, Sandhya KG, Srinivas K, SanilKumar V, Sabique L, Nherakkol A, KrishnaPrasad B, RakhiKumari JC, Kaviyazhahu K, RameshKumar M, Harikumar R, Shenoi SSC, Nayak S (2013) Performance of the ocean state forecast system at Indian National Centre for Ocean Information Services. Curr Sci 105(2):175–181
Nayak SR, Solanki HU, Dwivedi RM (2003) Utilization of IRS P4 ocean colour data for potential fishing zone – a cost benefit analysis. Indian J Mar Sci 32(3):244–248
NCAER (2010) Impact assessment and economic benefits of weather and marine services. National Council of Applied Economic Research, New Delhi, 104 p
Okubo A (1994) The role of diffusion and related physical processes in dispersal and recruitment of marine population. In: Sammarco P, Heron M (eds) The bio-physics of marine larval dispersal. American Association for the Advancement of Science/American Geophysical Union, Washington, DC, pp 5–32
Oliver MP, Shelton PA (1993) Larval fish assemblages of the Benguela current. Bull Mar Sci 53 (2):450–474
Pauly D, Watson R (2003) Counting the last fish. Sci Am 289:42–47
Pineda J (2000) Linking larval settlement to larval transport: assumptions, potentials, and pitfalls. Oceanogr East Pac I:84–105
Platt T, Sathyendranath S (2008) Ecological indicators for the pelagic zone of the ocean from remote sensing. Remote Sens Environ 112:3426–3436
Platt T, Yaco CF, Frank KT (2003) Spring algal bloom and larval fish survival. Nature 423:398–399
Platt T, Sathyendranath S, Fuentes-Yaco C (2007) Biological oceanography and fisheries management: perspective after 10 years. ICES J Mar Sci: J du Cons 64(5):863–869
Raghavan B, Raman M, Chauhan P, Kumar BS, Shylini S, Mahendra R, Nayak S (2006) Summer chlorophyll-a distribution in eastern Arabian Sea off Karnataka-Goa coast from satellite and in-situ observations. In: Frouin RJ, Agarwal VK, Kawamura H, Nayak S, Pan D (eds) Proceedings of SPIE, vol 6406, Remote sensing of the marine environment, volume B 269, Goa, India, p 64
Rao KH, Smitha A, Ali MM (2006) A study on cyclone induced productivity in south-western Bay of Bengal during November-December 2000 using MODIS (SST and chlorophyll-a) and altimeter sea surface height observations. Indian J Mar Sci 35(2):153–160
Rivonker CU, Verlencar XN, Reddy MPM (1990) Physico-chemical characteristics of fishing grounds off Mangalore, west coast of India. Indian J Mar Sci 19:201–205
Rogers DJ, Fish BC (1991) Marine algal lectins. Lectin Rev 1:129–142
Ryther JH (1969) Photosynthesis and fish production in the sea. Science 166(3901):72–76
Santos AMP, Fiuza AFG, Laurs RM (2006) Influence of SST on catches of swordfish and tuna in the Portuguese domestic longline fishery. Int J Remote Sens 27(15):3131–3152
Sathiadhas R (2006) Socio-economic scenario of marine fisheries in India – an overview. In: Kurup BM, Ravindran K (eds) Sustain fish. Paico Press, School of Industrial Fisheries, Cochin, pp 84–101
Sathyendranath S, Longhurst A, Caverhill CM, Platt T (1995) Regionally and seasonally differentiated primary production in the North Atlantic. Deep Sea Res I: Oceanogr Res Pap 42:1773–1802
Seki MP, Polovina JJ, Brainard RE, Bidigare RR, Leonard CL, Foley DG (2001) Observations of biological enhancement at cyclonic eddies tracked with GOES thermal imagery in Hawaiian waters. Geophys Res Lett 28(8):1583–1586

Sherman K, O'Reilly J, Belkin I, Melrose C, Friedland KD (2011) The application of satellite remote sensing for assessing productivity in relation to fisheries yields of the world's large marine ecosystems. ICES J Mar Sci 68:667–676

Simpson JJ (1987) Transport processes affecting the survival of pelagic fish stocks in the California Current. In: 10th annual larval fish conference: American Fisheries Society symposium 2. American Fisheries Society, pp 39–59

Smitha BR, Sanjeevan VN, Kumar KGV, Revichandran C (2008) On the upwelling off the southern tip and along the west coast of India. J Coast Res 24:95–102

Solanki HU, Raman M, Kumari B, Dwivedi RM, Narain A (1998) Seasonal trends in the fishery resources off Gujarat: salient observations using NOAA-AVHRR. Indian J Mar Sci 27:438–444

Solanki HU, Dwivedi RM, Nayak SR (2001) Synergistic analysis of SeaWiFS chlorophyll concentration and NOAA-AVHRR SST features for exploring marine living resources. Int J Remote Sens 22:3877–3882

Solanki HU, Dwivedi RM, Nayak SR, Gulati DK, John ME, Somavanshi VS (2003) Potential Fishing Zone (PFZs) forecast using satellite data derived biological and physical processes. J Indian Soc Remote Sens 31(2):67–69

Solanki HU, Pradhan Y, Dwivedi RM, Nayak S, Gulati DK, Somvamshi VS (2005) Application of QuikSCAT SeaWinds data to improve remotely sensed Potential Fishing Zones (PFZs) forecast methodology: preliminary validation results. Indian J Mar Sci 34(4):441–448

Stuart V, Platt T, Sathyendranath S (2011) The future of fisheries science in management: a remote-sensing perspective. ICES J Mar Sci: J Cons 68(4):644–650

Waluda CM, Rodhouse PG, Trathan PN, Pierce GJ (2001) Remotely sensed mesoscale oceanography and the distribution of Illex argentinus in the South Atlantic. Fish Oceanogr 10:207–216

Yentsch CS, Garside JC (1986) Patterns of phytoplankton abundance and biogeography. In: Pierrot-Bults AC, van der Spoel S, Zahranec BJ, Johnson RK (eds) Pelagic biogeography, UNESCO technical papers in marine science, 49. UNESCO, Paris, pp 153–163

Yoder JA, Atkinson LP, Lee TN, Kim HH, McLain CR (1981) Role of upwelling on phytoplankton patches on the outer southeast shelf. Limnol Oceanogr 26:1103–1110

Yokouchi K, Takeshi K, Matsumoto I, Fujiwara G, Kawamura H, Okudak K (2000) OCTS-derived chlorophyll-a concentration and oceanic structure in the Kuroshio frontal region off the Joban/Kashima Coast of Japan. Remote Sens Environ 73:188–197

Chapter 19
Identifying Suitable Sites of Shrimp Culture in Southwest Bangladesh Using GIS and Remote Sensing Data

Shak Md. Bazlur Rahaman, Khandaker Anisul Huq, and Md. Mujibor Rahman

Abstract Over the last decade, southwest coastal region of Bangladesh has gone through a rapid expansion in shrimp farming. Congenial conditions such as availability of coastal land and water, quality soil and water, successful transfer of hatchery technology and increased export demand led to this rapid growth in shrimp culture sector. Non scientific and unplanned development of shrimp culture has been accompanied by many controversies in the society and in other sectors of agriculture, and the sector has itself been affected adversely by environmental problems. Scientific and sustainable development of shrimp culture essentially needs better site selection. This study identified suitable sites of shrimp culture using GIS and remote sensing technology. Satellite imagery and GIS data- water and soil quality, shrimp culture area, method and production, source and seasonal availability of water, drainage system, water logging, disease outbreak, sanitation facility, road communication, electricity supply, land use pattern, land elevation, hazard frequency, fisheries statistics and population census data etc. were collected in site suitability study. Data were collected through in-situ measurement, laboratory analysis, questionnaire survey and secondary information. Satellite data were used in preparing base map of the study area and visual interpretation of the images was performed using the most important diagnostic characteristics. Decision rules for most important criteria of shrimp culture were followed during site suitability analyses. Study finds most suitable sites of shrimp aquaculture covering

S.Md.B. Rahaman (✉) • K.A. Huq
Fisheries and Marine Resource Technology Discipline, Khulna University, Khulna 9208, Bangladesh
e-mail: riti_rahaman@yahoo.com

Md.M. Rahman
Environmental Science Discipline, Khulna University, Khulna 9208, Bangladesh

C.W. Finkl and C. Makowski (eds.), *Remote Sensing and Modeling: Advances in Coastal and Marine Resources*, Coastal Research Library 9,
DOI 10.1007/978-3-319-06326-3_19,

9,309.40 ha area in Shyamnagar upazilla, Satkhira district where moderately and less suitable locations were identified as 19,256.92 and 9,103.90 ha respectively. Paikgacha, one of the shrimp culture dominated upazillas of Khulna got an estimated 9,954.20 ha land as most suitable with 8,919.40 and 6,775.80 ha as moderately and less suitable zone respectively. Study results indicate the existence of 6,403.20, 8,616.70 and 8,867.20 ha area as most suitable, moderately suitable and less suitable for shrimp culture industry in another coastal area-Rampal under Bagerhat district. The study will have utility in formulating shrimp culture policy for sustainable development of this promising sector in the country.

19.1 Introduction

Bangladesh, one of the developing countries of the world has many urgent needs and objectives, which can be mutually conflicting and the use of resources can be competitive. Shrimp culture in the coastal areas has been one of few economic activities which have globally witnessed high growth rates during the last decade. Though economic development yields some gains but often creates pollution, and becomes responsible for damages of natural habitats and depletion of natural resources. Much of the shrimp culture development in Bangladesh comes from the exploitation of natural resources such as mangroves and aquatic living organisms in the case of coastal and marine resources. The rapid development of shrimp culture has created so many ambiguities in our society and there is a growing demand whether we need to have a closer assessment at the environmental and socio-economic impacts of the culture system. The giant tiger shrimp, *Penaeus monodon*, provides a significant contribution to the total value of coastal aquaculture production in Bangladesh. Shrimp farming in Bangladesh has gone through a significant development in the last decade. Availability of coastal land and water, hatchery production of shrimp PL and rising export demand contributed to this unplanned and uncontrolled expansion of traditional shrimp culture in southwest Bangladesh. The ecological and social impacts of shrimp culture include large-scale degradation of coastal wetlands (especially mangroves), land subsidence, acidification, salinization of ground water, pollution of agricultural land and coastal waters by pond effluents, introduction of exotic species or pathogens into coastal waters, loss of wild larvae and subsequent loss of goods and services generated by natural resource system. Conversely shrimp aquaculture has itself been affected adversely by environmental problems such as poor water quality and disease leading to reduced productivity and in some cases abandonment of shrimp ponds. The ecological and commercial problems of shrimp culture are largely related to the removal of mangroves. Shrimp culture has adversely affected food security through the loss of agriculture lands and salinization of huge areas. Rice production has been affected by seepage of salt water and pollutants from the shrimp ponds (Hein 2000). The breaking of coastal embankments in search for seawater has led to salinization of rice lands and a drop in rice productivity such as occurred in Khulna, Bangladesh from 1.7 to 0.5 tones/ha in the last 10 years. The traditional shrimp farming practice in southwestern Bangladesh is based on brackish water salinity that led to salinization of adjacent land and waterways. Salinization reduces water supplies not only for agriculture but also for drinking and other domestic needs (Patil Pawan and Krishnan 1998). The environmental degradation caused by shrimp farming has often led to the deterioration of local livelihoods (Barraclough and Finger-Stich 1996) through conversion and privatization of mangroves and other lands, salinization of soil and water, marginalization of local population and food insecurity.

Within the context of sustainable shrimp culture development, it appears to be important and useful to address the factors affecting the sustainability of the environment. Sustainable development of aquaculture can contribute to the prevention and control of aquatic pollution since it relies essentially on good quality water resources. Maintenance of the productive potential of land resources and checking of land degradation is a fundamental element of sustainable land use. Shrimp farming activities are mostly carried out in mangroves and coastal regions in Bangladesh. Utilization of the coastal and marine ecosystems on a sustainable basis forms the heart of any policy for exploitation of these resources for which accurate and scientific data on the status, distribution, extent of use, exploitation trends, etc. are indispensable. Proper site selection with improved culture management practices is considered to be the key of scientific and sustainable development of shrimp culture. For a sustainable use of the land, it is essential that pragmatic planning and monitoring to be done. The root cause of many of the environmental and social problems resulting from, and affecting the shrimp aquaculture industry is the lack of planning of coastal land use (Jenkins et al. 1995). Poor site planning results in accumulation of pollutants, water quality deterioration, clearance of mangrove forests and death or reduction of shrimp productivity due to pollution from upstream agricultural or industrial activity. A principal concern of sustainable management is to ensure a rational development of an area with judicious use of its resources, which is consistent with the surrounding natural systems and environment. Planning activities to promote and monitor the growth of aquaculture in individual countries (or larger regions) inherently have a spatial component because of the differences among biophysical and socioeconomic characteristics from location to location. The spatial information needs for decision makers who evaluate such biophysical and socioeconomic characteristics as part of aquaculture planning efforts can be well served by GIS (Rahman and Subudhi 2004; Rahman and Hoque 2006; Rahman 2005, 2006). In recent years, several surveys had been conducted in Khulna district to evaluate biological, physical, social and economic conditions for sustainable aquaculture development. However, most of the surveys lack systematic spatial data. Planning activities to promote sustainable aquaculture in Khulna district require spatial analysis because of geographical variation in biophysical features and socio-economic status.

GIS and remote sensing database can play a vital role in sustainable shrimp culture by providing information on water bodies, land use pattern, water and soil quality parameters, productivity, tidal activity, coastal settlement, communication network etc. Potential site selection considering the impact of other land use activities like agriculture, industries, human uses, protected areas like sanctuaries, breeding and nursery grounds etc. that are part of the same ecosystem can help to maintain the sustainability of shrimp culture. GIS can be described as databases where the information is spatially referenced, what has made GIS so popular is the fact that the spatial referencing of information is related to maps. It is the manipulation and analysis of the spatial database and the display of maps with relative speed and ease is the trademark of GIS. Some works on carp and crab aquaculture opportunities by using GIS modeling have been done in southwest coastal areas in Bangladesh (Salam et al. 2002, 2005). For sustainable use of land resources, it is very much essential that we will have a scientific and systematic planning and monitoring. When a rational development of an area is ensured with judicious use of its resources, which is consistent with the surrounding natural systems and environment then it helps to maintain sustainable management. This research initiative would be supportive for sustainable shrimp culture in the region.

Biophysical characteristics may include criteria pertinent to water quality (e.g. temperature, dissolved oxygen, salinity, turbidity, and pollutant concentrations), water quantity (e.g. volume and seasonal profiles of availability), soil type (e.g. slope, structural suitability, water retention capacity and chemical nature) and climate (e.g. rainfall distribution, air temperature, wind speed and relative humidity). Socio-economic characteristics that may be considered in aquaculture development include administrative regulations, competing resource uses, market conditions (e.g. demand for fishery products and accessibility to markets), infrastructure support, and availability of technical expertise. The spatial information needs for decision makers who evaluate such biophysical and socioeconomic characteristics as part of aquaculture planning efforts can be well served by geographical information systems. These decision support needs cannot be effectively addressed without the use of GIS. Finally, the individual investor interested in aquaculture development also requires spatial information particularly at the time of site selection from among a range of alternative locations with different biophysical and socioeconomic characteristics. GIS is potentially a powerful tool for assisting this class of decision makers, and is already being effectively used for such purposes in some places. Development of a GIS and remote sensing database is a first step in formally evaluating the effects of local geographic and social conditions on the sustainable production of shrimp in these areas. With an adequate database, remote sensing and GIS can serve as a powerful analytic and decision making tool for sustainable shrimp culture development. This project proposes to develop a GIS and remote sensing database of southwest coastal areas of Bangladesh for sustainable shrimp culture, coupled with other physical, biological, and social attributes. GIS and remote sensing database is a first step in formally evaluating the effects of local geographic and social conditions on the sustainable shrimp culture.

19.2 Materials and Methods

Preliminary field surveys were conducted in Rampal, Paikgacha and Syamnagar upazilla of Bagerhat, Khulna and Satkhira district respectively to develop a strategic plan for collecting required ground data. Shrimp culture related data such as types of the farm, culture area, method and production, source of fry and water, seasonal availability of water; water quality parameters such as salinity, pH, temperature, hardness alkalinity, acidity; soil characteristics such as nutrient, pH, salinity, texture, consistency, permeability, reaction, association, moisture etc. were collected through questionnaire survey, in-situ measurement, laboratory analysis and secondary information. Data related to land use pattern, land elevation, hazard frequency, depot facility, electricity supply etc. were also collected. For laboratory analysis, water samples were taken from selected study points of each upazilla (Fig. 19.1). Ancillary data such as land reconnaissance and topographic map (scale: 1:50,000) prepared by Survey of Bangladesh (SOB), administrative map (scale: 1:1,000,000), soil map (scale: 1:50,000) and soil nutrient data supplied by Soil Resource Development Institute (SRDI), Bangladesh were collected. Meteorological data were obtained from Bangladesh Meteorological Department (BMD). Fisheries statistics such as historical fish production and shrimp culture area and population census were collected from Bangladesh Bureau of Statistics (BBS).

Remote sensing (RS) and Geographical Information System (GIS) software were used for the interpretation of satellite images. Base map of the study area was prepared using satellite imagery collected from Indian Institute of Remote Sensing (IIRS). Visual interpretation of the images was performed using the most important diagnostic characteristics. Land use map was recognized through digital process. Spatial database was made integrating satellite imagery and ancillary data. Necessary corrections were made after verification of field data and data collected from various sources. The major spatial data covered different maps of Rampal, Paikgacha and Syamnagar upazilla.

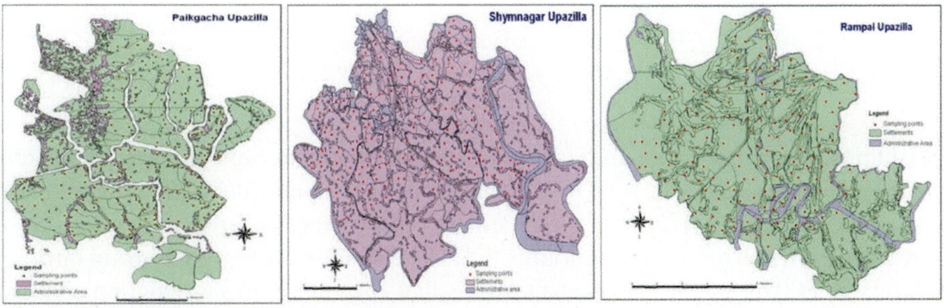

Fig. 19.1 Map of the study area and location of experimental sites

Collected data were analyzed using different data analyzing tools and rules. After the development of GIS and remote sensing database, potential sites were identified for shrimp culture following the decision rules.

19.3 Results and Discussion

19.3.1 Distribution of Water Bodies

Shyamnagar upazilla is situated in Satkhira district at the southwest coastal region of Bangladesh. The total area of Shyamnagar upazilla is 1,968.24 km^2. It lies between 22°33′06″ N latitude and 89°10′28″ E longitude. Table 19.1 indicates union wise distribution of various water bodies in Shyamnagar upazilla. The total area of Paikgacha upazilla (situated in Khulna District, south-west part of Bangladesh) is 411.19 km^2. It is located between 22°26′ and 22°44′ N latitude and 89°12′ and 89^{0}30′ E longitudes. Union wise shrimp farms and their areas of Paikgacha upazilla are shown in Table 19.2. The total area of Rampal upazilla, situated in Bagerhat district is 335.46 km^2. It's position in between 22°30′ and 22°42′ N latitude and 89°28′ and 89°48′ E longitudes. Table 19.3 indicates various water bodies of Rampal upazilla. A study using remote sensing techniques revealed that areas devoted to shrimp farming were encroached on agricultural lands, especially on paddy growing lands, in the south-western region, while in the Chokoria Sundarban of the south-eastern region of Bangladesh coast areas were expanding through destruction of the mangrove forest (Shahid et al. 1992). Alam et al. (1989) found that shrimp practice is rapidly modifying the existing land use pattern of south-western region. By contrast, although new sources of income have been created by shrimp culture in the region, there has been a significant reduction of livestock and poultry in the shrimp farming areas (Rahman et al. 1995; Khan and Hossain 1996).

19.3.2 Fisheries Production

Fisheries productions that come from different sources in Shyamnagar, Paikgacha and Rampal upazilla are shown in Table 19.4. It is found that highest production (5,509.35 MT) in Shyamnagar comes from shrimp ghers that cover nearly 14,659 ha area while fin fish production amounts to 1,613.59 MT in the region. Shrimp farming covers 16,800 ha area in Paikgacha upazilla that contributes to a production of about 5,270 MT. Table 19.4 indicates that crab farming stands for second highest

Table 19.1 Union-wise water bodies (ha) in Shyamnagar Upazilla

Union	River	Khal	Pond	Ditch	Gher	Seasonal Gher	Integration cannal	Low land	Total
Munsiganj	220.54	101.46	0.39	20.73	1354.67	731.01	18.07	0	2,446.87
Burigoalini	565.28	107.16	0.51	24.25	2008.08	322.1	6.86	21.7	3,055.94
Atulia	444.52	71.26	0.73	29.13	1,217.92	207.37	8.08	8.92	1,987.93
Shyamnagor	4.71	52.01	2.41	50.73	920.34	643.22	14.15	12.68	1,700.25
Koikhali	609.96	120.05	0.68	32.25	1,834.74	897.35	12.46	8.89	3,516.38
Ramjannagor	28.83	84.34	1.78	38.13	976.21	491.11	9.47	7.26	1,637.13
Issoripur	4.21	73.68	1.37	43.13	925.74	649.09	10.19	0.1	1,707.51
Kashimari	25.67	68.68	1.16	39.04	1,000.12	344.32	8.43	3.65	1,491.07
Nurnagor	15.69	94.68	0.93	47.25	1,204.23	223.42	7.88	2.18	1,596.26
Vurulia	3.25	62.24	1.29	48.97	980.46	342.2	10.2	4.39	1,453
Gabura	677.15	86.84	0.16	12.13	1807.45	640.24	6.48	6.73	3,237.18
Padmopukur	620.19	90.56	0.19	11.45	1,798.78	540.58	5.97	8.41	3,076.13
Sub-total	3,220	1,012.96	12	397.19	16,028.74	6,032.01	118.24	84.91	26,905.65

Table 19.2 Union-wise shrimp farms (ha) in Paikgacha Upazilla

Union	No. of farm	Area
Paikgacha Paurashava	88	230.87
Haridhali	222	546.44
Kapilmuni	272	1,660.82
Lata	447	3,580.70
Deluti	279	1,630.36
Raruli	103	899.28
Gadaipur	214	993.61
Chandkhali	242	679.11
Laskar	562	2,532.85
Soladana	1,121	3,502.81
Goruikhali	186	230.33
Sub total	3,736	1,6487.17

Table 19.3 Union-wise water bodies (ha) in Rampal Upazilla

Union	River	Khal	Pond	Ditch	Gher	Seasonal Gher
Hurka	97.55	32.98	45.15	12.44	467.48	700.81
Rampal	115.50	55.53	124.01	21.96	1,557.93	953.08
Rajnagar	179.97	61.24	66.58	15.44	1,302.80	974.64
Gaurambha	99.59	82.07	73.21	17.46	1,749.13	462.80
Ujalkur	48.49	93.14	101.84	18.23	176.01	444.27
Baintala	3.15	9.89	20.09	5.77	233.28	117.86
Banshtali	21.84	5.54	16.10	3.63	293.82	101.05
Grand total	566.08	340.39	446.99	94.94	5,780.45	3,754.51

Table 19.4 Fisheries production (MT) in Shyamnagar, Paikgacha and Rampal Upazilla

	Fisheries production (MT)		
Source	Shyamnagar	Paikgacha	Rampal
River	463.69	440.5	74
Khal	405.53	275	873
Pond	223.2	368.6	7,175.5
Shrimp gher	5,509.35	5,270	4,250
White fish culture	1,818.34	5.2	2,035
Prawn farm	6.3	8.2	765
Crab	1.89	5,000	237.50

production (5,000 MT) in fisheries sector of Paikgacha upazilla. In Rampal upazilla, shrimp culture is practiced in 14,811 ha area that come up with a production of 4,250 MT shrimp while white fish production in the region is estimated to be 2,035 MT.

19.3.3 Status of Water Quality

Various water quality parameters were recorded from ten unions of Shyamnagar upazilla during September–October, 2009 that is given in Table 19.5. Highest

Table 19.5 Water quality at different unions of Shyamnagar Upazila

Union	pH	Alkalinity (mg/l)	Hardness ($CaCO_3$ mg/l)	Temperature (°C)
Vurulia	7.0–7.4	100–237	720–1,020	25–27
Kashimari	7.0–7.1	100–150	830–1,730	26
Nurnagar	7.0–7.3	100–200	520–1,800	24–27
Shyamnagar	7.0–7.1	100–150	780–1,590	25–26
Atulia	7.0–7.2	75–150	940–1,700	26–27
Koikhali	7.0–7.1	75–200	570–1,800	26
Ramjannagar	7.0–7.1	87–150	890–1,610	25–27
Isshwaripur	6.9–7.2	75–150	865–1,595	25–26
Munshiganj	7.0–7.2	50–150	850–1,420	24–26
Burigoalini	7.0–7.1	75–150	900–1,520	25–26

temperature, pH, alkalinity, hardness and salinity were found 27 °C, 7.4, 237 mg/L, 1,880 mg/L and 26 ppt respectively while lowest figures for the parameters were counted as 24 °C, 6.9, 50 mg/L, 520 mg/L, 4.5 ppt respectively. Table 19.6 indicates month wise water pH of two rivers and two shrimp farms of Shyamnagar upazilla. The highest pH of river water was found from October to February within 8.0–8.2. Conversely, from March to September it was within 7.4–7.7. Month wise similar distribution pattern was found in shrimp farms though pH level was slightly lower than that of river water. Month wise water salinity of two major rivers and shrimp farms of two locations having significant number of shrimp farms are also shown in Table 19.6. It was found that water salinity gradually increased from November to June and salinity of shrimp farms was a bit higher than that of river water. From Table 19.6, it was observed that the lowest water temperature recorded in December-January and it ranged between 18–26 °C in river water and 15–26 °C in shrimp farms from January to December, 2009.

In Paikgacha (shown in Table 19.7), for pH and water temperature among the unions, no significant variation was found. For alkalinity, the maximum value was observed 286 mg $CaCO_3$/L and minimum was 83 mg $CaCO_3$/L. The highest value of hardness was recorded as 1,944 mg/L while the lowest figure 240 mg/L. Highest salinity was found in the month of June (19–21 ppt) and lowest in the months of October, November, December and January. Monthly pH values of river water and shrimp farm are depicted in Table 19.8. The pH value of river water ranged from 7.4 to 8.0 throughout the year. The maximum pH value of river water was 8.0 in the month of July and minimum in the month of December. In case of shrimp farms, pH value varied from 7.6 to 8.7. The maximum pH of shrimp farm was recorded in July and minimum in the month of January. Month wise salinity distribution of the Kapataska river is also shown in Table 19.8. Highest salinity was found in the month of June (19–21 ppt) and lowest in the months of October, November, December and January. River water temperature (shown in Table 19.8) was lowest in the month of January and rose gradually up to 32 °C in July–August. The characters of shrimp farm water temperature is as the river water which is lowest

Table 19.6 Water quality in river and shrimp farm of Shyamnagar Upazila

	pH				Salinity (ppt)				Water temperature (°C)			
Month	River (HM)	Shrimp farm (HM)	River (KB)	Shrimp farm (KB)	River (HM)	Shrimp farm (HM)	River (KB)	Shrimp farm (KB)	River (HM)	Shrimp farm (HM)	River (KB)	Shrimp farm (KB)
January	8.1	7.4	8.2	7.5	14	15	14	15	18	17	19	18.5
February	8	7.5	8	7.6	16	17	15	17	19	18	20	19
March	7.7	7.1	7.6	7	15	17	17	20	20	18	21	22
April	7.4	7	7.4	7.1	17	19	19	21	21	22	21	23
May	7.5	7.1	7.6	7.2	18	20	20	22	21	22	22	25
June	7.6	7.1	7.8	7.3	19	21	17	19	22	26	23	26
July	7.8	7.5	7.8	7.4	16	18	15	17	24	26	22	25
August	7.4	7.3	7.5	7.3	15	16	14	16	25	26	24	23
September	7.6	7.3	7.6	7.3	13	14	12	14	24	22	25	16
October	8.1	7.6	8	7.6	10	12	10	12	26	18	26	15
November	8	7.5	8	7.5	8	10	9	11	24	17	20	17
December	8.2	7.6	8.2	7.6	12	15	12	13	19	18	19	18.5

HM Harinagar Munshiganj, *KB* Kailbari Burigoalini

Table 19.7 Water quality at different unions of Paikgacha Upazila

Union	pH	Alkalinity (mg CaCO3/l)	Hardness (mg/l)	Temperature (°C)
Haridhali	6.8–8.1	83–200	360–1,420	28–32
Kapilmuni	6.8–8.3	100–266	240–1,173	29–30
Lata	6.8–8.1	100–233	340–1,386	31–32
Gadaipur	6.8–8.3	100–250	273–1,433	30–32
Chandkhali	6.8–7.5	100–200	433–1,433	31–32
Laskar	6.9–7.9	100–200	653–1,253	29–30
Soladana	6.6–8.1	100–286	373–1,944	29–32
Goruikhali	6.8–7.9	116–166	286–1,126	28–30

Table 19.8 Water quality in river and shrimp farm of Paikgacha Upazila throughout the year

	pH		Temperature (°C)		Salinity (ppt)
Month	River (Laxmikhula-Laskar)	Shrimp farm (Laxmikhula-Laskar)	River (Laxmikhula-Laskar)	Shrimp farm (Laxmikhula-Laskar)	Kapataska river
January	7.5	7.6	17	18	0
February	7.7	7.8	18	19	2–8
March	7.7	8	24	25	8–14
April	7.9	8.3	28	29	15–16
May	7.8	8.3	30	31	16–19
June	7.6	8.5	30	31	19–21
July	8	8.7	31	32	8–18
August	7.9	8.6	32	32	1–6
September	7.8	8.3	30	31	0–2
October	7.6	8.2	29	30	0
November	7.5	8	23	22	0
December	7.4	7.8	19	18	0

in December and January and highest in the month of July and August and represents as 18 and 32 °C respectively. In Rampal (shown in Table 19.9), highest range of salinity was recorded 15 ppt, while lowest was found to be 4 ppt. The highest (33 °C) and the lowest (16 °C) water temperatures were recorded in June–July and December–January respectively. The minimum and maximum pH was recorded as 7.5 and 10.0 respectively throughout the year. The highest and the lowest Dissolved Oxygen concentrations were 11 and 6 ppm respectively while maximum (22 cm) and minimum (12 cm) water transparency were recorded in January and August respectively. Das (1998) recommended the pH value for prawn and shrimp as 7.5–8.7. The best range of water pH for coastal aquaculture is 7–9 (Boyd and Fast 1992). Low pH reduces ammonia toxicity but mobilizes metals, including iron and aluminum, which can reduce yields in aquaculture ponds (Simpson et al. 1983). Chiu (1988b) noted that optimum range of pH is 6.8–8.7

Table 19.9 Month-wise water quality parameters in Rampal Upazila

Month	Salinity (ppt)	Temperature (°C)	pH	DO (ppm)	Transparency (cm)
January	4–6	16–20	7.8–8.0	7–8	20–22
February	6–7	18–20	8.0–8.5	8–9	16–18
March	7–8	25–28	7.5–8.5	9–10	14–16
April	10–14	30–31	9.2–10	9–10	16–17
May	10–14	30–32	9.0–9.2	9–10	14–15
June	10–15	32–33	7.9–8.9	8–11	14–16
July	9–11	32–33	8.5–9.0	9–11	16–17
August	8–10	30–31	8.5–9.0	8–10	12–16
September	8–10	30–31	8.0–8.8	8–10	12–17
October	7–8	29–30	8.0–8.7	8–10	16–18
November	5–6	22–26	8.0–8.5	6–7	16–18
December	4–5	18–20	8.5–9.0	6–8	18–20

for shrimp culture. The pH results of this study support the others work. Too high or too low salinity affects growth, and if extreme can be lethal. Fluctuations in salinity below 8 ppt or above 18 ppt have been reported to be retard for growth of penaeid shrimp (Boyd and Fast 1992). Chiu (1988b) mentioned that the optimum range of salinity for *P. monodon* farming would be 10–25 ppt. Subrahmanyan (1973), Verghese *et al.* (1975) and Liao (1977) have observed a direct influence of salinity on the growth of *P. monodon*. The salinity of Shyamnagar upazilla supports these work, on the other hand in Paikgacha and Rampal upazillas, a certain period of time showed the suitable level. But this result is supported by the statement of Hoq et al. (1996) who measured salinity ranged from 4 to 6 ppt in five shrimp farmers' ponds in Satkhira district. Temperature is one of the most important physico-chemical parameters (others are salinity, pH, water transparency and water depth) that have direct effect on the growth of marine shrimp (Lester and Pante 1992). Chiu (1988b) reported that the optimum temperature for *P. monodon* culture is 25–32 °C. Apud (1984) mentioned a range of water temperature from 25 to 30 °C would be favorable for *P. monodon* culture. For penaeid species, recommended range of temperature is about 25–30 °C (Anon 1986). Hoq et al. (1996) reported that water temperature ranging from 27.5 to 30.5 °C was suitable for the growth of Shrimp and prawn. During culture season the water temperature was within the suitable range in this study. Law (1988) suggested that there are long term effects of DO level on the growth of shrimp. Shingh (1984) considered 4–8 ppm or more to desirable range. Authors reported 7.0 mg^{-L} of dissolve oxygen of water body is good for productivity whereas that range from 3.0 to 5.0 mg^{-L} is unproductive for fish culture. The study supports the statements. Alkalinity is the buffering capacity of the pond water. The higher the alkalinity the better the stabilization of the pond system. For successful culture of *P. monodon* alkalinity is recommended to be >80 mg/L (Hansell 1993). The CP Aquaculture Pvt. Ltd. recommended the range of alkalinity from 80 to120 mg/L for shrimp culture. The suitable values of hardness/alkalinity in most of the farms were probably due to the application of lime. He stated that more carbonate was found when free CO_2 was drawn up from water by photosynthesis.

19.3.4 Types of Soil

Soil data of Table 19.10 represents that the major types of soils with maximum percentage were Bazoa (57.75 %), Jhalokati (54.17 %), Barisal (65.22 %), Kamalkati (89.80 %) in Shyamnagar upazilla. For Paikgacha (Table 19.11) upazilla, Barisal soil was found as majority in most of the unions and the figure stands as 80 %, 92.9 %, 60 %, 100 % and 62.5 % respectively. In Rampal (Table 19.12), Bazoa soil dominates than the other types. Gopalpur, Iswardi and Pirojpur soils are totally absent here. Barisal soil is found in significant amount (66 %) in comparison to other types.

19.3.5 Chemical Properties of Soil

Tables 19.13 and 19.14 show union wise chemical properties of soil in Shyamnagar upazilla. The highest soil pH, salinity, organic matter, acidity, total nitrogen, total phosphorus, sulphur, zinc, boron, potassium, calcium, magnesium, copper, iron and manganese were 7.8, 21.27 (ppt), 11.53 (%), 1.32 (mv/100 g soil), 0.207 (%), 39.74 (μg/g soil), 722.90 (μg/g soil), 2.86 (μg/g soil), 4.14 (μg/g soil), 2.8 (mv/100 g soil), 20 (mv/100 g soil), 21.5 (mv/100 g soil), 413 (μg/g soil), and 51.26 (μg/g soil) respectively. Whereas the lowest value was found 3.7, 4.7 (ppt), 1.79 (%), 0.08 (mv/100 g soil), 0.010 (%), 3.27 (μg/g soil), 41.50 (μg/g soil), 0.22 (μg/g soil), 0.35 (μg/g soil), 0.35 (mv/100 g soil), 1.90 (mv/100 g soil), 0.04 (mv/100 g soil), 19.0 (μg/g soil), and 2.40 (μg/g soil) respectively. For Paikgacha, (Tables 19.15 and 19.16) chemical properties of soil indicates that the highest soil pH, salinity, organic matter, total nitrogen, total phosphorus, sulphur, zinc, boron, potassium, calcium, magnesium, copper, iron and manganese were 8.3, 16.08 (ppt), 6.49 (%), 130 (%), 45.0 (μg/g soil), 632.0 (μg/g soil), 6.3 (μg/g soil), 2.46 (μg/g soil), 1.12 (mv/100 g soil), 17.35 (mv/100 g soil), 14.33 (mv/100 g soil), 15.8 (μg/g soil), 880 (μg/g soil), and 85.8 (μg/g soil) respectively. Whereas the lowest value was found 4.1, 0.36 (ppt), 1.03 (%), 20 (%), 2 (μg/g soil), 21.0 (μg/g soil), 0.7 (μg/g soil), 0.14 (μg/g soil), 0.23 (mv/100 g soil), 3.95 (mv/100 g soil), 2.67 (mv/100 g soil), 4 (μg/g soil), 32.0 (μg/g soil), and 9.3 (μg/g soil) respectively. The distribution of union wise chemical properties of soil in Rampal (Tables 19.17 and 19.18) shows that the highest value was found 8.2, 6.6 (ppt), 3.45 (%), 0.27 (%), 60 (μg/g soil), 720 (μg/g soil), 3.5 (μg/g soil), 3.94 (μg/g soil), 1.15 (mv/100 g soil), 11.0 (mv/100 g soil), 5.24 (mv/100 g soil), 6.4 (μg/g soil), 108.0 (μg/g soil), and 70.2

Table 19.10 Union-wise soil type in Shyamnagar Upazila

Union	Barisal (%)	Bazoa (%)	Ramgoti (%)	Jhalokati (%)	Doulatpur (%)	Kamalkati (%)	Kotalpoli (%)	Total (%)
Vurulia	8.25	57.75	0	33	0	0	0	100
Kashimari	37.13	11.23	0	46.23	0	5.41	0	100
Nurnagar	0	45.71	1	0	0	54.29	0	100
Shyamnagar	25.1	16.52	4.21	54.17	0	0	0	100
Atulia	65.22	4.33	0	21.74	0	8.71	0	100
Koikhali	10.2	0	1	0	0	89.8	0	100
Ramjannagar	17.64	35.29	0	29.41	5.92	5.87	5.87	100
Isshwaripur	57.14	14.29	0	7.15	0	21.42	0	100
Munshiganj	63.33	0	0	14.17	6.64	15.86	0	100
Burigoalini	20.01	19.98	0	26.68	0	33.33	0	100
Gabura	32.62	0	0	0	0	67.38	0	100
Paddopukur	43.14	14.2	0	0	0	42.66	0	100

Table 19.11 Union-wise soil type in Paikgacha Upazila

Union	Sara (%)	Gopalpur (%)	Iswardi (%)	Barisal (%)	Jhalokati (%)	Kamalkati (%)	Total (%)
Haridhali	16.67	16.67	66.66	0	0	0	100
Kapilmuni	0	11.1	11.1	22.2	0	55.6	100
Lata	0	0	0	80	20	0	100
Deluti	0	0	0	92.9	7.1	0	100
Raruli	20	0	20	60	0	0	100
Gadaipur	0	16.7	16.7	33.3	0	33.3	100
Chandkhali	0	0	0	100	0	0	100
Laskar	0	0	0	62.5	12.5	25	100
Soladana	0	0	0	45.5	0	54.5	100
Goruikhali	0	0	0	40	0	60	100

Table 19.12 Union-wise soil type in Rampal Upazila

Union	Barisal (%)	Bazoa (%)	Dumuria (%)	Gopalpur (%)	Iswardi (%)	Pirojpur (%)	Total (%)
Baintola	17	66	17	0	0	0	100
Banshtoli	0	60	40	0	0	0	100
Gourambha	41	29	15	0	0	15	100
Hurka	60	0	20	0	0	20	100
Rajnagar	66	0	34	0	0	0	100
Rampal	37	50	13	0	0	0	100
Ujalkur	50	50	0	0	0	0	100

(μg/g soil) respectively while the lowest value was figured out as 6.7, 1.8 (ppt), 1.14 (%), 0.05 (%), 6 (μg/g soil), 134 (μg/g soil), 1.5 (μg/g soil), 0.20 (μg/g soil), 0.58 (mv/100 g soil), 8.3 (mv/100 g soil), 3.1 (mv/100 g soil), 4.2 (μg/g soil), 11 (μg/g soil), and 6.0 (μg/g soil) respectively. The nature of a particular soil type is dependent on its physical properties and nutrient content (George et al. 2010). Soil quality is an important factor in fish pond productivity as it controls pond bottom stability, pH and salinity. It also regulates the quality of the overlying water (Hill 1976). The fish pond soil is made up of sediments, which are composed mainly of loose sand, silt and other soil particles that settle at the bottom (USEPA 2002). Textural and chemical properties of pond soils vary with pond depth, clay and organic matter accumulate in deeper areas of ponds (Boyd 1995). A clayey soil rich in organic matter encourages growth of benthic blue green algae, which along with the associated microorganisms form the main food of most of the brackishwater animals (Boyd 1995). High clay content reduces the seepage of water, thereby preventing or reducing the possible salinisation in nearby areas. If the soil pH is lower than 5.5, phosphorus is adsorbed on clays and humus particles by forming chemical bonds enabling its fixation and thus not available to the phytoplankton in the pond. Available phosphorus is that portion of the soil phosphorus readily absorbed by phytoplankton and other organisms (RPI 1985). Both of these are directly related to the amount of organic matter in the bottom soils. Some common

Table 19.13 Chemical properties of soil at different unions of Shyamnagar Upazilla

Union	pH	Salinity (ppt)	Org. Mat. (%)	Acidity (mv /100 g soil)	N (%)	P (micro gram/g soil)	S (micro gram/g soil)
Vurulia	4.9–7.8	7.41–26.21	1.72–11.53	0.09–0.23	0.080–0.195	3.67–15.22	253.82–717.64
Kashimari	6.7–7.1	7.35–34.26	1.79–3.96	0.21–0.93	0.079–0.157	3.76–17.82	96.55–483.75
Nurnagar	4.3–6.0	5.19–38.74	2.36–4.14	0.15–0.85	0.10–0.166	3.27–19.59	71.19–583.18
Shyamnagar	4.5–7.0	4.90–36.77	1.96–3.54	0.09–0.54	0.101–0.184	4.33–39.74	41.50-722-90
Atulia	3.8–6.4	13.05–43.1	3.12–6.57	0.09–2.28	0.068–0.207	6.11–32.92	114.41–378.27
Paddopukur	4.2–5.8	6.50–43.20	2.17–4.1	0.09–2.5	0.01–0.173	6.5–25.0	56.54–196
Koikhali	4.1–6.9	7.50–34.70	1.93–6.92	0.09–0.89	0.07–0.20	4.56–18.92	98.45–234.5
Ramjannagar	3.7–7.2	7.34–28.40	3.23–6.72	0.08–1.2	0.083–0.19	6.10–18.43	64.45–337.34
Isshwaripur	5.6–7.5	6.80–27.40	2.21–4.83	0.8–1.0	0.12–0.188	6.12–24.70	47.45–236.47
Munshiganj	3.8–6.8	4.7–37.5	1.87–3.98	0.15–1.32	0.018–0.193	4.70–31.60	58.34–287.47
Burigoalini	3.8–4.8	28.89–51.27	3.12–4.21	0.15–1.20	0.141–0.178	6.58–11.10	168.53–496.14
Gabura	4.1–5.0	7.09–40.53	2.19–3.9	0.08–1.21	0.010–0.153	6.40–19.39	56.76–178.21

Table 19.14 Chemical properties of soil at different unions of Shyamnagar Upazilla

Union	B (micro g/g soil)	K (mv/100 g soil)	Ca (mv/ 100 g soil)	Mg (mv/ 100 g soil)	Cu (micro g/g soil)	Fe (micro g /g soil)	Mn (micro g/g soil)
Vurulia	0.98–3.67	0.53–1.20	6.75–14.5	8.5–9.75	0.04–1.38	20.8–197.6	2.40–40.78
Kashimari	1.19–3.97	0.69–1.60	4.25–9.25	5.25–21.0	6.38–9.86	19.0–343.4	9.48–31.0
Nurnagar	0.55–2.97	0.60–1.2	4.0–11.75	7.75–11.0	4.02–10.08	36.8–190.0	10.2–51.26
Shyamnagar	0.58–4.01	0.35–2.1	4.25–15.25	4.0–10.25	4.34–16.92	20.0–224.2	11.98–14.76
Atulia	0.95–3.79	0.40–1.9	4.75–20.0	8.0–13.75	3.12–8.5	33.0–260.8	8.22–36.7
Paddopukur	0.41–2.47	0.64–1.7	1.90–12.0	5.7–14.2	4.53–8.74	73.2–207.4	16.7–45.6
Koikhali	0.36–4.10	0.45–2.7	3.75–17.5	3.75–16.5	4.23–8.23	32.56–286.23	4.34–40.54
Ramjannagar	0.83–3.9	0.43–2.8	3.98–16.5	7.0–16.5	3.56–15.67	73.45–227.48	7.27–50.32
Isshwaripur	0.76–3.8	0.73–1.8	4.25–14.5	4.5–14.75	5.30–16.50	27.87–326.25	6.32–46.6
Munshiganj	0.78–4.12	0.76–1.90	4.0–13.50	3.75–13.75	4.70–17.50	43.27–326.70	4.78–37.80
Burigoalini	1.25–4.14	0.72–2.30	4.25–9.25	11.0–21.5	6.18–17.44	219.4–413.2	23.6–50.0
Gabura	0.35–2.42	0.68–1.3	2.75–10.0	5.5–16.75	4.64–8.74	77.0–204.0	26.80–48.6

Table 19.15 Chemical properties of soil at different unions of Paikgacha Upazila

Union	pH	Salinity (ppt)	Org. Mat. (%)	N (%)	P (micro g/g soil)	S (μ/g soil)	Zn (μ/g soil)
Haridhali	5.4–8.3	0.68–7.01	1.6–2.76	20–50	2–25	16–368	1.3–2.41
Kapilmuni	4.3–6.47	0.36–14.3	1.88–5.5	20–40	5–45	21–555	1.5–2.8
Lata	4.3–6.8	4.95–12.29	2.21–3.80	30–90	5–23	179–632	0.7–6.3
Deluti	5.2–7.9	4.88–13.04	2.55–5.26	30–110	2–8	179–458	2.5–5.8
Raruli	6–8.3	4.35–8.38	1.03–2.82	20–40	7–28	26–400	1.2–2.6
Gadaipur	4.1–8.1	0.67–13.72	1.67–6.49	40–70	9–35	47–474	1.1–5.7
Chandkhali	5.2–5.5	7.06–8.43	2.31–3.6	30–40	5–9	395–447	2.8–4.5
Laskar	4.3–8	5.03–11.85	1.41–3.70	30–80	4–30	142–516	2.2–4.9
Soladana	4.6–6.9	4.87–16.08	2.2–5.7	30–130	8–25	253–632	1.7–4.4
Goruikhali	4.6–5.9	4.66–13.46	1.33–5	40–80	5–18	274–500	3.1–6

Table 19.16 Chemical properties of soil at different unions of Paikgacha Upazila

Union	Br ((μ/g soil)	K (mv/100 g soil)	Ca (mv/100 g soil)	Mg (mv/ 100 g soil)	Cu (μ/g soil)	Fe (μ/g soil)	Mn (μ/g soil)
Haridhali	0.14–0.71	0.29–0.76	12.5–17.35	2.67–10.25	5.0–14	39–394	14.4–35.4
Kapilmuni	0.73–2.17	0.63–0.72	7.60–16.20	6.67–10.25	6.0–14.5	39–290	26.6–77.5
Lata	0.35–1.76	0.55–1.04	6.0–9.25	6.33–10	13.0–15.8	64–728	15.4–81.9
Deluti	0.21–1.06	0.65–1.01	6.65–13.10	8.33–11.25	8.9–14.5	35–252	15–63.3
Raruli	0.21–1.41	0.23–0.74	8.5–9.90	8.5–9.7	4.0–13.9	45–227	18–45.4
Gadaipur	0.21–1.41	0.37–1.04	4.20–16.05	3.58–10.75	5.0–13.7	32–880	22.8–37.5
Chandkhali	0.42–1.06	0.55–0.58	6.75–9.55	7.5–14.33	9.2–11.5	151–227	24.9–36.3
Laskar	0.35–1.76	0.5–1.03	4.70–8.95	5.25–11.66	7.5–13.1	32–744	17.5–29.4
Soladana	0.28–2.46	0.66–1.10	6.4–9.75	7.5–10.83	8.0–13.8	84–446	16.4–85.8
Goruikhali	0.42–1.41	0.5–1.12	3.95–7.0	7.75–13	9.2–14.9	206–582	9.3–55.5

Table 19.17 Chemical properties of soil at different unions of Rampal Upazila

Union	pH	Salinity (ppt)	Org. Mat. (%)	N (%)	P (micro g/g soil)	S (micro g/g soil)	Zn (micro g/g soil)
Baintola	6.7–7.9	1.8–5.7	1.81–2.19	0.08–0.15	10–60	216–555	1.8–2.5
Banshtoli	7.6–7.8	2.6–4.8	1.3–2.3	0.05–0.10	36–48	265–715	1.5–2.8
Gourambha	7.3–8.1	2.3–4.8	1.7–2.7	0.07–0.15	09–60	170–476	1.6–3.3
Hurka	7.2–8.2	2.6–4.9	1.9–2.74	0.09–0.27	12–48	266–715	1.7–2.7
Rajnagar	7.6–8.0	3.5–4.8	1.19–2.57	0.08–0.15	07–33	134–636	1.7–3.5
Rampal	7.4–7.9	3.8–6.0	1.14–3.45	0.08–0.15	06–38	309–720	1.6–2.3
Ujalkur	7.5–8.0	2.7–6.6	1.88–2.6	0.05–0.10	25–48	366–647	1.5–2.1

soil problems that can hinder fish cultivation are acidity, alkalinity and excess organic matter (Dance and Hynes 1980). The range of some chemical parameters of the soil like pH, phosphorus (ppm), organic matter content (%), carbon content (%), nitrogen content (%) was found in a semi-intensive culture of shrimp, *Penaeus*

Table 19.18 Chemical properties of soil at different unions of Rampal Upazila

Union	B (micro g/g soil)	K (mv/100 g soil)	Ca (mv/100 g soil)	Mg (mv/100 g soil)	Cu (micro g/g soil)	Fe (micro g/g soil)	Mn (micro g/g soil)
Baintola	0.47–2.96	0.58–0.75	9.5–11.8	4.3–7.66	5.4–11.0	23–57	6.0–32.1
Banshtoli	0.73–2.17	0.63–0.72	10.5–17.5	4.2–5.6	6.4–7.5	23–33	12.1–32.6
Gourambha	0.76–2.08	0.59–0.85	8.3–13.2	3.1–6.33	4.7–9.3	21–108	12–46.6
Hurka	0.93–2.70	0.66–1.13	9.9–17.4	4.1–7.33	5.9–7.6	15–26	13.4–70.2
Rajnagar	0.73–3.94	0.81–1.15	9.5–14.5	5.24–9.66	5.8–7.6	11–30	9.7–43.5
Rampal	0.20–1.91	0.58–0.97	9.5–15.5	4.48–6.33	4.3–7.5	11–37	14.1–36.6
Ujalkur	1.04–1.46	0.61–0.89	11.0–12.5	5.0–5.3	4.2–5.8	20–21	10.0–28.7

monodon were, 6.7–7.20, 11.08–22.88, 1.3–2.74, 0.76–1.60, 0.07–0.13 respectively (Caipang et al. 2012). Chattopadhyay and Mandal (1980) reported that soils with organic carbon content below 0.5 % are poor and needs organic manuring. Chakraborti et al. (1985) suggested that available phosphorus range of 1.08–2.86 Cmol (p+)/kg was poor for *Penaeus monodon*. It has been reported that brackish-water with high amounts of calcium ions may react readily with water soluble phosphorus to render it into insoluble calcium phosphate. Alkaline water is more suitable for aquaculture rather than neutral or acidic. Acidic water restricted the growth of primary producer and also reduced feeding affinity of aquatic organisms (Boyd 1989).

19.3.6 Distribution of Drainage Condition, Hazard Frequency, Soil Nutrients, Soil Reactions, Soil Salinity, and Soil Depth

The distribution of drainage condition, hazard frequency, soil nutrients, soil reactions, soil salinity, and soil depth of Shyamnagar upazilla are shown in Fig. 19.2. The drainage condition of most of the areas of Shyamnagar upazilla was poorly defined while hazard frequency contained 6–20 % (north-west part) and nutrient rich soil dominated northwestern part of the upazilla. The soil reaction is the indication of pH level in soil. Northern part was characterized by a soil pH of 5.5–7.3 and the eastern part ranges between 4.5 and 5.5. The north and middle parts of the study area were characterized by high soil salinity (4–8 MMHOS/cm) while the south, southeast and southwest regions had soils with low salinity (<2 MMHOS/cm). The soil depth in majority parts of this area were covered predominantly 0.90–1.22 M.

Most of the parts of Paikgacha Upazilla (shown in Fig. 19.3) were poorly drained, only some western parts were mixed by imperfectly and poorly drained systems. Hazard frequency in most of the areas was ranged from 6 to 20 %.

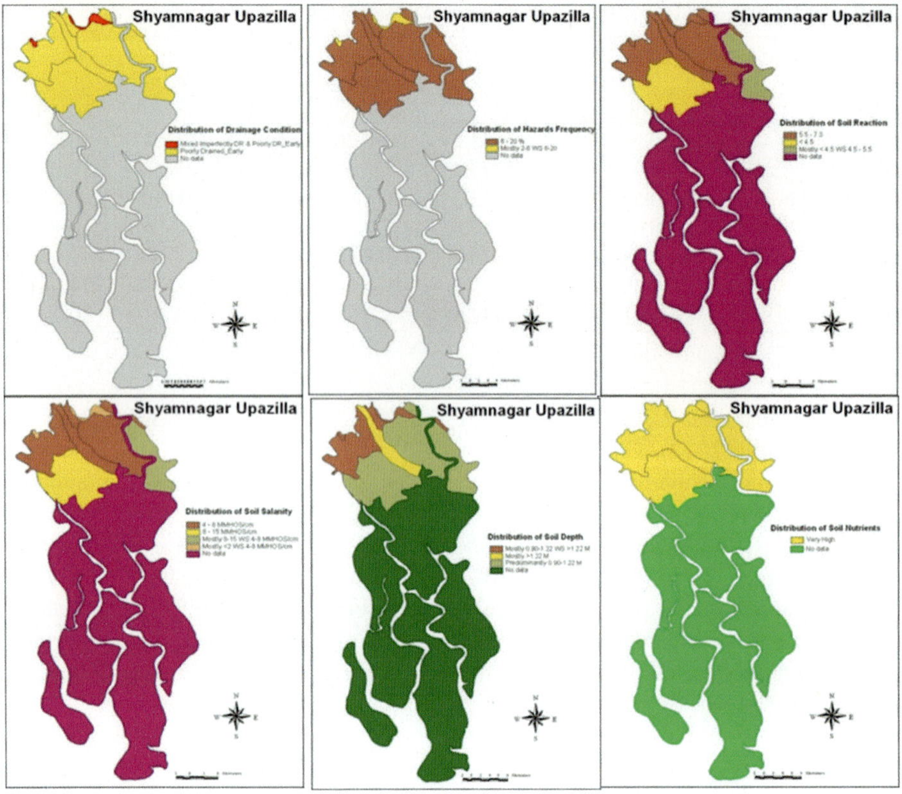

Fig. 19.2 Distribution of drainage system, hazard frequency, and soil characteristics in Shyamnagar Upazila

Northern and southern parts were characterized by a soil pH of 5.5–7.3, whereas north-eastern part had soil pH of 7.3–8.4. Soil salinity map indicated high soil salinity (4–8 MMHOS/cm) in the northeastern and southwestern parts of the study area while soil depth in majority parts was predominantly 0.90–1.22 M. Consistent soil was found to cover most of the areas of this upazilla.

Majority areas of Rampal Upazilla (shown in Fig. 19.4) were poorly drained and the distribution of soil nutrients in all parts was found to be low. The distribution of soil permeability in north, central and western parts were slow, whereas in southern part were mostly slow. Most of the areas were covered by mixed type soil pH (5.5–7.3 and 7.3–0.4). Southern part was characterized by a soil pH of 5.5–7.3. Soil pH of northern, middle and eastern parts ranged between 5.5–7.3 and 7.3–8.4. Northern and middle parts of the study area were characterized by high soil salinity (4–8 MMHOS/cm) while low salinity (<2 MMHOS/cm) soil was found in the south, southeast and southwest regions.

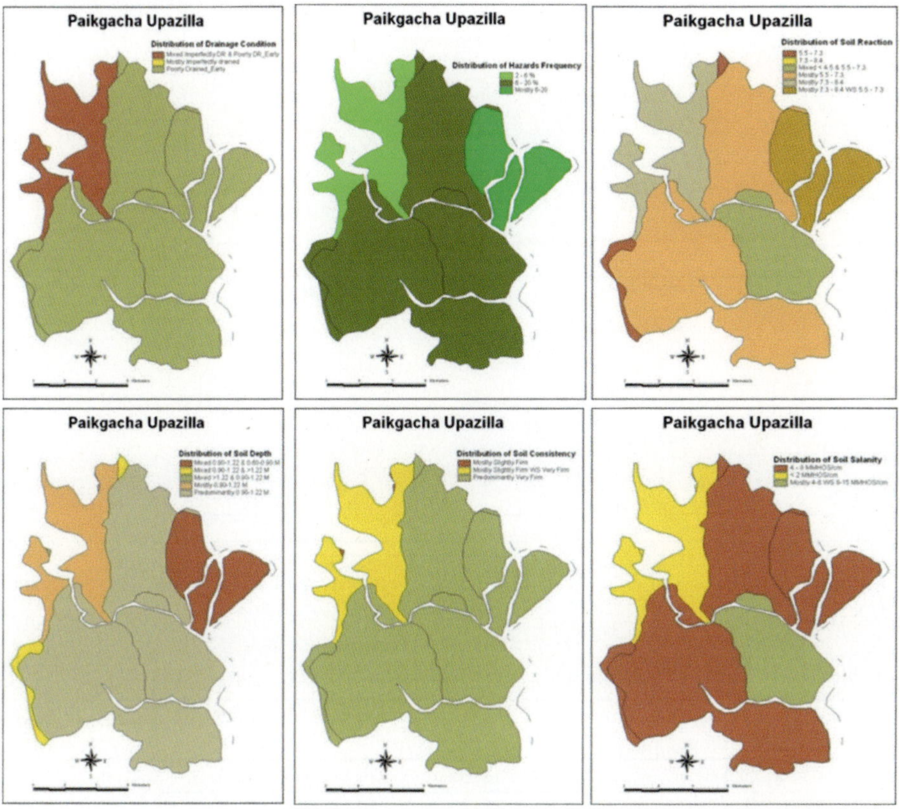

Fig. 19.3 Distribution of drainage system, hazard frequency, and soil characteristics in Paikgacha Upazila

19.3.7 Causes of Water Logging

In the study area water logging problems (shown in Table 19.19) mainly came from its low lying character (55.39 %). Absence of proper drainage facility also contributed to the problem (29.6 %).

19.3.8 Suitable Site Selection for Shrimp Culture

19.3.8.1 Decision Making

It is argued that quality of decision-making depends on sequence in which the activities are undertaken. Any decision-making process is composed of three major

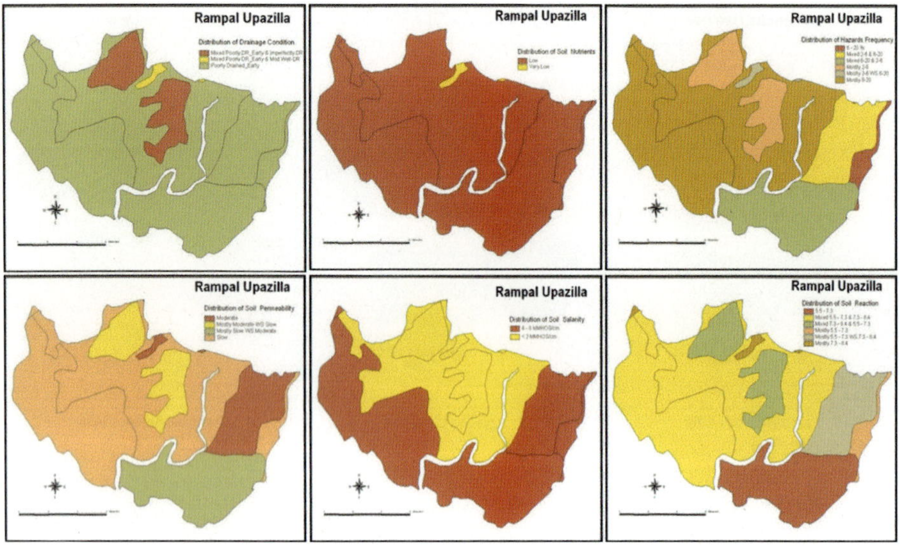

Fig. 19.4 Distribution of drainage system, hazard frequency, and soil characteristics in Rampal Upazila

Table 19.19 Causes of water logging in the study area

Low land (%)	Absence of drainage system (%)	Collapse of drainage system (%)	Gher (%)	Others (%)	Total (%)
55.39	29.6	12.26	2.33	0.42	100

phases- Intelligence, Design and Choice. Suitable sites for shrimp aquaculture were selected following three-phase decision making process.

19.3.8.2 Decision Rules and Knowledge Base

Decision rules for locating suitable sites for sustainable shrimp aquaculture has been prepared based on expert knowledge and multi criteria evaluation process. Table 19.20 shows the decision rules for site suitability for sustainable shrimp aquaculture.

19.3.8.3 Suitability Categories

Using satellite imagery and ancillary data shrimp culture related database was developed through the present research for sustainable shrimp culture in the

Table 19.20 Decision rules of site suitability for sustainable shrimp aquaculture

Criteria	Suitability class		
	Most suitable	Moderately suitable	Less suitable
Water salinity	10–25 ppt	5- < 10 ppt	3- < 5 ppt
Water source	River	Branch of river with water flow restricted	Man made narrow canal, underground tube well
Tidal activity	Prominent	Not so prominent	Absent
Soil type	Loamy	Clay loamy	Sandy loamy, pit
Land elevation	High	Medium	Low
Water pH	7.5–8.5	7.0–7.5	<7.0, >8.5
Soil pH	5.0–6.0	4.0–5.0	2.0- < 4.0
Water temperature	18–30 °C	12–18 °C	<12, >30 °C
Water alkalinity	125–200 mg/L	75–125 mg/L	40–75, >200 mg/L
Distance to road	Adjacent	100–500 m	500–1,000 m
Drainage condition	Well drained	Moderately drained	Poorly drained
Water logging	Free	Slightly present	Severely present
Hazard frequency	Hazard free	Hazard free	Hazard free
Electricity supply	Regular	Irregular	Absent
Distance to depot	Nearby	Far	Too far
Sanitation facility	Enough	Medium	Poor

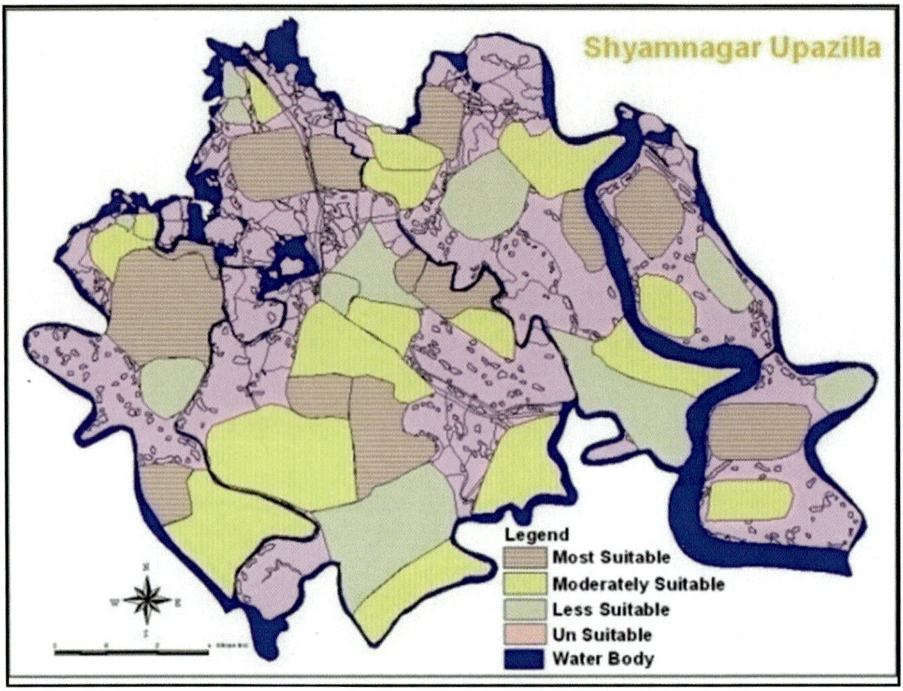

Fig. 19.5 Suitable areas of shrimp culture in Shyamnagar Upazila of Satkhira district

coastal districts of Bangladesh. Through site suitability study three different categories of sites were identified which are suitable for shrimp culture in Shyamnagar, Paikgacha and Rampal upazillas of Satkhira, Khulna and Bagerhat districts respectively. Site suitability analyses were conducted following the decision rules for most important factors of shrimp culture. Potential locations were classified as most suitable, moderately suitable and less suitable areas (in Figs. 19.5, 19.6 and 19.7). Present study indicates the most suitable sites of shrimp aquaculture covered 9,309.40 ha in Shyamnagar upazilla, Satkhira district. In the same upazilla, 19,256.92 and 9,103.90 ha areas were identified as moderately and less suitable for shrimp culture (Fig. 19.5). Most suitable areas in Paikgacha upazilla of Khulna district were estimated to be 9,954.20 ha while moderately suitable and less suitable regions of the upazilla were identified as 8,919.40 and 6,775.80 ha respectively (Fig. 19.6). Study results found (in Fig. 19.7) most suitable, moderately suitable and less priority in suitability areas of shrimp aquaculture as 6,403.20, 8,616.70 and 8,867.20 ha respectively in Rampal upazilla of Bagerhat district.

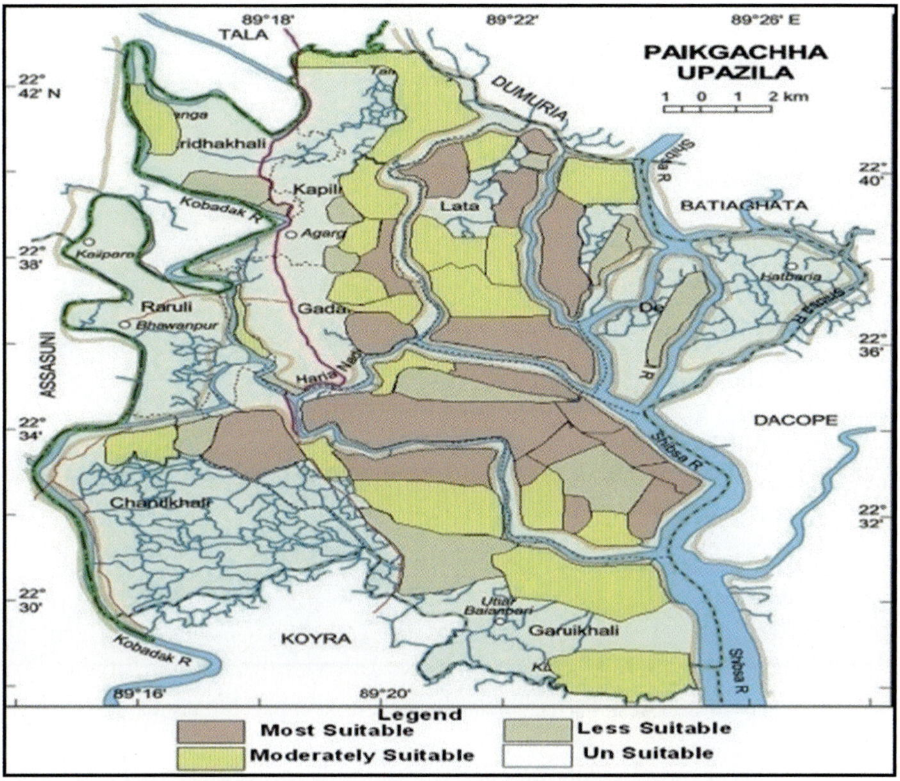

Fig. 19.6 Suitable areas of shrimp culture in Paikgacha upazila of Khulna district

19.4 Conclusion

Comprehensive field survey has been conducted in three coastal districts of Bangladesh namely Satkhira, Khulna and Bagerhat with an ultimate objective of developing a GIS and remote sensing database for sustainable shrimp culture in the region. In developing fisheries database different information were collected through field survey, laboratory analysis and from different GOs and NGOs. All these information were analyzed, interpreted and compiled to identify potential sites of shrimp culture in the study area. After suitability study the potential sites were classified into three types i.e. most suitable, moderate suitable and less suitable locations. GIS is potentially a powerful tool for spatial decision process and is already being effectively used for such purposes in many other countries. Development of a GIS and remote sensing database is a first step in formally evaluating the effects of local geographic and social conditions on the sustainable production of shrimp in these areas. With an adequate database, remote sensing and

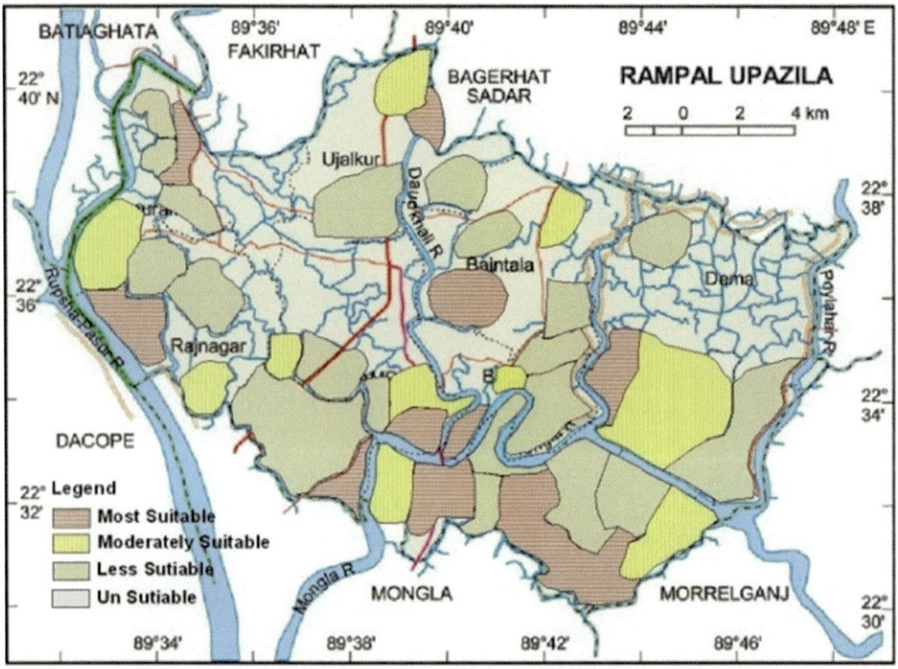

Fig. 19.7 Suitable areas of shrimp culture in Rampal upazila of Bagerhat district

GIS can serve as a powerful analytic and decision making tool for sustainable shrimp culture development. Through this research GIS and remote sensing database was developed for southwest coastal areas of Bangladesh for sustainable shrimp culture, coupled with other physical, biological, and social attributes. Shrimp culture industry itself can also be benefited through proper site selection and improved management system.

Acknowledgement The authors would like to acknowledge Krishi Gobeshona Foundation (KGF), Bangladesh for funding the research.

References

Alam MS, Elahi KM, Shamsuddin SD (1989) Remote sensing for coastal land use mapping: a case study on shrimp culture in paikgacha, Khulna. Department of Geography, Jahangirnagar University, Savar

Anon (1986) Shrimp culture: pond design operation and management. Selected pub. no. 2 for commemorating world food day. Network of Aquaculture Center in Asia, Bangkok, pp 15–48

Apud FD (1984) Extensive and semi-intensive culture of sugpo (*Penaeus monodon*) in the Philippines. In: Prawn industry development in the Philippines. SEAFDEC Aquaculture Department, Iioilo, pp 55–73

Barraclough S, Finger-Stich A (1996) Some ecological and social implications of commercial shrimp farming in Asia. Discussion paper. United Nations Research Institute for Social Development & World Wildlife Fund

Boyd CE (1989) Water quality management and aeration in shrimp farming. Fisheries and allied aquaculture development series no. 2, Alabama Agricultural Experiment Station, Auburn University, Alabana, 83 pp

Boyd CE (1995) Bottom soils, sediment and aquaculture. Chapman and Hall, New York

Boyd CE, Fast AW (1992) Pond monitoring and management. In: Fast AW, Lester LJ (eds) Marine shrimp culture: principles and practices. Elsevier, Amsterdam, pp 497–508

Caipang CMA, Fos MA, Golez MSM (2012) Bottom soil characteristics of brackishwater ponds after a culture period. AACL Bioflux 5(5):415–420

Chakraborti BK, Ravichandran P, Haider DD, Mandal SK, Sanfui D (1985) Some physico-chemical characteristics of Kakdwip brackishwater ponds and their influence on the survival, growth and production of Penaeus monodon. Indian J Fish 32(2):224–235

Chattopadhyay GN, Mandal LN (1980) Inorganic transformation of applied phosphorus in brackishwater fish pond soil under different water salinity levels. Hydrobiologia 17:125–130

Chiu YN (1988) Site selection for intensive prawn farms. In: Chiu YN, Santos LM, Juliano RO (eds) Technical consideration for the management and operation of intensive prawn farms. U.P. Aquaculture Society, Iioilo City, pp 25–28

Dance KW, Hynes HRN (1980) Some effects of agricultural land use on stream insect communities. Environ Pollut 22(Series A):14–28

Das B (1998) Chingri Chhash O Babosthapona (prawn culture and management), vol 1. Bangla Academy, Dhaka, pp 128–166

George ADI, Abowei JFN, Allison ME (2010) The sediment characteristics of Okpoka Creek, Niger Delta, Nigeria. Asian J Agric Sci 2(1):9–14

Hansell J (1993) Health management in shrimp ponds. Aquatic Animal Health Research Institute, Department of Fisheries, Kasetsart University Campus, Jatujak, 2 pp

Hein L (2000) Impact of shrimp farming on mangroves along east coast of India. Unasylva 51:48–54

Hill AR (1976) The environmental impacts of agricultural land drainage. J Environ Manage 4:251–274

Hoq ME, Islam MM, Hossain MM (1996) Polyculture of freshwater prawn (Macrobrachium rosenbergii) with Chinese and Indian carps in farmer's pond. J Trop Aquac 11:135–141

Jenkins S, White I, Phillips MJ, Tookwinas S (1995) Key researchable issues in sustainable coastal shrimp aquaculture in Thailand. Draft report to the Australian Centre for International Agricultural Research (ACIAR), 75 pp

Khan YSA, Hossain MS (1996) Impact of shrimp culture on the coastal environment of Bangladesh. Int J Ecol Environ Sci 22(2):145–158

Law AT (1988) Water quality requirements for *Penaeus monodon* culture. In: Proceedings of a seminar marine prawn farming in Malaysia, Malaysian Fisheries Society, Serdang, pp 53–65

Lester LJ, Pante MJR (1992) Penaeid temperature and salinity responses. In: Fast AW, Lester LJ (eds) Shrimp culture: principles and practices. Elsevier, Amsterdam, pp 515–534

Liao IC (1977) A culture study on grass prawn *Penaeus monodon* in Tafwan the patterns, the problems and the prospects. J Fish Soc Taiwan 5:11–29

Patil Pawan G, Krishnan M (1998) The social impacts of shrimp farming in Nellore district. India Aquac Asia 1:3–5

Rahman MM, Subudhi AP (2004) Knowledge based spatial expert support system in identifying suitable sites for urban development in Dehradun City, India. Plan Plus 1(2):13–25

Rahman MM (2005) Integration of GPS and GIS for mobile mapping. J Geo Environ 1(1):26–36

Rahman MM (2006) Spatial decision support system for planning location of additional village amenities considering isotropic & anisotropic cost distance. Plan Plus 4(2):33–45

Rahman MM, Hoque MA (2006) Site Suitability analysis for solid waste disposal using GIS: a case study on KCC area. J Geo Environ Dept Geogr Environ Stud Rajshahi Univ 4(1):45–65

Rahman MS, Malek MA, Matin MA (1995) Trend of pesticide usage in Bangladesh. Sci Total Environ 159(1):33–39

Research Planning Institute (RPI) (1985) Environmental baseline studies for the establishment of control criteria and standards against petroleum related pollution in Nigeria. Colombia South Carolina, USA, RPI/R/84/4/15–17

Salam MA, Ross LG, Malcolm CMC (2002) A comparison of development opportunities for crab and shrimp aquaculture in South-western Bangladesh, using GIS modelling. Aquaculture 220 (1–4):477–494

Salam MA, Khatun NA, Ali MM (2005) Carp farming potential in Barhatta Upazilla, Bangladesh: a GIS methodological perspective. Aquaculture 245:75–87

Shahid MA, Pramanik MAH, Ali S (1992) Remote sensing applications in coastal shrimp farming and agricultural areas. Asian Pac Remote Sens J 4(2):1

Shingh SP (1984) Site selection: soil quality considerations in the selection of sites for aquaculture farms. Aqua-Train/NACA/84-036, 19 pp

Simpson HJ, Ducklow HW, Deck B, Cook HL (1983) Brackish water aquaculture in pyrite-bearing tropical soils. Aquaculture 34:333–350

Subrahmanyan M (1973) Experimental studies on growth in *Penaeus monodon* (Fabricius). Proceedings of seminar on mariculture and mechanized fishing, pp 70–73

United States Environmental Protection Agency (USEPA) (2002) Water quality monitoring for Coffee Creek Porter County, Indiana. Retrieved from: http://www.usepa/research.htm. modecode=62-28-00-00

Verghese PU, Ghosh AN, Das PB (1975) On growth, survival and production of Jambo tiger prawn Penaeus monodon (Fab.) in brackish water ponds. Bull Dept Mar Sci Univ 7(4):781–789

Chapter 20
A Multi-criteria Approach for Erosion Risk Assessment Using a New Concept of Spatial Unit Analysis, Wave Model and High Resolution DEMs

Helena Granja, José Pinho, and João Mendes

Abstract In this work, field data and model outputs were integrated, processed and analyzed in a GIS environment in order to assess the vulnerability to erosion and to produce associated risk maps using a multi-criteria approach. Erosion risk assessment methodology was developed based on morphological, hydrodynamic, and meteorological indicators that were computed using data obtained from a short term monitoring program. It was applied to a stretch on the NW coastal zone of Portugal. Comparing this methodology to previously developed ones, there are three main distinctive aspects to be considered: (i) the coastal stretch was segmented for computation of indexes (susceptibility, exposure and risk), accordingly to specific coastal types (ii) a new set of vulnerability variables is proposed; and (iii) erosion impact indicators computed from detailed building and beaches data that was extracted from aerial photos.

20.1 Introduction

Roughly 40 % of the world population lives within 100 km of the coast. The amount of people in those increasingly highly populated areas is putting pressure over the necessary urban development to provide all of them the required accommodations

H. Granja (✉)
Interdisciplinary Centre of Marine and Environmental Research (CIIMAR/CIMAR), University of Porto, Rua dos Bragas, 289, P 4050-123 Porto, Portugal

Earth Sciences Department, University of Minho, Campus de Gualtar, 4710-057 Braga, Portugal
e-mail: hgranja@dct.uminho.pt

J. Pinho • J. Mendes
C-TAC/Department of Civil Engineering, University of Minho, Campus de Gualtar, 4710-057 Braga, Portugal

C.W. Finkl and C. Makowski (eds.), *Remote Sensing and Modeling: Advances in Coastal and Marine Resources*, Coastal Research Library 9, DOI 10.1007/978-3-319-06326-3_20, and services. One of the major aspects to have under consideration is the problematic of coastal erosion. The over-construction is increasing the degradation of the coastal regions, making them susceptible to sea advance, and thus losing the attractiveness they once had. This is especially serious along low clastic coasts facing storm surges. The need for an integrated coastal management is a necessity. The fact that the coast can, in some cases, present significant retreats rates per year is a troublesome fact that decision makers should have in mind.

It seems sounding that different coastal stretches should have different solutions: retreat, first priority for ICZM purposes; defend, just when a town or a high interest utility justify it; adapt, work with nature, finding less harmful solutions; do nothing, allowing the natural processes to evolve.

Coastal defense strategies include two main approaches: the hard one where heavy engineering interventions are held, by introducing static elements in a natural dynamic environment, thus changing mobility; and the soft, which mainly consists in artificial nourishment and other light solutions like fences, sand bags, drains, reprofiling, replanting vegetation, natural barriers for wave attenuation such as seaweeds.

The coast will evolve differently according to the taken choice. So, impacts and levels of risk are different from stretch to stretch.

The purpose of this paper is to present a methodology – based on data acquired with innovative techniques, considering a spatial segmentation based on administrative boundaries, human occupation and type of coast, and using a wave model – that can be used to classify a coastline according to its erosion risk, and be an important tool in coastal planning and management.

20.2 The Meaning of Risk

Risk can be considered the probability of occurrence of a hazardous process and the estimation of its consequences over people, goods and/or environment. It expresses direct and/or indirect damages. It is a product of probability of hazard and potential loss.

The probability of hazard is the chance of occurrence of a process with destruction potential (hazard). The process is characterized by its severity, which is understood as the capacity to produce damage as a function of magnitude, intensity, rank, velocity or other parameter.

Risk has not the same expression everywhere. It depends of susceptibility, exposure and location (Julião et al. 2009).

The susceptibility is the spatial incidence of hazard (function of location) and represents the predisposition of an area (exposed elements) to be affected by it. The exposure concerns the elements potentially affected by a hazardous process. The exposure is function of vulnerability that is represented by the degree of loss as a consequence of process action. Damage or loss over exposed elements is the consequence of impact of a hazardous process. It can be expressed as a product of vulnerability and value (economical value) and is a function of exposure of elements.

Several indicators can be used to assess all the above mentioned concepts.

20.3 Multi-criteria Approaches for Erosion Risk Assessment

Kumar et al. (2010) used a specific methodology to assess erosion risk, understood as the potential loss of the exposed elements or systems, recurring to a set of hazard and vulnerability indicators. This methodology assumed that a coastline could be categorized in terms of risk solely based on the physical hazard to which the coast was susceptible of experiencing.

A coastal vulnerability index (VI), similar to the one used by Pendleton et al. (2005), Thieler and Hammar-Klose (1999, 2000), was proposed and is quantified based on eight parameters. Six of those parameters were already considered by earlier researchers: shoreline change rate, sea-level change rate, coastal slope, mean significant wave height, mean tidal range and coastal regional elevation. Kumar et al. (2010) proposed yet the use of two other variables: coastal geomorphology and tsunami run-up.

The VI is calculated based on risk values assigned to input parameters using simple vector algebraic technique recurring to GIS tools. The VI values, thus generated for different segments of the coastline, are categorized into three classes – low, medium and high – vulnerable corresponding to, less then 25th percentile, between 25th and 50th percentile, and greater then 50th percentile, of the obtained distribution values, respectively.

Thieler and Hammar-Klose (1999) have also proposed a similar approach to the computation of the VI used by Gornitz et al. (1994), as well as to the sensitivity index employed by Shaw et al. (1998). The index concerns six physical variables related in a quantifiable manner: geomorphology, coastal slope, relative sea-level rise rate, shoreline erosion/accretion rate, mean tide range, mean wave height.

Del Rio and Garcia (2009) assumed that the evaluation of any type of risk should necessarily include two separate components that constitute the risk – the physical hazard and the socioeconomic impact, and for that purpose a hazard index and an impact index were combined to obtain a erosion risk index. It is a single numerical value obtained by means of a weighted average of both base indexes, according to the number of variables involved in their computation. The rationale behind this procedure was that a simple average to obtain the risk index would actually overestimate the individual weight of the impact index variables (6) against the hazard index variables (11) considered in their work. On the other hand, the risk index obtained by this procedure is expressed as a percentage of its maximum theoretical value in a similar way to the hazard and impact indexes. In this sense, it must be noted that, although risk is often defined in terms of probabilities, the percentage values of the obtained risk index by the proposed method do not bear a direct relationship with probabilities.

Ferreira (2006) developed vulnerability and risk assessment related to erosion phenomenon, since, either marine or continental, is one of the main processes responsible for environmental deterioration and the consequent loss of biodiversity in many places. The data used to perform the analysis of vulnerability was categorized in two distinct groups: continental and marine. The assessment of the vulnerability for these two groups was developed considering two distinct methods: (i) for the continental variables, a multi-criteria assessment was considered, based on spatial data, which allowed the ranking of information; (ii) due to the lack of data and detailed studies on the littoral dynamics of the studied area, for the marine variables, the analysis of the morphogenetic effects of a series of storm episodes, from December 1995 to January 1996, was chosen. To better characterize the storm episodes and their consequences, weather conditions and corresponding wave climate, as well as the field record of the erosion effects registered in the area, were studied. The wok showed that the simple and expedite approach of categorizing sections of coast with different variables allows a clear identification of problematic territorial units which are susceptible to greatest hazard and risk.

20.4 Methodology

The proposed methodology in our work is based on a multi-criteria analysis supported by reliable field data quantified in a high resolution grid and considering a set of indicators that were selected according to judicious criteria.

20.4.1 Spatial Units of Analysis

Coastal stretch analysis was done taking into consideration administrative boundaries, presence or absence of buildings, and type of coast (geomorphological features). The segmentation results from a spatial intersection of those geographical themes.

For the mapping of the indicators a baseline was defined, approximately corresponding to the border between not occupied and occupied areas. This is an artificial division because it is variable from stretch to stretch and along time. However, it can be easily represented by roads and paths present on aerial photos. This baseline should be kept, as much as possible, parallel to the shoreline and avoid abrupt changes of its distance to the shore. These conditions imply that in some stretches not occupied areas appear at both sides of the baseline, and therefore, these segments were distinguished and considered in the segmentation procedure.

Areas used to quantify indicators consist in a buffer extending landwards and seawards from that baseline, mostly obtained using GIS tools which includes aerial photo-maps. The buffer is comprised between 1 km inland and, at least, 5 km seawards.

Administrative boundaries can be chosen according to the scale of the study area. It depends on the administrative configuration of each country but desirably the smallest spatial administrative unit should be selected.

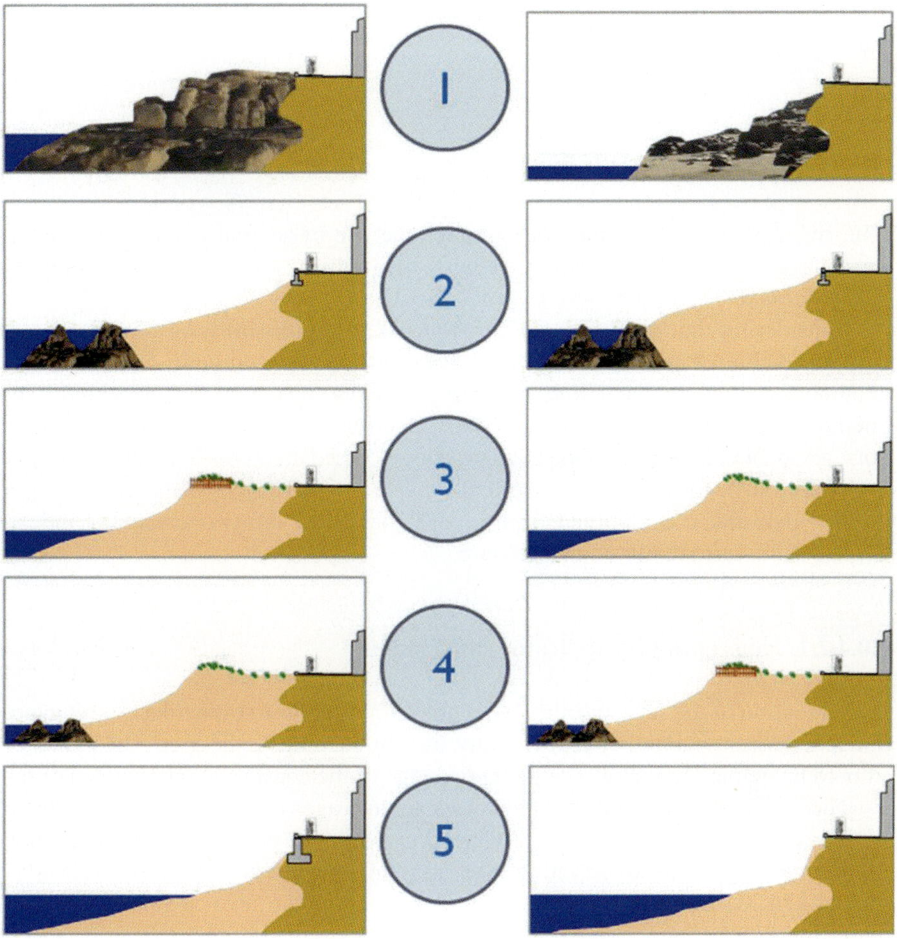

Fig. 20.1 Geomorphological types of urban coast

Another element considered for spatial segmentation is derived from a sound analysis of the urban coast and constitutes an attempt to classify the coast using its main geomorphological features, considering the presence of (i) rocky or sand beaches, (ii) outer rocky outcrops, (iii) dunes, (iv) dune fences and (v) adherent defense structures. Once considered several combinations of those features, five types of coast were identified (Fig. 20.1).

The first type of coast is a pure rocky beach or pocket beaches between rocky outcrops. The second is constituted by a sandy beach located between a seawall defense and outer rocky outcrops; the sandy beach in this type can either present a convex or concave profile. The third type of urban coast is a sandy beach with a vegetated area associated to frontal and/or embryonic dunes, that can be interventioned or not by the installation of dune fences. The fourth type is identical to the last one but with rocky outcrops in the sea border of the beach. And, finally, the fifth type is constituted by narrow and high slope sandy beaches delimited in the inner border by a road or a seawall.

20.4.2 *Risk Assessment*

The risk assessment methodology implies the use of several parameters, such as susceptibility, exposure, location, value, vulnerability and potential loss, as previously defined.

Considering that in the case of erosion risk, the probability of occurrence and also the value of potential losses are very difficult to compute, a multi-criteria methodology, as in the above mentioned works, has been implemented. This methodology was based on the judicious selection of a set of susceptibility indicators, giving a vulnerability index, and a set of exposed elements indicators giving an exposure index for the coastal segments.

The risk index is obtained from a combination of the two computed indexes related with vulnerability and exposure.

20.4.2.1 Susceptibility Indicators and Index

The selection of the susceptibility indicators should follow some rules: significance, independence, availability and friendly use. In each case, the group of indicators should be defined based on a sound knowledge of the coastal sector being analyzed.

The suggested susceptibility indicators are averages of: coastal elevation, beach slope, section width, sedimentary budget, wave climate and vegetation (Fig. 20.2).

The average coastal elevation was computed based on the digital elevation model (DEM) of the beach, on each segment, on the area seawards from the baseline (Fig. 20.2a).

For the same area the average beach slope can be obtained using GIS tools and a DEM of the beach (Fig. 20.2b). The digital elevation model can be obtained by aero-photogrammetric or laser topographic techniques.

The average section width is obtained using the polygon in front of the baseline, dividing its area by the extension of the baseline segment (Fig. 20.2c).

Sedimentary budgets are computed by comparison of DEMs obtained in different moments (Fig. 20.2d). For this purpose the volume of the terrain in front of the baseline is computed at different times and the bulk difference is used as a measure of the erosion/accretion trend. Desirably, these budgets should be based on high resolution topographical data and considering time intervals that can capture the resultant erosion/accretion major dynamics trend. This is not, of course, easy to obtain. When data from several monitoring campaigns are not available, a model, using specific software, should be used in order to predict the future evolution of the coastline, and, according to the results, the specific susceptibility factor can be estimated.

Wave climate is one of the most important forcing factors on the coastal environment. Susceptibility of erosion risk associated with this indicator is usually

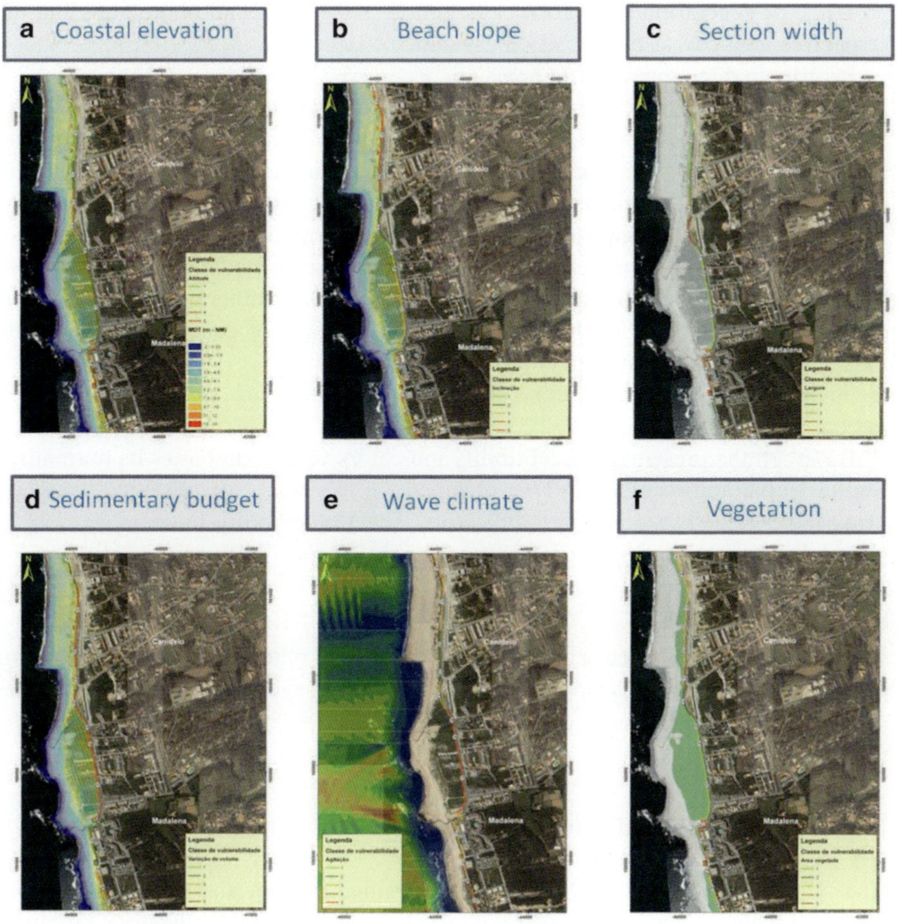

Fig. 20.2 Susceptibility indicators

expressed using significant wave heights derived from historical buoys data. In this work we use a new approach based on results obtained with a numerical model (Nawogu and Demirbilek 2001) for wave's propagation (Fig. 20.2e). This approach consists in propagate the same wave regime, for all coastal segments, from an outer boundary and, through the inner shelf, till the shore. The resulting wave energy in an ocean area located in front of the segment is estimated. Depending on the bathymetry, wave energy, when is propagating from the ocean to the shore, tends to dissipate and/or concentrate in some stretches. Significant wave heights that are related with that energy, in an area in front of each segment, were taken in order to measure that indicator. Bathymetrical data is needed to implement the wave model.

Behind the physical indicators previously presented, a biological indicator (Honrado et al. 2010) was also considered as a susceptibility indicator. The presence of vegetation (Fig. 20.2f) can be inferred from aerial photomaps and the respective area can also be easily measured.

In order to distinguish buffer areas not occupied with buildings at the moment of the analysis, an attribute (presence or absence) related with buildings is added to the baseline, which allows to segment it according to present occupation. So, erosion risk assessment is done considering the actual status and is not associated with potentially exposed elements derived from building coastal policy plans.

After the computation of all indicators, a transformation is required, so that all factors can be analyzed under a common numerical scale, and a classification from 1 (minimum vulnerability/exposure) to 5 (maximum vulnerability/exposure) is used for that purpose. For this classification, the limits of classes were based on the resulting distributions percentiles (20, 40, 60 and 80 %). Thus, any segment of the analyzed coastal sector can be classified from 1 to 5 for any of the considered indicators.

Based on the presented susceptibility indicators values, and after its normalization, a Vulnerability Index (VI) is obtained (in this work we use equal weights) using the obtained susceptibility indicators. The Vulnerability Index is given by:

$$VI_j = \sum\nolimits_{i=1}^{n} w_i S_{ij} \tag{20.1}$$

where, VI_j is the VI for the segment j, w_i is the weight of the susceptibility indicator i, S_{ij} is the value of the susceptibility indicator i in the segment j, and n is the number of susceptibility indicators.

20.4.2.2 Exposure Indicators and Index

Coastal erosion can potentially affect a diversified number and type of infrastructures and environmental systems. In our work we consider two indicators: buildings and sandy beaches (Fig. 20.3).

Exposed buildings were identified in a buffer of 100 m inland and 100 m seawards from the baseline (Fig. 20.3a). Their areas are computed for each segment giving a measure of exposed buildings in the buffer. Exposed beaches areas are also considered as a potential loss in the erosion process (Fig. 20.3b). On each segment, the beach area was computed considering the bathymetric −1 m (MSL) as the outer limit.

Based on the previous presented indicators an Exposure Index (EI) was computed. But, unlike the calculus of the VI, the weights for each indicator were assumed different. It was considered that the exposure indicator associated with building area has a significantly bigger importance, since the actual coastal policy defence privileges any type of build at risk relatively to the beaches themselves. Thus, a weight of 0.75 for the building area indicator and a value of 0.25 for the beach area indicator were attributed. The EI is given by:

$$EI_j = \sum\nolimits_{k=1}^{l} w_k E_{kj} \tag{20.2}$$

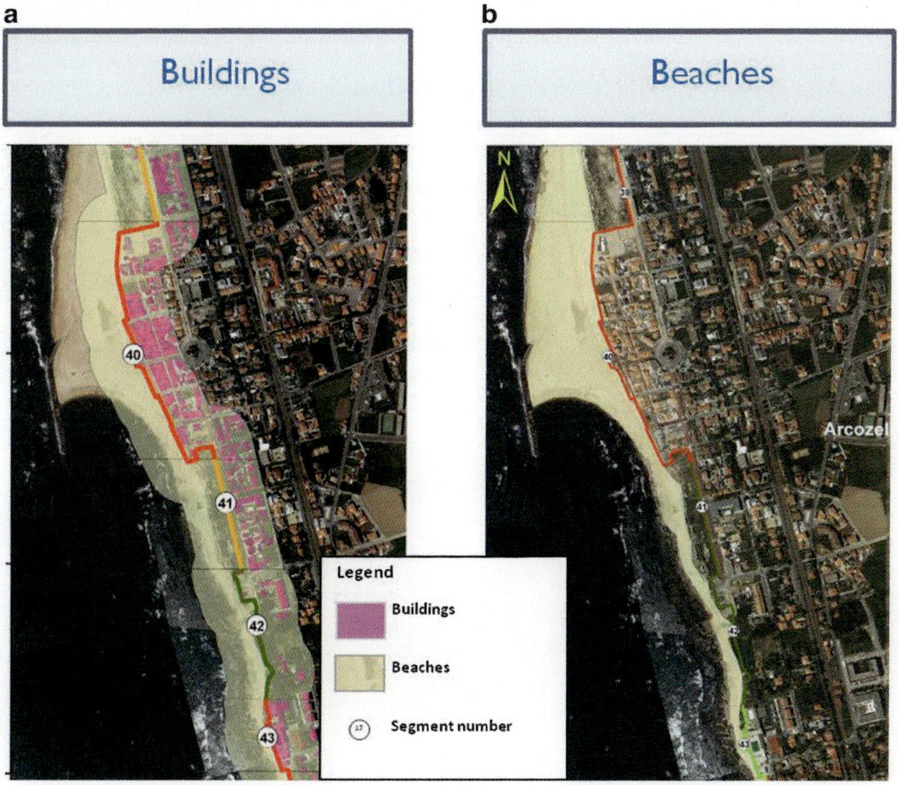

Fig. 20.3 Exposure indicators

where, EI_j is the EI for the segment j, w_k is the weight of the exposure indicator k, E_{kj} is the value of the exposure indicator k in the segment j, and l is the number of exposure indicators.

20.4.2.3 Erosion Risk Index

The erosion risk index ERI_j is obtained as a simple average of the VI and the EI results for each segment:

$$ERI_j = \left(VI_j + EI_j\right)/2 \tag{20.3}$$

After the calculus of both VI and EI for each segment, a relative classification was made considering three classes, whose limits are the 33 and 66 % percentiles of the corresponding distributions.

The final result should be expressed in a classification from 1 to 3, 1 being low risk and 3 high risk within the analyzed coastal sector.

20.5 Vila Nova de Gaia Case Study

The Municipality of Vila Nova de Gaia (Fig. 20.4) is located on Northern Portugal, on the coastal zone just south of Porto city and Douro River. It is one of the most populated municipalities of Portugal and is under a huge urban development pressure. One of the reasons that justify this work, besides the increasing urban growing, is the need for interventions in order to mitigate local impacts of coastal erosion, particularly in some sectors (Granja et al. 2010).

The erosion risk assessment on this coastal municipality was developed based on a comprehensive monitoring program that have collected relevant data to characterize the main morphodynamic patterns and evolutive trends. All the collected data was included in a GIS, specially implemented for this study.

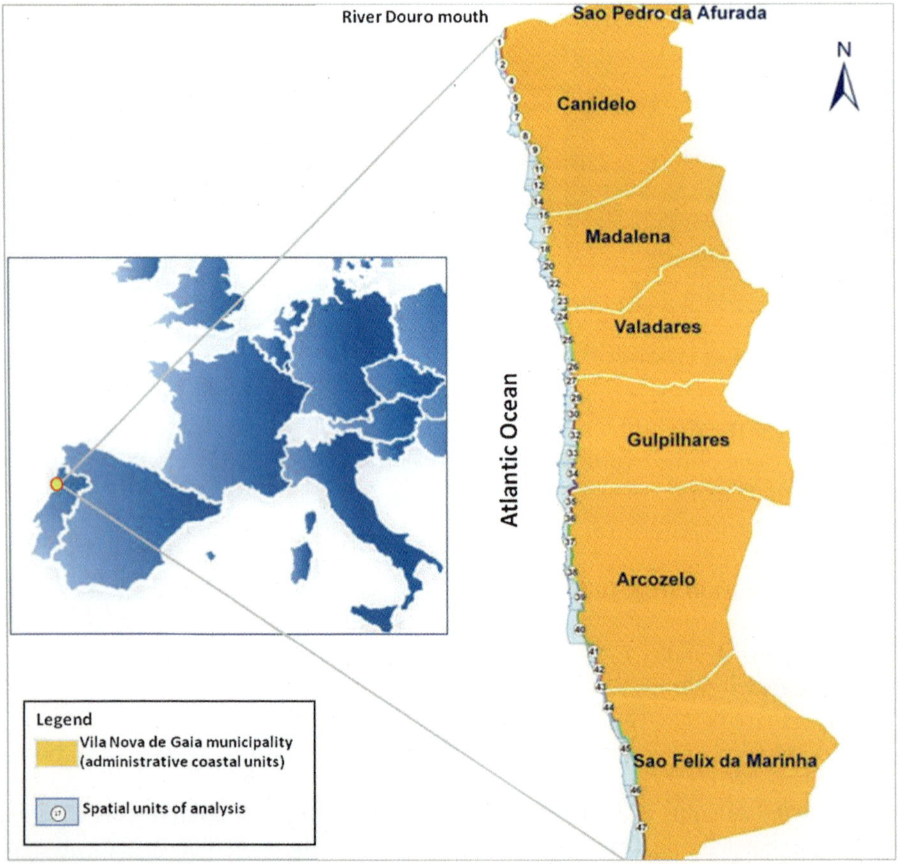

Fig. 20.4 Vila Nova de Gaia Municipality location

From the segmentation of the coastal zone of Vila Nova de Gaia municipality, and according to the previous detailed methodology, resulted 47 segments that were individually processed in order to compute the respective values of susceptibility and exposure indicators.

20.5.1 Field Data

Different methods were used to collect field data (Bio et al. 2011) during a 2 years monitoring program including: topographic surveys based on aerial photogrammetry and terrestrial videogrammetric mobile mapping; sub-tidal bathymetry with sonar imagery using an autonomous surface vehicle (ASV); as well as field observations, with sediment sampling and beach profiling. Accuracies of the topographic survey methods ranged from few centimeters to decimeters for terrestrial and aerial surveys, respectively.

20.5.2 Results and Discussion

Results of the VI, EI and ERI indexes, for each segment, and its risk classes are summarized in Table 20.1 and the segment classification concerning coastal risk is represented in Fig. 20.5.

The presented results must be interpreted in the context of the coastal zone of Vila Nova de Gaia municipality for which they were obtained. They allow the establishment of a spatial differentiation based on human and natural characteristics of that coastal zone. Results should be carefully interpreted taking into consideration the set of chosen indicators. Some other forcing factors, such as tsunamis, storms and other human interventions, were not considered in this study.

Besides those remarks, a differentiation between the northern and southern sectors (the boundary approximately between the segments 33 and 34) can be clearly identified. The northern sector is dominated by low and medium coastal erosion risk and the southern sector by high and medium coastal erosion risk.

However, on the northern sector, particularly on locations coincident with creek mouths, coastal erosion risk is high. This fact can be related with the frequent migration of creek mouths and the low elevation of these areas.

Moreover, on the northern sector, two sets of coastal segments with different orientation can be identified – one presenting a main N-S orientation (maps 1 and 3, Fig. 20.5) and another mainly NNW-SSE oriented (maps 2 and 4, Fig. 20.5) – presenting different erosion risk trends. Having in mind that the prevailing wave approach at this location is from NW, the more exposed segments set is the one that reveals higher erosion risk indexes values.

On the southern sector, the obtained results can be mainly explained by the high density of building, its proximity to the shore, the presence of an important railway dangerously close to the shore (segments 43 and 44), and the probable impact of a small harbor on the segment 40.

Table 20.1 Indexes and risk classes results

Segment	VI	EI	ERI	Risk class	Segment	VI	EI	ERI	Risk class
1	53	3	29	Low	**25**	73	28	50	High
2	83	6	45	High	**26**	41	5	23	Low
3	100	3	52	High	**27**	62	7	35	Medium
4	99	19	59	High	**28**	65	2	33	Medium
5	80	3	42	High	**29**	47	34	40	Medium
6	77	4	40	Medium	**30**	48	13	30	Low
7	58	17	38	Medium	**31**	44	4	24	Low
8	75	29	52	High	**32**	36	15	26	Low
9	54	4	29	Low	**33**	41	26	34	Medium
10	50	9	29	Low	**34**	51	38	44	High
11	43	21	32	Low	**35**	71	38	55	High
12	39	18	28	Low	**36**	81	1	41	Medium
13	59	6	32	Low	**37**	80	27	53	High
14	57	4	31	Low	**38**	75	12	44	High
15	45	11	28	Low	**39**	31	44	37	Medium
16	37	5	21	Low	**40**	62	100	81	High
17	40	39	40	Medium	**41**	49	17	33	Medium
18	70	27	49	High	**42**	69	10	33	Medium
19	63	0	31	Low	**43**	75	13	44	High
20	58	4	31	Low	**44**	78	73	76	High
21	56	4	31	Low	**45**	26	99	63	High
22	77	16	46	High	**46**	22	50	36	Medium
23	52	10	31	Low	**47**	0	79	40	Medium
24	57	13	35	Medium					

The obtained results also show some coherence with the observed behavior of the coastal zone of this municipality, that is influenced by river Douro discharge at north and have high vulnerability at south on the vicinity of a location where erosion problems are well known for a long time (more than a century).

20.6 Conclusions

A multi-criteria approach for erosion risk assessment considering high spatial resolution data, obtained with advanced techniques, is presented. The proposed methodology can be a powerful tool in the planning and management of coastal areas. The computation of the Risk Index through two simple indexes, that can be easily obtained – VI and EI – is of most importance for an adequate establishment of spatial planning policies.

This methodology was successfully applied to a coastal zone giving results that, in a first evaluation, reveal to be coherent with the previous knowledge about its morphodynamic behavior.

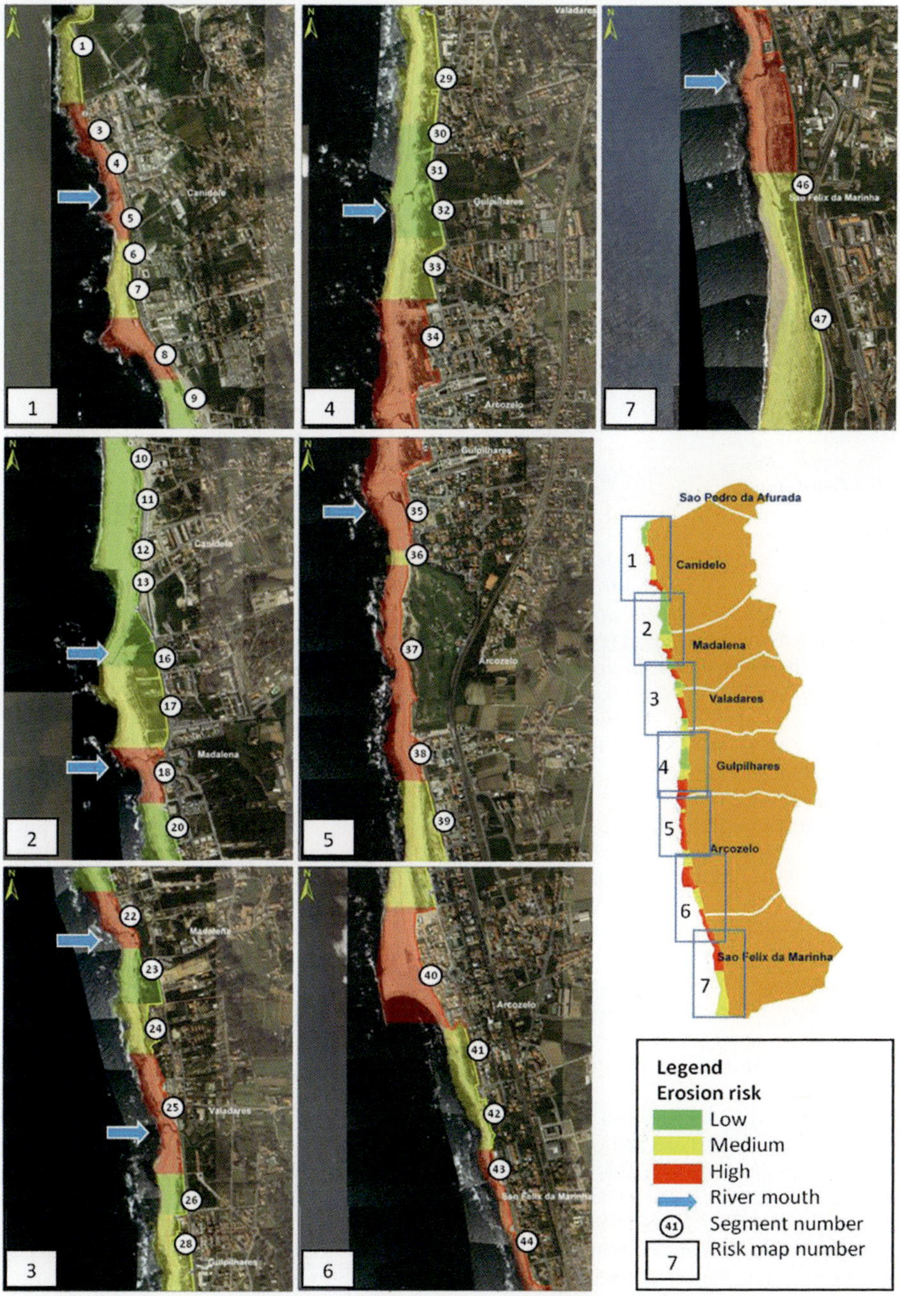

Fig. 20.5 Coastal erosion risk maps for Vila Nova de Gaia

Soft engineering solutions, like dune fences that have been installed on this coastal stretch, can be prioritized according to the erosion risk results obtained in this study. Moreover, other temporary solutions, like artificial nourishment of sandy beaches using local sources, can be defined considering the obtained results for the erosion risk.

The new proposed segmentation procedure, based on a sound set of geomorphological features, allows a more reliable evaluation of the set of indicators than a methodology based on a division considering equidistant segments.

Acknowledgments The authors would like to thank to Biological Park of Gaia for the financial support to develop this coastal monitoring program at Vila Nova de Gaia.

This research was partially supported by the European Regional Development Fund (ERDF) through the COMPETE – Operational Competitiveness Programme and national funds through FCT – Foundation for Science and Technology, under the project "PEst-C/MAR/LA0015/2013".

References

Bio A, Bastos L, Silva, Pinho JL, Granja H (2011) Dynamics of the Douro estuary sand spit before and after breakwater construction. Estuar Coast Shelf Sci 109:53–69

Del Rio L, Garcia F (2009) Erosion risk assessment of active coastal cliffs in temperate environments. Geomorphology 112:82–95

Ferreira JC (2006) Coastal zone vulnerability and risk evaluation: tool for decision-making (an example in the Caparica Littoral – Portugal). J Coast Res SI 39 (Proceedings of the 8th international coastal symposium), 1590–1593. Itajaí, SC, Brazil, ISSN 0749-0208

Gornitz VM, Daniels RC, White TW, Birdwell KR (1994) The development of a coastal risk assessment database: vulnerability to sea-level rise in the U.S. southeast. J Coast Res, Special Issue No. 12:327–338

Granja H, Bastos L, Pinho JLS, Gonçalves J, Henriques R, Bio A, Magalhães A (2010) Small harbours risks: lowering for fishery and increasing erosion. The case of Portinho da Aguda (NW Portugal). In: Proceedings of littoral 2010, London, September 2010

Honrado J, Vicente J, Lomba A, Alves P, Macedo JA, Henriques R, Granja H, Caldas FB (2010) Fine-scale patterns of vegetation assembly in the monitoring of changes in coastal sand-dune landscapes. Web Ecol 10:1–14

Julião R, Nery F, Ribeiro J, Castelo Branco M, Zêzere J (2009) Guiametodológico para a produção de cartografia municipal de risco e para a criação de sistemas de informaçãogeográfica (SIG) de base municipal. Ed.: AutoridadeNacional de Protecção Civil. Co-Ed.: Direcção-Geral do Ordenamento do Território e DesenvolvimentoUrbano e InstitutoGeográficoPortuguês. Lisboa., 91 p

Kumar TS, Mahendra RS, Nayak S, Radhakrishnan K, Sahu KC (2010) Coastal vulnerability assessment for Orissa State, east coast of India. J Coast Res 26(3):523–534

Nawogu G, Demirbilek Z (2001) BOUSS-2D: a Boussinesq wave model for coastal regions and harbors. Report 1. Theoretical background and user's manual. US Army Corps of Engineers, Engineer Research and Development Center, USA

Pendleton EA, Thieler ER, Williams SJ (2005) Coastal vulnerability assessment of national park of American Samoa to sea-level rise. U.S. Geological Survey open-file report 2005-1055

Shaw J, Taylor RB, Forbes DL, Ruz M-H, Solomon S (1998) Sensitivity of the Canadian coast to sea-level rise. Geol Surv Can Bull 505:114

Thieler ER, Hammar-Klose ES (1999) National assessment of coastal vulnerability to sea-level rise, U.S. Atlantic coast. U.S. Geological Survey open-file report 99–593, 1 sheet

Thieler ER, Hammar-Klose ES (2000) National assessment of coastal vulnerability to sea-level rise, U.S. Pacific coast: U.S. Geological Survey open-file report 00-178, 1 sheet

Index

C.W. Finkl and C. Makowski (eds.), *Remote Sensing and Modeling: Advances in Coastal and Marine Resources*, Coastal Research Library 9,
DOI 10.1007/978-3-319-06326-3, © Springer International Publishing Switzerland 2014

S

T

Printed by Printforce, the Netherlands